Photonics and Lasers

Photonics and Lasers
An Introduction

Richard S. Quimby
Department of Physics
Worcester Polytechnic Institute
Worcester, MA

WILEY-INTERSCIENCE

A Wiley-Interscience Publication
JOHN WILEY & SONS, INC.

Published by John Wiley & Sons, Inc., Hoboken, New Jersey.
Published simultaneously in Canada.

For general information on our other products and services or for technical support, please contact our Customer Care Department within the United States at (800) 762-2974, outside the United States at (317) 572-3993 or fax (317) 572-4002.

Wiley also publishes its books in a variety of electronic formats. Some content that appears in print may not be available in electronic format. For information about Wiley products, visit our web site at www.wiley.com.

Library of Congress Cataloging-in-Publication Data is available.

ISBN-13 978-0-471-71974-8
ISBN-10 0-471-71974-9

Printed in the United States of America.

10 9 8 7 6 5 4 3 2 1

To my parents
who have given me much more
than these few words can say

Contents

Preface

This book grew out of a series of courses that I developed and taught over many years in the areas of lasers, optoelectronics, and photonics. Although these courses have been taught in the physics department, I have made a conscious effort to keep them accessible to nonphysics majors, especially sophomores and juniors from engineering or the other sciences. These students are typically looking for a survey course to introduce them to the basic elements of photonics and lasers, often to fulfill a science "distribution requirement." It has always been difficult to find an appropriate textbook for such a course because the existing books in these areas are aimed at either too high a level or too low a level, or they cover only a portion of the topics that are needed. In teaching these classes, I came to rely mostly on my lecture notes as the reading material for the course. This need for a truly introductory level book, covering a wide range of topics in photonics and lasers, was my motivation for writing this book.

In deciding what material to include, and how to present it, I have kept two distinct audiences in mind. One is the college undergraduate described above, and the other is the working professional who wants to "come up to speed" in the photonics area by learning the fundamentals in an accessible format. Both of these audiences, I believe, can benefit from the level of treatment given here. The reader's physics background is expected to include the usual freshman-level sequence of courses in mechanics, electricity and magnetism, waves, and modern physics. Knowledge of differential and integral calculus is assumed, including simple ordinary differential equations, but no knowledge of partial differential equations is needed. Although I do present and discuss certain solutions of Maxwell's equations that are relevant for photonics (such as the Gaussian beam), I do not derive these solutions here. Similarly, I discuss topics relating to quantum mechanics at the de Broglie wave and "particle in a box" levels, without ever writing down the Schrödinger equation. Readers with a more advanced physics background will better appreciate some of the points that are made, but the bulk of the material should be understandable by those with only a modest physics background.

My goal throughout has been to make sure that the mathematics does not get in the way of the physical concepts. I've tried whenever possible to give physical arguments that lead to an intuitive understanding, while including sufficient mathematical detail to make that understanding quantitative as well. This is a tough balancing act, and necessarily results in trading off rigor versus accessibility. I have deliberately avoided the temptation to be "comprehensive," choosing instead to limit the discussion when appropriate to certain limiting cases that are mathematically simple. This not only makes the discussion easier to follow for the beginning student, but also brings out the fundamental concepts more clearly. To further aid the student who is just learning to think symbolically, I have written some equations in words as well as symbols.

The topical coverage in this book is somewhat unusual, in that it treats two subjects—photonics and lasers—that are usually found in separate books. One reason they are included together here is that there is a natural synergy between them. On the one hand, understanding the operating principle of certain lasers requires knowledge of photonics concepts such as waveguiding, while on the other hand, understanding the principles of optical communications (an important photonics system) requires some knowledge of lasers. An additional benefit to treating them together is a consistency of notation, which is very helpful to the beginning student. An annotated list of symbols is provided at the end of the book.

Because of the combined coverage of photonics and lasers, it is probably unrealistic to try to cover the entire book in a one-semester course. If a course emphasizes photonics, a suggested list of chapters to cover would be 1–7, 10–15, and 24. Chapters 8 and 9 are additional options, should time permit. A course emphasizing lasers might cover Chapters 2, 10–11, and 15–23, with Chapter 9 optional if time permits. Different combinations of chapters or parts of chapters can certainly be used, depending on the emphasis of the particular course.

I would like to thank the many students who have taken my classes over the years, for their questions and comments. You have been my inspiration, and your struggles with the course material have helped me to sharpen my presentations, ultimately making this a better book. Thanks are also due to the reviewers commissioned by Cambridge University Press and Wiley who took the time to make helpful comments on the manuscript. And, finally, thanks to my wife Julie and daughters Claire and Grace, for putting up with the many long hours that took me away from family life. This project could not have been completed without your patience and understanding.

R. S. Quimby

Worcester Polytechnic Institute
Worcester, MA
August 2005

Chapter 1

Overview

1-1. PHOTONICS DEFINED

During the twentieth century, the electronics industry has revolutionized the way we work and play. The vacuum tube made practical the transmission of information over long distances through radio and television. Vacuum tubes were also used in the first electrical computers for the processing of information. From these first steps, the trend has been toward smaller and faster electronic devices, first with transistors as discrete components, essentially replacing vacuum tubes, and later with integrated circuits, in which thousands and then millions of transistors were incorporated onto the same semiconductor chip. This miniaturization has given rise to many of the conveniences that we have become accustomed to today, including personal computers, cell phones, stereo music systems, television, and camcorders, to name just a few.

Today, at the dawn of the twenty-first century, there is a similar revolution underway. In this new revolution, it is not the electrons of the now mature electronics industry that are being put to work, but rather the photons of the nascent photonics industry. The word "photonics" will be taken, for the purpose of this book, to mean phenomena and applications in which light (consisting of photons) is used to transmit or process information, or to physically modify materials. Perhaps the most important example to date is fiber optic communications, in which light traveling down long lengths of ultraclear optical fibers now carries the bulk of telephone traffic across and between the continents. These same optical fibers serve as the backbone of high-speed data transmission networks, allowing Internet users to access not only text and single-frame graphics, but also multimedia content.

Photonics, as defined above, also includes optical data storage, such as CDs and DVDs for audio, video, or computer storage. These applications, although under continual development, are becoming mature technologies. Less well developed are applications in optical switching and optical image processing, also considered within the realm of photonics. Optical computing may be considered to be the final goal of photonics research, in which information is processed and stored mostly optically. This could result in extremely fast and efficient computers, since light signals travel very fast and there is the possibility of efficient parallel processing. However, the practical realization of optical computers remains today, as it has all along, a rather distant goal.

Optical sensors can be considered to be photonic devices, since they optically detect and transmit information about some property such as temperature, pressure, strain, or the concentration of a chemical species in the environment. Such devices have applications as diverse as biosensors for the human body and strain sensors for bridges. Applications such as laser surgery or laser machining are also considered photonic in nature, since they rely on a stream of high-intensity photons.

Photonics and Lasers: An Introduction. By Richard S. Quimby

1-2. FIBER OPTIC COMMUNICATIONS

Although fiber optic communications is just one aspect of the broader topic of photonics, we will emphasize it in this book since it is a well-established and increasingly important technology. The beginnings of optical communications can be traced to the inventor of the telephone, *Alexander Graham Bell.* In 1880, Bell invented a device he termed the photophone, which allowed information to be transmitted through air on a beam of modulated sunlight. The modulated light was detected by the photoacoustic effect, in which a sound wave is produced inside a closed gas cell when modulated light is absorbed inside the cell. Although this was a clever device, it was much less practical than the telephone, and was not developed further.

It was not until the 1960s that optical communications was considered seriously again, this time motivated by two parallel developments. The laser had been developed at the beginning of the decade, and this provided a light source that was powerful and highly directional, both valuable characteristics for sending a light signal over long distances. Sending a laser beam through the air, however, has obvious limitations as a practical communications source over long distances. What was needed was a way to guide light over a controlled path for distances measured in miles rather than feet. It was proposed in 1966 by Kao and Hockham that glass, if sufficiently purified, could form such a light guide by confining the light to the central region of an optical fiber through the principle of total internal reflection (TIR). Although this theoretical paper suggested the possibility of optical fiber communications, the attenuation of light in the glasses available at that time was too great to make the scheme practical.

To quantify the degree of light attenuation in glass, we digress from the historical development to define the *decibel,* or dB, which is commonly used to characterize attenuation. If light of power P_{in} is incident on a length of fiber, and light of power P_{out} exits the far end of that fiber, then the dB loss is defined as

$$\text{dB loss} = 10 \log_{10}\left(\frac{P_{in}}{P_{out}}\right) \tag{1-1}$$

which can also be written in the form

$$\frac{P_{out}}{P_{in}} = 10^{-(\text{dB loss}/10)} \tag{1-2}$$

From this definition, you can see that a factor of 10 drop in power corresponds to a 10 dB loss, a factor of 100 drop corresponds to a 20 dB loss, and so on. In an electrical circuit, power is proportional to the square of the voltage, so in electrical engineering the dB loss is often defined in terms of a voltage ratio as

$$\text{dB loss} = 10 \log_{10}\left(\frac{V_{in}^2}{V_{out}^2}\right) = 20 \log_{10}\left(\frac{V_{in}}{V_{out}}\right) \tag{1-3}$$

The utility of the decibel concept becomes apparent when loss elements are cascaded. Suppose there are two fiber lengths, as shown in Fig. 1-1, with losses of $(\text{dB loss})_1$ and $(\text{dB loss})_2$ respectively. Light of power P_1 enters the first fiber, and light of power P_2 exits this fiber. The light power P_2 then enters the second fiber, and exits the second fiber with power P_3. The dB losses for the individual fiber sections are

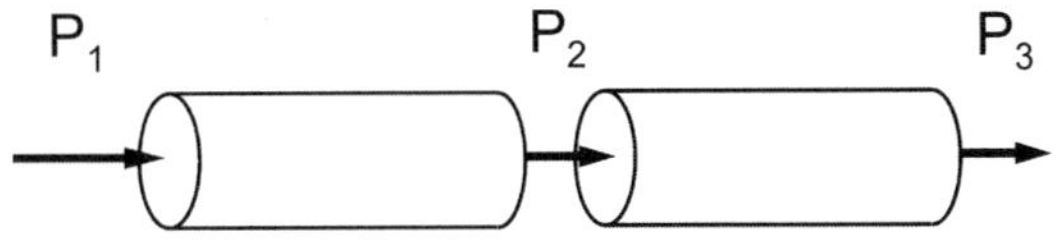

Figure 1-1 The dB concept is useful for cascaded losses.

$$(\text{dB loss})_1 = 10 \log_{10}\left(\frac{P_1}{P_2}\right)$$
$$(\text{dB loss})_2 = 10 \log_{10}\left(\frac{P_2}{P_3}\right) \qquad (1\text{-}4)$$

The overall loss for the combination is

$$\text{dB loss} = 10 \log_{10}\left(\frac{P_1}{P_3}\right) = 10 \log_{10}\left(\frac{P_1}{P_2}\cdot\frac{P_2}{P_3}\right) \qquad (1\text{-}5)$$

or, using Eqs. 1-4 and the properties of logarithms,

$$\text{dB loss} = (\text{dB loss})_1 + (\text{dB loss})_2 \qquad (1\text{-}6)$$

The advantage of using decibels to describe attenuation is that we just need to add and subtract when combining elements. In complex systems, this is more convenient than multiplying and dividing by transmission factors such as P_2/P_1 and P_3/P_2. In a similar way, optical power is often expressed logarithmically in terms of dBm, according to the definition

$$\text{optical power in dBm} = 10 \log_{10}\left(\frac{P}{1\text{ mW}}\right) \qquad (1\text{-}7)$$

This gives the optical power relative to 1 mW on a logarithmic scale, so that, for example, an optical power of –20 dBm is 0.01 mW, whereas an optical power of +20 dBm is 100 mW. This is convenient because a loss measured in dB can then simply be subtracted from the original optical power (measured in dBm) to obtain the new optical power.

Since the dB loss is additive for cascaded fiber lengths, the total attenuation in dB for a fiber is proportional to the length of the fiber. This allows us to characterize the fiber loss in units of dB/km, the attenuation per unit length. Typical loss coefficients for glass available in the 1960s were on the order of 10^3 dB/km, or 1 dB/m. For a practical fiber optic communications system, it was estimated that the loss would have to be reduced to the order of 10 dB/km. Using Eq. 1-2, you can gain an appreciation for the extreme transparency of such a fiber. This glass would be so clear that after light propagated through it for 1000 feet, approximately 50% of the optical power would still remain. The challenge in obtaining such glass was to remove the impurities from the glass, which were largely responsible for the high attenuation.

A breakthrough occurred in 1970, when a team of researchers from Corning Inc. found a way to dramatically decrease the fiber loss by depositing highly purified SiO_2 vapor on the inside of a glass tube. After heating and drawing the tube into a fiber, this process (now referred to as "inside vapor deposition") resulted in losses below 20 dB/km. From this point on, development of low-loss optical fibers was rapid, achieving 0.5 dB/km in 1976, 0.2 dB/km in 1979, and 0.16 dB/km in 1982. This loss was now approaching the theoretical limit for silica fibers, for reasons that will be discussed in Chapter 5. The transparency of these later-generation fibers was incredible—50% transmission through 10 miles of fiber. With such low-loss fibers available, optical fibers began to replace copper wires for most long-haul telecommunications.

There are a number of advantages that led to the widespread replacement of copper wires by fiber optic cables. Optical fibers can transmit data at a higher rate, over longer distances, and in a smaller volume compared with copper wires. The fiber optic cable is lightweight and flexible, and can carry information further before the signal needs to be amplified. The natural resources needed to make fiber are not scarce—mostly silicon and oxygen, which make up a large part of the sand found on beaches. A further advantage of optical transmission is its insensitivity to electrical interference. Optical signals can maintain their high quality, even over the vast distances between the continents.

1-3. OVERVIEW OF TOPICS

Although the field of photonics is broader than fiber optic communications, the various components of a fiber optic system provide a convenient framework for presenting the basic principles of photonics. The approach we will take in this book is to frame the discussion around the elements of an optical communications system, recognizing that these same components have a number of applications in the general area of photonics.

Figure 1-2 gives a schematic overview of the components of a fiber optic communications system. Source data, in the form of audio, video, or computer data, is converted into a digital data stream, and this is used to modulate (see Chapter 9) the intensity of a light source, typically a light-emitting diode (LED) (see Chapter 11) or a laser (see Chapters 15–23). This light is coupled into a fiber (see Chapter 12), and the light propagates (see

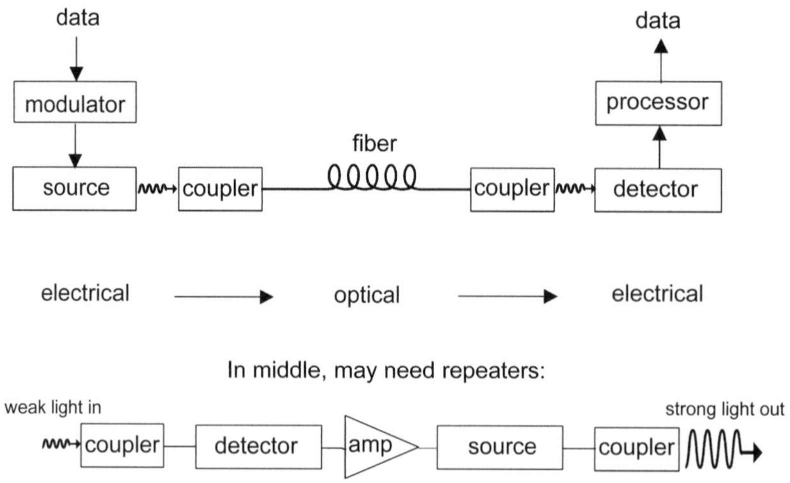

Figure 1-2 Components of an optical communications system.

Chapters 3–6) along the fiber to the receiving location, where it is decoupled from the fiber and converted into an electrical signal by a photodetector (see Chapter 13), typically a photodiode (see Chapter 14). The electrical signal from the photodetector is then decoded and converted back into a replica of the original source data.

During the propagation of light along the fiber, the signal strength decreases due to scattering and absorption losses (see Chapter 5), and amplification is necessary. In early fiber optic systems, this amplification was accomplished by converting the weak light signal into an electrical signal, amplifying the electrical signal with conventional electronic amplifiers, and then regenerating a strong light signal from this amplified electrical signal. Devices that accomplish this task are called repeaters, and they were a major part of the cost of early fiber optic systems. Since the mid 1990s, these repeaters have largely been supplanted by optical amplifiers (see Chapters 19 and 24), in which the weak light signal is directly amplified within a fiber by the process of stimulated emission. Optical amplifiers allow light signals of different wavelengths to be amplified simultaneously with high efficiency. The transmission of many distinct information channels down a single fiber, each at a slightly different wavelength, is known as wavelength division multiplexing (WDM) (see Chapter 24). This technology has rapidly expanded the capacity of fiber optic systems, allowing the transmission of bandwidth-hungry multimedia content over the Internet. Future developements will likely include the practical implementation of new types of optical waveguides, such as photonic crystals (see Chapter 8). In these new materials, light is confined to a region of space by novel interference effects in a nanostructured material. The possibilities seem endless, and it may be no exageration to say that, just as the 20th century was the age of electronics, so the 21st century will be the age of photonics.

Chapter 2

Review of Optics

2-1. THE NATURE OF LIGHT

In this chapter, we will review those aspects of optics that are most relevant to the study of photonics. It is natural to begin with the fundamental question, What is light? Historically, light has at times been considered to be in the form of particles, or corpuscles, a point of view favored by Isaac Newton. The view of light as a wave was promoted in the 17th century by Christiaan Huygens, among others, and came to dominance after the experiments of Thomas Young on light interference in the early 19th century. Our modern view of light arose during the early part of the 20th century with the advent of quantum mechanics. In this view, light must be considered to be both a particle and a wave, in the same way that material particles such as electrons have both a particle and wave nature. Generally, the classical, or wave nature of light is appropriate when light is propagating from one point to another, whereas the quantum, or particle nature of light manifests itself when light is absorbed or emitted by atoms. During absorption or emission, light acts like a stream of particles or packets of energy called *photons.* Each photon contains energy equal to

$$E_{\text{photon}} = h\nu = \frac{hc}{\lambda} \tag{2-1}$$

where $h = 6.63 \times 10^{-34} J \cdot s$ is Planck's constant, ν and λ are the frequency and wavelength of the light wave, respectively, and c is the speed of light in a vacuum. In most situations other than absorption and emission, light can be treated as a wave, consisting of oscillating electric and magnetic fields. The variation of these two fields in space and time is governed by Maxwell's equations, the treatment of which are outside the scope of this book. We will, however, quote certain results from Maxwell's equations from time to time and use these results to explain various phenomena relevant to photonics. The interested reader is referred to the bibliography for more advanced treatments that show how these results follow from Maxwell's equations.

One simple solution to Maxwell's equations in a uniform medium is that of a *plane wave,* in which the electric field is constant everywhere along a plane (at a particular instant in time), and varies sinusoidally in a direction perpendicular to that plane. For example, if the electric field varies in the x direction, then

$$\mathbf{E} = \mathbf{E}_0 e^{i(kx-\omega t)} \tag{2-2}$$

where $k \equiv 2\pi/\lambda$ is the wave vector magnitude or *wavenumber,* $\omega \equiv 2\pi\nu$ is the angular frequency (measured in radians per second), and the quantity $\phi = (kx - \omega t)$ is the *phase* of the wave. Here and throughout the book, it will often be convenient to use the complex exponential notation for waves and oscillations, with the understanding that the real part of the expression corresponds to the physical oscillation. Using Euler's identity $e^{i\theta} = \cos(\theta) + i\sin(\theta)$, the wave in Eq. (2-2) is then equivalent to

$$\mathbf{E} = \mathbf{E}_0 \cos(kx - \omega t) = \mathbf{E}_0 \cos \phi(t) \tag{2-3}$$

The electric field amplitude $\mathbf{E}_0$ is a vector in the y–z plane. If $\mathbf{E}_0 = E_0\,\hat{y}$, the wave is said to be polarized in the y direction, whereas if $\mathbf{E}_0 = E_0\hat{z}$, it is polarized in the z direction. Any other direction for $\mathbf{E}_0$ can be described by a linear combination of polarizations in the y and z directions, so we say in general that there are two distinct polarizations for a given plane wave. Figure 2-1 shows the variation of E_y with x and t for y-polarized light. The value of E_y depends on the phase ϕ of the wave at a particular x and t. When $\phi = 0$, E_y is at a positive maximum, and when $\phi = \pi/2$, $E_y = 0$. A phase $\phi = \pi$ gives a negative maximum in E_y, and $\phi = 2\pi$ gives again a positive maximum. The wave is therefore periodic in phase with period 2π. It is periodic in space with wavelength λ and periodic in time with period T.

The light wave contains not only an electric field, but also an oscillating magnetic field. As indicated in Fig. 2-2, the magnetic field has the same dependence on time and space as the electric field, but is perpendicular to both the electric field and the direction of propagation. The relative orientation of $\mathbf{E}$ and $\mathbf{B}$ is always such that the cross product $\mathbf{E} \times \mathbf{B}$ is in the direction of wave propagation. For an arbitrary wave direction, Eq. (2-2) can be generalized to

$$\mathbf{E} = \mathbf{E}_0 e^{i(\mathbf{k}\cdot\mathbf{r} - \omega t)} \tag{2-4}$$

In this case, the planes of constant phase are perpendicular to the wave vector $\mathbf{k}$, which specifies the direction of wave propagation. The wavelength is related to the wave vector by $k = |\mathbf{k}| = 2\pi/\lambda$.

The wave in Eq. (2-2) is characterized by planes of constant phase at $x = \omega t/k$ where the amplitude is a maximum. As time advances, these planes of constant amplitude propagate in the $+x$ direction with a speed

$$v_p = \frac{\omega}{k} = \nu\lambda \tag{2-5}$$

which is referred to as the *phase velocity* of the wave. For electromagnetic waves in a vacuum, this phase velocity is $v_p = c$, where $c = 3 \cdot 10^8$ m/s is the speed of light. In a material

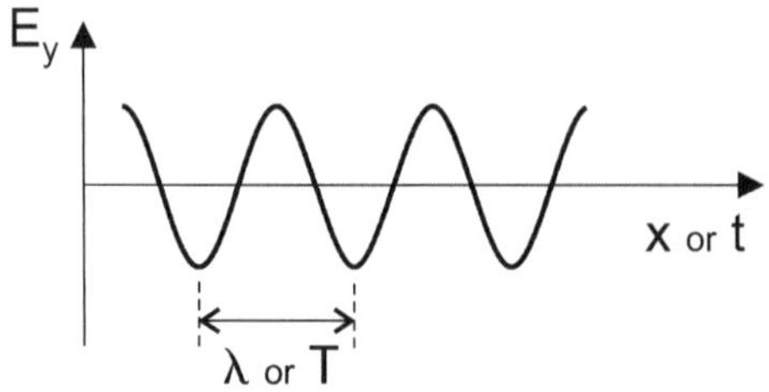

Figure 2-1 Electric field oscillation in time and space.

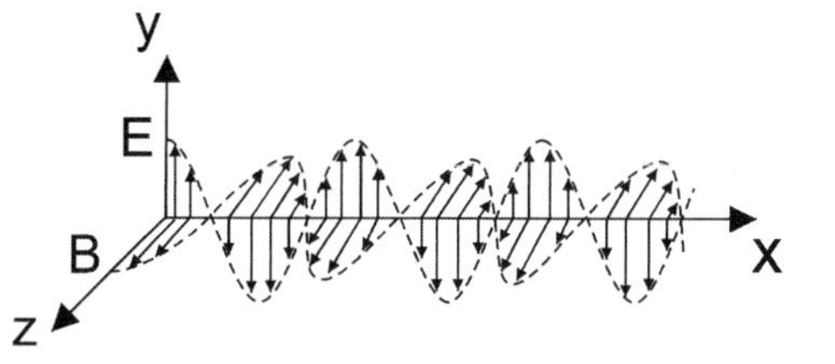

Figure 2-2 Transverse electromagnetic wave.

medium, the atoms interact with the light, and the phase velocity of the wave is changed to

$$v_p = \frac{c}{n} = \frac{\omega}{nk_0} \tag{2-6}$$

where n is the *index of refraction* and we have defined the free-space *wavenumber* $k_0 = 2\pi/\lambda_0$ in terms of the free-space wavelength $\lambda_0 = c/\nu$. The effect of a higher refractive index is to slow the wave down and to decrease the wavelength to $\lambda = \lambda_0/n$. Table 2-1 gives the index of refraction for a few materials.

Phase and Group Velocity

Generally, the index of refraction is greater than 1, so the speed of light in a medium is less than the speed of light in a vacuum. This is in accordance with special relativity, which indicates that speeds greater than c are not allowed because causality would be violated. However, in certain cases it is possible to have $n < 1$, which implies $v_p > c$. At first glance, this would seem to be inconsistent with relativity, since we have something moving faster than the speed of light. The reason that this is not a problem is that the wave of Eq. (2-2) conveys no information, since it is infinite in extent—it has no beginning and no ending. In order to transmit information, you must modulate this wave, that is, turn it on and off to create a pulse, as shown in Fig. 2-3. Such a pulse of light can be represented by a linear superposition of infinite plane waves having some distribution of frequencies, in

Table 2-1 Refractive index of selected materials at the specified wavelength. Variation with wavelength is given in Palik (1985).

Material	Index	λ (μm)
air	$\cong 1$	all
water	1.33	0.65
fused silica (SiO_2)	1.45	1
silicate glass	≈ 1.5	1
sapphire (Al_2O_3)	1.76	0.83
$Al_{0.3}Ga_{0.7}As$	3.4	0.88
Si	3.45	2
GaAs	3.6	0.88
InAs	3.5	4
Ge	4.0	4-10

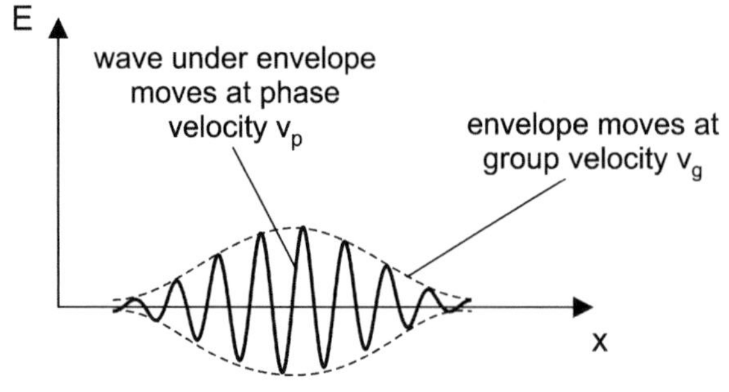

Figure 2-3 The peak of the envelope function moves at the group velocity.

the manner of Fourier synthesis (see Appendix B). The speed with which the pulse propagates is given by the *group velocity:*

$$v_g = \frac{d\omega}{dk} \tag{2-7}$$

If the index of refraction is independent of frequency, the medium is said to be dispersionless, and the two velocities are the same since $d\omega/dk = \omega/k$. Generally, there is dispersion (index varies with frequency) and $v_g \neq v_p$. In this case, it is the group velocity that determines how fast information can be transmitted. This applies not only to plane waves, but also to waveguide modes in an optical fiber.

Energy in a Light Wave

A propagating electromagnetic wave carries energy in both its electric and magnetic field components. The energy per unit volume stored in the electric field component is

$$\rho = \tfrac{1}{2}\,\varepsilon_r \varepsilon_0 E^2 \qquad \text{(energy density in } E \text{ field)} \tag{2-8}$$

where $\varepsilon_0 \simeq 8.85 \times 10^{-12}$ F/m is the *permittivity* of free space, a fundamental constant of nature, and ε_r is the relative dielectric constant, related to the index of refraction by $\varepsilon_r = n^2$. For a plane wave, it turns out that the electric and magnetic components carry an equal amount of energy, so the total energy density should be twice that of Eq. (2-8). However, the fields oscillate in time as $\cos \omega t$, and averaging $\cos^2 \omega t$ over a complete cycle gives an additional factor of $\frac{1}{2}$. As a result, Eq. (2-8) is valid as well for the total energy in a light wave, with E the peak electric field.

It is useful to characterize the energy in a light beam not just by the energy density ρ, but also by the rate at which energy flows across a surface. The *intensity* I is defined as the energy passing through a surface per unit time per unit surface area when the surface is oriented perpendicular to the beam. I and ρ can be related by considering a beam with intensity I and cross-sectional area A that propagates for time Δt, filling a cylinder of length $L = (c/n)\Delta t$ with light energy. The energy inside the cylinder can be computed from either the intensity or energy density, as shown in Fig. 2-4, giving the desired relationship

$$I = \frac{c}{n}\rho = \frac{1}{2} c n \varepsilon_0 E^2 \tag{2-9}$$

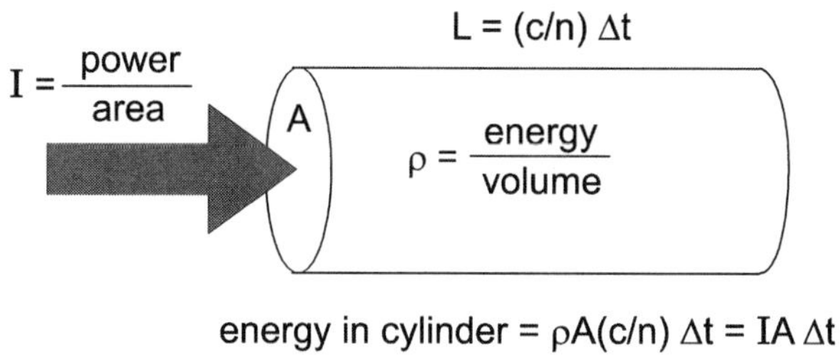

Figure 2-4 Light energy inside cylinder can be calculated either from (energy density) × (volume) or (intensity) × (area) × (time interval).

2-2. LIGHT AT A BOUNDARY

The discussion so far has been for plane waves in an infinite, uniform medium. We consider now what happens to plane waves at the boundary between two semiinfinite media having different indices of refraction.

Snell's Law

Figure 2-5 shows an incident plane wave propagating toward such a boundary from medium 1, the side with index of refraction n_1. At the boundary, some of the light energy is reflected back into medium 1. The remainder is transmitted into medium 2 (with index n_2), undergoing a change in direction known as *refraction.* The direction of each wave is specified by the angle between its **k** vector and the normal to the boundary (indicated by the dotted line). The law of reflection states that the angle of reflection is equal to the angle of incidence, just as for a mirror. The angle of refraction θ_2 is related to the angle of incidence θ_1 by *Snell's law,* which can be written as

$$n_1 \sin \theta_1 = n_2 \sin \theta_2 \tag{2-10}$$

If the second medium has a higher index than the first ($n_2 > n_1$), Snell's law says that the angle of refraction is smaller than the angle of incidence ($\theta_2 < \theta_1$). In such a case (pic-

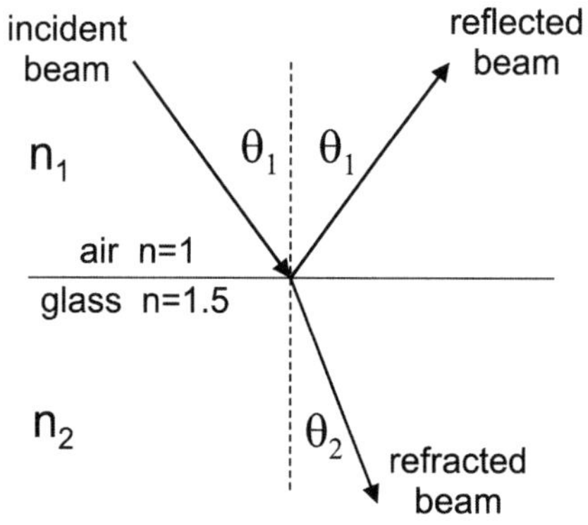

Figure 2-5 Refraction of light at the dielectric boundary.

tured in Fig. 2-5), we say that the refracted beam is bent toward the normal. Similarly, if $n_2 < n_1$, then $\theta_2 > \theta_1$ and the beam is bent away from the normal. Generally, materials that are more dense have a greater index of refraction, so we can summarize this by saying that in going from less dense to more dense materials, the refracted beam is bent toward the normal, and vice versa.

The change in direction of the refracted beam can be understood intuitively by considering how the wave fronts (planes of constant phase) behave at the boundary. Fig. 2-6 shows the wave fronts of the plane wave as they pass through the boundary. All parts of the wave front move at the same speed while in medium 1, but when part of the wave front passes into medium 2 it moves with a slower speed (assuming, for example, that $n_2 > n_1$). Since the part of the wave front still in medium 1 is moving faster than the part now in medium 2, the wave fronts must bend as shown, resulting in a change in the beam's direction once it is entirely in medium 2. If you pursue this argument quantitatively, you arrive at Snell's law.

Fresnel's Equations

Snell's law tells us the allowed directions for any reflected or refracted beams, but it does not tell us what fraction of the incident beam is reflected and what fraction is transmitted. This information is provided by Fresnel's equations, which are derived by requiring that the plane wave solutions on either side of the boundary satisfy certain boundary conditions. For example, the component of electric field parallel to the interface must be continuous as we cross the boundary. Since this component will in general depend on the polarization of the light, there will be different Fresnel's equations for the different polarizations.

Figure 2-7 shows the notation commonly used to describe plane wave reflection and refraction at a planar boundary. We define the *plane of incidence* as the plane formed by the incident, reflected, and refracted rays (they must all be in the same plane for a uniform boundary, due to symmetry). When the electric field of the incident wave is in the plane of incidence, the light is said to be p polarized, or TM (transverse magnetic). The TM no-

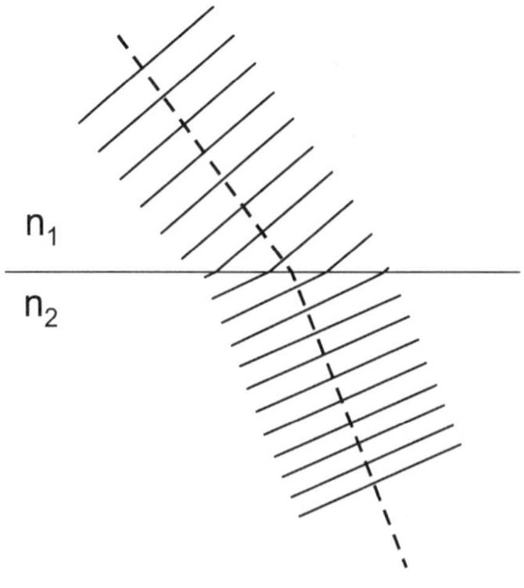

Figure 2-6 Bending of wavefronts in refraction, shown for $n_2 > n_1$.

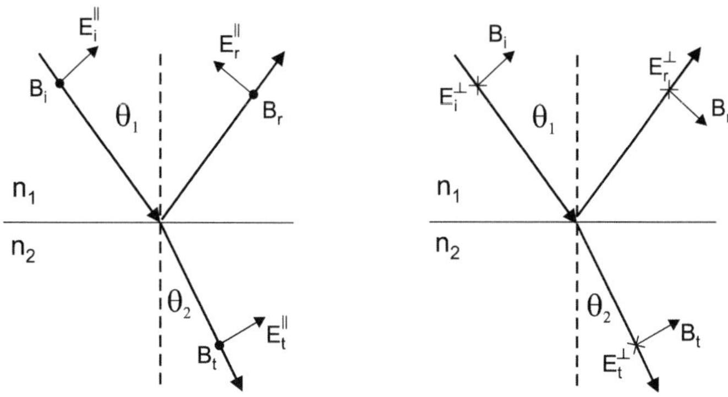

Figure 2-7 Polarization notations for Fresnel reflection. Dots and crosses denote fields out of or into the page, respectively.

tation refers to the fact that the magnetic field is perpendicular, or transverse, to the plane of incidence. The electric field of p polarized light will be denoted as $E^{\|}$. Light with the electric field perpendicular to the plane of incidence is said to be s polarized, or TE (transverse electric). The electric field of s polarized light will be denoted as $E^{\perp}$.

The *Fresnel equations* for the reflected and transmitted E fields in p polarization (TM) can be written as

$$\left(\frac{E_r}{E_i}\right)_{\|} = \frac{n_1 \cos\theta_2 - n_2 \cos\theta_1}{n_1 \cos\theta_2 + n_2 \cos\theta_1} \tag{2-11a}$$

$$\left(\frac{E_t}{E_i}\right)_{\|} = \frac{2n_1 \cos\theta_1}{n_1 \cos\theta_2 + n_2 \cos\theta_1} \tag{2-11b}$$

and for s polarization (TE) they are

$$\left(\frac{E_r}{E_i}\right)_{\perp} = \frac{n_1 \cos\theta_1 - n_2 \cos\theta_2}{n_1 \cos\theta_1 + n_2 \cos\theta_2} \tag{2-12a}$$

$$\left(\frac{E_t}{E_i}\right)_{\perp} = \frac{2n_1 \cos\theta_1}{n_1 \cos\theta_1 + n_2 \cos\theta_2} \tag{2-12b}$$

To determine the fraction of incident light power reflected and transmitted by the boundary, we use the fact that the energy carried by a light wave is proportional to the square of its electric field amplitude. The power reflection and transmission coefficients R and T can then be found from

$$\begin{aligned} R &\equiv \frac{I_r}{I_i} = \left(\frac{E_r}{E_i}\right)^2 \\ T &\equiv \frac{I_t}{I_i} = \frac{n_2 \cos\theta_2}{n_1 \cos\theta_1}\left(\frac{E_t}{E_i}\right)^2 \end{aligned} \tag{2-13}$$

where the intensity $I \propto nE^2$ from Eq. (2-9) has been used. Note that E_t/E_i can be > 1, although the fraction T of light power that is transmitted is always < 1. It can easily be verified from Eqs. (2-11), (2-12), and (2-13) that $R + T = 1$, which is consistent with conservation of energy.

The Fresnel equations are simplified considerably in the case of normally incident light, where $\theta_1 = \theta_2 = 0$. The fraction of light reflected then becomes

$$R = \frac{(n_1 - n_2)^2}{(n_1 + n_2)^2} \tag{2-14}$$

a result valid for both polarizations. Note that the reflectivity remains the same when n_1 and n_2 are interchanged, which means that at normal incidence the reflectivity is the same from either side of the boundary.

EXAMPLE 2-1

Determine the fraction of light transmitted at normal incidence through a pane of window glass in air, assuming that the glass has an index of refraction of $n = 1.5$.

Solution: The fraction of light transmitted through the first interface (going from air to glass) is

$$T_1 = 1 - R_1 = 1 - \frac{(1.5 - 1.0)^2}{(1.5 + 1.0)^2} = 0.96$$

In going through the second interface (glass to air), the transmission factor is the same, $T_2 = 0.96$. The total transmission through the combination is therefore $T = T_1 T_2 = 0.922$. The fractional reflection loss at each interface is 4% and the total reflection loss is approximately twice that. This calculation ignores multiple reflections, which only become important when the reflection coefficient is much higher.

Brewster's Angle

Another special case in which the Fresnel's equations are simplified is when the numerator in Eq. (2-11a) goes to zero. Under these conditions, there will be no reflected beam and all of the light energy is transmitted from medium 1 into medium 2. This will occur at a particular angle of incidence θ_1 known as *Brewster's angle,* which can be found by setting

$$n_1 \cos\theta_2 - n_2 \cos\theta_1 = 0$$

To solve for the Brewster's angle, we combine this with Snell's law to obtain the two equations

$$\begin{aligned} n_1 \cos\theta_2 &= n_2 \cos\theta_1 \\ n_1 \sin\theta_1 &= n_2 \sin\theta_2 \end{aligned} \tag{2-16}$$

It is clear by inspection that these two equations will both be satisfied simultaneously when $\theta_1 + \theta_2 = 90°$, since in that case $\cos\theta_2 = \sin(90 - \theta_2) = \sin\theta_1$, and similarly, $\cos\theta_1 = \sin\theta_2$. From Snell's law we then have

$$\sin \theta_1 = \frac{n_2}{n_1} \sin \theta_2 = \frac{n_2}{n_1} \cos \theta_1$$

which yields the Brewster's angle $\theta_B = \theta_1$:

$$\tan \theta_B = \frac{n_2}{n_1} \tag{2-17}$$

This Brewster's angle was found for p polarized light [Eq. (2-11a)]. It is left as an exercise for the reader to show that there is no Brewster's angle for s polarized light.

There is a simple way to understand physically why there is no reflected light at the Brewster's angle. At an atomic level, the reflection of light from a solid surface can be thought of as the radiation of light from electrons in the solid that are accelerated sinusoidally by the electric field of the light wave. Electrons that are accelerated in a particular direction radiate light preferentially in a direction perpendicular to the acceleration vector, as shown in Fig. 2-8. There is no radiated light along the direction of the electron's acceleration. The emission pattern is similar to that from a half-wave dipole antenna, in which the transmission (or reception) is most efficient perpendicular to the wire. Now consider p polarized light incident on the interface from medium 1, as depicted in Fig. 2-9. Electrons in medium 2 are sinusoidally accelerated in the direction of the electric field in medium 2, and radiate light with the angular distribution shown in Fig. 2-8. At Brewster's angle the transmitted and reflected beams would be at right angles and there would be no light energy radiated into the reflected beam direction because this direction is parallel to the electron's acceleration. For s polarized light, on the other hand, the E field is always perpendicular to the reflected beam direction and there is no Brewster's angle.

Fig. 2-10 shows the power reflection coefficients R as a function of angle of incidence θ_1 for $n_2/n_1 = 1.5$, calculated using Eqs. (2-10), (2-11), (2-12), and (2-13). The two polarizations have the same reflectivity at normal incidence ($\theta_1 = 0$), as discussed previously. The p polarized reflectivity decreases with increasing angle of incidence, going to zero at the Brewster's angle, and then increases with a further increase in incident angle up to a maximum of unity at $\theta_1 = 90°$. The s polarized reflectivity, on the other hand, increases monotonically with increasing θ_1 up to the same maximum of unity at $\theta_1 = 90°$. In each case, the fraction of incident light transmitted can be found from $T = 1 - R$.

Total Internal Reflection

When going from a less dense to a more dense medium, as in Fig. 2-10, there is a transmit-

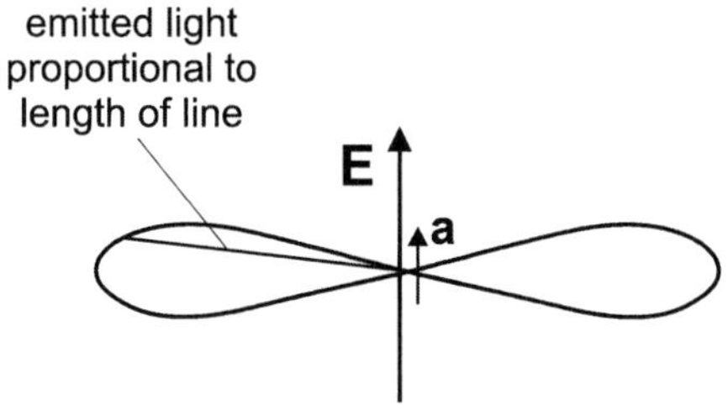

Figure 2-8 The electric dipole radiation pattern is directed perpendicular to the electron's acceleration vector **a**.

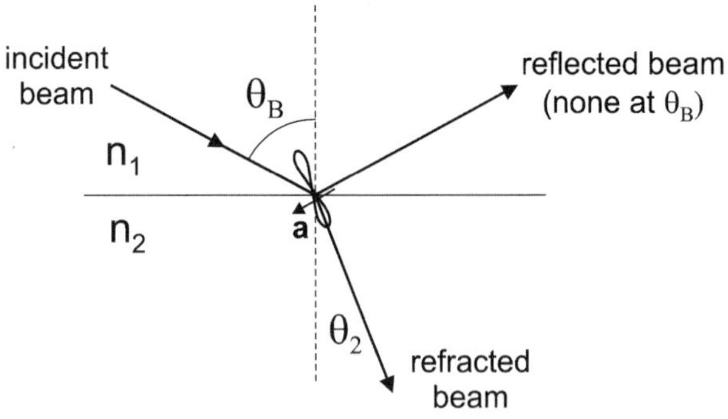

Figure 2-9 At Brewster's angle, there is no dipole radiation in the direction of the reflected beam and, hence, no reflected light.

ted beam for all angles of incidence θ_1. In going from a more dense to a less dense medium, however, there is a restricted range of incident angles that allow a transmitted beam. To see why, consider a beam with angle of incidence θ_1 in a medium with refractive index n_1, passing into a medium with refractive index $n_2 < n_1$, as shown in Fig. 2-11a. Snell's law (Eq. 2-10) dictates that the beam gets bent away from the normal, $\theta_2 > \theta_1$. As θ_1 increases, the angle of refraction θ_2 also increases until $\theta_2 = 90°$. This occurs at an angle of incidence $\theta_1 = \theta_c$, where θ_c is the *critical angle.* At the critical angle, Snell's law becomes

$$n_1 \sin \theta_c = n_2 \sin(90°)$$

or

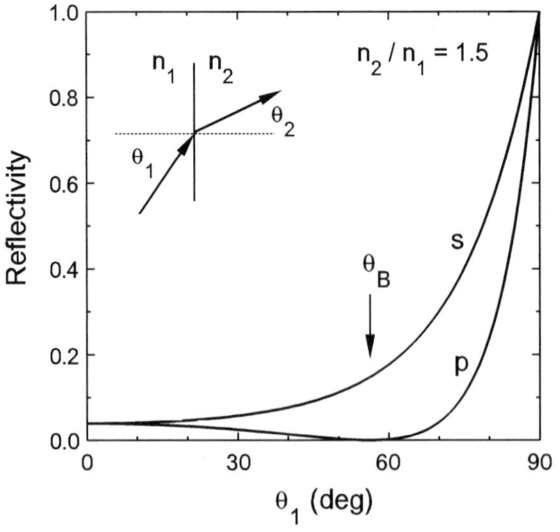

Figure 2-10 Variation of reflectivity with angle of incidence for a dielectric interface having n_2/n_1 = 1.5. Reflectivity goes to zero at Brewster's angle for p polarization, but not for s polarization.

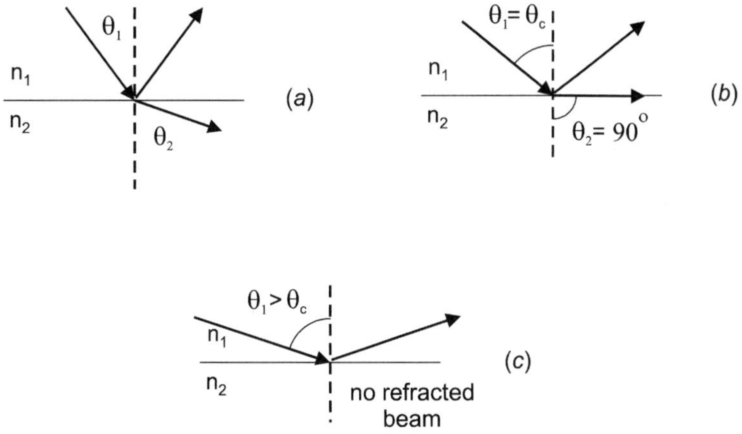

Figure 2-11 Total internal reflection occurs when the incident angle θ_1 exceeds the critical angle θ_c.

$$\sin \theta_c = \frac{n_2}{n_1} \tag{2-18}$$

For angles of incidence greater than the critical angle ($\theta_1 > \theta_c$), the light is completely reflected back into medium 1 and there is no transmitted beam. The interface acts like a perfect, lossless mirror with angle of incidence equal to angle of reflection. This situation (see Fig. 2-11c) is referred to as *total internal reflection,* and is the basis for long-distance propagation of light down optical fibers. It is also used to make low-loss mirrors for directing high-power laser beams (see Problem 2.9).

Although there is no propagating wave in medium 2 under conditions of total internal reflection, there is still an electric field (and magnetic field) which penetrates into

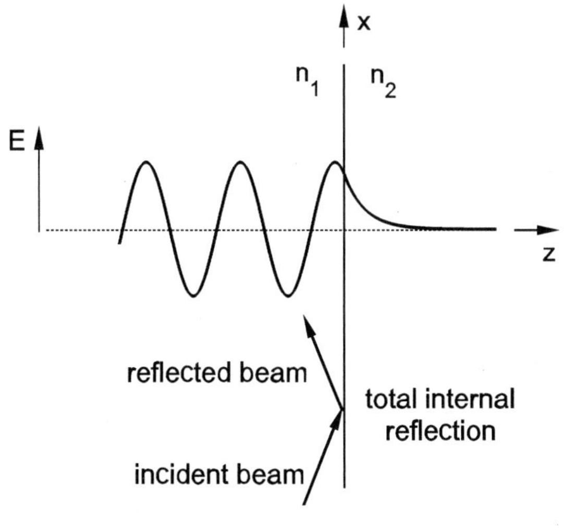

Figure 2-12 An evanescent wave decays exponentially in the lower index medium n_2 during total internal reflection. The wave shown corresponds to grazing incidence $\theta_1 \approx 90°$ and $n_1/n_2 = 1.5$.

the second medium. This field decays exponentially as a function of distance into medium 2 and is referred to as an *evanescent field.* Figure 2-12 shows how the z dependence of the E field changes from oscillatory in the higher index medium to exponentially decaying in the lower index medium. The variation of field with z in the second medium is given by

$$E(z) = E_0 e^{-\alpha z} \tag{2-19}$$

where

$$\alpha = k_0 \sqrt{n_1^2 \sin^2 \theta_1 - n_2^2} \tag{2-20}$$

and z is measured from the boundary. Using $k_0 = 2\pi/\lambda_0$, we find from Eqs. (2-18), (2-19), and (2-20) that the E field decays by a factor $1/e$ at a distance $\delta = 1/\alpha$ from the interface, where

$$\delta = \frac{\lambda_0}{2\pi n_1 \sqrt{\sin^2 \theta_1 - \sin^2 \theta_c}} \tag{2-21}$$

It can be seen that in general the evanescent field decays to a negligible value after a very small distance from the interface, on the order of the wavelength. The exception to this is for incident angles close to the critical angle. When $\theta_1 \approx \theta_c$, the penetration depth δ can become much greater than the wavelength. This result will prove to be useful in understanding certain properties of optical waveguides.

We have emphasized that under total internal reflection there is no steady-state propagation of energy into the second medium. There is, however, some propagation of energy within the second medium *parallel to the interface.* The path of energy flow for a ray of light with finite lateral extent is indicated in Fig. 2-13. The apparent lateral displacement of the beam along the interface is known as the *Goos–Haenchen Shift,* and is generally quite small (less than a wavelength). The phase of the reflected wave is also shifted with respect to that of the incident wave. Taking the time dependence of the incident wave to be

$$E_{\text{incident}} = E_0 e^{i\omega t}$$

the time dependence of the reflected wave is

$$E_{\text{reflected}} = E_0 \, e^{i(\omega t + \phi_r)} \tag{2-22}$$

where ϕ_r is the phase shift upon reflection. For TE polarization, this phase shift is given by

$$\tan \frac{\phi_r}{2} = \frac{\sqrt{\sin^2 \theta_1 - (n_2/n_1)^2}}{\cos \theta_1} \tag{2-23}$$

and for TM polarization it is

$$\tan \frac{\phi_r}{2} = \left(\frac{n_1}{n_2}\right)^2 \frac{\sqrt{\sin^2 \theta_1 - (n_2/n_1)^2}}{\cos \theta_1} \tag{2-24}$$

Note that the phase shift goes to zero at the critical angle, and goes to 180° as $\theta_1 \rightarrow 90°$. These expressions will be useful in the analysis of modes in planar waveguides.

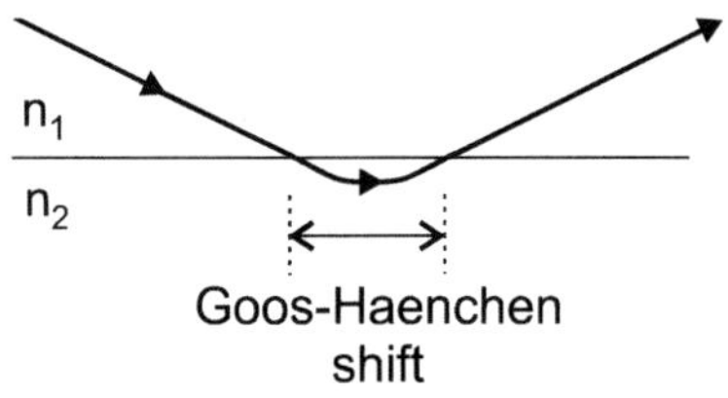

Figure 2-13 Lateral shift in position of reflected beam in total internal reflection.

2-3. LIGHT PASSING THROUGH APERTURES

The two phenomena of diffraction and interference are closely related, and can be understood by considering how light behaves once it has passed through one or more apertures.

Diffraction

The preceding discussion of reflection and refraction has assumed plane waves of infinite extent perpendicular to the direction of propagation **k**. In practice, a beam of light has a finite lateral width, and this causes the beam to spread out as it propagates, a process called *diffraction*. As indicated in Fig. 2-14, a beam with initial diameter D will diverge into a cone of half-angle

$$\theta \sim \frac{\lambda}{D} \tag{2-25}$$

where λ is the wavelength of light in the medium. The boundary of this cone is not sharp but is defined so that most of the light energy is contained within the cone. If the beam intensity at the beam waist (smallest-diameter region) falls off gradually away from the beam axis, then the light distribution near the cone edge is also smooth, falling off monotonically to zero. If the intensity distribution at the beam waist is sharp, however, then the the light intensity near the cone edge will be oscillatory, with dark and bright fringes appearing around a central bright spot. This would be the case for an infinite plane wave

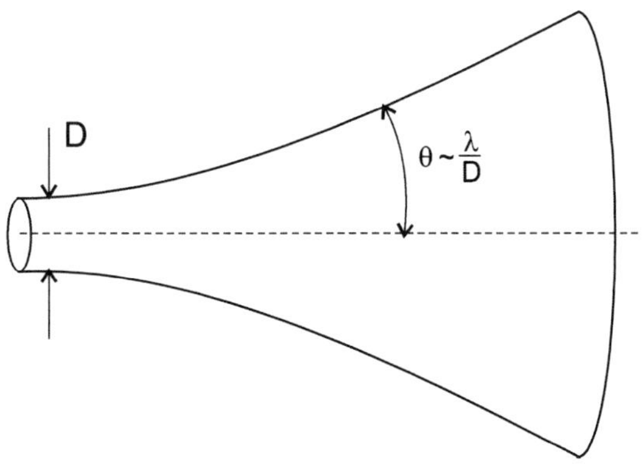

Figure 2-14 Diffraction of light through an aperture of dimension *D*.

passing through a sharp circular aperture of diameter D. The exact version of Eq. (2-25) for this situation is

$$\theta = 1.22\,\frac{\lambda}{D} \tag{2-26}$$

where now the cone edge is defined by the angle θ at which the first dark ring appears in the diffraction pattern (the factor of 1.22 comes from the zero of a Bessel function). The appropriate numerical factor to be put in the proportionality of Eq. (2-25) depends on how sharply the intensity falls off at the beam waist, and also on how the cone edge is defined. Other common definitions for the cone edge are the angles at which the diffracted light intensity falls to 1/2 or $1/e^2$ of the value at the center. The latter definition is used for laser beams that have a Gaussian distribution (see Chapter 17), where it is found that

$$\theta = \frac{\lambda}{\pi w_0} \qquad \text{(Gaussian beam divergence)} \tag{2-27}$$

Here, w_0 is the initial beam radius, defined by the $1/e^2$ intensity point.

Interference

The diffraction of light can be viewed as a special case of the general phenomenon of light *interference.* To understand the essence of interference, consider an infinite plane wave of wavelength λ incident on a mask containing two small pinhole apertures, as shown in Fig. 2-15. According to Huygen's wavelet principle, the light field inside each pinhole can be considered to be a new source of radiated light, emitting a spherical light

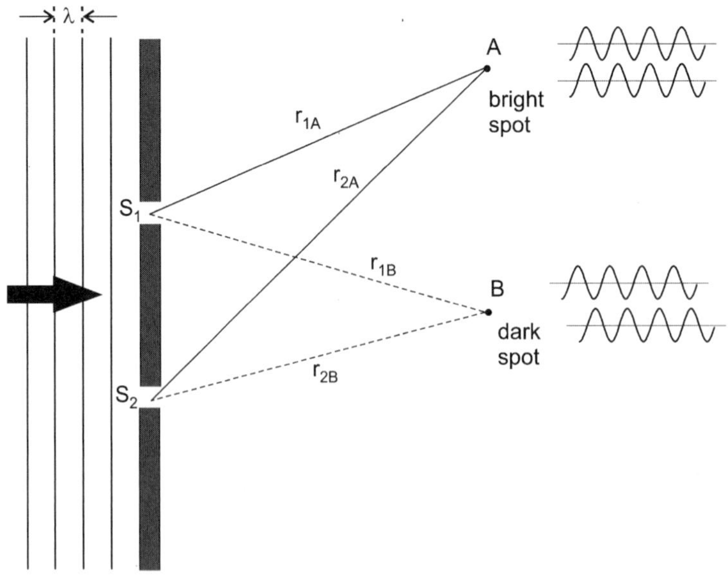

Figure 2-15 Interference of light from two pinhole sources, S_1 and S_2. At point A, the waves from the two sources are in phase, whereas at point B they are out of phase.

wave centered about that pinhole. At some point A in space (or on a screen), the electric fields of the light waves from the two sources S_1 and S_2 may arrive in phase, which means that the electric fields are both a maximum at the same time. This will occur when the distance r_{2A} differs from r_{1A} by an integer number of wavelengths, $r_{2A} = r_{1A} \pm m\lambda$. We call this *constructive interference* because the two component waves add together to give a larger total electric field, resulting in a bright spot at point A.

At another point, B, the electric fields from the two sources may be 180° out of phase, such that the positive maximum of one field arrives at the same time as the negative maximum of the other. This will occur when the two distances r_{2B} and r_{1B} differ by a half-odd integer number of wavelengths, $r_{2B} = r_{1B} \pm (m + \frac{1}{2})\lambda$. This is called *destructive interference* and results in a dark spot at point B. Points at which the phase difference between the two component waves is between 0 and 180° will have an intermediate intensity, proportional to the square of the total electric field magnitude.

The concept of light interference just described can be generalized to include more than two point sources. For example, the diffraction of a light beam (Fig. 2-14) can be understood by considering the beam waist to be composed of an infinite number of point sources, each radiating a spherical wave of wavelength λ. The cone angle of the diffracted beam is determined by finding the angle θ for which destructive interference occurs when adding the contributions from all these point sources.

Another example is that of the diffraction grating, shown in Fig. 2-16, in which a beam of light is diffracted by an array of parallel slits. We will assume here that the incident light beam is collimated (has planar wave fronts), and that the diffraction pattern is observed very far away from the grating. In that case, the two rays labeled 1 and 2 can be considered to be approximately parallel, both before and after the grating. Defining the angles of incidence (θ_i) and transmission (θ_t) as in Fig. 2-16, the extra distance that ray 2 has to go (the *optical path difference*) is $d(\sin\theta_i + \sin\theta_t)$, where d is the slit spacing: For these two rays to interfere constructively, giving a bright spot in the diffraction pattern, this optical path difference must be an integer number of wavelengths. The condition for diffraction maxima then becomes

$$d(\sin\theta_i + \sin\theta_t) = m\lambda \tag{2-28}$$

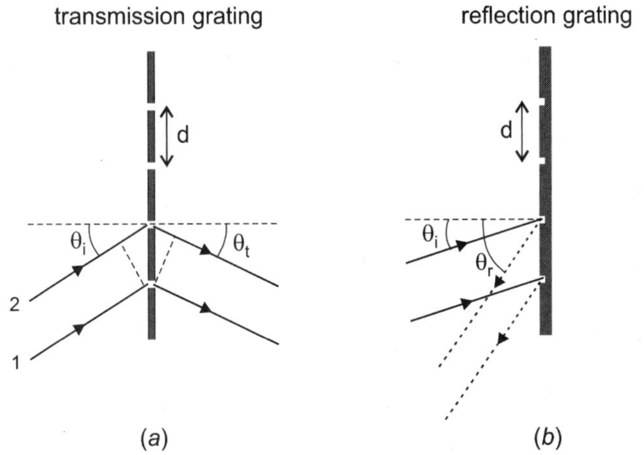

Figure 2-16 (a) Diffraction geometry for a thin transmission grating; incident and diffracted angles θ_i and θ_t are related by the grating equation, Eq. (2-28). (b) Geometry for a reflection grating, where incident and reflected beams are on same side. Eq. (2-28) still applies, with θ_t replaced by θ_r.

where the integer m specifies the order of the diffraction peak. Note that for 0th order, $\theta_t = -\theta_i$, which corresponds to the light passing straight through undeflected. For each value of the incident angle θ_i between –90° and +90°, the transmitted beam can take on a finite number of possible angles θ_t, corresponding to the various orders. Diffraction gratings are often used in spectroscopy to separate out the different wavelength components of light, because according to Eq. (2-28), the angle of diffraction depends on the wavelength.

The diffraction grating of Fig. 2-16 is a *thin grating,* since it is a two-dimensional mask with negligible thickness in the third dimension. More generally, we can have a *thick grating* that extends into the third dimension. Fig. 2-17 shows a cross-sectional view of a thick grating, with a regular array of partially reflecting planes oriented perpendicular to the page. As light propagates through this structure, some light is reflected as it encounters each plane, and the rest is transmitted to be incident on the next plane. After passing through many such planes, most of the light has been reflected. In order for the light reflected from the various planes to interfere constructively, giving rise to a bright spot in the diffraction pattern, the optical path difference between any two rays must be an integer number of wavelengths, just as for a thin grating. If we adopt the same definition of incident and transmitted angles used for thin gratings, then Eq. (2-28) also applies to thick gratings. The difference in the case of thick gratings is that the two angles θ_i and θ_t must be equal, since this is a reflection process and the angles must obey the law of reflection. Putting $\theta_i = \theta_t = \theta$ in Eq. (2-28) yields

$$2d \sin \theta = m\lambda \tag{2-29}$$

where again m is an integer specifying the order of the diffraction.

Equation (2-29) is known as the *Bragg condition,* and was first developed by Lawrence Bragg in 1912 to describe the diffraction of X-rays by the periodic arrays of atoms in crystals. It also describes the diffraction of light by periodic planes in a solid with different index of refraction, and has applications in a number of areas of photonics. For example, a volume hologram is essentially a thick grating formed by two interfering light waves, and can be used for high-density optical storage of data. Another example is acoustooptic diffraction of light, in which a moving refractive index grating is created by propagating a high-intensity sound wave through a solid. The pressure oscil-

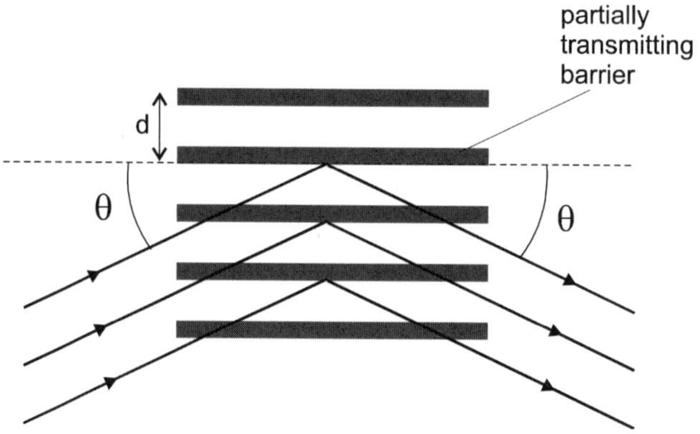

Figure 2-17 Diffraction geometry for a thick grating; incident and diffracted angles are the same, and are given by the Bragg condition.

lations of the sound wave modify the refractive index by changing the local density of the material (recall that higher density leads to higher refractive index). Acoustooptic diffraction can be used for light-beam deflection and for fast switching of a laser pulse (see Chapter 9).

Another application of thick gratings that has become important for photonics applications is that of the fiber optic *Bragg grating.* It was found in 1978 that the refractive index inside an optical fiber can be modified periodically to form a thick refractive index grating. These gratings can be highly reflecting for a particular wavelength, and highly transmitting for other wavelengths. The high reflectivity and wavelength selectivity inherent in Bragg gratings has made them essential elements in devices such as fiber lasers and multiplexers for WDM systems. The fiber Bragg grating is discussed in detail in Chapter 8.

2-4. IMAGING OPTICS

When lenses are used to form an image, the geometrical optics treatment is usually adequate. Figure 2-18 shows how the location and size of an image can be determined by tracing rays from a point on the object to a point on the image. Rays traveling parallel to the optical axis before the lens pass through the focal point after the lens, and rays passing through the focal point before the lens become parallel to the axis after the lens. Rays passing through the lens center are undeflected. In the *paraxial approximation,* the rays make small angles with the optical axis, and these three rays (and any others drawn from the same point on the object) will converge to a common point after the lens, forming an image.

The relationship between object and image distances and sizes can be obtained by using the geometry of the similar triangles containing angle θ to write

$$\tan\theta = \frac{h_1}{s_1} = \frac{h_2}{s_2} \qquad (2\text{-}30)$$

Note that the image will be larger when it is further from the lens, and vice versa. Another useful relation is obtained by considering the similar triangles containing angle ψ:

$$\tan\psi = \frac{h_1}{f} = \frac{h_2}{s_2 - f} \qquad (2\text{-}31)$$

Combining this with Eq. (2-30) gives, after a few steps of algebra,

$$\frac{1}{s_1} + \frac{1}{s_2} = \frac{1}{f} \qquad \text{(lens equation)} \qquad (2\text{-}32)$$

This relation is called the *lens equation,* and is one of the most widely used equations in geometrical optics. It is valid for any position of the image, according to the sign convention that s_1 is positive to the left of the lens, and s_2 is positive to the right of the lens. The focal length f is taken as positive for a converging lens (the kind drawn in Fig. 2-18), and negative for a diverging lens.

Curved mirrors have focusing properties similar to those of lenses. Figure 2-19 shows two rays incident on a concave mirror, one through the mirror's center of curvature and the other offset from the center but parallel to the optical axis. At the mirror's surface, the

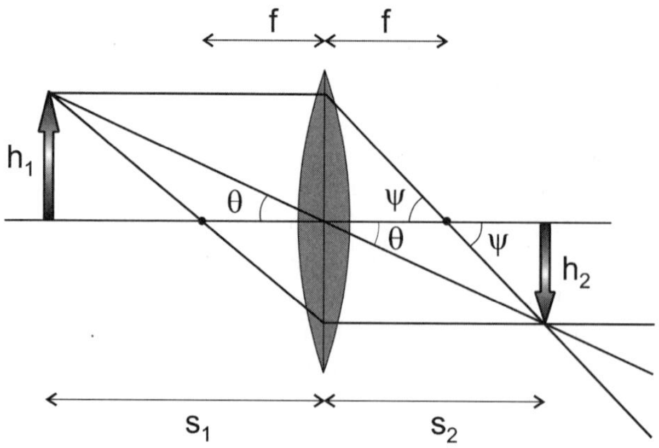

Figure 2-18 Light rays from a point on an object are refracted by a lens of focal length *f* and converge to a point on the image.

angles α that the incident and reflected offset rays make with the normal to the surface are equal by the law of reflection. The ray through the center is incident perpendicular to the mirror surface, and is therefore reflected back along its original direction. These two rays meet at the focal point, a distance f from the mirror surface. Using the geometry shown for the small right triangle having one vertex at the center of curvature, we can write

$$\frac{R}{2} = (R - f) \cos \alpha \qquad (2\text{-}33)$$

which in the paraxial approximation ($\alpha \ll 1$) becomes

$$f \simeq \frac{R}{2} \qquad \text{(mirror focal length)} \qquad (2\text{-}34)$$

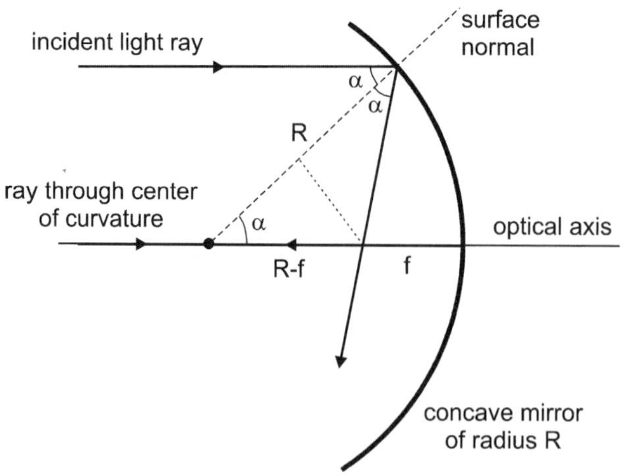

Figure 2-19 Parallel rays incident on a mirror of radius of curvature *R* are reflected, converging at a focal point a distance $f \approx R/2$ from the mirror.

Using this focal length for the mirror, the relation between object and image locations and sizes can be found using Eqs. (2-30) and (2-32), just as for a lens. The difference is in the sign convention, which for a mirror takes both s_1 and s_2 as positive for objects and images to the left of the mirror, and negative to the right. A positive f corresponds to a concave mirror, as shown, whereas negative f corresponds to a convex mirror.

The sign conventions that we have mentioned for lenses and mirrors assume that the incident beam is propagating from left to right. In some applications, however, the beam gets turned around and propagates from right to left. In such cases, the meaning of phrases such as "to the left of the mirror" should be generalized to "on the side from which the ray is incident." When the sign conventions are rephrased in this way, they apply to all situations.

PROBLEMS

2.1 (a) Show that there is no angular deflection of a beam passing through a dielectric plate with parallel faces. (b) If the two faces are not perfectly parallel, so that one face makes a small angle α with the other face, determine the resulting angular deviation of a beam that is incident from air with an angle θ (also assumed small) from the normal to one of the faces. Write your result in terms of the angles α and θ, and the index of refraction n of the plate.

2.2 The wavelength dependence of the refractive index for silica glass can be expressed as

$$n^2(\lambda) = 1 + \sum_{i=1}^{3} \frac{a_i \lambda^2}{\lambda^2 - b_i^2}$$

in the range $0.3 < \lambda < 2.5$, where λ is the free-space wavelength in units of μm. The constants are

$$(a_1, a_2, a_3) = (0.50716, 0.59707, 0.69879)$$

and

$$(b_1, b_2, b_3) = (0.04014, 0.11359, 8.81674)$$

(a) Determine the phase and group velocities for light traveling in silica glass for $\lambda =$ 1.30 μm. (b) Repeat for $\lambda = 500$ nm.

2.3 A laser beam is incident on the side of a rectangular fish tank with angle θ_1 from the normal to the glass surface. The beam enters the water and strikes the surface of the water. For what range of angles θ_1 does the beam undergo total internal reflection at the water–air interface?

2.4 A He–Ne laser beam has power 1 mW and beam diameter 1 mm. Determine the electric field amplitude in the light wave, assuming that the light intensity is uniform across the beam profile.

2.5 A light wave is incident from air on a thick glass slab of index 1.8, with angle of incidence 30°. Determine the fraction of light reflected from and transmitted through the air–glass interface, and verify that these two fractions add to unity. Assume s polarization.

2.6 Show that there is no Brewster's angle for s polarization

2.7 Light passes through a glass slab with parallel faces. Show that if light is incident at Brewster's angle on the first (air–glass) interface, then there will also be no reflection at the second (glass–air) interface.

2.8 Light is incident on a glass–air interface from the glass side, and researchers want to use the evanescent field on the air side to excite molecules adhered to the surface. It is desired that the evanescent field extend a distance $\delta = 20$ μm into the air side when using light of free-space wavelength 1 μm. (a) How close to the critical angle must the incident beam be? (Give the difference $\Delta\theta = \theta_1 - \theta_c$.) (b) Considering that a beam of finite width contains rays with a spread of angles due to diffraction, how wide must the beam be so that the angular spread is just equal to the difference in angle found in part a?

2.9 A laser beam is incident perpendicular to the surface of one of the short faces of a 45–45–90 prism. If the refractive index of the glass is 1.5, show that the light undergoes total internal reflection when it strikes the long face of the prism. This type of reflector is often used to redirect high-power laser beams, because little heat is deposited in the device.

2.10 Collimated laser light of wavelength 632.8 nm is incident on a mm-scale ruler at grazing incidence (light nearly parallel to ruler axis). The light is diffracted off the mm-spaced lines and strikes a screen 2 m away. Determine the angular deflection δ of the beam (with respect to the original beam direction) for each diffraction order, in terms of the angle α between the original beam direction and the ruler axis. Sketch the pattern of diffracted spots seen on the screen, for $\alpha = 1°$, identifying the diffraction order of each spot and the corresponding vertical position on the screen. If the wavelength of the incident light were unknown, one could use this method to "measure the wavelength of light with a ruler."

2.11 A transmission diffraction grating with grating spacing $d = 3$ μm is originally oriented perpendicular to a collimated beam of wavelength 0.5 μm. (a) Determine the angular position of the first two diffracted orders. (b) The grating is now tilted by an angle of 40° about an axis parallel to the grating grooves. Determine the angular deflection with respect to the original beam direction for the same diffracted orders considered in the previous part. Is there still symmetry in the diffraction pattern for positive and negative orders?

2.12 A compact disk can be used to diffract light, because the spirals of data are evenly spaced and act like a diffraction grating of groove spacing $d \approx 1.5$ μm. A CD is oriented at 45° to the direction of an incident He–Ne laser beam ($\lambda = 632.8$ nm), and diffracted spots reflected from the CD are observed on a screen parallel to and 30 cm distant from the incident beam. Determine the position of the spots on the screen for all observable diffraction orders.

2.13 Derive Eq. (2-32) using Eqs. (2-30) and (2-31).

2.14 An LED (light-emitting diode) has an emitting surface of diameter 0.5 mm. Light power of 1.5 mW is collected by a lens with focal length 25 mm and diameter 10 mm, placed 80 mm from the LED. Determine the position, the diameter, and the light intensity for the image of the LED.

2.15 Light from a light bulb with filament height 2 mm is coupled into an optical fiber of core diameter 50 μm, using a lens of focal length f. If the bulb is 20 cm from the end of the fiber, determine the value of f and the required location of the lens so that the image of the filament just fits inside the fiber core. If the lens diameter is at most equal to the focal length, what does this say about the efficiency with which light from a filament can be coupled into a fiber?

2.16 For the mirror in Fig. 2-19, rays only converge at the focus when $\alpha \ll 1$ (in radians). If the incident beam diameter is D, determine the value of D/R for which the focal point becomes spread out along the optical axis by $0.05\,f$.

Chapter 3

Planar Waveguides

This chapter treates the propagation of light between parallel planes having different indices of refraction. We start with the planar geometry because it is easier to treat mathematically than the cylindrical geometry of optical fibers. This material has direct application to integrated optic and semiconductor devices, and will also allow us to develop an intuitive feeling for optical modes and dispersion that can be carried over into the later chapters on optical fibers.

3-1. WAVEGUIDE MODES

Consider the planar dielectric waveguide shown in Fig. 3-1, with medium 1 of refractive index n_1 sandwiched between two semiinfinite media with index of refraction n_2. If $n_2 < n_1$, a ray of light propagating in medium 1 will undergo total internal reflection at the upper boundary, provided that the angle of incidence θ_1 on the boundary is greater than the critical angle $\theta_c = \sin^{-1}(n_2/n_1)$. The angle with the boundary is preserved upon reflection, so the ray will then strike the lower boundary at the same angle of incidence, assuming parallel surfaces. A ray with a well-defined initial direction will continue to propagate down the waveguide in this zigzag path without loss from the reflections. It is this perfect reflection of light energy by the waveguide boundaries that allows light to propagate such great distances down optical fibers.

The view of light as a narrow ray with a well-defined direction is called the geometric or ray optics picture, and is valid when the wavelength is much smaller than the width of the ray. In this limit, there is no restriction on the beam direction in the waveguide (other than $\theta_c < \theta_1 < 90°$), and the concept of a waveguide mode does not apply. However, when the wavelength is larger, we must use the wave optics picture of light, according to which an initially collimated beam of diameter D will eventually spread out due to diffraction (see Chapter 2). Because of this diffraction, any light beam of finite width inside a waveguide that starts out at a particular angle θ_1 will spread out into other angles, and the angular distribution will change as the light propagates down the waveguide. What we would like to find is a pattern of light distribution that remains constant along the waveguide. Such a pattern is referred to as a *mode.*

It is important to understand the concept of a mode, because we will refer to modes a lot in this book. An intuitive view of a mode can be obtained by picturing two people holding a rope that is stretched between them. If one person shakes the rope in just the right way, a stable pattern of oscillations will be seen, and this corresponds to a vibrational mode of the rope. If the rope is shaken the "wrong" way, then it still vibrates, but there is no stable pattern. The essential feature of a mode is that there is a pattern that is stable in time.

Photonics and Lasers: An Introduction. By Richard S. Quimby

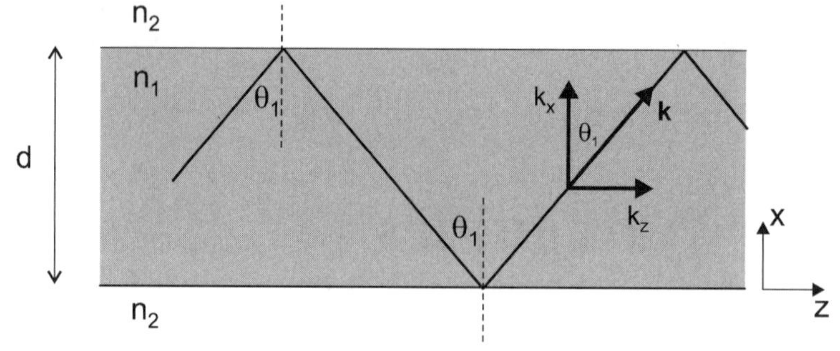

Figure 3-1 A single ray propagating down a planar waveguide. Superposition of two such rays with opposite k_x constitutes a waveguide mode.

To find a mode for the waveguide, we recall from Chapter 2 that a wider beam suffers less diffraction than a narrower beam. A plane wave that is infinitely wide would not diffract at all, so this is a candidate for our mode. However, being infinitely wide, it would undergo repeated reflections from the top and bottom waveguide boundaries, so that at a particular point inside the waveguide there would be plane waves moving both up and down with the same angle θ_1. If we let the vectors $\mathbf{k}_1 = k_x\hat{\imath} + k_z\hat{k}$ and $\mathbf{k}_2 = -k_x\hat{\imath} + k_z\hat{k}$ be the propagation vectors for these two waves, the total electric field inside the waveguide can be written as

$$
\begin{aligned}
E_{\text{mode}} &= E_0 e^{i(\omega t - \mathbf{k}_1 \cdot \mathbf{r})} + E_0 e^{i(\omega t - \mathbf{k}_2 \cdot \mathbf{r})} \\
E_{\text{mode}} &= E_0 e^{i(\omega t - k_x x - k_z z)} + E_0 e^{i(\omega t + k_x x - k_z z)}
\end{aligned}
\tag{3-1}
$$

$$
\begin{aligned}
E_{\text{mode}} &= E_0 e^{i(\omega t - k_z z)}[e^{ik_x x} + e^{-ik_x x}] \\
E_{\text{mode}} &= 2E_0 \cos(k_x x) e^{i(\omega t - k_z z)}
\end{aligned}
\tag{3-2}
$$

According to Eq. (3-2), the distribution of electric field in the x direction is given by $\cos(k_x x)$ and does not vary with time. There is also a sinusoidal variation of field in the z direction, but the z dependence changes with time, in the manner of a traveling wave. This combination of a traveling wave along the waveguide with a stationary wave in the perpendicular direction is the characteristic feature of a waveguide mode. More generally, we could write

$$
E_{\text{mode}} = E_0\, g(x, y) e^{i(\omega t - \beta z)} \tag{3-3}
$$

where the transverse distribution $g(x, y)$ is a function that must be consistent with Maxwell's equations and the boundary conditions at the waveguide boundaries, and β is the *propagation constant* for the waveguide mode. Determining $g(x, y)$ and β is easy for a planar waveguide because a simple sum of two plane waves gives a solution of the correct form, as in Eq. (3-2). For two-dimensional waveguides such as optical fibers, waveguide modes still have the form of Eq. (3-3), but with a more complicated function $g(x, y)$.

The field in Eq. (3-2) will represent a true mode of the waveguide only when it satisfies the appropriate boundary conditions. Each plane wave component of the mode can be thought of as undergoing multiple reflections from the top and bottom boundaries as it propagates down the waveguide. In order for the waves to reinforce each other after many reflections, the total round-trip phase change for propagation in the transverse (x) direction must be an integer multiple of 2π. For a waveguide of thickness d, the total round-trip distance in the x direction is $2d$, resulting in a phase shift of $-k_x(2d)$. If the phase shift upon reflection is ϕ_r, there is an additional contribution of $2\phi_r$ to the total round-trip phase shift. The condition for self-reinforcing fields then becomes

$$-k_x(2d) + 2\phi_r = \pm m2\pi \tag{3-4}$$

where the integers $m = 0, 1, 2, 3 \ldots$ label the different modes allowed in the waveguide.

Equation (3-4) specifies the allowed values of k_x for the waveguide modes. Each value of k_x in turn corresponds to a different propagation angle θ given by $k_x = k_1 \cos\theta$ (see Fig. 3-1), where k_1 is the propagation constant of one of the plane wave components in medium 1. This can be written in terms of wavelength as

$$k_x = \frac{2\pi}{\lambda}\cos\theta = \frac{2\pi n_1}{\lambda_0}\cos\theta \tag{3-5}$$

where λ and λ_0 are the wavelengths in the medium and free space, respectively. Combining Eqs. (3-4) and (3-5) gives

$$\cos\theta_m = \frac{(\pm m\pi + \phi_r)\lambda_0}{2\pi n_1 d} \tag{3-6}$$

where θ_m is angle θ for mode number m. What we have found is that not all angles of light propagation are possible in a waveguide. Instead, only those discrete angles θ_m that satisfy Eq. (3-6) will propagate in a way that is self-reinforcing.

In addition to the restriction on mode angles θ_m given in Eq. (3-6), there is also the condition that total internal reflection be satisfied, $\theta_m > \theta_c$, where $\sin\theta_c = n_2/n_1$ (see Chapter 2). As the mode number m increases from zero, the mode angle given by Eq. (3-6) decreases from 90°, eventually becoming less than the critical angle θ_c. There will therefore be a maximum mode number, which we designate as p, such that $\theta_p = \theta_c$. The range of possible mode numbers is then the finite set $m = 0, 1, 2, 3, \ldots p$, with the angle for the pth mode given by

$$\cos\theta_p = \sqrt{1 - \sin^2\theta_c} = \sqrt{1 - \left(\frac{n_2}{n_1}\right)^2}$$

When combined with Eq. (3-6), this becomes

$$\frac{2\pi d}{\lambda_0}\sqrt{n_1^2 - n_2^2} = p\pi + \phi_r \tag{3-7}$$

which gives the maximum mode number p for a waveguide of given thickness and refractive index.

It is convenient to define

$$V_p \equiv \frac{2\pi d}{\lambda_0}\sqrt{n_1^2 - n_2^2} \tag{3-8}$$

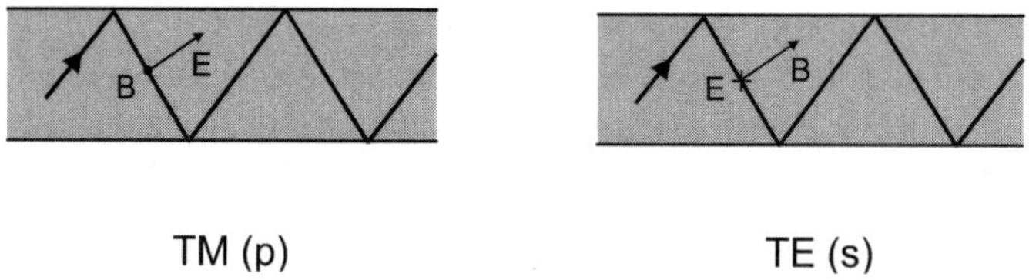

Figure 3-2 Polarization definitions for a planar waveguide.

where V_p is a dimensionless parameter sometimes referred to as the *normalized film thickness.* Using this definition, the maximum number of modes in a planar waveguide is p + 1, where

$$p = \text{int}\left(\frac{V_p}{\pi} - \frac{\phi_r}{\pi}\right) \tag{3-9}$$

The integer function int(x) truncates x to the integer value below it; for example, int(5.27) = 5.

The phase shift ϕ_r varies between 0 and π radians, depending on the waveguide angle θ. For the pth mode, the waveguide angle is near the critical angle, which results in $\phi_r = 0$ according to Eqs. (2-23) and (2-24). The number of waveguide modes can then be written simply as

$$\text{number of modes} = p + 1 = \text{int}\left(\frac{V_p}{\pi}\right) + 1 \tag{3-10}$$

In the case of a thick waveguide where $V_p \gg 1$, the number of modes is well approximated by V_p/π. For each of these modes, there are two possible polarizations, as shown in Fig. 3-2. For TM polarization the E field is in the plane formed by the zigzagging ray, whereas for TE polarization the E field is perpendicular to this plane. If the different polarizations are considered to be different modes, then the total number of modes in the planar waveguide is $\approx 2V_p/\pi$. The allowed values of k_x for the thick waveguide modes can be approximated by

$$k_x \simeq \frac{m\pi}{d} \tag{3-11}$$

where Eq. (3-4) has been used with the approximation $m\pi \gg \phi_r$. This approximation is most valid for the higher-order modes, where m is large.

EXAMPLE 3-1

Assuming light of free-space wavelength 1 μm, determine the number of modes (a) in a microscope slide of thickness 1 mm, immersed in water, and (b) in a soap film in air of thickness 2 μm. Take the refractive index of glass as 1.5 and that of water as 1.33.

Solution: (a) The normalized film thickness is

$$V_p = \frac{2\pi(1 \times 10^{-3})}{1 \times 10^{-6}} \sqrt{(1.5)^2 - (1.33)^2}$$

$$V_p = 4.7 \times 10^3$$

The number of waveguide modes is then $\approx 4700/\pi \approx 1500$, not including different polarizations. When including different polarizations, there are about 3000 modes.

(b) The soap film is mostly water, with refractive index 1.33, so

$$V_p = \frac{2\pi(2 \times 10^{-6})}{1 \times 10^{-6}} \sqrt{(1.33)^2 - 1^2}$$

$$V_p = 11$$

The number of waveguide modes is then $\text{int}(11/\pi) + 1 \simeq \text{int}(3.5) + 1 = 4$, not including different polarizations. When including different polarizations, there are eight modes.

Effective Index

There are a number of parameters that can be used to specify a particular waveguide mode. For example, there are the transverse wave vector k_x and the waveguide angle θ, which are related by Eq. (3-5). There is also the longitudinal wave vector component k_z, which is similarly related to θ by

$$k_z = k_1 \sin\theta = \frac{2\pi n_1}{\lambda_0} \sin\theta \tag{3-12}$$

where $k_1 = 2\pi/\lambda$ is the propagation constant in medium 1 for one of the plane wave components of the mode. The wavevector components satisfy $k_x^2 + k_z^2 = k_1^2$, as can be verified from Eqs. (3-5) and (3-12). Any one of the three parameters k_x, k_z, or θ can be used to specify the mode, with the other two parameters then determined by Eqs. (3-5) and (3-12).

A fourth parameter that is often used to specify the mode is the *effective index of refraction,* defined by

$$k_z \equiv \frac{2\pi n_{\text{eff}}}{\lambda_0} \tag{3-13}$$

This is similar to the relation $k = (2\pi n)/\lambda_0$ for a single plane wave propagating through a medium with index of refraction n, except that the plane wave propagation constant k has been replaced by the waveguide mode propagation constant k_z. According to Eq. (3-2), the guided wave propagates at the phase velocity given by $\omega t - k_z z = \text{constant}$, or

$$v_p = \frac{\omega}{k_z} = \frac{2\pi c}{\lambda_0 k_z} = \frac{c}{n_{\text{eff}}} \tag{3-14}$$

The effective index thus governs the phase velocity of waveguide modes in the same way that the ordinary index does for plane waves (see Eq. 2-6). Equations (3-13) and (3-14) can be applied quite generally to waveguides in either planar or circular geometry. The propagation constant k_z is commonly denoted as β in optical fibers, and referred to as the *axial wave vector*

Mode Velocities

For a planar waveguide, the effective index of refraction can be written as

$$n_{\text{eff}} = n_1 \sin\theta \tag{3-15}$$

using Eqs. (3-12) and (3-13). The phase velocity of the waveguide mode can then be written using Eqs. (3-14) and (3-15) as

$$v_p = \frac{c}{n_1 \sin\theta} = \frac{v_1}{\sin\theta} \tag{3-16}$$

where $v_1 = c/n_1$ is the phase velocity of a plane wave in medium 1. Since $\sin\theta_c < \sin\theta < 1$ for a guided mode, then, using $\sin\theta_c = n_2/n_1$, we have

$$n_2 < n_{\text{eff}} < n_1 \tag{3-17}$$

and

$$v_2 > v_p > v_1 \tag{3-18}$$

This says that the phase velocity of the waveguide mode (which is confined to medium 1) is greater than the speed of a plane wave in medium 1. At first glance, this does not seem to make sense, since one would expect light propagating in a waveguide mode at an angle θ to move more *slowly* down the waveguide than light moving in a straight line as a plane wave.

The resolution of this apparent paradox can be found in the distinction between phase and group velocities, as discussed in Chapter 2. Information is sent down the waveguide in the form of pulses, which move at the group velocity given by

$$v_g = \frac{d\omega}{d\beta} \tag{3-19}$$

where we have defined $\beta \equiv k_z$. The functional relationship between ω and β can be found from Fig. 3-3, which shows the wave vector $\mathbf{k}_1$ for one of the plane wave components of the waveguide mode. The x and z components of $\mathbf{k}_1$ satisfy

$$(n_1 k_0)^2 = k_x^2 + \beta^2 \tag{3-20}$$

where we have used $k_1 = n_1 k_0 = n_1 2\pi/\lambda_0$. For high-order modes in which Eq. (3-11) applies, this can be written as

$$\left(\frac{n_1}{c}\right)^2 \omega^2 = \left(\frac{m\pi}{d}\right)^2 + \beta^2 \tag{3-21}$$

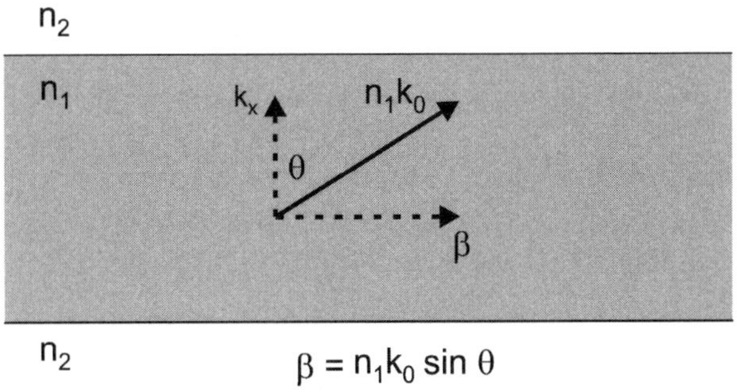

Figure 3-3 The wave vector for the ray of magnitude n_1k_0 can be broken down into its longitudinal and transverse components, β and k_x.

where $\omega = ck_0$ has been used (see Eq. 2-6). Taking the derivative with respect to β on both sides of Eq. 3-21 yields

$$2\omega \frac{d\omega}{d\beta} = \left(\frac{c}{n_1}\right)^2 2\beta$$

Combining this with Eqs. (3-14) and (3-19) gives

$$v_g v_p = v_1^2 \tag{3-22}$$

This is the desired relation between the group and phase velocities for a waveguide mode. We see that if the phase velocity is greater than v_1, then the group velocity is less than v_1 by the same factor. Using Eq. (3-16) for the phase velocity, the group velocity becomes

$$v_g = v_1 \sin \theta \tag{3-23}$$

so that $v_g < v_1$, as expected. In fact, referring to Fig. 3-3, we see that v_g can be interpreted as the component of the plane wave's velocity vector along the z axis. This is physically satisfying, since we would expect energy in a mode to be transmitted down the waveguide at the speed with which the individual plane waves making up the mode travel down the waveguide. Although our analysis has assumed a high-order mode, the same result applies more generally.

Our conclusion from the previous discussion is that the group velocity is the relevant parameter for determining how fast information propagates down the waveguide. However, we are still left with the question: What is the physical interpretation of the phase velocity? Consider Fig. 3-4, which shows two wave fronts (1 and 2) propagating down the $+z$ axis to form a waveguide mode. Plane wave 1 is propagating up and to the right, and plane wave 2 is propagating down and to the right. The lines of constant phase for each wave are shown by solid lines at $t = 0$ and by dotted lines at $t > 0$. As wave fronts 1 and 2 move to positions 1′ and 2′, the point at which the waves intersect moves to the right as shown. This intersection point will be a point of constant phase for the combined waves, since it is a point of constant phase for each of the constituent waves 1 and 2. It is clear from the geometry that the intersection point can move down the waveguide faster than the speed of either wave individually. The speed with which points of constant phase propagate is by definition the

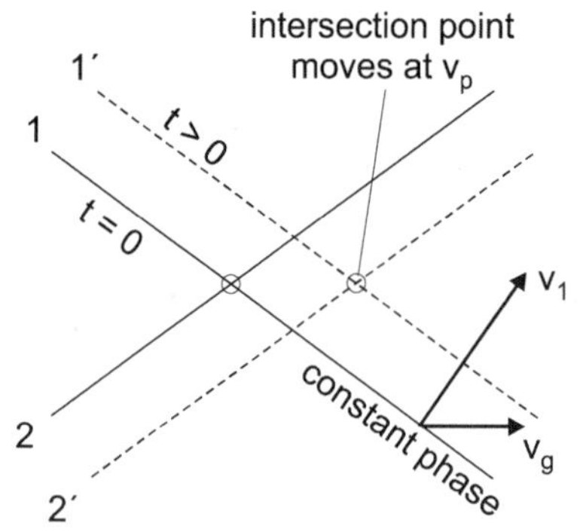

Figure 3-4 The motion of wave fronts for two component rays with opposite k_x illustrates the difference between group and phase velocity.

phase velocity, so we identify the phase velocity here with the speed of this intersection point. With this picture in mind, it is clear that information or energy does not propagate at the phase velocity, and there is no difficulty having $v_p > v_1$.

3-2. MODE CHART

A waveguide mode can be specified by any one of the parameters n_{eff}, θ, k_x, or $k_z = \beta$. For a waveguide of given thickness d, an approximate expression for the allowed mode angles θ_m can be obtained from Eq. (3-6):

$$\cos \theta_m \simeq \frac{m\lambda_0}{2\, n_1 d} \tag{3-24}$$

where it has been assumed that $m\pi \gg \phi_r$. This expression can be combined with Eqs. (3-15), (3-5), and (3-12) to obtain expressions for the allowed n_{eff}, k_x, and k_z values, valid for high-order modes. To obtain a result valid for all modes, we must incorporate the proper expression for ϕ_r into Eq. 3-6. Solving Eq. (3-6) for ϕ_r and substituting into Eq. (2-23), we have for TE polarization

$$\tan\left[\frac{\pi n_1 d \cos\theta}{\lambda_0} - m\,\frac{\pi}{2}\right] = \frac{\sqrt{1 - \left(\dfrac{n_2}{n_1}\right)^2 - \cos^2\theta}}{\cos\theta} \tag{3-25}$$

A similar expression can be obtained for TM polarization using Eq. (2-24). The results for TE and TM polarizations will be nearly the same if $n_1 \approx n_2$.

For a fixed value of d, Eq. (3-25) is a transcendental equation for the mode variable cos θ, and can be solved graphically. Figure 3-5 shows the left-hand side (LHS) and right-hand side (RHS) of Eq. (3-25) plotted as a function of cos θ. The LHS has multiple curves corresponding to different m, spaced evenly along the cos θ axis by $\lambda_0/(2n_1 d)$. The RHS goes to infinity as cos θ goes to 0, and goes to zero at the critical angle θ_c. The intersections of the LHS and RHS give the solutions to Eq. (3-25) and correspond to the allowed waveguide modes. The modes are finite in number and approximately spaced by $\lambda_0/(2n_1 d)$, in accordance with Eq. (3-24).

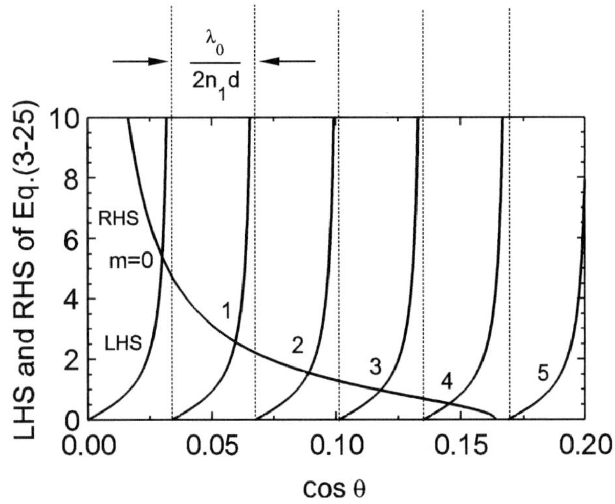

Figure 3-5 Graphical solution of Eq. (3-25) for modes in a planar waveguide, with n_1 = 1.48, n_2 = 1.46, and d/λ_0 = 10. These parameters lead to five allowed modes, which correspond to the five line crossings.

As the waveguide thickness d is decreased, the modes become more widely separated and fewer in number, until at some point there is only one allowed mode. The condition for such a *single-mode waveguide* is

$$\frac{\lambda_0}{2n_1 d} > \cos\,\theta_c$$

which can be written using Eq. (2-18) as

$$d < \frac{\lambda_0}{2\sqrt{n_1^2 - n_2^2}} \tag{3-26}$$

The condition for a single-mode waveguide can also be written in terms of the V_p parameter (Eq. 3-8) as

$$V_p < \pi \tag{3-27}$$

This is in agreement with Eq. (3-10), which gives the number of modes as 1 when $V_p < \pi$. There are advantages to single-mode waveguides, which we will discuss later in this chapter. It should be noted that no matter how small the film thickness d, there is always one crossing between curves for m = 0 in Fig. 3-5, and therefore the waveguide always has at least one mode.

Equation (3-25) is a transcendental equation for the variable cos θ and cannot be solved analytically. However, if cos θ is taken as given, then the equation *can* be solved analytically for the waveguide thickness d. After some manipulation (Problem 3.2) we find

$$d_m = d_0 + \frac{m\lambda_0}{2n_1 \cos\,\theta} \tag{3-28}$$

where d_m is the thickness for mode number m and

$$d_0 = \frac{\lambda_0}{\pi n_1 \cos\theta} \cos^{-1}\left(\frac{n_1 \cos\theta}{\sqrt{n_1^2 - n_2^2}}\right) \tag{3-29}$$

Equations (3-28) and (3-29) can be used to develop a graph of allowed modes by computing the values of d for a fixed θ, and then repeating this for different values of θ. It is customary in such a graph to characterize the modes by the effective index n_{eff} rather than θ, where

$$n_1 \cos\theta = \sqrt{n_1^2 - n_{eff}^2} \tag{3-30}$$

using Eq. (3-15).

An example of such a *mode chart* is shown in Fig. 3-6. Although the mode chart is constructed by computing d for fixed values of n_{eff}, in practice we use the chart by fixing the value of d for a particular waveguide and going up vertically from the d/λ_0 axis to find the n_{eff} values for the various allowed modes. When $d/\lambda_0 < 1/(2\sqrt{n_1^2 - n_2^2})$ there is only one allowed mode, in accordance with Eq. (3-26).

Field Distribution in a Mode

Each mode of the waveguide has a characteristic variation of electric field with position, which is quite different inside and outside the waveguiding region. Inside the waveguide (in the higher index n_1), the field is oscillatory both in x and z, with the form

$$E(x, z, t) = E_{max} \cos(k_x x) \cos(\omega t - \beta z) \tag{3-31}$$

The field can also vary as $\sin(k_x x)$, but we consider only $\cos(k_x x)$ for simplicity. Here the definition $\beta \equiv k_z$ has been used, and we have taken the real part of the complex exponential in Eq. (3-2). For large d/λ_0 where $m \gg 1$, we can use $k_x \approx m\pi/d$ from Eq. (3-11). Out-

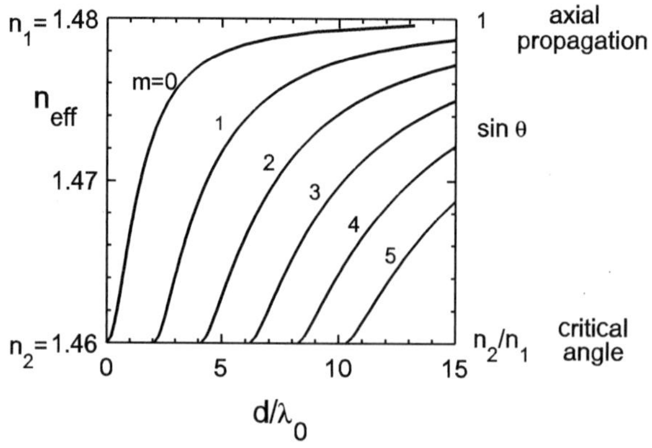

Figure 3-6 Mode chart calculated from Eqs. 3-28 and 3-29, using n_1 = 1.48 and n_2 = 1.46 as in Fig. 3-5. Values of n_{eff} for the various modes are obtained by drawing vertical lines and looking for curve crossings.

side the waveguide (in the lower index n_2), the field is still oscillatory in the z direction, but decays exponentially in the x direction.

The transverse (x) variation of E at one instant in time is illustrated in Fig. 3-7, for some representative positions z along the waveguide. As time increases, the whole pattern shifts to the right with the phase velocity $v_p = \omega/\beta$. At any fixed point in the waveguide, the E field will vary sinusoidally with time, with an amplitude given by $E_{max} \cos(m\pi x/d)$. There will thus be certain values of x for which the E field is zero at all times. The locus of such zero-field points are called the *nodal lines,* one of which is shown in Fig. 3-7 by the dotted line. For a planar waveguide mode with mode number m, there are m nodal lines, resulting in $m + 1$ lines of maximum intensity. The lowest-order mode, with $m = 0$, has just a single intensity maximum and no nodal lines.

If one could excite just a single mode of higher order m, the intensity distribution observed at the end of the waveguide would consist of $m + 1$ bright lines with m dark lines in between. In practice, it is quite difficult to excite just one mode, especially in a thick waveguide with many possible modes. When many modes are excited simultaneously, the peak of one mode tends to fill in the node of another, resulting in a more uniform intensity distribution. The uniformity of the resulting light distribution depends on the coherence of the light, which we will discuss in Chapter15.

3-3. DISPERSION

In the previous sections, we have found that light propagates in one or more waveguide modes, each mode being characterized by a different propagation angle θ. Energy or information in these modes propagates down the waveguide at the group velocity, which varies with θ according to Eq. (3-23). The time it takes for a pulse of light to propagate a distance L down the waveguide will then vary with θ as

$$t = \frac{L}{v_g} = \frac{Ln_1}{c \sin \theta} \tag{3-32}$$

If the energy in a light pulse is spread out among the various waveguide modes, the parts of the energy in different modes will arrive at the far end of the waveguide at differ-

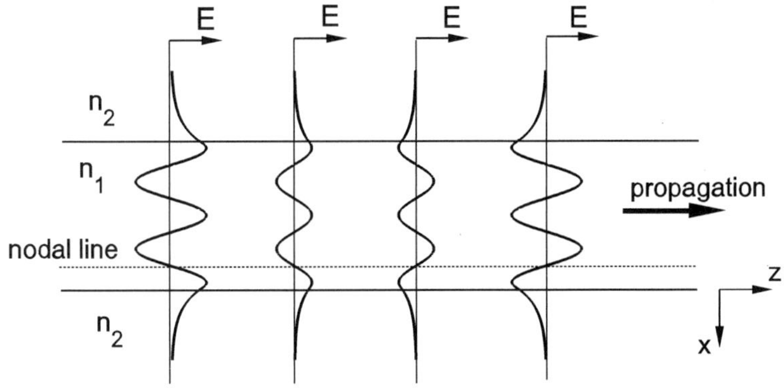

Figure 3-7 Transverse spatial distribution of the E field at one instant in time for four positions along the z axis of a planar waveguide. Parameters are those of Fig. 3-5, with mode number $m = 4$.

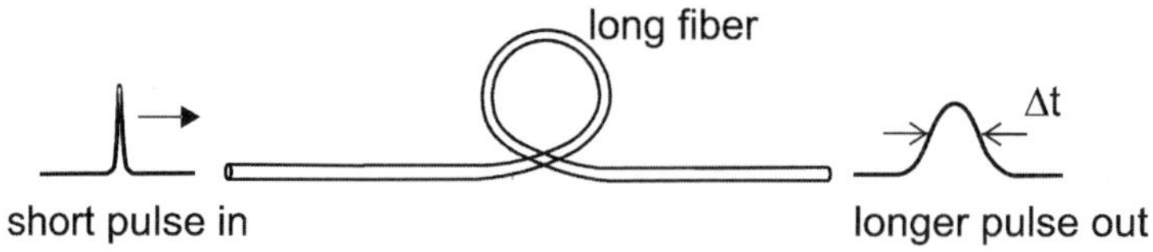

Figure 3-8 Dispersion causes a pulse to broaden in time as it propagates down a long waveguide.

ent times. This means, as indicated in Fig. 3-8, that an initially short pulse will broaden in time by an amount Δt when it reaches the far end, where

$$\Delta t = \frac{Ln_1}{c}\left(\frac{1}{\sin\theta_{\min}} - \frac{1}{\sin\theta_{\max}}\right) \tag{3-33}$$

The limits on θ are $\theta_c < \theta < 90°$, where θ_c is the critical angle for total internal reflection. Using $\sin\theta_c = n_2/n_1$, the limits on θ become

$$\frac{n_2}{n_1} < \sin\theta < 1 \tag{3-34}$$

Combining Eqs. (3-33) and (3-34) yields

$$\Delta t = \frac{Ln_1}{cn_2}(n_1 - n_2) \simeq \frac{L}{c}(n_1 - n_2)$$

where we have assumed $n_1 \approx n_2$, generally a good assumption for optical fibers. This can also be written as

$$\Delta t \simeq \frac{Ln}{c}\Delta \tag{3-35}$$

where we have defined the fractional index difference Δ as

$$\Delta \equiv \frac{n_1 - n_2}{n_1} \tag{3-36}$$

Equation (3-35) gives the spreading in time of a light pulse due to propagation in different modes, referred to as *intermodal dispersion.* It occurs in optical fibers as well as planar waveguides, and is most important when sending pulses a great distance. Note that the dispersion does not depend on the waveguide thickness, although it is implicitly assumed that the waveguide is thick enough to support several modes. Since Δt is proportional to L, it is customary to specify the degree of dispersion as $\Delta t/L$, in units of ns/km.

EXAMPLE 3-2

Determine the intermodal dispersion of an optical fiber with a core index of 1.5 and a fractional index difference of 0.01.

Solution: Using Eq. (3-35) with $L = 1$ km, we have

$$\Delta t = \frac{(10^3 \text{ m})(1.5)}{3 \times 10^8 \text{ m/s}}(10^{-2}) = 50 \text{ ns}$$

The intermodal dispersion is thus approximately 50 ns/km.

The spreading out of a light pulse in time sets an upper limit on the rate of data transmission in optical communications. Consider a digital data stream as shown in Fig. 3-9a, consisting of pulses of width $T/2$ separated in time by T. After propagating along the fiber for a distance L, each pulse spreads out by Δt, so that the pulses start to overlap as shown in Fig. 3-9b. When the width of the pulses $(T/2) + \Delta t$ is much greater than the separation T, the individual pulses cannot be distinguished, and reliable transmission of information is not possible. The criterion for distinguishable pulses is then

$$T > \frac{T}{2} + \Delta t$$

which can be written as

$$\text{BR} \equiv \frac{1}{T} < \frac{1}{2\Delta t} \tag{3-37}$$

where BR is the number of pulses per second or the *bit rate*. The bit rate is often specified in units of Mb/s or Gb/s—mega (10^6) or giga (10^9) bits per second. Using Eq. (3-37) with the intermodal dispersion value of 50 ns/km from Example 3-2, we find a maximum bit rate of 1/100 ns = 10 Mb/s for $L = 1$ km. For $L = 2$ km the spreading Δt is twice as great, leading to $\text{BR}_{\text{max}} = 5$ Mb/s. In general, the product of fiber length and maximum bit rate is a constant, so the effect of dispersion can be characterized by the *length × bit rate product* ($L \times BR$), which for Example 3-2 is $L \times BR = 10$ km Mb/s.

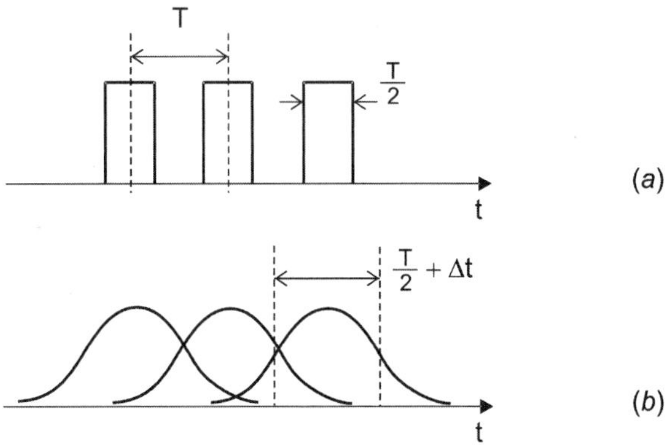

Figure 3-9 Pulses are broadened by Δt after propagating a distance L, which limits the rate at which data can be transmitted down a long waveguide.

For planar waveguides, propagation lengths are typically on the order of centimeters rather than kilometers, and intermodal dispersion is not usually significant. It is more important in optical fibers, and will be discussed further in Chapters 5 and 6.

PROBLEMS

3.1 Show that $k_x = k_0\sqrt{n_1^2 - n_{\text{eff}}^2}$ for a planar waveguide, where $k_0 = 2\pi/\lambda_0$ and n_1 is the index of the center region of the waveguide. Show further that for any mode near cutoff, $k_x \simeq k_0\sqrt{n_1^2 - n_2^2}$.

3.2 Derive Eqs. (3-28) and (3-29) from Eq. (3-25).

3.3 A waveguide has refractive indices $n_1 = 3.6$, $n_2 = 3.4$, and thickness $d = 5$ μm. If light of free-space wavelength 1.3 μm is coupled into the waveguide, how many TE modes can propagate?

3.4 For the waveguide in Problem 3.3, consider the mode with $m = 2$. Determine θ, n_{eff}, and β for this mode.

3.5 Sketch the electric field distribution $E(x)$ for the mode considered in Problem 3.4.

3.6 Calculate the phase and group velocities for the mode considered in Problem 3.4.

3.7 Consider the waveguide of Problem 3.3, except that now the thickness d can be varied. For what range of d do modes with $m = 2$ exist?

3.8 A waveguide has $n_1 = 3.6$ and $n_2 = 3.4$. For what range of d does the waveguide support only one TE mode at a free-space wavelength 1.5 μm, and only two TE modes at a free-space wavelength 1.3 μm?

3.9 A waveguide has $n_1 = 3.6$, $n_2 = 3.4$, and thickness 1.2 μm. For what range of free-space wavelengths does it support only three TE modes?

3.10 A multimode optical fiber has $n_1 = 1.48$ and $\Delta = 0.015$. Determine the range of propagation angles θ for the various modes in the fiber. Express this also as a range of angles ψ that the rays make with the fiber axis.

3.11 For the fiber described in Problem 3.10, determine the time spread in an optical pulse after propagating a distance 2.5 km along the fiber. What is the maximum bit rate that can be transmitted over this fiber for this distance?

3.12 A fiber is characterized by an intermodal dispersion of 40 ns/km. If the core refractive index is $n_1 = 1.49$, what is the cladding index n_2?

Chapter 4

Cylindrical Waveguides

The previous chapter treated propagation of light in a planar waveguide, in which the E field varies in only one transverse dimension (1-D), for example the x direction. We now extend this to two dimensions (2-D), in which the field varies in both the x and y transverse directions. An important special case is that of the optical fiber, which has (usually) cylindrical symmetry about the fiber axis (z axis). A full treatment of the 2-D waveguide modes is beyond the scope of this book. However, many of the features of light propagation in an optical fiber can be understood, at an intuitive and semiquantitative level, by simple extensions of the 1-D treatment to 2-D.

4-1. ACCEPTANCE ANGLE AND NUMERICAL APERTURE

Consider the cylindrical dielectric waveguide shown in Fig. 4-1, with a solid cylindrical core of refractive index n_1 surrounded by a concentric cladding shell of refractive index n_2. The medium outside the fiber will be taken to have index n_0. A ray of light that enters the fiber end at an angle α from the fiber axis will be refracted upon entering, striking the core-cladding boundary at an angle of incidence θ. Total internal reflection will occur at the core-cladding boundary if $n_2 < n_1$, provided that the internal waveguide angle θ is greater than the critical angle $\theta_c = \sin^{-1}(n_2/n_1)$. As with the planar waveguide, the angle with the boundary is preserved upon reflection, and the ray will continue to propagate without reflection loss. The waveguide modes corresponding to such rays are termed *guided modes* or *propagating modes,* since they are guided in a near-lossless propagation down the fiber. Losses other than reflection, such as absorption and scattering, will be considered in Chapter 5.

As the entrance angle α is increased from the value shown in Fig. 4-1, the angle of incidence θ on the core-cladding boundary will decrease, until at some α_{max} the critical angle is reached, $\theta = \theta_c$. Rays having $\alpha > \alpha_{max}$ will still enter the fiber, but the reflection at the core-cladding boundary will only be partial, as shown in Fig. 4-2. After a short distance down the fiber, the light will mostly have been lost from the core, and the modes corresponding to such rays are termed *unguided modes* or *nonpropagating modes.* The fiber will therefore accept light into the guided modes only for entrance angles within the range $0 < \alpha < \alpha_{max}$. In three dimensions for a cylindrical fiber, this corresponds to an acceptance angle cone of half-angle α_{max}, as shown in Fig. 4-3. Light incident on the fiber core within this range of angles is accepted into guided modes, whereas light incident outside of this range goes into unguided modes. This same cone angle applies for light leaving the end of the fiber.

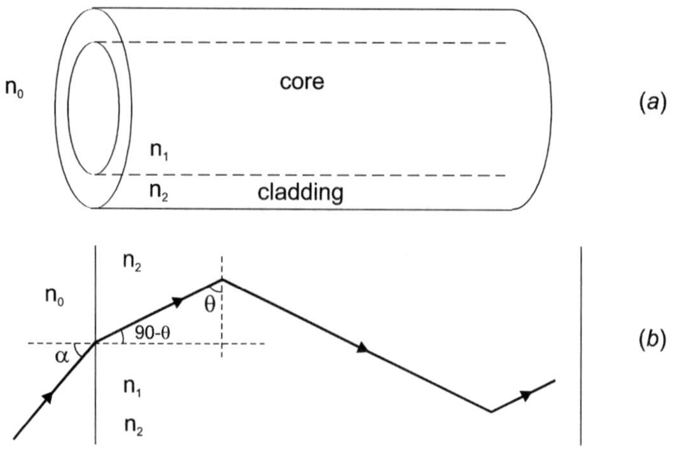

Figure 4-1 (a) Perspective view of optical fiber core-cladding structure. (b) Side view of optical fiber, showing path of a light ray that enters the fiber end.

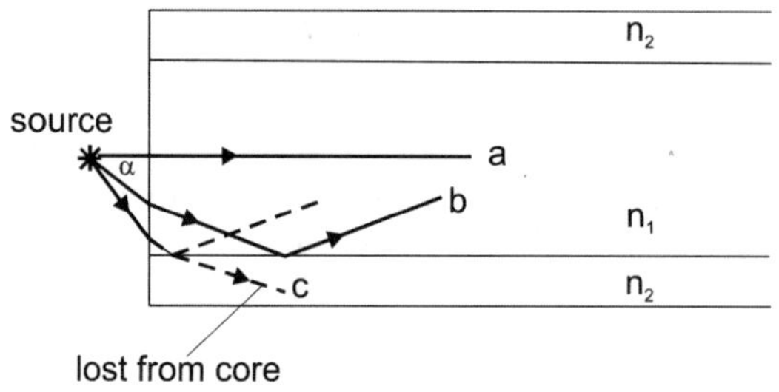

Figure 4-2 Rays incident over some range of angles α are coupled into propagating modes (a and b). Other rays (c) are attenuated by partial transmission at core–cladding boundary.

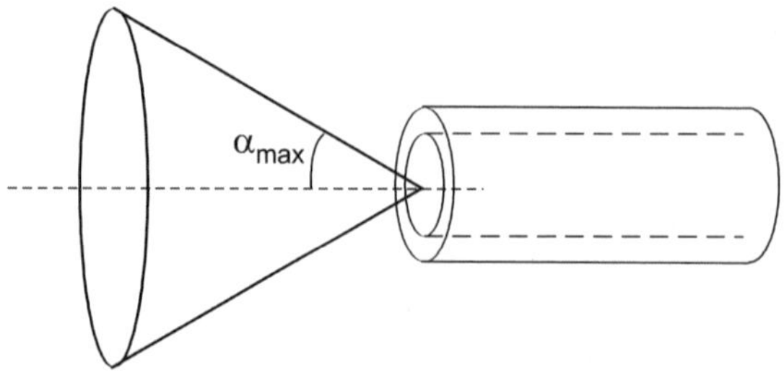

Figure 4-3 Light enters or exits the fiber within a cone of half-angle α_{max}.

The acceptance angle α_{max} can be related to the refractive indices of the core and cladding by applying Snell's law at the fiber entrance:

$$n_0 \sin \alpha_{max} = n_1 \sin(90 - \theta_c) \tag{4-1}$$

Using $\sin(90 - \theta_c) = \cos(\theta_c)$, $\cos(\theta_c) = \sqrt{1 - \sin^2\theta_c}$, and $\sin \theta_c = n_2/n_1$, this can be written as

$$n_0 \sin \alpha_{max} = \text{NA} \tag{4-2}$$

where the *numerical aperture* NA has been defined here as

$$\text{NA} \equiv \sqrt{n_1^2 - n_2^2} \tag{4-3}$$

The numerical aperture is widely used in optical systems to specify the maximum acceptance angle for light to enter the system, and is most generally defined by Eq. (4-2). For example, the spatial resolution Δx of an optical microscope is related to the NA of the microscope objective by

$$\Delta x = 0.61 \frac{\lambda}{\text{NA}} \tag{4-4}$$

where λ is the wavelength of light. Optimum resolution in a microscope requires not only short wavelengths, but also high numerical apertures.

For optical fibers, the NA is defined by Eq. (4-3) rather than by Eq. (4-2). The two definitions are equivalent in the case of wide core diameters, which support many transverse modes (so-called *multimode fibers*). For small core diameters that support only one transverse mode (*single-mode fibers*), however, the angular distribution of light entering or leaving the fiber is influenced by diffraction effects, which were not considered in deriving Eq. (4-2). Although Eq. (4-2) does not apply to single-mode fibers, it is still true that a smaller NA as defined in Eq. (4-3) leads to a smaller acceptance angle for incident light (see Problem 4.13). The definition of NA in Eq. (4-3) is therefore most useful for optical fibers since it can apply to both multimode and single-mode fibers.

In the previous chapter, the fractional index difference Δ was defined [see Eq. (3-36)] for planar waveguides as $\Delta = (n_1 - n_2)/n_1$ for a medium of index n_1 sandwiched in between two media of index n_2. In the case of optical fibers, a similar definition is made, with n_1 and n_2 now the index of the core and cladding, respectively. This parameter is usually small, a typical value for telecommunications fiber being $\Delta \sim 0.01$. For such small values of Δ, the indices of the core and cladding are nearly the same, so we can define a single approximate index $n \approx n_1 \approx n_2$. In this case, the numerical aperture can be approximated by the simple expression (see Problem 4.1):

$$\text{NA} \simeq n\sqrt{2\Delta} \tag{4-5}$$

Germanium is often added to the core glass in an optical fiber to raise its refractive index. Adding 20% Ge by weight gives $\Delta n \approx 0.025$.

EXAMPLE 4-1

Determine the numerical aperture and acceptance angle for a multimode fiber with core index 1.5 and fractional index difference 0.01, assuming that light is incident on the fiber from air. Repeat if the fiber is immersed in water (index 1.33).

Solution: Using Eq. (4-5), we have for the fiber in either air or water

$$\mathrm{NA} \simeq (1.5)\sqrt{(2)(0.01)} = 0.21$$

Although the NA is the same, the acceptance angle is different in air and water:

$$\alpha_{\max} = \sin^{-1}(0.21) = 12° \qquad \text{(in air)}$$

$$\alpha_{\max} = (1/1.33)\sin^{-1}(0.21) = 9° \qquad \text{(in water)}$$

4-2. CYLINDRICAL WAVEGUIDE MODES

The problem of determining the allowed modes in a cylindrical geometry is similar in principle to that of the planar waveguide, but the mathematical treatment is much more complex. Fig. 4-4 shows the fiber geometry for a multimode fiber with core radius a. A rigorous solution to the problem involves solving Maxwell's equations in the core and cladding regions, and applying appropriate boundary conditions at the core–cladding boundary. For example, one might look for solutions in the form of Eq. (2-3), which can be written with cylindrical coordinates as

$$E_{\mathrm{mode}} = E_0 g(r, \phi) e^{i(\omega t - \beta z)} \tag{4-6}$$

The solution for $g(r, \phi)$ turns out to be in the form of Bessel functions in the radial (r) direction and sinusoidal functions in the azimuthal (ϕ) direction.

Number of Modes

To gain some physical insight into the nature of the modes, without being overwhelmed with the mathematics, let us first consider the fiber as a rectangular waveguide of width $d = 2a$, as shown in Fig. 4-5.

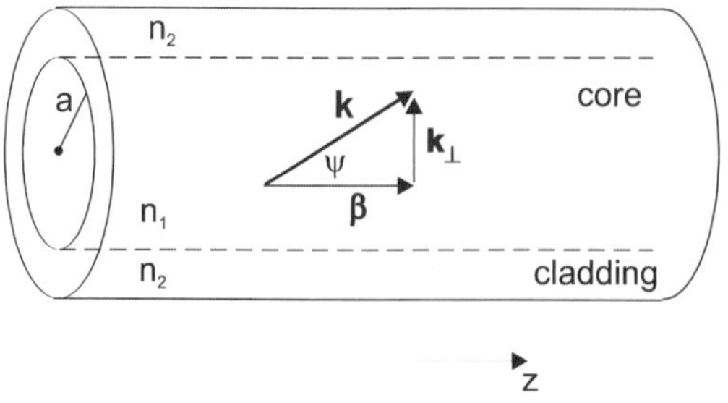

Figure 4-4 Fiber geometry with core radius a, core index n_1, and cladding index n_2.

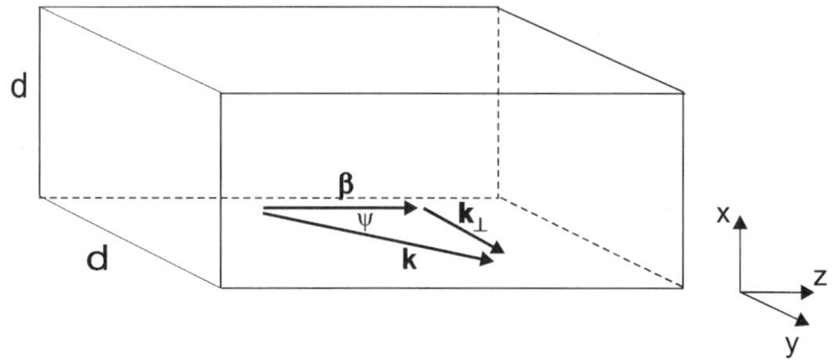

Figure 4-5 Rectangular approximation for fiber, with width $d = 2a$.

In rectangular geometry, we look for solutions in the form of a plane wave with wave vector $\mathbf{k} = \mathbf{k}_{\|} + \mathbf{k}_{\perp}$ with

$$\mathbf{k}_{\|} = \beta \hat{k}$$

$$\mathbf{k}_{\perp} = k_x \hat{i} + k_y \hat{j}$$

where $\hat{i}$, $\hat{j}$, and $\hat{k}$ are the usual unit vectors. Ignoring the phase shifts upon reflection (a valid approximation for multimode waveguides with high-order modes), the boundary conditions require that the round-trip phase change in either the x or y directions be a multiple of 2π. The restrictions on k_x and k_y are then determined from

$$\begin{aligned} k_x 2d &= m2\pi \\ k_y 2d &= l2\pi \end{aligned} \tag{4-7}$$

where m and l are integers. The allowed values of k_x and k_y,

$$k_x = m\frac{\pi}{d}, \qquad k_y = l\frac{\pi}{d} \tag{4-8}$$

are evenly spaced with increment π/d, and the waveguide modes can be represented as

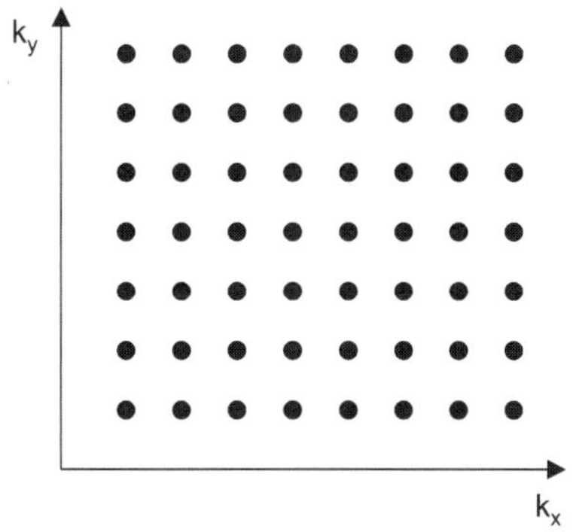

Figure 4-6 Allowed modes for rectangular waveguides are uniformly spaced in k_x–k_y space.

points in the two-dimensional k space shown in Fig. 4-6. The upper limits on m and l are found by requiring that total internal reflection occur at the core–cladding boundary.

For a mode with $k_y = 0$, the treatment reduces to that of Chapter 3 for a planar waveguide. In our approximate treatment, the fiber diameter $2a$ corresponds to the 1-D waveguide thickness d. It is conventional for optical fibers to define a V parameter that is similar to the V_p for planar waveguides [Eq. (3-8)], except that the waveguide thickness d is replaced by the fiber core radius a:

$$V \equiv \frac{2\pi a}{\lambda_0}\sqrt{n_1^2 - n_2^2} \tag{4-9}$$

where λ_0 is the free-space wavelength. The two definitions for V are related by $V_p = 2V$ for our approximate treatment.

The number of guided modes with $k_y = 0$ can then be given by Eq. (3-10), which for large V reduces to $\approx 2V/\pi$. Similarly, for modes with $k_x = 0$ there are $\approx 2V/\pi$ guided modes. The total number of guided modes for any combination of k_x and k_y is therefore expected to be the product of these two numbers, or $\approx (2V/\pi)^2$. This simple analysis gives the essential feature that the number of modes is ~ V^2 for a 2-D waveguide, rather than ~ V for a 1-D waveguide.

For fiber geometry, the above calculation overestimates the number of allowed modes, because modes with $k_\perp = \sqrt{k_x^2 + k_y^2} > V/a$ are not guided. Correcting for this yields a factor of $\pi/4$ (see Problem 4.2), which gives for the estimated number of modes

$$\#\text{ modes (rough estimate)} \approx \frac{2}{\pi}V^2 \tag{4-10}$$

where the result has also been multiplied by 2 to account for two polarizations for each spatial mode. This estimate is quite close to the often-quoted result (Senior 1992) for the number of modes (including both polarizations) in a multimode fiber,

$$\#\text{ modes (actual value)} \leqslant \frac{1}{2}V^2 \tag{4-11}$$

and is also close to the result $(4/\pi^2)V^2$ obtained in a more rigorous treatment (Saleh 1991). Since generally only an estimate is needed in practice, we will use the usual simple expression $V^2/2$ for calculations.

EXAMPLE 4-2

How many modes can propagate in a step-index fiber with a 100 μm diameter core and $\Delta = 0.03$? Take the core index of refraction as 1.5 and the free-space wavelength as 1.00 μm.

Solution: The core radius is $a = D/2 = 50$ μm, and the V parameter is

$$V \simeq \frac{2\pi a n\sqrt{2\Delta}}{\lambda_0} = \frac{2\pi(50 \cdot 10^{-6})(1.5)}{10^{-6}}\sqrt{2(0.03)}$$

$$V \simeq 115$$

$$\#\text{ modes} \simeq \frac{1}{2}V^2 = 6{,}660$$

Mode Patterns

The mode pattern for a rectangular waveguide would consist of standing waves in the x and y directions, with propating waves in the z direction. Larger mode numbers m and l mean a more rapid variation in intensity with x and y, giving rise to $\approx m$ maxima in the x direction and $\approx l$ maxima in the y direction.

The situation is qualitatively the same for fibers with cylindrical symmetry, the difference being the symmetry and shape of the resulting modes. Fig. 4-7 shows representative mode patterns for low-order modes in an optical fiber. The modes are designated as LP (for "linearly polarized"), and labeled with integer subscripts l and m. There are m maxima in the mode intensity along a radial direction, and $2l$ maxima along the circumference of a circle around the fiber center.

In the ray picture, m corresponds to rays making different angles with the fiber axis, as indicated in Fig. 4-8. Likewise, the integer l corresponds to the helicity (tightness of the spiral) of the ray as it corkscrews down the fiber. Rays that pass through the fiber axis are termed *meridonial rays,* and have $l = 0$ or zero helicity. Rays not passing through the fiber axis are termed *skewed rays,* and have $l \neq 0$.

Single-Mode Fibers

It is often of interest to have a fiber that allows only a single propagating mode. For example, this will eliminate the intermodal dispersion that was discussed in the previous chapter and allow a higher data rate for optical communications. From the analysis of planar waveguides we found [Eq. (3-27)] that only a single mode was allowed when $V_p < \pi$. Setting $V_p = 2V$ as before, we might expect the single-mode condition for a fiber to be $V \sim \pi/2$. The actual result of a rigorous treatment is close to this:

$$V < 2.405 \qquad \text{(single-mode condition in fiber)} \qquad (4\text{-}12)$$

where V for the fiber is given by Eq. (4-9). It should be kept in mind that when we say "single-mode" fiber, we mean a single spatial mode. There are two distinct polarizations possible (**E** along either x or y), and therefore always at least two modes if different polarizations are considered as different modes.

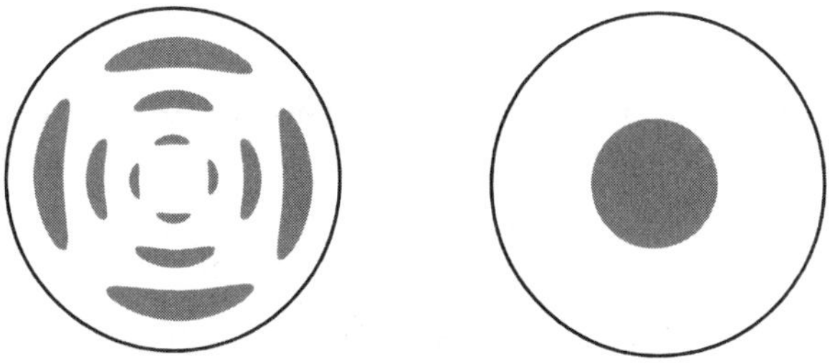

(*a*) LP_{23} (*b*) LP_{01} (HE_{11})

Figure 4-7 Typical mode patterns for the LP_{lm} modes.

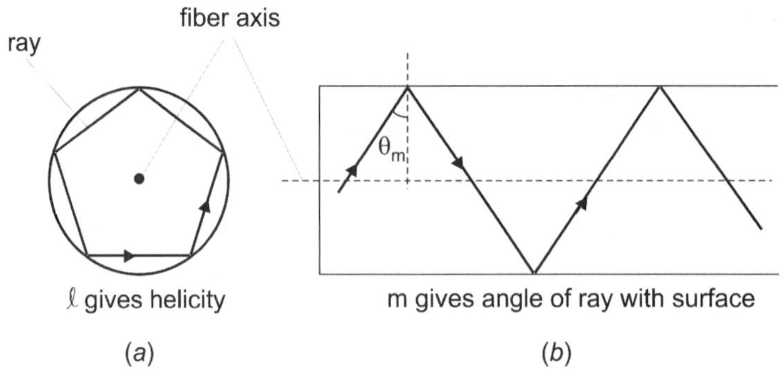

Figure 4-8 (a) End view of fiber, showing spiraling of skewed rays around the fiber axis. (b) Side view, showing propagation of meridonial ray.

Eq. (4-12) can be written in the form

$$\frac{2\pi a}{\lambda_0}\,\mathrm{NA} < 2.405 \qquad \text{(single-mode condition)} \tag{4-13}$$

using Eqs. (4-3) and (4-9). The condition for single-mode propagation is thus seen to depend on three parameters: the core radius a and numerical aperture NA of the fiber, and the optical wavelength λ_0. In principle, either a or NA could be reduced to achieve single-mode operation. Making the NA too small, however, restricts the acceptance angle for incident light and reduces source-to-fiber coupling efficiency (see Chapter 12). In practice, NA ~ 0.20 is typical for fibers used in optical communications.

The dependence on wavelength in Eq. (4-13) implies that any fiber will be single-mode for a sufficiently long wavelength of light. Of course, not all wavelengths will propagate efficiently in the fiber, due to attenuation processes such as absorption and scattering (see Chapter 5). The wavelength at which the fiber just becomes single-mode is termed the *cutoff wavelength* λ_c, defined by

$$2.405 = \frac{2\pi a}{\lambda_c}\,\mathrm{NA}$$

which can be written

$$\lambda_c = \frac{2\pi a\,\mathrm{NA}}{2.405} \qquad \text{(cutoff wavelength)} \tag{4-14}$$

For wavelengths $\lambda > \lambda_c$, the fiber will be single-mode, whereas for $\lambda < \lambda_c$ it will be multimode.

EXAMPLE 4-3

a) How small must the core be if only one mode is to propagate in a fiber with $\Delta = 0.01$? Take the core index of refraction as 1.5 and the free-space wavelength as 1.00 μm.

Solution: For $\Delta \ll 1$, we use the approximation $\mathrm{NA} \simeq n\sqrt{2\Delta} = 0.212$. From Eq. (4-13), the core radius must be

$$a < \frac{(2.405)\lambda}{2\pi \mathrm{NA}} = \frac{(2.405)(1\ \mu\mathrm{m})}{2\pi(0.212)} = 1.8\ \mu\mathrm{m}$$

The core diameter must therefore be less than 3.6 μm.

b) A fiber with the same Δ and n has a core diameter of 4.4 μm. For what range of wavelengths will the fiber be single-mode?

Solution: Using Eq. (4-14), the cutoff wavelength is

$$\lambda_c = \frac{2\pi a \mathrm{NA}}{2.405} = 1.22\ \mu\mathrm{m}$$

This fiber would therefore be single-mode for $\lambda > 1.22\ \mu$m.

Mode Chart

In Chapter 3, it was shown that the modes in a planar waveguide can be described by an effective refractive index:

$$n_{\mathrm{eff}} = \frac{\lambda_0}{2\pi}\beta \tag{4-15}$$

where β is the axial wave vector and λ_0 is the free-space wavelength of light. The effective index varies with waveguide thickness d in the manner of the mode chart shown in Fig. 3-6. Similarly, one can determine the effective index for an optical fiber, defined by Eq. (4-15), as a function of the core radius a. The calculations are much more difficult for the fiber due to the 2-D cylindrical geometry. The resulting variation in n_{eff} for a fiber is shown in Fig. 4-9, plotted versus the dimensionless V parameter [Eq. (4-9)]. Note that the abscissa in Fig. 3-6 would correspond to V_p if rescaled by the factor 2πNA.

Qualitatively, the mode charts for the fiber and planar waveguide are similar. In each case, there is one mode that can propagate for an arbitrary small waveguide width. As the waveguide width increases, other modes (which were "cut off" for smaller V) become propagating as well. The value of V for which a second mode just becomes allowable in a fiber waveguide turns out to be 2.405, which corresponds to the single-mode condition given in Eq. (4-12). The number of propagating modes is found by drawing a vertical line at a particular value of V and counting the number of mode lines that are crossed. Some of the modes are "degenerate," in the sense that they have the same value of n_{eff} for a given V. For example, for $2.6 < V < 3.8$ there are four mode line crossings, for a total of eight modes (including two polarizations for each spatial mode).

The nomenclature for the modes in a circular waveguide depends on the level of approximation. Most optical fibers can be considered to be "weakly guiding," with $\Delta \ll 1$. The modes in an exact treatment are labeled TE, TM, HE, and EH. The TE and TM modes are analogous to the transverse electric and transverse magnetic modes discussed in Chapter 3 for a planar waveguide, and have $E_z = 0$ (TE) or $B_z = 0$ (TM). The HE and EH modes are *hybrid modes* in which both E_z and B_z are nonzero. In the optical fiber there are no modes that are truely TEM (transverse electric and magnetic), with $E_z = B_z = 0$. However, because of the weak guiding, light propagation in the fiber is *paraxial* (close to the fiber axis), and the axial components E_z and B_z are small.

The next level of approximation is to neglect these axial fields and consider the modes

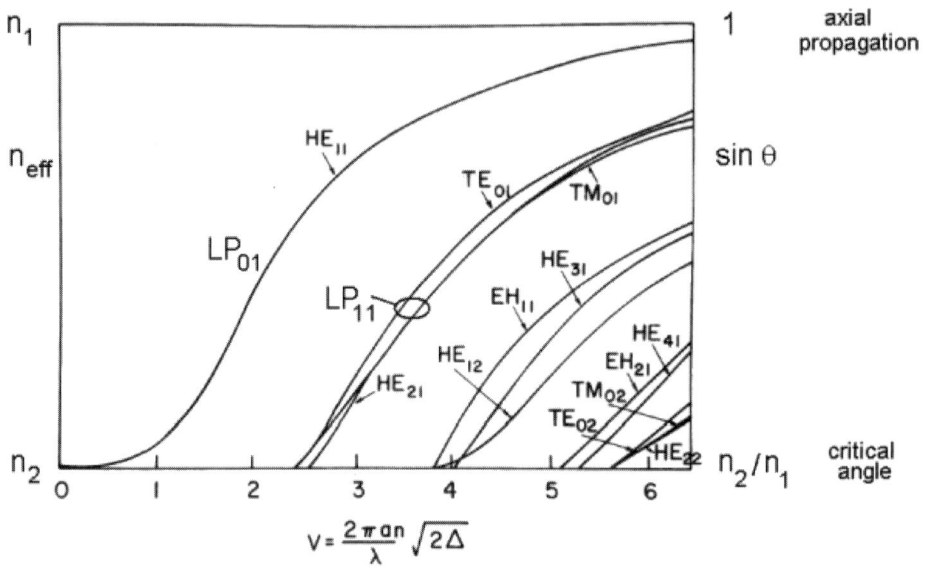

Figure 4-9 Mode chart for optical fiber (after Keck, 1981).

to be TEM. These are the *linearly polarized modes,* LP_{lm}, the patterns for which were shown in Fig. 4-7. The lowest order mode is LP_{01}, which corresponds to the HE_{11} hybrid mode. The next-highest mode is LP_{11}, which corresponds to the three modes HE_{21}, TE_{01}, and TM_{01}. Figure 4-9 shows that these three levels are nearly degenerate, being designated as a group by LP_{11}. Similar groupings occur for the higher-order modes. For most applications in which $\Delta \ll 1$, the LP approximation is adequate.

Gaussian Mode Approximation

The lowest-order mode LP_{01} in a single-mode fiber is found to be approximately *Gaussian,* with electric field varying with radial distance r from the fiber axis as

$$E(r) = E_0 e^{-(r/\omega)^2} \qquad \text{(Gaussian profile)} \tag{4-16}$$

The parameter w is the *mode waist size,* and $2w$ is the *mode field diameter,* which characterizes the spread of the optical field, as illustrated in Fig. 4-10. Since the optical intensity I varies as the square of E,

$$I(r) = I_0 e^{-2(r/\omega)^2} \tag{4-17}$$

For $1.2 < V < 2.4$, the mode waist size is approximately given by

$$w \simeq a\left(0.65 + \frac{1.619}{V^{1.5}} + \frac{2.879}{V^6}\right) \tag{4-18}$$

[see (Marcuse 1977) and (Jeunhomme 1990)], where a is the fiber core radius.

It can be seen from Eq. (4-18) that as V decreases, say by decreasing the core radius,

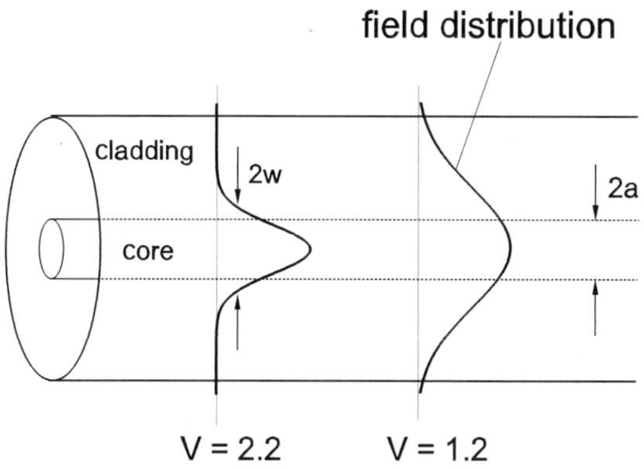

Figure 4-10 Gaussian modes with mode field diameter $2w$ in fiber of diameter $2a$. Shown is a tightly confined mode with $V = 2.2$, and a loosely confined mode with $V = 1.2$.

the mode waist size increases. This can be understood fundamentally as a diffraction phenomenon—as the core is made smaller to confine the light, diffraction becomes more important and acts to resist that confinement. The steady-state mode distribution given by Eq. (4-16) is a result of the balance between the tendency of the high-index core to confine the beam and the tendency of diffraction to spread the beam out.

A fiber with very small V has a large ratio of w/a, and the resulting mode is referred to as *weakly guided.* The beam may become so spread out that the majority of the mode's energy is contained in the cladding region rather than the core. This behavior can be useful for various devices such as fiber sensors and fiber couplers, but is detrimental for low-loss communications fiber. For best confinement of the mode the fiber V is often chosen to be not much below the cutoff value of 2.405.

A commonly used telecommunications fiber designed for 1550 nm has parameters $2a = 8.3$ μm and $\Delta = 0.0036$. The numerical aperture according to Eq. (4-5) is then NA $\simeq$ 0.13, and $V \simeq 2.19$ from Eq. (4-9). Using Eq. (4-18), we then obtain a mode field diameter of $2w = 9.8$ μm. This is only 18% higher than the actual core diameter, which means that the mode is well confined by the core.

PROBLEMS

4.1 Show how Eq. (4-5) follows from Eq. (4-3) under the approximation $n_1 \approx n_2$.

4.2 Show that for a ray in a cylindrical waveguide to correspond to a guided mode, it must have a transverse wavenumber $k_\perp = \sqrt{k_x^2 + k_y^2} < V/a$. Show further that this introduces a factor of $\pi/4$ in obtaining the estimated number of modes given in Eq. (4-10).

4.3 An optical fiber has core index $n_1 = 1.495$, cladding index $n_2 = 1.485$, and core diameter 50 μm. Light of (free-space) wavelength 1.3 μm is coupled into this fiber. Determine the fiber V parameter and the maximum number of modes that can propagate. Consider different polarizations as different modes.

4.4 For the fiber and wavelength of Problem 4.3, determine the maximum angle ψ that rays make with the optical axis. (Hint: use the result of Problem 4.2.)

4.5 Sketch the mode field pattern for the LP_{32} and LP_{21} modes.

4.6 A step index fiber has core index $n_1 = 1.48$ and $\Delta = 0.013$, and is single-mode only for free-space wavelengths $\lambda > 1.25$ μm. Determine the radius of the fiber core.

4.7 For the fiber of Problem 4.6, determine the range of wavelengths for which the fiber supports eight modes (and no more than eight). Consider different polarizations as different modes.

4.8 A step index fiber with core diameter 7 μm has more than one mode only for $\lambda <$ 1400 nm. Determine the fiber's numerical aperture.

4.9 The cutoff wavelength for a step index fiber is 1400 nm. How many modes can propagate at a signal wavelength of (a) 1550 nm, (b) 1100 nm, and (c) 750 nm. Consider different polarizations as different modes.

4.10 The fiber in Problem 4.6 is used with a signal wavelength of 1.7 μm. Use the mode chart of Fig. 4-9 to determine n_{eff} and β for the fiber mode.

4.11 For the fiber and wavelength of Problem 4.10: (a) determine the mode field diameter, and (b) determine the radial distance from the fiber axis at which the light intensity is reduced to 5% of the peak (on-axis) value.

4.12 Consider a standard telecommunications fiber designed for operation at 1550 nm, with core diameter 8.3 μm, core index 1.48, and fractional index difference 0.0036. What is the mode field diameter when the operational wavelength is 1630 nm? If this fiber is used at a wavelength of 1310 nm, will it still be single-mode?

4.13 Using the definition of NA in Eq. (4-3), along with Eqs. (4-9) and (4-18), show that a smaller NA in a single-mode fiber leads to a smaller divergence angle for light emitted from the fiber core.

Chapter 5

Losses in Optical Fibers

Historically, the success of fiber optic communications depended critically on the development of low-loss optical fiber, as discussed in Chapter 1. In an optical fiber, there are three fundamental loss mechanisms: absorption, scattering, and bending loss, as illustrated in Fig. 5-1. Absorption results in the loss of a propagating photon, the photon's energy generally being converted into heat. In a scattering process, the photon does not disappear, but its direction (and possibly its energy) is changed. Absorption and scattering are fundamental materials properties, occurring both in fibers and in bulk glass (large uniform sections of glass). The third loss mechanism, bending loss, is unique to the fiber geometry, and relates to the requirement of total internal reflection (TIR) for lossless transmission down the fiber. In this chapter, each of these three loss mechanisms will be discussed in turn.

5-1. ABSORPTION LOSS

Consider light propagating from left to right through a length L of uniformly absorbing material, as shown in Fig. 5-2. We define the *attenuation coefficient* α as the fractional loss in light power per unit length of propagation. The amount of power lost in a thin slice of thickness dz is then $P\alpha dz$, where P is the power incident on the slice. If the power entering the material from the left is P_{in}, it is straightforward to show (see Problem 5.1) that the power exiting the right side is

$$P_{out} = P_{in}e^{-\alpha L} \tag{5-1}$$

which is known as *Beer's law.* When the attenuation of light is predominantly due to absorption, α is also referred to as the *absorption coefficient.* Since αL is dimensionless, the units for α are reciprocal length, often given in cm^{-1}.

The attenuation coefficient α is an alternative to the decibel concept discussed in Chapter 1. Using Eq. (1-1), we have

$$\text{dB loss} = 10 \log_{10}\left(\frac{P_{in}}{P_{out}}\right) = 10 \log_{10}(e^{\alpha L}) = 10\, \alpha L \log_{10} e \tag{5-2}$$

or,

$$\text{dB loss} = 4.34\, \alpha L \tag{5-3}$$

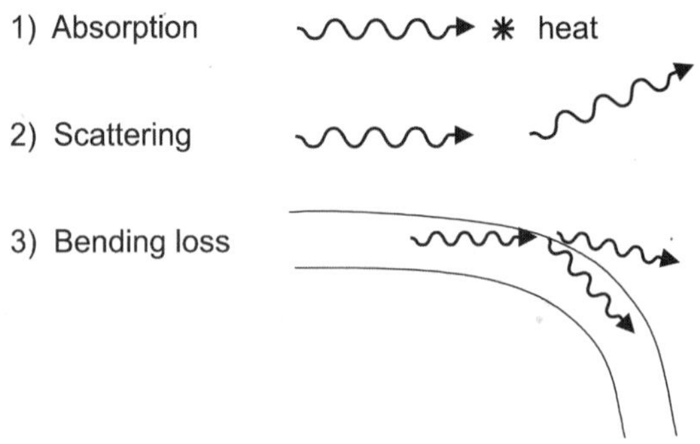

Figure 5-1 Loss mechanisms for light propating in optical fiber.

Note that the dB loss per unit length and the attenuation coefficient are the same, apart from a scaling factor. The equivalence can be expressed by a conversion factor for the two units:

$$\begin{aligned} 1\ \text{cm}^{-1} &= 4.34 \times 10^5\ \text{dB/km} \\ 1\ \text{dB/km} &= 2.303 \times 10^{-6}\ \text{cm}^{-1} \end{aligned} \tag{5-4}$$

In practice, the dB/km unit is usually used to describe losses in optical fiber systems, whereas the cm^{-1} unit is used when relating propagation losses to fundamental physical processes.

Figure 5-3 summarizes the various absorption processes that can lead to attenuation in an optical fiber. These processes can be separated into two fundamentally different types: those that promote an electron from a lower to a higher energy state (an *electronic transition*), and those that increase the vibrational energy of some group of atoms (a *vibrational transition*). Electronic transitions are generally of higher energy than vibrational transitions because of the small mass of an electron compared to the mass of an atom. At the shortest wavelengths (typically $\lambda < 400$ nm), the photon energy is sufficient to promote an electron from the valence band to the conduction band of the host glass (energy bands are discussed in Chapter 10). At the longest wavelengths (typically $\lambda > 2\ \mu$m) the photon energy matches the vibrational energy of the host glass, and vibrational transitions become efficient.

In between these two wavelengths, the absorption loss can be very small— there is a "window of transparency" in the visible and near-infrared regions between $0.4 < \lambda < 2$ μm. However, the presence of impurities introduces additional electronic and vibrational absorption, which can reduce the transparency in this window. Typical impurities include transition metal ions such as Cu^{2+}, Fe^{2+}, and Cr^{3+}, which introduce electronic transitions,

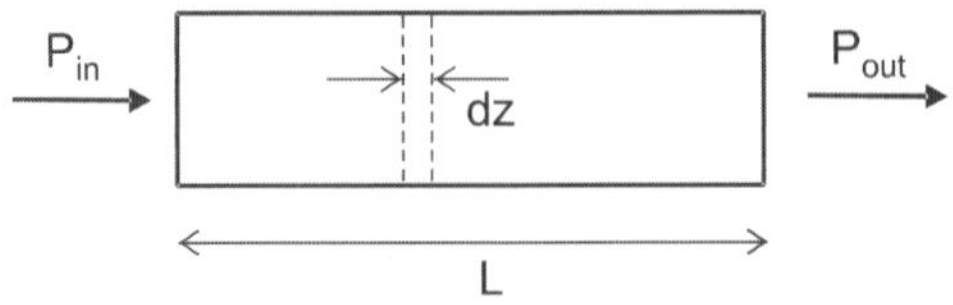

Figure 5-2 A fraction $\alpha\, dz$ of light power P is absorbed in slice of thickness dz.

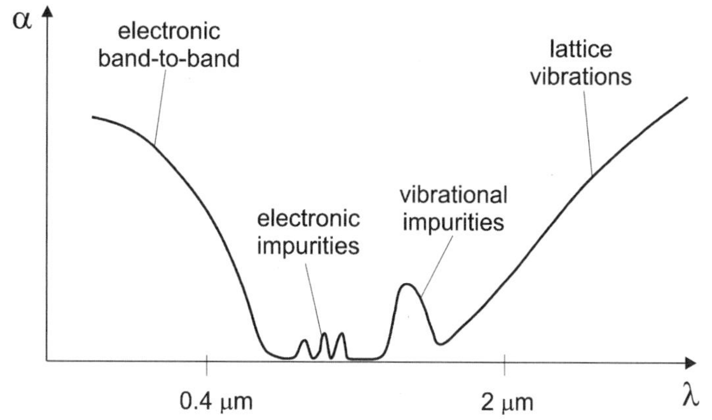

Figure 5-3 Absorption coefficient versus wavelength for optical fiber, showing electronic and vibrational loss mechanisms.

and the hydroxyl ion OH^-, which introduces strong vibrational transitions at 1.4 and 2.8 μm. The transition at 1.4 μm is especially detrimental, being close to the important telecommunications wavelength 1.5 μm where the attenuation in silica fiber is a minimum. For the lowest-loss fibers, it is important to keep water out during the manufacturing process, to minimize the OH content.

5-2. SCATTERING

In an ideal crystal at zero temperature, light can propagate without scattering. Real crystals have defects and impurities that interupt the perfect crystalline order and give rise to some degree of scattering. Noncrystalline materials such as glasses and liquids (and air!) have an inherent disorder that results in some minimum level of light scattering, even in the absence of impurities.

Rayleigh Scattering

The most important scattering loss in glass fibers is *Rayleigh scattering,* in which the wavelength of the scattered light remains unchanged. Rayleigh scattering arises from the interaction of the light wave with stationary fluctuations Δn in the index of refraction n. These fluctuations occur due to random thermal motion when the glass is in a liquid state, and are frozen in place when the glass makes the transition from liquid to solid at temperature T_F. The scattering process can be thought of as equivalent to the scattering of light from small spheres of diameter d and index $n + \Delta n$, embedded in a uniform medium of index n. If $d \ll \lambda$ (a good approximation here), the attenuation coefficient is found to be

$$\alpha_R \propto \frac{\langle(\Delta n)^2\rangle}{\lambda^4} \propto \frac{k_B T_F}{\lambda^4} \qquad \text{(Rayleigh scattering)} \qquad (5\text{-}5)$$

where $\langle(\Delta n)^2\rangle$ is the average square of the refractive index fluctuation and k_B is Boltzmann's constant.

The λ^{-4} dependence on wavelength in Eq. (5-5) is the most important characteristic of Rayleigh scattering and explains, for example, why the sky is blue (because light at shorter wavelengths, i.e. blue, is more strongly scattered into our eyes). For fiber optic communications the important consequence is that longer signal wavelengths will experience less loss, with less amplification and signal regeneration required. For the silica (SiO_2) glass typically used in communication-grade fibers,

$$\alpha_R \simeq (0.8)\left(\frac{1\ \mu\text{m}}{\lambda}\right)^4 \text{dB/km} \qquad \text{(silica fiber)} \tag{5-6}$$

with λ in μm and α_R in dB/km. If special attention is given to processing parameters during fiber manufacture, the prefactor in Eq. (5-6) can be made 0.7 or even 0.6, but not much lower. The earliest optical communications systems used wavelengths around 850 nm, the so-called *first telecommunications window,* in which Rayleigh scattering losses are $\approx$ 1.5 dB/km. Increasing the operating wavelength to 1.3 μm (the *second telecommunications window*) reduced Rayleigh scattering losses to $\approx$ 0.28 dB/km, a five-fold improvement. Further development of the *third telecommunications window,* around 1.55 μm, reduced the losses another factor of two to $\approx$ 0.14. Each factor of two reduction in α_R was important in the developing success of optical communications, effectively halving the number of regeneration stations required for long-haul optical networks.

One might ask whether the losses can be further reduced for wavelenths longer than 1.55 μm. The answer for silica fiber is no, because at longer wavelengths the absorption of light by vibrational transitions of the host glass becomes more important than Rayleigh scattering. Figure 5-4 shows a typical loss spectrum for a silica fiber, along with the Rayleigh scattering and lattice absorption contributions. The combination of Rayleigh scattering at shorter wavelengths and lattice absorption at longer wavelengths results in a V-shaped curve that is characteristic for a particular type of glass. In addition, there is a pronounced OH absorption peak at 1.4 μm, which creates local minima in the attenuation

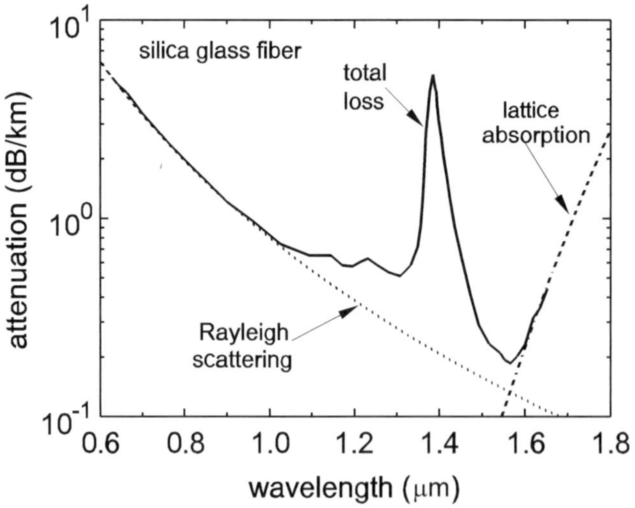

Figure 5-4 Typical attenuation spectrum for silica glass fiber, showing contributions from Rayleigh scattering and lattice absorption that result in a V-shaped curve.

around 1.3 and 1.5 μm. As optical communications systems evolved toward longer wavelengths, it was only natural to choose operating wavelengths in these windows of transparency. There are other advantages as well to the 1.3 μm window, to be discussed in the next chapter. Only recently has the technology for manufacturing optical fiber improved to the extent that the OH peak can be practically eliminated. Although low-OH fiber can now be made, most of the communications fiber that is currently installed and in use does have an OH peak.

For pure silica fiber, the minimum attenuation obtainable is ≈ 0.15 dB/km. To reduce the attenuation further, alternative types of glass could be tried, but to date there has not been significant improvement. For example, the Rayleigh scattering can be reduced by adding small amounts of dopants such as Na_2O to the silica host, but the reduction is only a modest 20% (Saito et al. 1997).

Another strategy for lower attenuation is to reduce the lattice absorption, which would shift the minimum in the V curve to longer wavelengths and smaller minimum attenuation. For a time in the late 1980s there was much interest in ZrF_4-based, heavy-metal fluoride glasses for this purpose, since they have reduced absorption in the infrared compared with silica glass. The reduced absorption arises from the need to simultaneously create a greater number of lattice vibrational quanta (*phonons*) in the fluoride glass, due to the lower vibrational frequency of the Zr–F bond compared with that of the Si–O bond. However, the lowest loss so far in fluoride glass fibers is ~ 1 dB/km, due to problems with crystallization and other sources of loss.

It seems at present that for light propagating in a glass fiber, the ultimate practical minimum loss will be ~ 0.1 dB/km. To reduce the loss further would require that the light propagate not in glass, but in air. Recent developments that make this possible will be discussed in Chapter 8.

Brillouin Scattering

Light will generally be scattered by any nonuniformity in a material's index of refraction. In the case of Rayleigh scattering in glass, the nonuniformity consists of stationary density and compositional fluctuations which were "frozen in" when the liquid cooled into a solid. The index of refraction can be thought of as nonuniform in a "lumpy" sort of way. Another way that the index of refraction can be nonuniform is via a sound (acoustic) wave, in which the density and pressure vary periodically inside the material. The varying density causes a varying refractive index, resulting in "waves" of changing index of refraction, propagating at the speed of sound v_s in the material. The separation between planes of maximum index will be $d = v_s/f_a$, where f_a is the frequency of the acoustic wave.

When light is incident upon these "index waves," it can be scattered as shown in Fig. 5-5, a process termed *Brillouin scattering*. The situation actually corresponds to Bragg diffraction from a thick grating, as discussed in Chapter 2 (see Fig. 2-17). The Bragg condition [Eq. (2-28)] then applies, giving the allowed angles θ for efficient scattering. Since the index grating is in motion, the scattered light waves will undergo a *Doppler shift,* just as for light reflected from a moving mirror. The optical frequency ν will increase or decrease depending on the component of the acoustic wave's velocity along the direction of the incident light wave. The change in optical frequency $\Delta\nu$ is

$$\frac{\Delta\nu}{\nu} = 2\,\frac{v_{\text{along ray}}}{c/n} = \frac{2n}{c}\,v_s \sin\theta \tag{5-7}$$

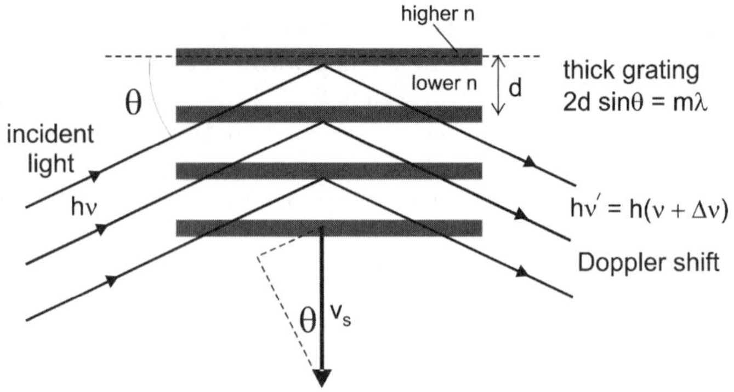

Figure 5-5 Brilloin scattering from acoustic waves.

where n is the average refractive index of the material and θ is the incident angle defined in Fig. 5-5. Combining Eq. (5-7) with Eq. (2-28) yields the simple result (see Problem 5.4) that the change in optical frequency is just equal to the acoustic frequency, $\Delta\nu = f_a$. In the quantum picture of light, this can be interpreted to mean that the photon energy $h\nu$ has changed by a unit of the vibrational energy hf_a (the phonon energy), implying that a phonon has been either created or destroyed in the scattering process.

The magnitude of the frequency shift in Brillouin scattering can be estimated from Eq. (5-7) by putting in typical values for glass ($n = 1.5$ and $v_s = 5 \times 10^3$ m/s) and setting $\sin\theta = 1$. This gives a maximum fractional frequency shift $\Delta\nu/\nu \approx 5 \times 10^{-5}$, which for 1500 nm light corresponds to $\Delta\nu = 10$ GHz or $\Delta\lambda = 0.075$ nm. The intensity of the scattered light is very weak for thermally generated acoustic waves. However, for externally applied sound waves of large amplitude, this scattering process can be efficient, and forms the basis for a practical way of deflecting laser beams, the acoustooptic deflector. In fibers, Brilluoin scattering is an important source of loss only when it becomes nonlinear (see Chapter 9). This occurs primarily for narrowband light, with spectral width $\Delta\nu < 10$ MHz.

Raman Scattering

In the previous section, we saw that light could scatter off acoustic waves in a process called Brillouin scattering. These acoustic waves consist of the collective motion of a large number of atoms, with nearby atoms moving in nearly the same direction. Other types of vibrations are possible as well, including *localized vibrations* in which neighboring atoms are moving in opposite directions. Figure 5-6 illustrates such a vibrational mo-

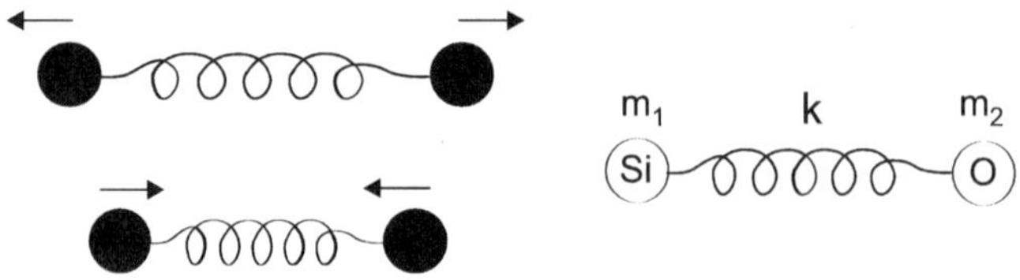

Figure 5-6 Molecular vibrations involved in Raman scattering can be modelled by masses and springs.

tion for adjacent silicon and oxygen atoms in SiO_2 glass. The vibration can be modeled as a molecule of two masses (the Si and O atoms) connected by a spring of spring constant k. For small-amplitude motion, this results in simple harmonic motion with vibrational frequency f_v given by

$$f_v = \frac{1}{2\pi}\sqrt{\frac{k}{m_r}} \tag{5-8}$$

where $m_r = m_1 m_2/(m_1 + m_2)$ is the reduced mass of the system.

When light is incident on the vibrating molecule, it can be scattered as shown in Fig. 5-7, a process termed *Raman scattering*. Energy is conserved in Raman scattering, just as for Brillouin scattering, and the new (scattered) photon energy $h\nu'$ is

$$h\nu' = h\nu \pm hf_v \qquad \text{(Raman shift)} \tag{5-9}$$

When the scattered light is decreased in frequency (often called *Stokes scattering*), the molecule is left in a more highly excited vibrational state after the scattering process. The converse process (*anti-Stokes scattering*) takes vibrational energy out of the molecule to increase the frequency of light. At finite temperature, there is some probability that the molecule is initially in the ground vibrational state, in which case no energy can be extracted. Therefore, the ratio of anti-Stokes to Stokes scattering probabilities is less than one, and is temperature dependent.

The magnitude of the frequency shift $\nu' - \nu$ is much greater for Raman scattering than for Brillouin scattering because the localized vibrational frequency f_v is much larger than the typical acoustic frequency f_a. Typically, $f_v \sim 10 - 30$ THz, whereas $f_a \sim 10$ GHz. Fundamentally, this can be understood by considering the motion of adjacent atoms in the glass, as in Fig. 5-8. For localized vibrations, the atom–atom separation d undergoes large oscillations in time. In an acoustic wave, however, where the atoms move collectively, the distance between adjacent atoms undergoes only small oscillations. Since the restoring force arises from changes in the spacing between adjacent atoms, the effective spring constant is reduced, and so is the oscillation frequency according to Eq. (5-8).

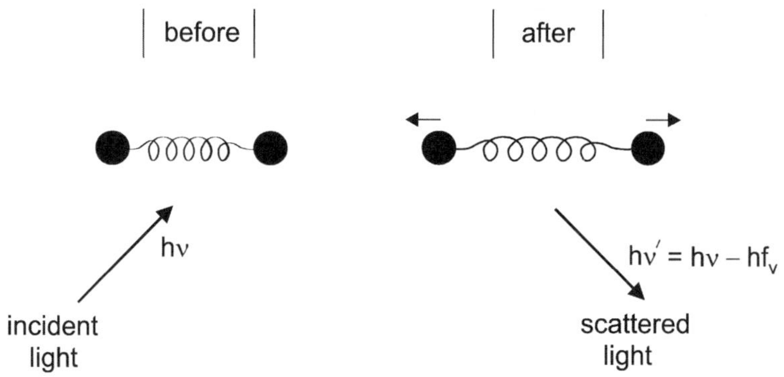

Figure 5-7 Some energy from the photon is transferred to molecular vibrational energy in Raman scattering.

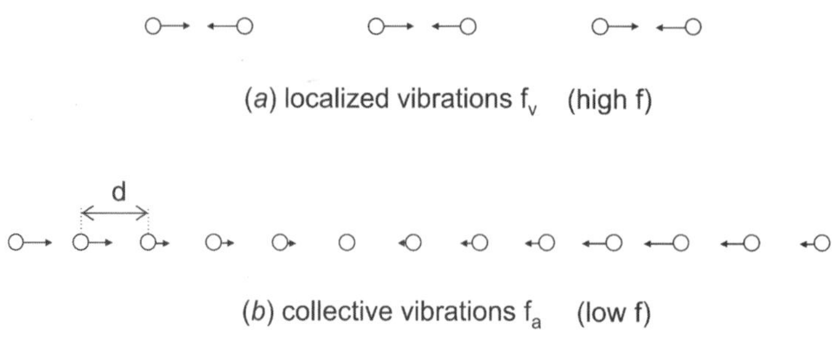

Figure 5-8 Vibrational patterns for (a) Raman scattering and (b) Brillouin scattering. Arrows show the relative displacement of the atoms.

EXAMPLE 5-1

Light of free-space wavelength 1500 nm is incident on silica glass. Determine the frequency and wavelength of the Stokes-shifted, Raman-scattered light, assuming f_v = 15 THz.

Solution: The optical frequency of the incident light is

$$\nu = \frac{c}{\lambda} = \frac{3 \times 10^8 \text{ m/s}}{1.5 \times 10^{-6} \text{ m}} = 2 \times 10^{14} \text{ Hz}$$

From Eq. (5-9), the frequency of scattered light is

$$\nu' = (20 - 1.5) \times 10^{13} = 1.85 \times 10^{14} \text{ Hz}$$

which corresponds to

$$\lambda' = \frac{c}{\nu'} = \frac{3 \times 10^8 \text{ m/s}}{1.85 \times 10^{14} \text{ Hz}} = 1620 \text{ nm}$$

Like Brillouin scattering, the Raman effect tends to be very weak and difficult to detect when the scattering occurs off the thermally generated vibrations naturally present in the glass. However, the effect can be significantly enhanced when the vibrations are large, leading to practical devices such as lasers and optical amplifiers. These devices are based on nonlinear phenomena, which will be discussed in Chapter 9. Losses due to Raman scattering become important in single-mode fibers when nonlinear effects set in, typically at power levels > 500 mW.

5-3. BENDING LOSSES

When an optical fiber is bent, light that was originally guided in the core may become unguided, resulting in a loss of guided light power. The light might also shift from one guided mode to another guided mode, a process known as mode coupling. These processes can be understood from either a geometrical optics or physical optics point of view.

Geometrical Optics View

Consider a light ray that is initially propagating at point A along the axis of a multimode fiber of radius a, as shown in Fig. 5-9. If the fiber is bent into a circular arc of radius R, then the ray will strike the core–cladding boundary at B, making an angle θ with the surface normal at that point. If R is too small, θ will become smaller than the critical angle θ_c, and the ray will no longer be guided. It is straightforward to show (see Problem 5.10) that this occurs when

$$R < \frac{a}{\Delta} \qquad \text{(significant bending loss)} \tag{5-10}$$

where $\Delta \ll 1$ has been assumed, as is typical for optical fiber. For example, if the core diameter is 100 μm and $\Delta = 0.01$, the bending loss will be significant for $R < 5$ mm. A smaller-diameter core would allow the fiber to be bent in an even smaller radius of curvature, although the fiber may break before the losses become significant.

The derivation of Eq. (5-10) assumed that the ray was initially propagating along the fiber axis, a so-called *low-order mode.* If, instead, the ray was initially already making a steep angle with the fiber axis, but was still guided (a *high-order mode*), it could become unguided with a much more gentle bending than that of Eq. (5-10). The degree of bending loss therefore depends not only on the bend radius, but also on which modes are propagating. The low-order modes are more stable and resistant to bending losses, whereas the high-order modes are only marginally stable, and prone to significant loss from even small bends.

Physical Optics View

The loss of propagating light in a bent fiber can also be understood from the wave optics point of view. Consider a low-order mode propagating in a fiber bent into a circular arc with center at O and radius of curvature R, as shown in Fig. 5-10. As the wave moves along the arc, different parts of the wave must move at different speeds, in order for the wave to maintain the same shape as it propagates. The situation is similar to that of a line of ice skaters, joining hands and pivoting about one end, so that the skaters at the far end

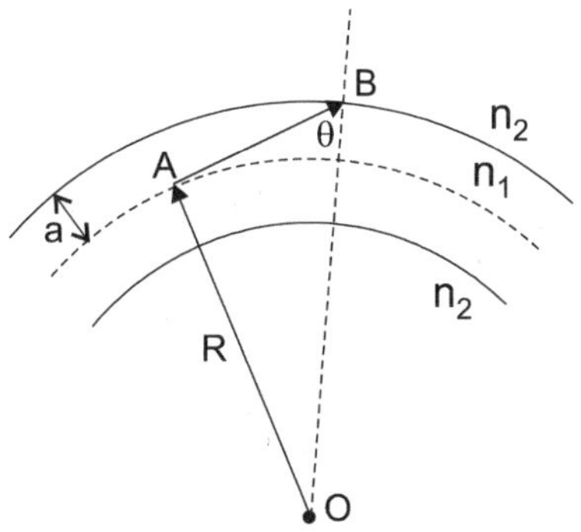

Figure 5-9 Ray optics picture of light loss due to fiber bending.

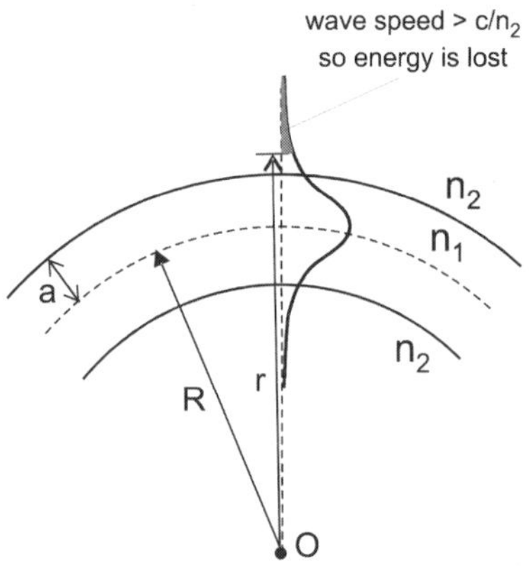

Figure 5-10 Wave optics picture of light loss due to fiber bending.

are moving very fast compared to those near the pivot. For the optical wave, this implies that at some distance r_{max} from the pivot, the evanescent wave in the cladding must be moving at a speed greater than the speed of light in the cladding material, c/n_2. Since this is not allowed by Maxwell's equations, the energy contained in the evanescent wave for $r > r_{max}$ will then be lost from the wave.

From this physical optics viewpoint, the degree of bending loss is seen to depend on how far the evanescent field extends into the cladding. High-order modes with internal waveguide angle θ close to the critical angle θ_c extend the furthest into the cladding, according to Eq. (2-21). Low-order modes, in contrast, are more tightly confined to the core, as shown in Fig. 5-11. High-order modes are then expected to be more lossy than low-order modes, a conclusion in accord with the ray optics result. The physical optics approach has the advantage that it applies to single-mode as well as multimode fibers.

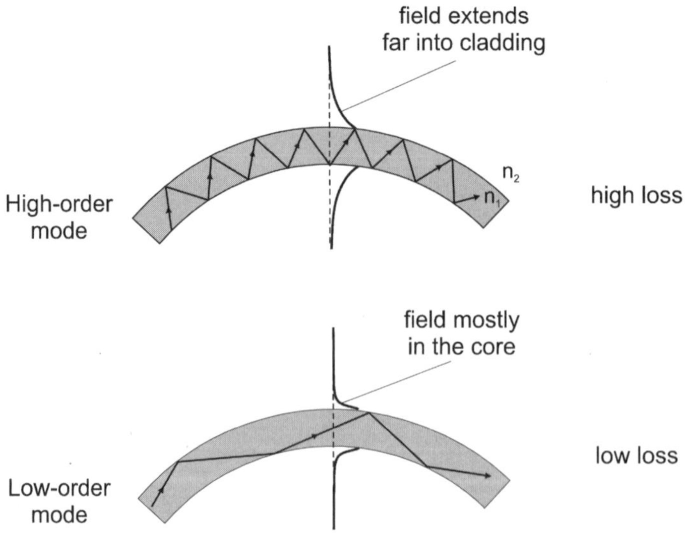

Figure 5-11 High-order modes are more lossy because more of the mode's energy is in the evanescent wave, where energy is lost due to bending.

Length Scale for Bending Loss

Bends in a fiber can be on any scale of length, but it is useful to classify them into two broad categories: macrobending and microbending. *Macrobending* losses are those caused by bends with R in the centimeter to meter range, and occur when the fiber is deliberately bent around a corner, for example. In practice, these losses can be minimized and are usually small, affecting mostly the higher-order modes in a multimode fiber.

Microbending losses are more difficult to control, arising from bends on the μm length scale. These microbends can be introduced by anything that crimps or stresses the fiber, including the packaging material that houses the fiber. Figure 5-12 shows two schemes for jacketing an optical fiber that minimize microbending losses. On the left is the *tight buffering* scheme, which contains the solid glass core and cladding within a soft plastic. This soft plastic buffering is solid enough to hold the core in a well-defined position, while being soft enough to relieve stresses and minimize microbends. The surrounding hard plastic adds mechanical protection to the fiber. On the right is the *loose buffering* scheme, in which the core and cladding are contained loosely within a larger hollow tube. The protection from microbending is better with loose buffering, since there is more "wiggle room" for the fiber. Each of these types of packaging can be incorporated into fiber cables containing large numbers of individual fibers in a variety of geometries (Hecht, J. 2002).

Mode Coupling

In our discussion to this point, we have considered light scattering and fiber bending to be loss mechanisms, in the sense that photons are lost from a signal beam. However, these processes do not always result in a loss of light power, because light propatating in one guided mode can be scattered into another guided mode, as illustrated in Fig. 5-13. This process of transferring energy from one mode to another is termed *mode coupling*. It is convenient to visualize the distribution of modes as in Fig. 5-14, with the axial component of the wave vector β taking on discreet values between the minimum n_2k_0 and maximum $n_1\, k_0$. Modes with the highest β values are *low-order modes,* whereas those with the lowest are *high-order modes.* Light that is no longer guided ($\beta < n_2k_0$) is said to be in a *radiation mode,* although this is not a true mode of the fiber.

Modest bending of a fiber results in only small changes in β, so the mode coupling takes place mostly between adjacent modes in β space. The low-order modes are fairly

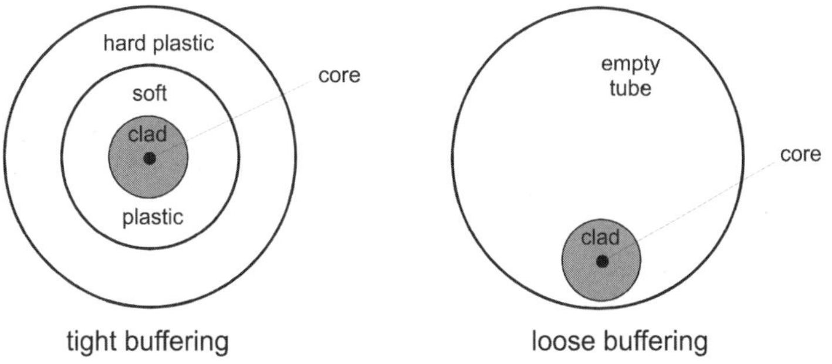

Figure 5-12 Two typical fiber-jacketing schemes. The method of containing the fiber in the cable can influence the degree of microbending loss.

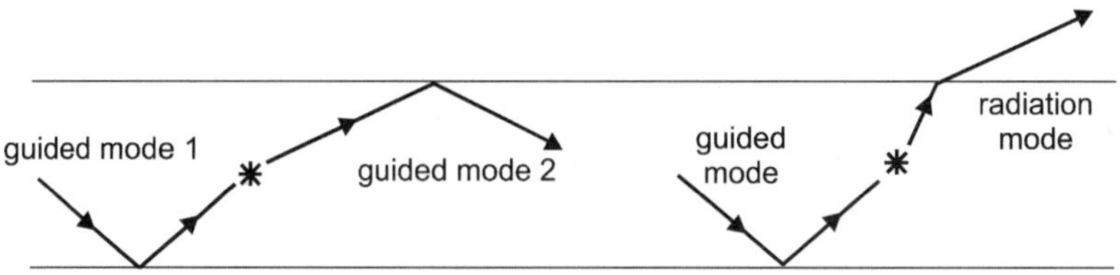

Figure 5-13 Scattering or fiber bending can couple light from one mode to another.

stable, coupling mostly with other guided modes, whereas the high-order modes are lossy, coupling efficiently with the radiation modes. After light has propagated over a long length of fiber, the distribution of light among the modes reaches a steady state that is skewed toward high values of β. In effect, the higher-order modes have been filtered out by the process of bending-induced mode coupling. As the light propagates, energy is continually hopping between adjacent modes, with a gradual leaking away of total energy as the high-order modes are coupled into radiation modes.

This mode coupling has implications for the intermodal dispersion discussed in Chapter 3. The spreading in time for a pulse in a multimode fiber was calculated in Eq. (3-35) to be $\Delta t \simeq Ln\Delta/c$, assuming that the pulse energy was distributed equally among all modes. This result would be valid if light launched initially into a particular mode stayed in that mode. Mode coupling, however, causes the light to switch back and forth between faster and slower modes as it propagates, resulting in a more nearly uniform arrival time for the different parts of the pulse. After propagating ~ 1 km, it is found that $\Delta t \propto \sqrt{L}$ rather than the linear dependence on L predicted by Eq. (3-35).

For some applications, it is desirable to create an equilibrium distribution of modes in a short fiber length, for example in fiber loss measurements (see Chapter 7). The coupling of modes can be enhanced by forcing the fiber through a series of tight bends, as illustrated in Fig. 5-15. Such a device is termed a *mode mixer,* and can be realized by simply sandwiching the fiber between pieces of sandpaper.

Cladding Modes

Light that is lost into radiation modes is no longer guided by the core, but it can still propagate some distance along the fiber in what is known as a *cladding mode.* Fig. 5-16 illus-

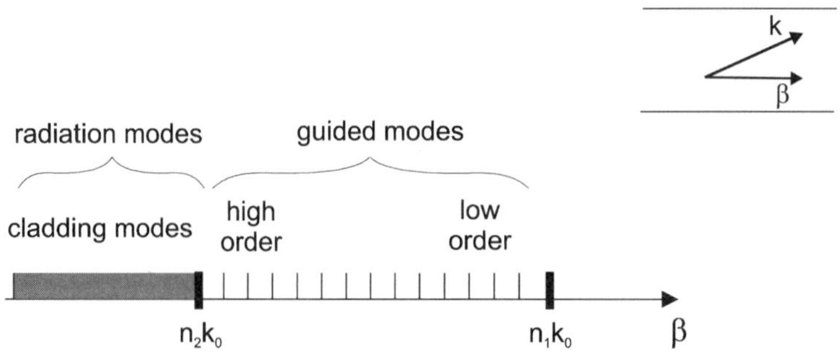

Figure 5-14 Distribution of fiber modes in β space, with discreet guided modes in the range $n_2k_0 < \beta < n_1k_0$, and continuous radiation modes for $\beta < n_2k_0$.

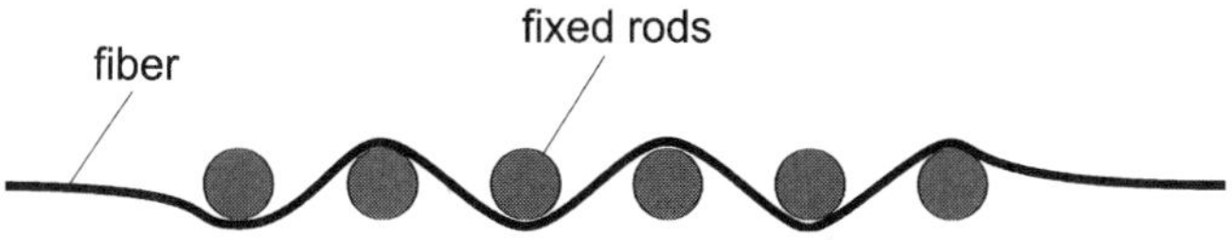

Figure 5-15 A mode mixer creates an equilibrium modal distribution.

trates the process, in which light entering the cladding is partially reflected at the interface with the surrounding jacket material. The light can propagate in the cladding for some distance, losing some of its energy with each bounce. These "modes" are highly lossy, and when excited directly by end-pumping the fiber, they can complicate measurements of the fiber attenuation coefficient.

One way to remove light from the cladding modes is to employ a *mode stripper* (see Fig. 5-16), in which the protective coating is removed from a section of the fiber, leaving just the bare cladding exposed. The fiber is then immersed in a liquid that nearly matches the index of refraction of the cladding, so there is little reflection of light at the cladding boundary. The mode stripper removes light from the cladding modes, leaving only true guided modes carrying the light energy. As the light propagates, some guided modes will continue to feed energy into the cladding modes by mode coupling. However, this is generally much less of a problem than the light that was initially injected into the cladding modes.

PROBLEMS

5.1 Derive the Beer's law expression in Eq. (5-1) by integrating the power lost per unit distance over the length L, as indicated in Fig. 5-2.

5.2 Light with wavelength 1.3 μm is coupled into a long silica fiber. (a) Determine the attenuation coefficient in units of cm^{-1}, assuming that Rayleigh scattering is the predominant loss mechanism. (b) If the optical power at the beginning of the fiber is 5 mW, determine the optical power at a distance 2.5 km down the fiber. (c) Determine the power a distance 25 km down the fiber.

5.3 In a certain (nonsilica) fiber, the loss due to Rayleigh scattering is 6 dB/km at λ = 800 nm. What would be the corresponding Rayleigh scattering loss at λ = 600 nm?

5.4 Use Eq. (5-7) along with Eq. (2-28) to show that the change in optical frequency in Brillouin scattering is equal to the acoustic frequency f_a.

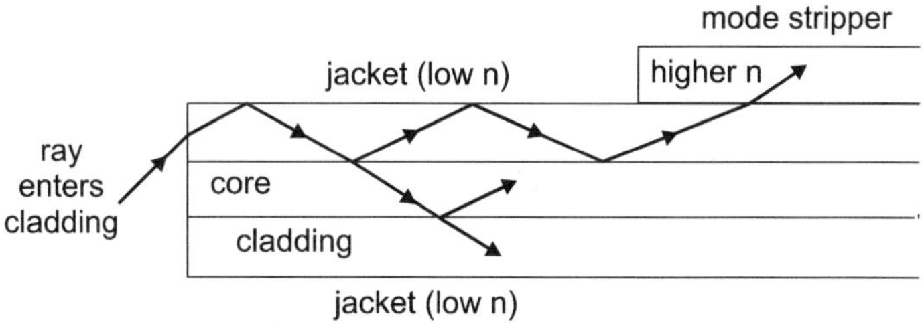

Figure 5-16 Cladding modes can be removed with a mode stripper.

5.5 An acoustooptic deflector uses sound waves of frequency 200 MHz to deflect light of free-space wavelength 1053 nm. Assuming a sound velocity of 5000 m/s and refractive index of 1.5, determine the angle by which the beam is deflected from its original direction in first order. Also calculate the wavelength of the deflected light.

5.6 Incident light of wavelength 1064 nm is Raman scattered in a glass with localized vibrational frequency $f_v = 20$ THz. Determine the wavelength of the scattered light.

5.7 Raman scattering from the CO_2 molecule occurs on the symmetric stretch mode (see Fig. 23-22), which vibrates at 40 THz. If the usual isotope of oxygen (8 protons, 8 neutrons) is replaced by one with 8 protons and 10 neutrons (^{18}O), the mass that governs the vibrational frequency in Eq. (5-8) will be increased by the factor $\approx$ 18/16. (a) Determine the molecular vibrational frequency for CO_2 with the ^{18}O isotope. (b) If the incident light has wavelength 800 nm, calculate the wavelength of Raman-scattered light for both the ^{16}O and ^{18}O isotopes.

5.8 In Raman scattering from CO (carbon monoxide) molecules, the principal stretching vibration occurs at a frequency of $\approx$ 65 THz. Taking the mass of carbon as 12 u, and the mass of oxygen as 16 u (where 1 u $\simeq 1.66 \times 10^{-27}$ kg is the atomic mass unit), calculate the effective "spring constant" for the C–O chemical bond.

5.9 Argon laser light at 514.5 nm is incident on a gas cell with some unknown molecules, and Raman scattering is observed at 579 nm. Are the unknown molecules CO_2 or CO? (See Problems 5.7 and 5.8 for data.)

5.10 Use the geometry of Fig. 5-9 to show that the minimum bend radius is given by Eq. (5-10).

5.11 Show that in Fig. 5-10, the radius at which light becomes lost from the waveguide mode is

$$r_{\text{max}} \simeq R(1 + \Delta)$$

Also show that if the criterion for significant light loss is taken to be $r_{\text{max}} < R + a$, we obtain the same result as in Eq. (5-10). Explain why this is a reasonable criterion.

5.12 Determine the bend radius at which bending losses become significant, for (a) multimode fiber with NA = 0.25 and core diameter 50 μm, and (b) single-mode fiber with NA = 0.18 and core diameter 8 μm. Assume core index = 1.5.

Chapter 6

Dispersion in Optical Fibers

As a pulse of light propagates down a long fiber, it will generally broaden in time, a phenomenon known as *dispersion.* In multimode fibers, the dispersion is largely due to the different propagation speeds for the various modes, which is known as *intermodal dispersion.* Typical values of intermodal dispersion are ~ 50 ns/km (see Chapter 3), which limits the useful propagation range for a 100 Mb/s signal to ~ 100 m.

There are two basic approaches to reducing dispersion. The first is to design the fiber core so that different modes have a more nearly equal transit time down the fiber. This can be accomplished with a *graded-index fiber,* as discussed in the next section. The second approach is to eliminate all modes but one, that is, use a single-mode fiber. Although single-mode fibers have no intermodal dispersion, they have other sources of dispersion, which are the subject of much of this chapter. The dispersion that occurs for propagation in a single mode is termed *intramodal dispersion.*

6-1. GRADED INDEX FIBER

Consider a fiber with an index of refraction that varies radially in the core, as shown in Fig. 6-1. The index is highest on the fiber axis, and decreases smoothly to the core-cladding boundary. This is called a *graded-index fiber,* and is in contrast to the usual *step-index fiber,* in which the core index is constant right up to the core–cladding boundary. A ray such as b (Fig. 6.1) that initially makes an angle with the fiber axis will be deflected into a curved path as it propagates, because the speed of the wavefront $v = c/n$ is higher in the lower-index medium than in the higher-index medium. The physical understanding of this deflection is similar to that of refraction through a dialectric interface, shown in Fig. (2-6).

To see how a graded-index fiber can reduce intermodal dispersion, consider two rays, taking the paths marked a and b (Fig. 6.1). A ray taking the path b has a longer distance to go, but it moves along a path where the average speed is higher. If the radial index variation $n(r)$ is chosen properly, the time taken for light to propagate along the two paths will be nearly the same. It turns out that a quadratic variation,

$$n(r) = n_1[1 - (r/a)^2\Delta]$$

accomplishes this, where $\Delta = (n_1 - n_2)/n_1$, n_1 and n_2 are the (maximum) core index and cladding index, respectively, and a is the core radius. Theoretically, this variation in $n(r)$ reduces the dispersion by a factor of $\Delta/8$ compared to an equivalent step-index fiber (Senior 1992). For example, if $\Delta = 0.01$, then the 50 ns/km intermodal dispersion for a step-

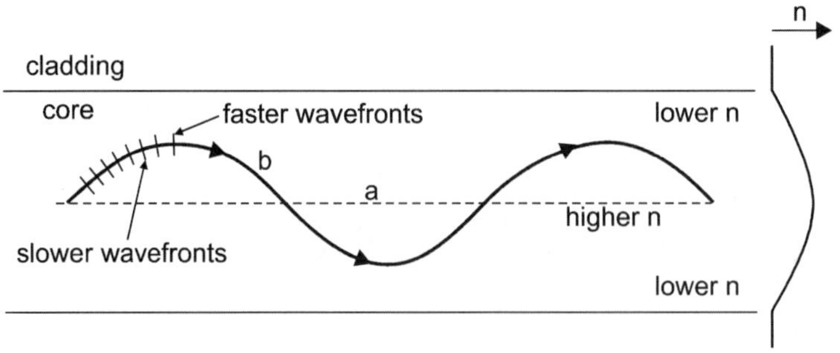

Figure 6-1 Ray propagation along two paths a and b in a graded-index fiber.

index fiber becomes only $\approx$ 50 ps/km for the corresponding graded-index fiber. The improvement is very sensitive to the exact index profile, however, and care must be taken to ensure that $n(r)$ is close to quadratic.

6-2. INTRAMODAL DISPERSION

In a single-mode fiber, there would be no dispersion for monochromatic light, that is, light with a single precise wavelength. However, light with any finite spectral width $\Delta\lambda$ will suffer from dispersion because the different wavelength components of the light travel at different speeds. Since this dispersion occurs within a single mode, it is referred to as *intramodal dispersion,* and because it depends on the spectral distribution of the light, it is also called *chromatic dispersion.* Chromatic dispersion arises from two distinct processes, as discussed in the following two sections.

Material Dispersion

One cause of chromatic dispersion is the variation of refractive index $n(\lambda)$ with wavelength. The refractive index governs the speed of the wave, so a varying index will result in a varying speed for the different wavelength components. This is called *material dispersion* because it depends only on a property of the material—the index of refraction. It applies equally to light propagating in a fiber geometry or in a large, homogeneous medium such as a glass rod, a liquid, or the atmosphere.

We can characterize the material dispersion quantitatively by calculating the time t for light of a particular wavelength λ to travel a distance L down the fiber. Recognizing that the group velocity given by Eq. (3-19) is the appropriate speed to use, we have

$$t = \frac{L}{v_g} = L\frac{d\beta}{d\omega} = L\frac{d\beta}{d\lambda}\cdot\frac{d\lambda}{d\omega} \tag{6-1}$$

where λ here will be taken as the free-space wavelength. The derivative $d\lambda/d\omega$ can be determined from the relations $\lambda = c/\nu = 2\pi c/\omega$ to be

$$\frac{d\lambda}{d\omega} = -2\pi\frac{c}{\omega^2} = -\frac{\lambda^2}{2\pi c} \tag{6-2}$$

The derivative $d\beta/d\lambda$ in Eq. (6-1) can be evaluated easily if we assume that the light propagates as a plane wave. In this case,

$$\beta = nk_0 = n\frac{2\pi}{\lambda} \qquad \text{(plane wave assumption)} \tag{6-3}$$

where the wavelength dependence appears both explicitly and through the index $n(\lambda)$. Using the rule for the derivative of a product, we have

$$\frac{d\beta}{d\lambda} = \frac{2\pi}{\lambda}\frac{dn}{d\lambda} - \frac{2\pi}{\lambda^2}n = -\frac{2\pi}{\lambda^2}\left(n - \lambda\frac{dn}{d\lambda}\right) \tag{6-4}$$

Substituting Eq. (6-2) and Eq. (6-4) into Eq. (6-1) then yields for the propagation time

$$t = \frac{L}{c}\left(n - \lambda\frac{dn}{d\lambda}\right) \tag{6-5}$$

Notice that the propagation time for a given wavelength is not found by simply using the wavelength-dependent index $n(\lambda)$ in $t = Ln/c$, but rather depends on the variation of index with wavelength $dn/d\lambda$. This comes from the use of the group velocity rather than the phase velocity. The group velocity is given by

$$v_g = \frac{c}{n - \lambda\dfrac{dn}{d\lambda}} \tag{6-6}$$

whereas the phase velocity is $v_p = c/n$. If the index of refraction is a constant, then $dn/d\lambda = 0$, and the simple result $t = Ln/c$ is obtained. The group and phase velocities are the same in this special case.

If a light pulse has a spectral width $\Delta\lambda$, the different wavelength components in the pulse will propagate with different delay times according to the $n(\lambda)$ and $dn/d\lambda$ for each wavelength. The spread in arrival times Δt for the different wavelength components will then be

$$\Delta t = \frac{dt}{d\lambda} \cdot \Delta\lambda \tag{6-7}$$

The derivative in Eq. (6-7) can be evaluated from Eq. (6-5) as

$$\begin{aligned}\frac{dt}{d\lambda} &= \frac{L}{c}\left[\frac{dn}{d\lambda} - \frac{dn}{d\lambda} - \lambda\frac{d^2n}{d\lambda^2}\right] \\ &= -\frac{L}{c}\lambda\frac{d^2n}{d\lambda^2}\end{aligned} \tag{6-8}$$

which, after combining with Eq. (6-7), yields

$$\Delta t = -\frac{L}{c}\lambda\frac{d^2n}{d\lambda^2}\,\Delta\lambda \qquad \text{(material dispersion)} \tag{6-9}$$

According to Eq. (6-9), there are two ways to minimize material dispersion. The first approach is to use a nearly monochromatic light source (small $\Delta\lambda$). This is a fundamental property of laser light, making the laser an ideal source for high-speed optical communications.

The second approach is to choose a wavelength where $d^2n/d\lambda^2$ becomes very small. The variation of refractive index with wavelength arises from the interaction of the light with the electronic and vibrational transitions in the glass. For wavelengths in the visible and near infrared regions, the balance between these two types of transitions causes n to decrease with increasing λ in the fashion shown in Fig. 6-2. At some wavelength λ_0, there is an inflection point in the graph of $n(\lambda)$, at which point $d^2n/d\lambda^2 \to 0$. In silica glass, this occurs at $\lambda_0 \approx 1300$ nm, which is another reason (in addition to low attenuation) that the 1300 nm region was chosen for the second telecommunications window.

Since the dispersion is proportional to L and $\Delta\lambda$, it is convenient to define a *material dispersion* coefficient,

$$D_m \equiv \frac{1}{L}\frac{\Delta t}{\Delta\lambda} = -\frac{\lambda}{c}\frac{d^2n}{d\lambda^2} \quad \text{(material dispersion coefficient)} \qquad (6\text{-}10)$$

which is only dependent on the material property $d^2n/d\lambda^2$. The bottom graph in Fig. 6-2 shows that D_m is negative for $\lambda < \lambda_0$ and positive for $\lambda > \lambda_0$. A negative D_m implies that longer wavelengths have a shorter arrival time, that is, they travel faster. This is termed *normal dispersion* since it is what is normally encountered in the visible and near IR spec-

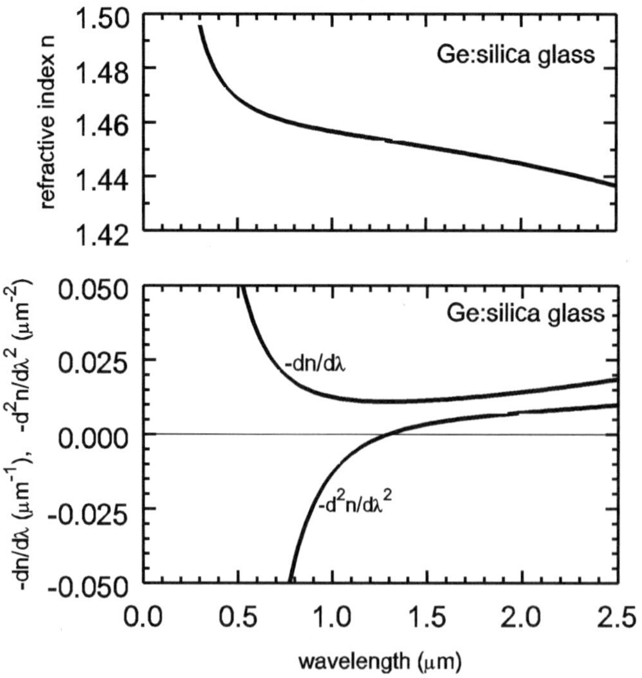

Figure 6-2 Variation of refractive index with wavelength for silica glass in the visible and near IR regions. The curvature $d^2n/d\lambda^2$ goes to zero at $\lambda_0 \approx 1300$ nm.

tral regions. It also happens to agree with how the phase velocity changes with wavelength: longer λ means decreasing n, which gives a greater $v_p = c/n$.

A positive value of D_m implies that longer wavelengths have longer arrival times, and thus travel slower. This is termed *anomalous dispersion,* and occurs in silica glass for $\lambda >$ 1300 nm. In this spectral region, the group velocity decreases with increasing λ, but the phase velocity increases, since n continues to decrease with increasing λ. This difference in group and phase velocity behavior is one sense in which the dispersion is "anomalous," and has applications in soliton propagation in fibers, to be discussed in Chapter 9.

EXAMPLE 6-1

Calculate the dispersion coefficient D_m in ps/(nm · km) and the dispersion per unit length in ns/km or ps/km for each of the following:

(a) LED (light-emitting diode) light with spectral width of 40 nm, operating at wavelength 800 nm, where $d^2n/d\lambda^2 = 4 \times 10^{10}\ \text{m}^{-2}$.

(b) Laser light with spectral width of 0.2 nm, operating at wavelength 1500 nm, where $d^2n/d\lambda^2 = -2.7 \times 10^9\ \text{m}^{-2}$.

Solution: (a) For $\lambda = 800$ nm, the material dispersion coefficient is

$$D_m = -\frac{800 \times 10^{-9}}{3 \times 10^8}(4 \times 10^{10}) = -107 \times 10^{-6}\ \text{s/m}^2$$

or

$$D_m = -107\ \frac{\text{ps}}{\text{nm} \cdot \text{km}}$$

The dispersion per unit length (dropping the minus sign) is

$$\frac{\Delta t}{L} = \left(107\ \frac{\text{ps}}{\text{nm} \cdot \text{km}}\right) \cdot (40\ \text{nm}) = 4.3\ \frac{\text{ns}}{\text{km}}$$

This can be compared with the ~ 50 ns/km typical for intermodal dispersion (see Example 3-2).

(b) For $\lambda = 1500$ nm,

$$D_m = -\frac{1500 \times 10^{-9}}{3 \times 10^8}(-2.7 \times 10^9) = 13.5 \times 10^{-6}\ \text{s/m}^2 = 13.5\ \frac{\text{ps}}{\text{nm} \cdot \text{km}}$$

The dispersion per unit length is

$$\frac{\Delta t}{L} = \left(13.5\ \frac{\text{ps}}{\text{nm} \cdot \text{km}}\right) \cdot (0.2\ \text{nm}) = 2.7\ \frac{\text{ps}}{\text{km}}$$

The dramatic reduction in dispersion for the laser source illustrates its importance for high-speed optical communications.

Waveguide Dispersion

In the previous section, the plane wave assumption $\beta = n2\pi/\lambda$ was made. Although this is a reasonable assumption for multi-mode fibers, it is not as valid for single-mode fibers. The proper treatment for single-mode fibers is to replace n by n_{eff}, as in Eq. (4-15). The effective index n_{eff} varies with wavelength not only because of material dispersion, but also because n_{eff} varies with V (see Fig. 4-9), and V, in turn, varies with wavelength. This implicit variation of $n_{\text{eff}}[V(\lambda)]$ with λ gives rise to the second cause for intramodal dispersion, which is termed *waveguide dispersion.*

Waveguide dispersion can analyzed quantitatively by making the substitution $n \to n_{\text{eff}}$ in each of Eqs. (6-3)–(6-10). The resulting total chromatic dispersion is then

$$D_c \equiv \frac{1}{L}\frac{\Delta t}{\Delta\lambda} = -\frac{\lambda}{c}\frac{d^2 n_{\text{eff}}}{d\lambda^2} \qquad \text{(total chromatic dispersion coefficient)} \qquad (6\text{-}11)$$

To evaluate this, it is useful to define the waveguide contribution to the index as $\delta = n_{\text{eff}} - n_2$.* Eq. (6-11) then becomes

$$\begin{aligned} D_c &= -\frac{\lambda}{c}\frac{d^2 n_2}{d\lambda^2} - \frac{\lambda}{c}\frac{d^2\delta}{d\lambda^2} \\ &= D_m + D_w \end{aligned} \qquad (6\text{-}12)$$

where D_w is the *waveguide dispersion coefficient.* The waveguide dispersion coefficient can be rewritten using a change of variables (see Problem 6.6) as

$$D_w \simeq -\frac{V}{\lambda c}\frac{d^2(V\delta)}{dV^2} \qquad \text{(waveguide dispersion coefficient)} \qquad (6\text{-}13)$$

with V given by Eq. (4-9). In writing Eq. (6-13), we are assuming that the wavelength dependence of V is $V \propto 1/\lambda$, which ignores the wavelength dependence of NA in $V = 2\pi a$ NA$/\lambda$. Relaxing this assumption gives rise to another (usually minor) contribution to D_c called *profile dispersion.*

The waveguide dispersion given by Eq. (6-13) can be positive or negative, depending on the curvature of $V\delta$ versus V. Fig. 4-9 shows that the curvature of δ versus V is positive for small V, and becomes negative for large V. A plot of $V\delta$ versus V exhibits a similar behavior, except that the point of zero curvature is shifted to larger V. In the range $V < 2.405$ relevant for single-mode fibers, the curvature is positive, making D_w negative.

The dependence of D_w on wavelength can be understood by referring to Eq. (6-13) and Fig. 4-9. Longer λ corresponds to a smaller V, where the curvature of $V\delta$ versus V is larger. This leads to a larger negative value of D_w for increasing λ. Fig. 6-3 illustrates this variation of D_w, along with the material dispersion D_m and the combined dispersion $D_c = D_m + D_w$. The effect of waveguide dispersion is to shift the wavelength λ_{min} for which the dispersion is zero from λ_0 to a longer wavelength. In conventional single-mode fiber optimized for the 1300 nm band (now referred to as "legacy fiber"), this shift is small.

The shift in the zero-dispersion wavelength can be made larger by modifying the refractive index profile in the fiber core. For example, if the core radius is made very small,

*A related constant $b = \delta/(n_1 - n_2)$ is defined by some authors, such that $0 < b < 1$.

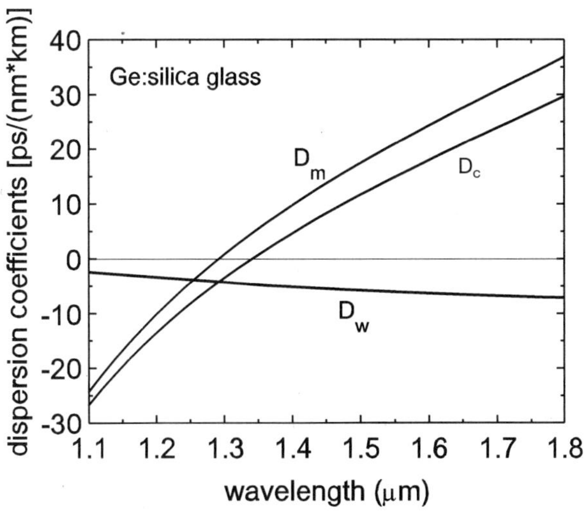

Figure 6-3 Dispersion coefficient D_c for single-mode fiber, showing contributions D_m from material dispersion and D_w from waveguide dispersion (core radius $\approx$ 4 μm).

the fiber V parameter is also small, leading to larger negative values of D_w. For step-index, single-mode fiber of core radius a it is found that $D_w \propto a^{-2}$. Fiber for which the zero-dispersion wavelength $\lambda_{\min}$ is significantly shifted from λ_0 is called *dispersion-shifted fiber,* as illustrated in Fig. 6-4b. When $\lambda_{\min}$ is shifted to ~ 1550 nm, the wavelength of lowest loss coincides with the wavelength of minimum dispersion. This allows the maximum possible data transmission rate for long optical fiber communications links operating at a single wavelength.

One limitation in using dispersion-shifted fiber is that the system is optimized for only a single wavelength. When multiple wavelengths are simultaneously sent down the fiber (called *wavelength division multiplexing,* or WDM), the different wavelengths will have vastly different dispersion, complicating the system engineering. Another complication occurs when nonlinear effects in the fiber core mix different wavelengths in a WDM signal (see Chapter 24). The mixing process is most efficient when different wavelengths propagate with equal speeds, that is, when the dispersion is very small.

For the above reasons, there is interest in fibers having a dispersion that is small but nonzero, and fairly constant over the wavelengths used in long-haul telecommunications. An example of this is the so-called "W fiber," which has a double-cladding structure, as shown in Fig. 6-4c . This type of refractive index profile gives rise to *dispersion-flattened fiber,* which has the desired dispersion characteristics for WDM. Other multiple-cladding structures can be used to fine-tune and optimize fibers for dispersion management (Senior 1992).

Polarization-Mode Dispersion

So far, we have considered the spreading in time of a light pulse due to intermodal dispersion (light distributed among several modes) and chromatic dispersion (light distributed over a range of wavelengths). A third type of time-spreading mechanism, that of *polarization mode dispersion* (PMD), is a consequence of the light being distributed over different

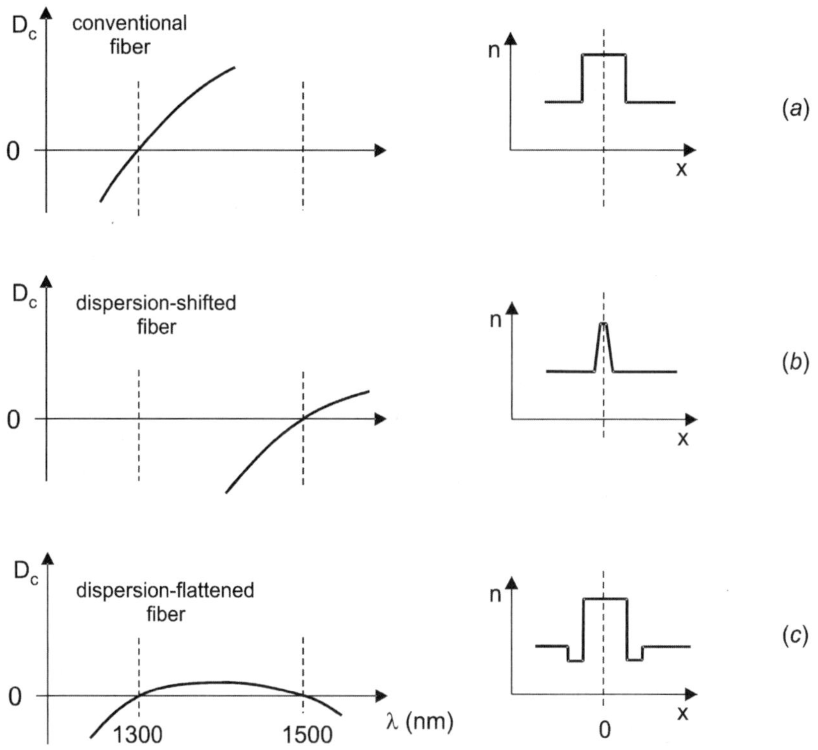

Figure 6-4 The refractive index profile can be tailored to produce different dispersion curves.

polarizations. For a perfectly uniform and symmetrical fiber, the propagation speed would be independent of polarization, and light of all polarizations would arrive at the far end of the fiber at the same time. In real fibers, however, there are always small stresses on the fiber that make the refractive index slightly different for light of two orthogonal polarizations, and the arrival time, therefore, depends slightly on polarization. The resulting dispersion tends to be small, because light of one polarization is rather easily coupled (usually within a few meters) into the orthogonal polarization by fiber bends and irregularities. This gives rise to a fairly uniform propagation speed, with statistical fluctuations from the average that increase with fiber length. The time spread of a pulse due to PMD is found to obey

$$\Delta t_{\mathrm{PMD}} \simeq D_{\mathrm{PMD}}\sqrt{L} \qquad \text{(polarization mode dispersion)} \qquad (6\text{-}14)$$

where D_{PMD} is the *polarization mode dispersion coefficient.* Typical values for communications fiber are 0.2-2 ps/$\sqrt{\mathrm{km}}$.

The $\sqrt{L}$ dependence for PMD is similar to that noted earlier (see p. 66) for intermodal dispersion when $L > 1$ km. The origin of the $\sqrt{L}$ dependence lies in the statistical nature of the processes. Light is coupled randomly from mode to mode, or from one polarization to another, resulting in statistical fluctuations from an average. This corresponds to the well-known "random walk problem" in statistics, which is characterized by a distribution with a width proportional to the square root of the number of steps.

Total Fiber Dispersion

To determine the total time spread of a pulse, we must combine the various sources of fiber dispersion. The way that we combine them depends on whether they are correlated or uncorrelated. For example, material dispersion D_m and waveguide dispersion D_w are correlated because they both depend in a specified way on the wavelength. In this case, we add these dispersions directly, as in Eq. (6-12), to obtain the total chromatic dispersion, D_c. However, intermodal dispersion, chromatic dispersion and PMD do not share any common origin, and are therefore uncorrelated. In this case, we must add the time spreads "in quadrature":

$$\Delta t_{\text{fiber}} = \sqrt{\Delta t^2_{\text{modal}} + \Delta t^2_{\text{chromatic}} + \Delta t^2_{\text{PMD}}} \quad \text{(total fiber dispersion)} \qquad (6\text{-}15)$$

In many situations, one of these three terms dominates and the others can be neglected. For example, in step-index multimode fiber, the $\Delta t^2_{\text{modal}}$ term usually dominates; it is zero, however, in single-mode fiber. The contribution from PMD can often be neglected, but can be significant in long fiber spans when very monochromatic light sources are used.

PROBLEMS

6.1 A light pulse with wavelength 850 nm passes through a single-mode silica fiber. (a) Determine the time spread of the pulse per unit length due to material dispersion if the spectral width is 20 nm. (b) Repeat if the spectral width is 2 nm. (From Fig. 6-2, take $d^2n/d\lambda^2 = 3 \times 10^{10}$ m^{-2}.)

6.2 A light pulse with wavelength 1550 nm passes through a single-mode silica fiber. Using Fig. 6-3, determine the time spread of the pulse per unit length due to material dispersion if the spectral width is 2 nm.

6.3 Determine the maximum bit rate for digital modulation using the results in Problems 6.1 and 6.2 for fiber lengths of 100 m and 10 km.

6.4 Pulses with a 2 nm spectral width at two discrete wavelengths 850 and 860 nm are coupled simultaneously into a long step-index, single-mode fiber with core radius 4 μm. (a) Which of these pulses reaches the far end of a 5 km long fiber first? (b) What is the time delay between the two pulses at the fiber end?

6.5 Repeat Problem 6.4 if the two wavelengths are 1550 and 1560, each with spectral width 2 nm.

6.6 Using $V = 2\pi a \text{NA}/\lambda$, show that Eq. (6-13) is equivalent to the second term in Eq. (6-12). Assume that NA is independent of λ.

6.7 Using Fig. 6-3, estimate the total chromatic dispersion at 1550 nm for a step-index silica fiber having core radius 3 μm. Repeat for a core radius 2 μm.

6.8 Show that the intermodal pulse spread per unit length in a fiber is approximately given by $\Delta t/L \simeq \text{NA}^2/(2nc)$, where NA is the fiber numerical aperture, and the core and cladding are assumed to have indices both close to n.

6.9 Light from a GaAs LED (center wavelength 850 nm, spectral width 50 nm) is sent down a step-index fiber with core diameter 60 μm. At this wavelength, the silica glass in the core has material dispersion $d^2n/d\lambda^2 = 3 \times 10^{10}$ m^{-2}. What NA is needed for this fiber so that the $\Delta t/L$ for intermodal dispersion will be equal to that of material dispersion?

6.10 Chromatic dispersion in a long fiber link can be "undone" by periodically passing the dispersed pulses through a special dispersion compensation fiber (DCF), which has a chromatic dispersion coefficient of opposite sign. If the DCF has $D_c = -300$ ps/(nm km) at the operating wavelength of 1550 nm, and a length of 1 km, determine the distance required between the DCF insertions along the fiber link. Assume that the fiber link uses standard silica telecommunications fiber of core radius 4 μm.

6.11 Single-mode fiber in a communications link is dispersion-flattened with $D_c = 2$ ps/(nm km) at the operating wavelength of 1550 nm, and has polarization mode dispersion coefficient $D_{\text{PMD}} = 2$ ps/$\sqrt{\text{km}}$. The linewidth of the light source is 1 nm. Determine the total time spread of light for fiber lengths of (a) 500 m and (b) 5 km. The chromatic dispersion can be reversed using dispersion-compensating fiber (see Problem 6.10). Will this work for polarization dispersion?

Chapter 7

Fiber Connections and Diagnostics

To use fibers for photonics applications, light must usually be coupled from one fiber to another. The different ways of accomplishing this will be discussed in this chapter. The losses involved in making fiber connections will be considered, as well as diagnostic techniques for determining the degree of loss.

7-1. FIBER CONNECTIONS

Connections between fibers can be classified into three types: a splice, a connector, or a coupler. In this section, each of these is briefly described.

Fiber Splice

A *fiber splice* is a permanent connection between fibers and is the optical equivalent of soldering wires together. In a *mechanical splice,* the fibers are typically held together in a "V groove" arrangement or in a tightly fitting capillary tube, as shown in Fig. 7-1. Index-matching fluid can be inserted between the fiber ends to reduce Fresnel reflection losses, and glue can be used to secure the alignment. An *elastomeric splice* is a type of mechanical splice having some elasticity in the capillary tube, which allows fibers of somewhat different dimensions to be joined.

The other type of splice is the *fusion splice,* in which the fiber ends are fused (melted) together to form a seamless junction. The fusion is accomplished with a *fusion splicer,* which is essentially an optical arc welder. In this process, a high-current, pulsed electrical discharge is created between closely spaced electrodes, which raises the temperature of the fiber to above the melting point of the glass. The duration of the pulse is carefully controlled to melt just emough glass to make the join, but not so much as to degrade the core-cladding structure. Slight misalignment of the two fibers tends to be self-correcting in this process, because the surface tension of the liquified fiber region acts to bring the fiber cores into proper alignment. Losses in a fusion splice can be quite low, in the 0.1 dB range.

Fiber Connector

A *fiber connector* is a connection intended to be repeatedly made and broken, as in an electrical plug. The two fibers are held securely in a connector housing, and are joined by simply "butt coupling" the two fiber ends, as in Fig. 7-2, so that the cores match up.

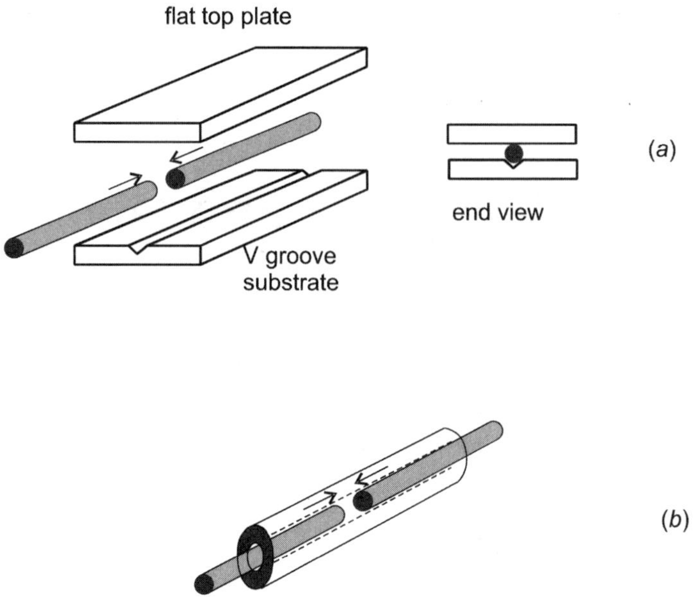

Figure 7-1 Two ways of holding fibers together for splicing. (a) Fiber sandwiched between V-groove substrate and flat glass top plate. (b) Fiber inserted into tight-fitting capillary tube.

Several different connector housing types have become industry standards, including the SC connector, which snaps in place, the ST connector, which twists on with a bayonet latch, and the FC connector, which twists onto a threaded mount. (Hecht, J. 2002)

Fiber Coupler

In a *fiber coupler,* light is combined or divided between three or more fibers. The *Y coupler,* for example, couples power P_1 and P_2 from two input fibers into a power P_3 in a single output fiber, as shown in Fig. 7-3a. In an ideal Y coupler, any light incident from port 1 will be coupled entirely into the output port 3, and any light incident from port 2 will also be coupled entirely into port 3. For multimode fibers, the throughput efficiency $P_3/(P_1 + P_2)$ can be reasonably high, but for single-mode fibers there are fundamental limits to this efficiency. The limitations on coupling light from two sources into a single-mode fiber can be understood based on the principle of *optical reciprocity.* This principle is based on the symmetry of Maxwell's equations with respect to the replacement $t \rightarrow (-t)$, and asserts that any allowed passive optical process (such as propagation of light) has an equally allowed counterpart that is identical in every respect but reversed in time.

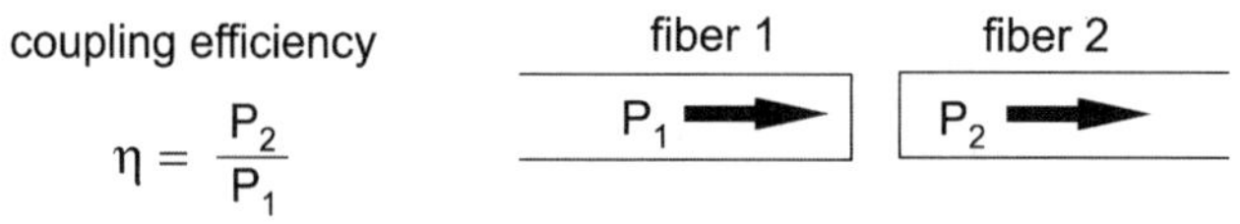

Figure 7-2 In butt-coupling geometry, the cores of two optical fibers are pressed directly together.

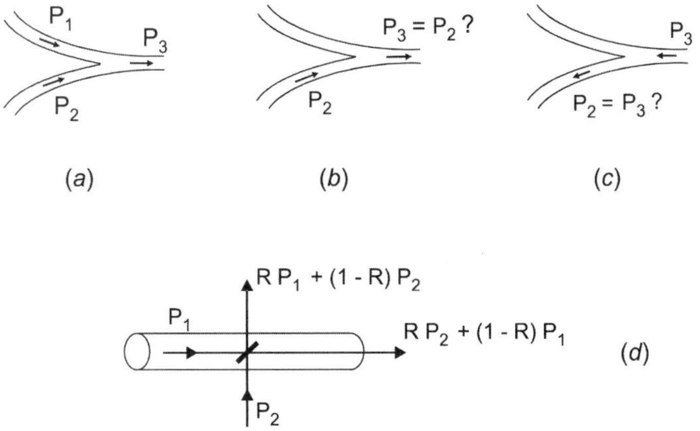

Figure 7-3 Combining two beams in a (a) Y coupler and (d) an in-fiber beam splitter. Perfect coupling of each input port into the output (b) is not allowed by time-reversal arguments (c).

Applying the reciprocity principle to the Y coupler, say that there is power P_2 incident in port 2 and no power incident in port 1, as shown in Fig. 7-3b, and it is desired to extract the entire power P_2 out of port 3. In this case, the time-reversed propagation of Fig. 7-3c would also be allowed, which means that a power P_3 coupled into the port 3 would be entirely output into port 2. However, we could equally well have started with light only coupled into port 1, in which case the conclusion would have been that light coupled into port 3 is entirely output into port 1. Since in a single-mode fiber there is only one way for the light to propagate after being coupled into port 3, the two conclusions are inconsistent. This means that our assumption that Fig. 7-3b is an allowed solution must be wrong; light entering port 2 (or port 1) must go somewhere in addition to port 3.

Further insight into this problem of limited coupling efficiency can be obtained by considering the specific coupling scheme shown in Fig. 7-3d. A small, partially reflective surface is placed at an angle of 45° inside the fiber, and light incident from the side is injected into the fiber core by reflection off this surface. If a fraction R of light is reflected, then an amount of light power RP_2 will be injected into the core, with $(1 - R)P_2$ passing through the fiber and lost. At the same time, however, the light already in the fiber core will be partially reflected out of the fiber, an amount RP_1 being lost. The total amount of light remaining in the fiber after the coupling will be $RP_2 + (1 - R)P_1$, which is less than the total incident light $P_1 + P_2$. The difference between total incident and total coupled light is lost into the upward-going beam, which constitutes a de facto fourth port for this device.

The limits on coupling efficiency discussed above apply only when *combining* two beams into one single-mode fiber. It is important to note that it does not apply to the *splitting* of energy from a single beam into two beams or fibers going in different directions. A device that performs this task is termed a *directional coupler,* and can have high efficiency. Examples of directional couplers would be Fig. 7-3a with the arrows reversed (P_3 is input beam), or Fig. 7-3d without the beam P_2 incident from the side.

The efficiency limits for combining two beams also only applies to beams of precisely the same wavelength. Beams of different wavelengths can be more efficiently coupled by means of wavelength-selective reflective elements such as dichroic mirrors or diffraction gratings. This has important implications for wavelength division multiplexing (WDM), as will be discussed in Chapter 22.

An important fiber coupler that can be used as a beam splitter or combiner is the *fused biconical taper coupler.* This four-port device is constructed by twisting together two bare-clad fibers, then applying heat and stretching the fibers to create a region in which the fibers become thinner, as shown in Fig. 7-4.

In the tapered region, the core radius of each fiber becomes very small, and the resulting small V parameter gives a large mode waist size w according to Eq. (4-18). The mode field from each fiber core extends far enough into the cladding to overlap significantly with the core of the adjacent fiber, as illustrated in Fig. 7-4b. As a result, light propagating in one of the cores will gradually "leak into" or "couple with" the other core. After propagating a distance known as the *coupling length,* the light energy has moved completely from one core to the other. As the light continues to propagate, the energy is transferred back and forth between the two cores in a periodic fashion. The coupling length will be shorter for cores that are closer together, since the modes then interact more strongly. By adjusting the ratio of the physical taper length to the coupling length, one can design the coupler so that a given fraction of light entering port 1 will be exit in port 3, and the remainder in port 4. This device plays the role of an all-fiber beam splitter, and has many applications in photonics.

7-2. LOSSES IN FIBER CONNECTIONS

Light can be lost in a fiber connection in a number of ways, including scattering or absorption of light at the fiber surfaces, misalignment of the two fiber cores, or a difference in numerical aperture (NA) when connecting different types of fiber. Scattering and absorption can be minimized by sufficiently polishing and cleaning the fiber surface. The coupling loss due to NA mismatch will be considered in Chapter 12. In this section, we consider the various types of misalignment and the associated coupling loss for multimode and single-mode fiber.

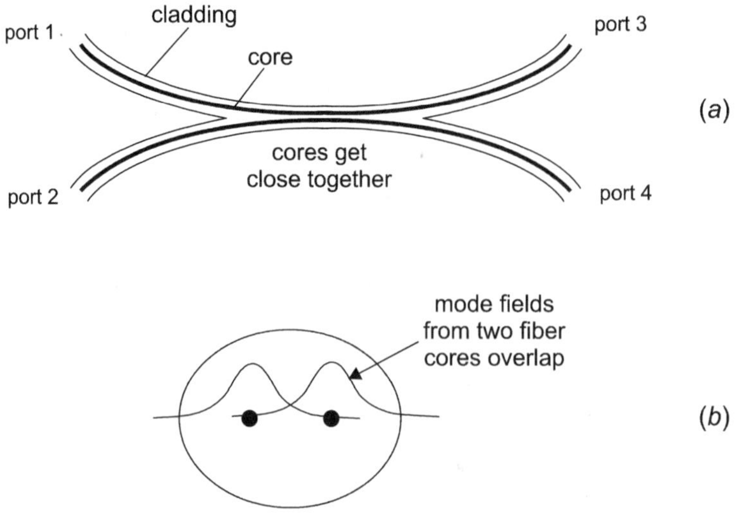

Figure 7-4 Fused biconical taper coupler in (a) side view, (b) cross section.

Multimode Fiber

There are three types of misalignment that can cause loss of coupled power, as indicated in Fig. 7-5. The greatest loss usually comes from a lateral offset δ in the fiber cores. For multimode fibers, the cores can be considered to be uniformly filled with light. The coupling efficiency is then determined by the overlap of core areas, and can be shown to be (see Problem 7.1)

$$\eta_{\text{lat}} \equiv \frac{P_2}{P_1} = \frac{1}{\pi}\left\{2\cos^{-1}\left(\frac{\delta}{2a}\right) - \left(\frac{\delta}{a}\right)\left[1-\left(\frac{\delta}{2a}\right)^2\right]^{1/2}\right\} \tag{7-1}$$

where a is the core radius, and P_1 and P_2 are the optical powers before and after the connection. The dB loss due to lateral offset is then

$$\text{dB loss} = 10\log_{10}(1/\eta_{\text{lat}}) \qquad \text{(lateral offset loss)} \tag{7-2}$$

using Eq. (1-1).

If there is a gap between the fibers, there will be an additional loss due to the Fresnel reflection at each fiber end. This will cause the transmitted light to be reduced by the factor (see Problem 7.4)

$$\eta_{\text{Fres}} = \frac{16(n_1/n_0)^2}{[1+(n_1/n_0)]^4} \qquad \text{(Fresnel transmission)} \tag{7-3}$$

where n_1 and n_0 are the refractive indices of the core and coupling medium, respectively. The combined coupling efficiency for both lateral offset and Fresnel losses is then $\eta = \eta_{\text{lat}}\eta_{\text{Fres}}$.

Another source of coupling loss is angular misalignment, as shown in Fig. 7-5b. If one multimode fiber is tilted by an angle θ from the other, it can be shown (Tsuchiya et al. 1977) that the coupling efficiency is

$$\eta_{\text{ang}} \approx 1 - \frac{n_0\theta}{\pi \text{NA}} \qquad \text{(angular offset)} \tag{7-4}$$

where θ is measured in radians, NA is the numerical aperture, n_0 is the refractive index of the medium between fibers, and small deviation $\theta \ll 1$ is assumed. Note that since the cone angle for light entering or leaving a fiber is $\alpha_{\text{max}} \simeq \text{NA}/n_0$, the fractional decrease in coupling efficiency due to angular misalignment is $\approx \theta/(\pi\alpha_{\text{max}})$.

Still another (usually minor) source of coupling loss is a longitudinal separation z between the fiber ends, as in Fig. 7-5c. For small separations $z \ll a$, the coupling efficiency due to longitudinal offset is (Tsuchiya et al. 1977)

$$\eta_{\text{long}} \approx 1 - \frac{z\text{NA}}{4an_0} \qquad \text{(longitudinal offset)} \tag{7-5}$$

The total coupling efficiency for any combination of small offsets is given by

$$\eta_{\text{tot}} \approx \eta_{\text{Fres}}\eta_{\text{lat}}\eta_{\text{ang}}\eta_{\text{long}} \qquad \text{(total coupling efficiency)} \tag{7-6}$$

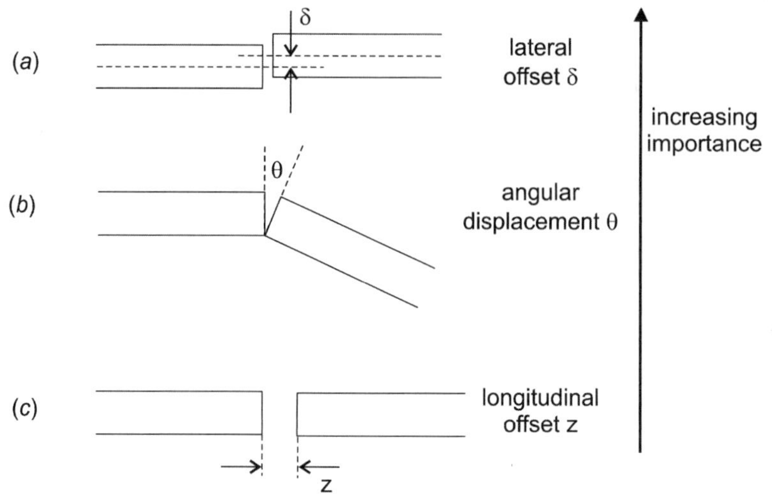

Figure 7-5 Types of misalignment in fiber–fiber coupling.

The corresponding total dB loss is found by adding the dB losses for for the individual loss mechanisms.

Single-Mode Fiber

For single-mode fibers, it is not a good assumption that the fiber core is uniformly filled with light. Rather, the mode field is approximately Gaussian in distribution, as given by Eq. (4-16). Therefore, when calculating the loss due to lateral offset it is not the core area overlap that must be calculated, but instead the overlap of the two (offset) Gaussian mode fields. The coupling efficiency is then found (Marcuse et al. 1979) to be

$$\eta_{\text{lat}} = e^{-(\delta/w)^2} \qquad \text{(single-mode lateral offset)} \qquad (7\text{-}7)$$

where w is the mode waist size given by Eq. (4-18). One difference between the single-mode and multimode results (see Problems 7.2 and 7.5) is that for small δ the loss is linear with δ for the multimode case, but quadratic with δ for the single-mode case. This helps to alleviate the sensitivity of single-mode fibers to lateral offset, but the very small w typical of single-mode fibers still makes precise alignment necessary.

Angular misalignment losses are also different for single-mode fibers, because the angular spread of light exiting the fiber is determined by diffraction rather than by the numerical aperture. The divergence angle is thus [see Eq. (2-25)] $\alpha_{\text{max}} \sim \lambda/(2w)$, and the coupling loss should then depend on the ratio $\theta/\alpha_{\text{max}}$. A detailed calculation bears this out (Marcuse et al. 1979), with the result

$$\eta_{\text{ang}} = e^{-(\pi n_1 w\theta/\lambda)^2} \qquad \text{(single-mode angular offset)} \qquad (7\text{-}8)$$

where, again, w is the mode waist size and n_1 is the core index of refraction. Note that for small angular offset, the fractional loss varies quadratically with angle as $\sim (\theta/\alpha_{\text{max}})^2$, in contrast to the linear dependence for multimode fibers.

There is also some loss from longitudinal offset, but in practice it is usually small compared with the other two offsets, and can be neglected. The total coupling efficiency can then be obtained by multiplying the efficiencies for lateral and angular offsets and reflection losses (or adding the corresponding dB losses), provided the losses are small.

The results in Eqs. (7-1)–(7-8) serve as a useful guideline for estimating losses in real multimode and single-mode fiber systems. However, it should be kept in mind that irregularities in the fiber geometry or modal distribution can cause deviations from these calculations in practice, and the equations are more useful for estimating trends than for making detailed quantitative predictions.

7-3. FIBER LOSS DIAGNOSTICS

It is important in practice to be able to evaluate the losses occuring in a fiber optic system. Say, for example, that we have a splice connection between two fibers, and want to know the loss due to the splice. One possibility, as indicated in Fig. 7-6, would be to measure and compare the powers P_{in} incident on the fiber and P_{out} exiting the fiber. There are several difficulties with this approach, however. First, if the fiber link is long, the powers would need to be read on different power meters, requiring careful calibration. Also, the amount of power actually coupled into the fiber is not the same as the incident power, due to the (unknown) coupling efficiency. Even if these problems were overcome, it would not be possible with a single measurement to determine whether the measured loss were due to the splice or to intrinsic losses within the fibers themselves. And finally, if there were more than one splice, it would not be possible to determine the contribution from each splice separately.

In this section, two important techniques are discussed that address one or more of the above problems.

Cutback Method

If the primary goal is to accurately measure the propagation loss in a long length of fiber, the *cutback method* can be used. As illustrated in Fig. 7-7, light is coupled into the long fiber, which is typically coiled up on a drum in the laboratory, and the power exiting the far end is measured with a power meter. After the power is measured (designated P_B), the fiber is cut at point A, which is close to the beginning of the fiber. The power exiting the end of the (now very short) fiber is again measured with the same power meter, and designated P_A. Since the input coupling efficiency has not changed, and since the collection efficiency remains the same, the only reason for a difference in the two power readings

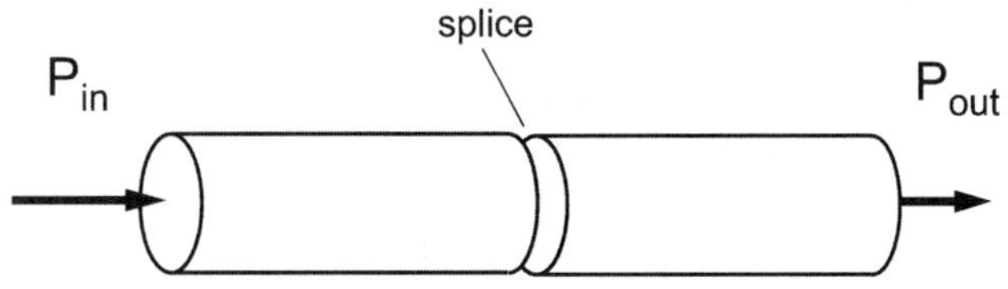

Figure 7-6 Measuring input and output power to determine fiber loss.

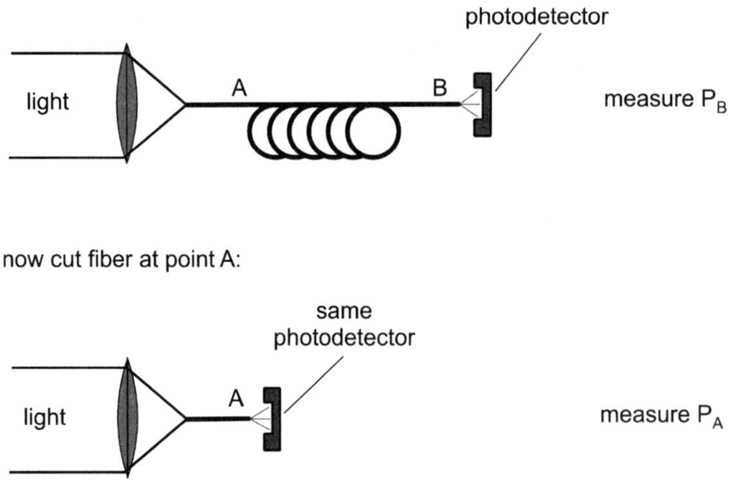

Figure 7-7 The cutback method for determining fiber loss.

would be propagation loss in the fiber under test. The attenuation coefficient α in the fiber is then determined from

$$P_B = P_A e^{-\alpha L} \tag{7-9}$$

where L is the length of fiber that was cut off.

For meaningful loss measurements, it is important that light is propagating only in the core, and (for multimode fiber) is uniformly distributed among the modes. This can be accomplished with a mode stripper (Fig. 5-16) and mode scrambler (Fig. 5-15) inserted in the fiber before point A. The cutback method is a destructive technique, since a small amount of the long fiber is removed each time a measurement is made. It is mostly used in laboratory work for examining the quality of an optical fiber, rather than as a diagnostic technique in the field.

Optical Time-Domain Reflectometer

The *optical time-domain reflectometer,* or OTDR, is a general purpose instrument for characterizing fiber losses, and is used both in the laboratory and in the field. Like the cutback method, it corrects for the unknown coupling loss by taking ratios of measured signals. Instead of taking the ratio of signals at different locations, however, it takes the ratio of reflected signals arriving at different times. In this way, the OTDR can separate out multiple sources of loss in a fiber link, determining the location and magnitude of splice losses as well as the attenuation coefficients of the fibers in the link.

As illustrated in Fig. 7-8, an OTDR operates by sending a short pulse of light through a beam splitter, and coupling it into the fiber under test. As the pulse propagates in the fiber, a certain fraction is scattered (point A) or reflected (point B) from splices or other irregularities, and a fraction of this scattered light is coupled back into the fiber core going in the reverse direction. When this backward-going light exits the front end of the fiber, a portion reflects off the beam splitter and into a photodetector, which converts the light

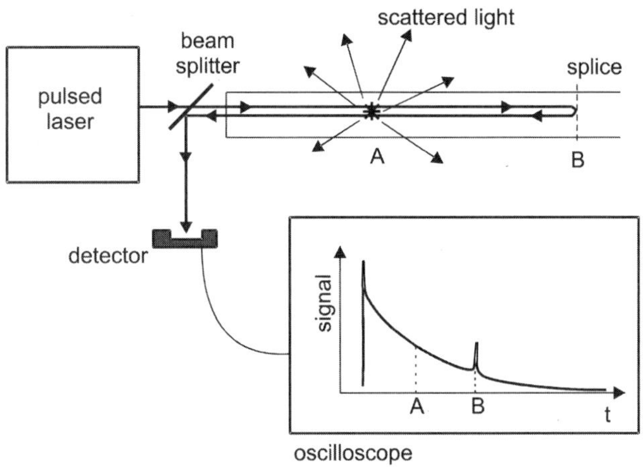

Figure 7-8 The optical time domain reflectometer (OTDR) uses the time dependence of reflected light to analyze fiber losses.

into an electrical signal. The resulting time-dependent photodetector signal is displayed on an oscilloscope or waveform averager, and constitutes the OTDR trace that is used to determine losses in the fiber link.

There are two distinct sources of the OTDR signal: reflection at discrete points at which there is a break or end in the fiber, and Rayleigh scattering along the entire length of fiber. In a reflection process, the angle that the ray makes with the boundary is preserved, and a large fraction of the backward-reflected light remains coupled in the fiber. In a scattering process, however, the direction of the scattered light is randomized, and only a small fraction (~ $\Delta/2$) is trapped by the fiber core in the backward-going direction. Fiber breaks, therefore, appear as large spikes on top of a continuous background from Rayleigh scattering, as shown in the inset of Fig. 7-8.

Three important aspects of fiber loss can be analyzed quantitatively from a single OTDR trace.

1. Location of Splice or Break

The fundamental idea of the OTDR is that it maps the spatial distribution of defects along the fiber into the time domain. The pulse of light must travel for a time $t = z/v$ to reach a defect (such as a splice) that is a distance z down the fiber, as illustrated in Fig. 7-9. Here, v is the group velocity given by Eq. (6-6), with n replaced by n_{eff}. Although the differences between v_p, v_g, and v_{eff} are important when considering dispersion (Chapter 6), they are insignificant when determining the propagation times to different parts of a fiber, and it is a good approximation to let $v = c/n$, where $n = n_1$ is the core index. The propagation time for the pulse to reach the defect is then $t = z/(c/n)$. After reflecting or scattering from the defect, the scattered light must then propagate for another time $t = z/(c/n)$ to reach the front end of the fiber, for a total "delay time" of

$$t_{\text{delay}} = \frac{2nz}{c}$$

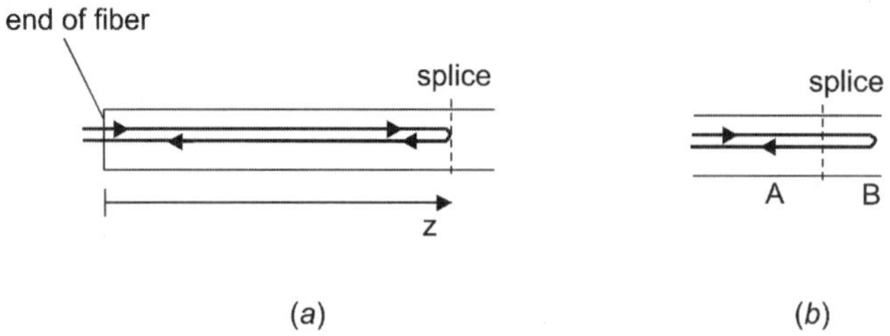

Figure 7-9 An OTDR can be used to: (a) locate the position of a splice by the travel time of reflected light, (b) determine the transmission loss of a splice.

Putting it another way, if a spike is observed on the OTDR trace at a delay time of t_{delay}, the corresponding defect is located at a distance

$$z = \frac{c}{2n} t_{delay} \qquad \text{(position of defect)} \tag{7-10}$$

The reference time for measuring t_{delay} is taken to be the reflection spike from the front end of the fiber, which is generally quite large due to the high refractive index contrast between the incident medium (usually air) and the fiber core. Compared to the delay from propagating hundreds of meters in the fiber, delays associated with propagation from the beamsplitter to detector are quite small and can usually be neglected.

2. Magnitude of Splice Loss

In addition to measuring the location of a splice, an OTDR can also determine the magnitude of loss in that splice. The splice loss is determined by comparing the Rayleigh scattering OTDR signal from points before the splice (point A in Fig. 7-9b) and after the splice (point B). If the one-way transmission through the splice is T, then a fraction T of the forward-propagating pulse energy at point A remains at point B to undergo scattering. Of the light that is scattered at point B, only a fraction T remains after traversing the splice again in the reverse direction. The OTDR signals S_A and S_B due to light scattered at points A and B are then related by

$$S_B = T^2 S_A \tag{7-11}$$

and the corresponding dB drop in OTDR signal is

$$\text{dB}_{\text{drop}} = 10 \log_{10}\left(\frac{S_A}{S_B}\right) = 20 \log_{10}\left(\frac{1}{T}\right) \tag{7-12}$$

Note that the one-way dB loss through the splice is half that given by Eq. (7-12).

3. Loss Coefficient of Fiber

The third important fiber characteristic that can be studied with the OTDR is the attenuation coefficient of the fiber. According to Beer's law (Eq. 5-1), the intensity of light decreases by the factor $\exp(-\alpha z)$ after propagating a distance z down the fiber. The amount of Rayleigh-scattered light generated at z is, therefore, also decreased by the factor $\exp(-\alpha z)$ compared with that generated at the beginning of the fiber ($z = 0$). The light scattered at position z must propagate back to the beginning of the fiber, decreasing the measured signal by another factor of $\exp(-\alpha z)$. The signal $S(z)$ due to light that was scattered at position z is then

$$S(z) = S(0)e^{-2\alpha z} \tag{7-13}$$

where $S(0)$ is the signal from light scattered near the beginning of the fiber ($z = 0$), and α is the attenuation coefficient. Light scattered at position z will be detected at time t given by Eq. (7-10), so the time dependence of the OTDR signal will be

$$\begin{aligned} S(t) &= S(0)e^{-2\alpha(ct/2n)} \\ &= S(0)e^{-(\alpha c/n)t} \end{aligned} \tag{7-14}$$

which is a decaying exponential. Taking the log of Eq. (7-14) gives

$$\log_e\left[\frac{S(z)}{S(0)}\right] = -\alpha\left(\frac{c}{n}\right)t \tag{7-15}$$

which is a straight line with slope $-\alpha c/n$. The attenuation coefficient α can, therefore, be determined by plotting $\log_e [S(z)/S(0)]$ versus t, and dividing the slope of this line by c/n.

The types of information that can be obtained from an OTDR trace are illustrated in Fig. 7-10. Straight-line sections correspond to distributed Rayleigh scattering, with steeper slope implying higher loss. A splice or break between two fiber sections is marked by a sudden vertical shift in the trace, the time of the shift giving the location of the splice and the magnitude of the shift giving the dB loss in the splice. The spatial resolution for locating defects is limited by the time resolution Δt, given by the pulse width or the detector response time, whichever is larger. Using Eq. (7-10), the spatial resolution Δz is

$$\Delta z = \frac{c}{2n}\Delta t \tag{7-16}$$

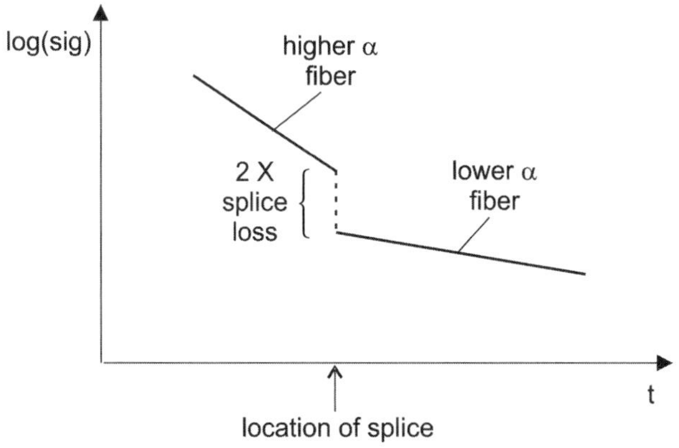

Figure 7-10 Fiber attenuation is determined from the slope of the OTDR trace on a semilog plot.

For example, a pulse width of 10 ns results in a spatial resolution of 1 m, assuming $n = 1.5$. Pulse widths in the picosecond range can give cm–mm scale resolutions, provided that the detector response is sufficiently fast (see Chapter 14).

Although the OTDR was developed for fiber optic communications diagnostics, the basic principle of time-resolved detection has some other potentially important applications. For example, optical fiber sensors sense changes in the fiber's environment via the changing stresses, strains, and associated losses, and find applications in monitoring the integrity of buildings and bridges into which a network of optical fibers has been embedded. These sensors can be made into distributed fiber sensors by monitoring the time-dependent reflections from the fiber network upon pulsed excitation. With such a monitoring system in place, the user is not only notified that there is a defect in the structure, but is also given information about the location of that defect.

PROBLEMS

7.1 Derive Eq. (7-1) by calculating the area overlap of two circles, both of radius a, with centers offset by δ.

7.2 Develop an approximate expression for the fraction of light lost due to lateral offset in a multimode fiber by expanding Eq. (7-1) in powers of δ/a. Assume $\delta/a \ll 1$ and keep just the lowest-order terms in δ/a.

7.3 Two multimode fibers with $a = 25$ μm are joined together, with a lateral offset $\delta = 1.5$ μm. Use Eq. (7-1) to calculate the coupling efficiency and the fraction of light lost. Compare this with the result obtained using the approximate expression derived in Problem 7.2.

7.4 Use the Fresnel reflection coefficient for normal incidence given in Eq. (2-14) to derive Eq. (7-3). Also derive an approximate expression for the fraction lost when $|n_0 - n_1| \ll 1$.

7.5 Develop an approximate expression for the fraction of light lost due to lateral offset in a single-mode fiber by expanding Eq. (7-7) in powers of δ/w. Assume $\delta/w \ll 1$ and keep just the lowest-order terms in δ/w.

7.6 For a multimode fiber, how small must the fractional displacement δ/a be in order to keep the coupling loss below 0.2 dB? Repeat this calculation for the δ/w ratio of a single-mode fiber. It is convenient here to use the approximate expressions derived in Problems 7.2 and 7.5.

7.7 A single-mode fiber has a mode-field diameter of 8.5 μm. Determine the lateral offset δ that will produce a transmission loss of 0.5 dB.

7.8 Two multimode fibers are being connected with an air gap between them. The air gap is now filled with an index-matching fluid that is not perfect, as it has a refractive index of 1.4 whereas the fiber core has an index of 1.48. Compute the transmission loss in dB, and also the reflected power level (in dB) with respect to the incident power level.

7.9 A lab technician measures the optical power transmitted through a long spool of fiber, and then cuts off 2.2 km of fiber from the spool to measure the attenuation coefficient. If the two measurements yield powers of 3 mW and 10 mW, what is the attenuation coefficient, expressed both in cm^{-1} and in dB/km?

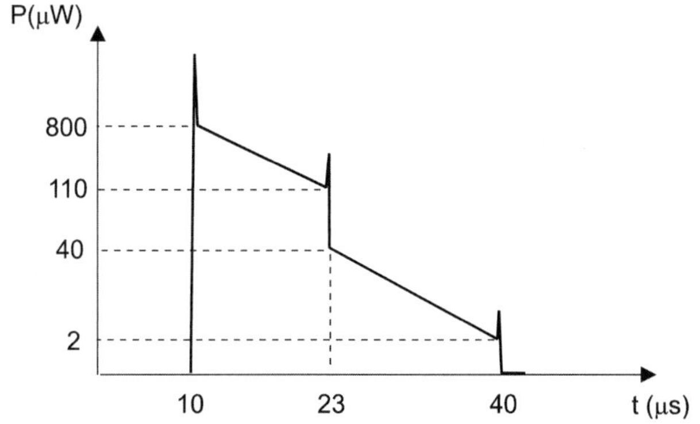

Figure 7-11 Reflected power measured in OTDR for Problem 7.10. The vertical axis is log scale; the horizontal axis is linear.

7.10 A field technician is diagnosing the losses in a fiber optic link consisting of two fibers, A and B, connected together with a mechanical splice. She sends in 800 nm light at the accessible end of fiber A, and looks at the reflected light signal from that same port using an OTDR. The measured time dependence is shown in Fig. 7-11. From this data, how long is the fiber link, and where along the fiber link is the splice located? Assume an index of 1.5.

7.11 For the fiber link of Problem 7.10, what is the one-way transmission loss through the splice, expressed in dB? Also give the reflected power from the splice expressed in dB relative to the power incident on the splice, assuming that the splice loss is due to reflection (rather than, e.g., absorption or scattering).

7.12 For the fiber link of Problem 7.10, determine the attenuation coefficients for the two fibers A and B. Express your answer both in cm^{-1} and in dB/km.

Chapter 8

Photonic Crystal Optics

The conventional method for guiding light along a particular path in space is to utilize total internal reflection (TIR) so that light propagating in a higher index core reflects with 100% efficiency off the boundary with a lower index cladding. This fundamental principle underlies the operation of nearly all fiber optic and planar waveguide devices in use today. Recently, however, there has been emerging an alternative and promising paradigm for controlling the flow of light. In this new scheme, light is confined to the core not by TIR, but instead by modifying the microstructure of the cladding region so that light cannot propagate there. The cladding material is modified so that the refractive index varies periodically in space, with a repetition distance on the order of the wavelength of light. In three dimensions, the resulting microstructure can be likened to that of a crystalline solid, with regions of high index where the atoms would be, and regions of lower index in between. Because of this analogy with crystalline lattice structure, a microstructured material of this type is termed a *photonic crystal.*

A photonic crystal can be periodic in one, two, or three dimensions. The Bragg grating, for example, which was discussed briefly in Chapter 2, is an example of a one-dimensional (1-D) photonic crystal. We will begin this chapter by treating Bragg gratings in further detail, not only because they are quite useful in and of themselves, but also because their fundamental properties are easy to understand, and this can be used to develop an intuitive understanding of the more complex two- (2-D) and three-dimensional (3-D) photonic crystals. We then consider 2-D and 3-D photonic crystals, emphasizing their relation to simple 1-D structures, and pointing out those applications that seem most promising. Progress in constructing and utilizing 2-D and 3-D photonic crystals is proceeding at a rapid pace, and new developments will certainly need to be included in any future overview. However, the fundamental principles are now well established, and the introductory treatment given here is intended to provide an intuitive foundation for understanding future advances in this exciting field.

8-1. 1-D PHOTONIC CRYSTALS

Step-Index Grating

The simplest one-dimensional photonic crystal consists of an array of N uniformly spaced parallel slabs, as shown in Fig. 8-1. We define the refractive index of the slabs as n_2, and that of the medium between slabs as n_1, with the index difference $\Delta n \equiv n_2 - n_1$. The center-to-center spacing of the slabs is denoted by Λ, so the total length of the photonic crystal in the x direction is $L = N\Lambda$. The thickness of each slab will be taken as $\Lambda/2$, so that on average there is an equal amount of material with indices n_1 and n_2. This not only simpli-

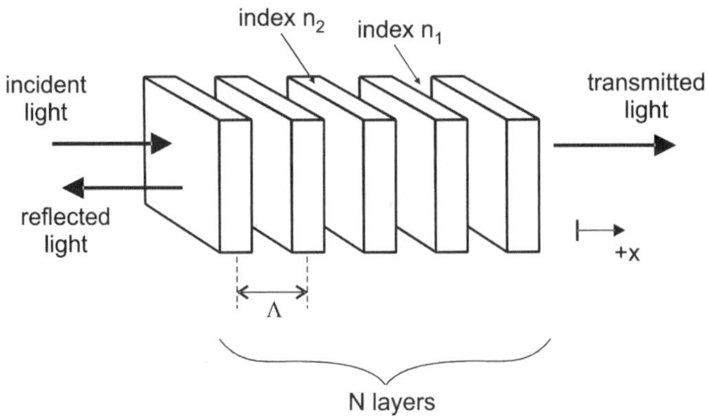

Figure 8-1 A simple 1-D photonic crystal consists of uniform parallel slabs with refractive index n_2, separated by a medium with refractive index n_1.

fies the analysis, but is also the appropriate condition for making comparisons with a sinusoidal grating.

When light is incident from the left, it is partially reflected from each interface in the series of slabs. To analyze the reflection quantitatively, we take the incident light wave's electric field to be of the form

$$\begin{aligned} E_i(x, t) &= A \cos(\omega t - kx) \\ &= A\Re e^{i(\omega t - kx)} \end{aligned} \tag{8-1}$$

where $\Re$ designates the "real part" of the following expression, ω is the angular frequency of the light, $k = 2\pi\, n/\lambda_0$ is the wave vector magnitude, and λ_0 is the free-space wavelength. In the following, we will assume $\Delta n \ll 1$, so $n_1 \approx n_2 \approx n$. The complex exponential notation is useful here because the phase shift of light reflected from different interfaces is then easily accounted for. Taking the left-most interface to be at $x = 0$, the E field of light reflected from the jth interface can be written in the form

$$E_{rj}(0, t) \equiv \Re \tilde{E}_{rj} e^{i\omega t} \tag{8-2}$$

where $\tilde{E}_{rj}$ is a complex amplitude that includes relative phase as well as magnitude. The complex amplitude of the incident light at $x = 0$ is $\tilde{E}_i = A$, a real number.

The total electric field of the reflected light is found by adding the contribution from each interface, as indicated in Fig. 8-2. The reflections can be separated into two groups: those in which the light is incident on the interface from index n_1, and those in which the light is incident from index n_2. For reflections of the first type, the Fresnel equations (Eqs. 2-11 and 2-12) at normal incidence ($\theta = 0$) give an amplitude reflection coefficient:

$$r = \frac{E_r}{E_i} = \frac{n_1 - n_2}{n_1 + n_2} \simeq -\frac{\Delta n}{2n} \tag{8-3}$$

where E_i and E_r are evaluated just before and after reflection. Reflections of the second type obey the same relation, except that n_1 and n_2 are interchanged, so that $r \simeq \Delta n/2n$. Assuming that $|r| \ll 1$, the total reflected field at $x = 0$ due to reflections of the first type is

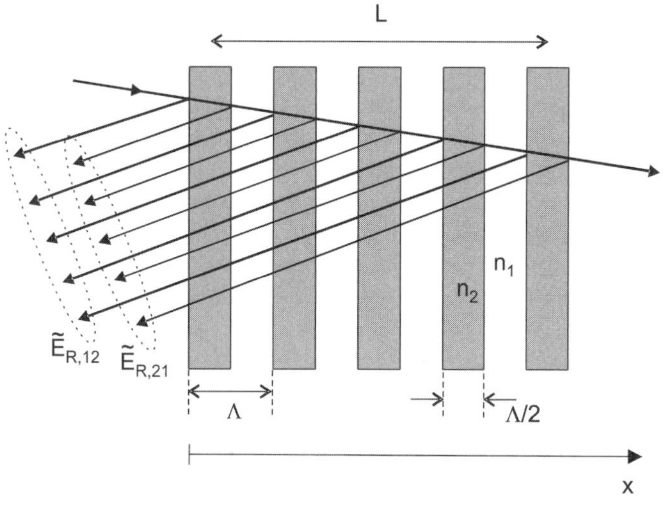

Figure 8-2 Light incident on the array of slabs is partially reflected at each boundary at which the refractive index changes. Reflected waves from all boundaries at which the index changes from n_1 to n_2 combine to give a total reflected complex field amplitude $\tilde{E}_{r,12}$. The corresponding reflections from $n_2 \rightarrow n_1$ boundaries give a reflected amplitude $\tilde{E}_{r,21}$.

$$\tilde{E}_{r,12} = -\frac{\Delta n}{2n} A[1 + e^{-i\delta} + e^{-i2\delta} + \ldots + e^{-i(N-1)\delta}] \tag{8-4}$$

where

$$\delta = k(2\Lambda) = \frac{2\pi n}{\lambda_0}(2\Lambda) = \frac{4\pi n \Lambda}{\lambda_0} \tag{8-5}$$

is the phase delay due to propagation over the round-trip distance 2Λ between slabs. For reflections of the second type, the corresponding total reflected field is

$$\tilde{E}_{r,21} = \frac{\Delta n}{2n} A e^{-i\delta/2} [1 + e^{-i\delta} + e^{-i2\delta} + \ldots + e^{-i(N-1)\delta}] \tag{8-6}$$

The extra phase change of $-\delta/2$ in this last expression comes from the additional round-trip propagation distance $2(\Lambda/2)$ for each type 2 reflection compared with the corresponding type 1 reflection.

The simplest way to add the terms inside the square brackets is to visualize them as vectors in the complex plane. Each vector has the same magnitude, but the angle from the real axis is 0, $-\delta$, -2δ, and so on, for the successive terms. The terms will add to give a maximum resultant when the vectors all point in the same direction, as depicted in Fig. 8-3a. In this case, there is constructive interference of the various reflected waves, and the incident light undergoes efficient *Bragg reflection*. This will occur when $\delta = m2\pi$, where m is an integer giving the order of diffraction. For first-order Bragg diffraction ($m = 1$), light will be efficiently reflected when the wavelength satisfies Eq. (8-5) with $\delta = 2\pi$. This gives

$$\lambda_B = 2n\Lambda \qquad \text{(Bragg wavelength)} \tag{8-7}$$

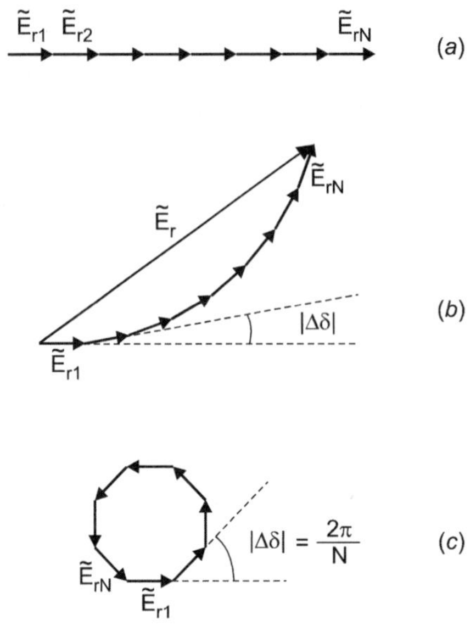

Figure 8-3 The complex amplitudes of the various reflected waves can be added by treating them as vectors in the complex plane. (a) Bragg resonance, all vectors in phase. (b) Slightly off resonance, small difference in phase between vectors. (c) Further off resonance, vectors add to zero, giving destructive interference.

where λ_B is the *Bragg wavelength.* (Note that this is the free-space wavelength, not the wavelength in the medium.) The Bragg wavelength can also be evaluated using Eq. (2-29), which gives the same result as Eq. (8-7) if we use $\theta = 90°$, $m = 1$, and Λ in place of d for the grating spacing.

When the Bragg condition $\delta = 2\pi$ is satisfied, $\exp(-i\delta/2) = -1$, and $\tilde{E}_{r,21} = \tilde{E}_{r,12}$. In this case, the total reflected complex amplitude from all interfaces is

$$\begin{aligned}\tilde{E}_r &= \tilde{E}_{r,12} + \tilde{E}_{r,21}\\ &= -\frac{\Delta n}{n} A[1 + e^{-i\delta} + e^{-i2\delta} + \ldots + e^{-i(N-1)\delta}]\end{aligned} \tag{8-8}$$

Since each of the N terms in the square brackets is unity for $\delta = 2\pi$, the reflected amplitude becomes simply

$$\tilde{E}_r = -\frac{\Delta n}{n} AN$$

and the power reflectivity is

$$R_{\max} = \left|\frac{\tilde{E}_r}{\tilde{E}_i}\right|^2 = \left|-\frac{\Delta n}{n} N\right|^2 = N^2\left(\frac{\Delta n}{n}\right)^2 \tag{8-9}$$

where we have used $\tilde{E}_i = A$. Since the grating length is $L = N\Lambda$, this can be written as

$$R_{\max} = \left(\frac{L}{\Lambda}\frac{\Delta n}{n}\right)^2 = \left(\frac{2\Delta n L}{\lambda_B}\right)^2 \tag{8-10}$$

where $\lambda_B = 2n\Lambda$ from Eq. (8-7) has been used. Defining the parameter $\kappa \equiv 2\Delta n/\lambda_B$, this can be written more compactly as

$$R_{\max} = (\kappa L)^2 \qquad \text{(peak reflectivity, weak grating)} \tag{8-11}$$

The constant κ (Greek letter kappa, not to be confused with the propagation constant k) is a measure of how strongly the light is attenuated as it propagates through the Bragg grating.

The above is only an approximate result, based on certain assumptions. It is assumed not only that $\Delta n/n \ll 1$, but also that $\kappa L \ll 1$. This amounts to the restriction $R_{\max} \ll 1$. Another assumption is that the Bragg condition is satisfied exactly. If we relax this assumption and allow the incident wavelength to be slightly detuned from λ_B, the terms in Eq. (8-8) will have phases that become progressivly further apart. The phase difference ϕ between successive terms is $\phi = -\delta$, and if the wavelength changes by $\Delta\lambda$, this phase difference changes by $\Delta\phi = -\Delta\delta$. This is illustrated graphically in Fig. 8-3b, which shows the addition of N complex amplitude vectors, each differing from the next by a phase difference $\Delta\phi$. When $N\Delta\phi = 2\pi$, as depicted in Fig. 8-3c, the vectors add to give a resultant of zero. The grating reflectivity, therefore, goes from a maximum to zero over a range of wavelengths $\Delta\lambda$ such that $\Delta\phi = 2\pi/N$. The wavelength interval $\Delta\lambda$ is the spectral half-width of the Bragg resonance. The full width would be $2\Delta\lambda$.

To evaluate $\Delta\lambda$, we use Eq. (8-5) to write

$$\Delta\phi = -\Delta\delta = -\frac{d\delta}{d\lambda}\Delta\lambda = \frac{4\pi n\Lambda}{\lambda_0^2}\Delta\lambda \tag{8-12}$$

Since $\lambda_0 \approx \lambda_B$ near the Bragg resonance, we use $\lambda_B = 2n\Lambda$ from Eq. (8-7) and the condition $\Delta\phi = 2\pi/N$ to obtain

$$\frac{2\pi}{N} = \frac{2\pi\lambda_B}{\lambda_0^2}\Delta\lambda$$

or

$$\frac{\Delta\lambda}{\lambda_0} = \frac{1}{N} \qquad \text{(spectral half-width, weak grating)} \tag{8-13}$$

This remarkably simple result says that the fractional width of the Bragg resonance is the reciprical of the number of grating planes. The width of a resonance is often characterized by the *quality factor* Q, defined as the center frequency divided by the frequency width. In terms of wavelength, this is equivalent to $\lambda/\Delta\lambda$. The quality factor for the Bragg grating is therefore $Q \approx N$.

If the incident wavelength is detuned from λ_B by more than $\Delta\lambda$, the vectors in Fig. 8-3c continue to curl around in the complex plane. This leads to a secondary maximum in the resultant field when $N\Delta\phi \simeq 3\pi$, and another zero when $N\Delta\phi = 4\pi$. When this second zero occurs, the vectors have wrapped around in two complete circles. This pattern continues with increasing detuning, and results in an oscillatory dependence of reflectivity on wavelength like that shown in Fig. 8-4.

Sinusoidal Index Grating

The step-index grating discussed in the preceeding section is useful for obtaining an intuitive view of Bragg reflection. In practice, however, the refractive index in a Bragg grat-

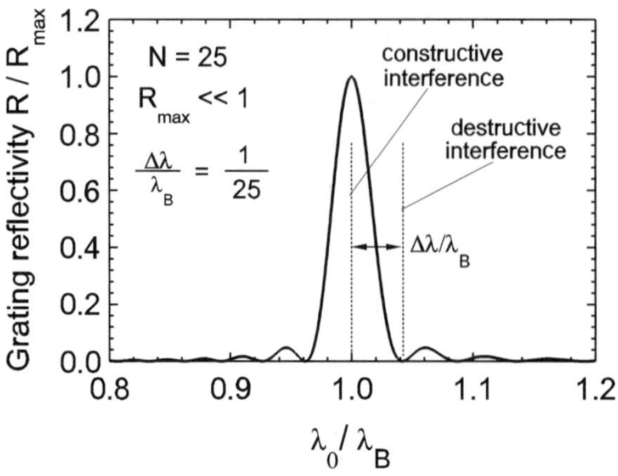

Figure 8-4 Reflectivity of 25 layer step index Bragg grating versus wavelength for $R_{max} \ll 1$. Reflectivity is normalized to the maximum value $R_{max} = (\kappa L)^2$, and free-space wavelength is normalized to the Bragg wavelength λ_B. The fractional half-width is $1/N$.

ing often varies sinusoidally with position, rather than in a step-wise fashion. Perhaps the most important example is that of the *fiber Bragg grating,* in which the refractive index is made to vary periodically along the core of an optical fiber. Figure 8-5 shows how this can be accomplished by the interference of two intersecting pump beams of (free-space) wavelength λ_p. The interference of two beams crossing at an angle was analyzed when we developed the concept of waveguide modes [see Eq. (3-2) and related discussion]. Using similar arguments here, it is straightforward to show that the pump light intensity in Fig. 8-5 varies sinusoidally along the fiber axis, and has a constant value along planes perpendicular to the axis. If the angle between the beams is 2θ, the separation Λ between planes of maximum intensity can be shown (see Problem 8.5) to be

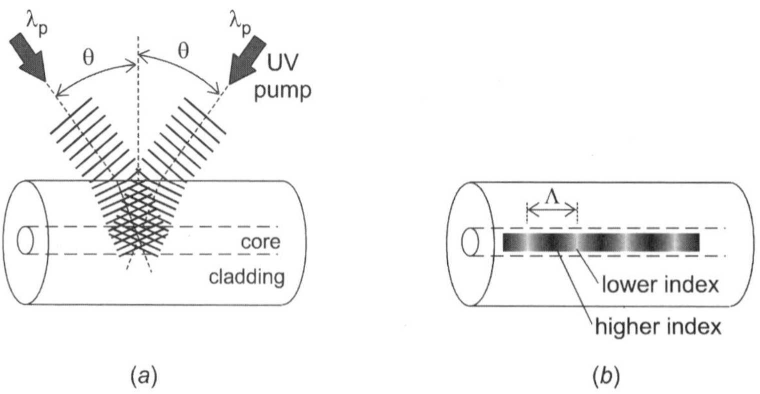

Figure 8-5 (a) The holographic scheme for fabricating a fiber Bragg grating utilizes the inteference fringes created by crossed laser beams. (b) The resulting index of refraction varies sinusoidally along the fiber core.

$$\Lambda = \frac{\lambda_p}{2 \sin \theta} \tag{8-14}$$

This relation shows that for a fixed pump wavelength λ_p, the grating spacing Λ can be conveniently adjusted by choosing the proper angle θ.

The spatial variation of light intensity produced by the crossed pump beams will create a corresponding spatial variation of refractive index if the fiber exhibits the property of *photosensitivity*. Photosensitivity in optical fibers was discovered by Ken Hill and coworkers in 1978, at the Canadian Communication Research Center, and has been extensively studied since then. For certain glasses, especially those with dopants such as germanium, it is found that the refractive index increases with increasing pump intensity and exposure time, up to some limiting value at which the response saturates. In silica fiber, for example, the maximum index change $(\Delta n)_{max}$ varies from ~ 3×10^{-5} for standard telecommunications fiber (3 mole% GeO_2 in core) to ~ 2.5×10^{-4} for high-germanium fiber (20 mole% GeO_2 in core). The effect is largest for pump wavelengths in the UV (~ 240 nm), where the large photon energy rearranges chemical bonds in the glass. The rearrangement of these bonds modifies the glass structure, and this causes a change in the refractive index.

If the product of pump intensity and exposure time is short enough so that Δn has not yet reached saturation, the resulting index variation may be expressed by the sinusoidal form*

$$n(x) = \bar{n} + \Delta n \cos\left(\frac{2\pi x}{\Lambda}\right) \tag{8-15}$$

where $\bar{n}$ is the average index and Λ is the spacing between index maxima (the grating spacing). The detailed analysis of reflections from this type of grating is more complicated than that of the step-index grating, and we do not present it here. However, many of the results for the step-index grating apply equally well to the sinusoidal grating. For example, the wavelength λ_B for Bragg reflection is still $\lambda_B = 2n\Lambda$, in agreement with Eq. (8-7). In the low-reflectivity limit, the resonance half-width is still $\Delta\lambda = \lambda_0/N$, and the peak reflectivity is $R_{max} \approx (\kappa L)^2$, as in Eqs. (8-13) and (8-11). One small difference is that the attenuation constant for the sinusoidal grating is given by

$$\kappa \equiv \frac{\pi \Delta n}{\lambda_B} \qquad \text{(attenuation constant, sinusoidal grating)} \tag{8-16}$$

rather than $\kappa = 2\Delta n/\lambda_B$ for the step-index grating.

In the discussion so far, the weak reflection limit has been assumed, so that $\kappa L \ll 1$. Since high-reflectivity gratings are needed for many applications, the more general result for arbitrary κL is of interest. The analysis for this more general case involves the use of "coupled mode theory," which accounts for the exchange of energy between forward- and backward-propagating beams. Using these methods, it is found [see, for example, (Kashyap 1999)] that the peak reflectivity is

$$R_{max} = \tanh^2(\kappa L) \qquad \text{(peak reflectivity, arbitrary } \kappa L\text{)} \tag{8-17}$$

*Nonsinusoidal gratings can be expressed as a Fourier series of terms like this with multiples of the spatial frequency $2\pi/\Lambda$.

where $\tanh u = (e^u - e^{-u})/(e^u + e^{-u})$ is the hyperbolic tangent function. In the limit $\kappa L \ll 1$, this reduces to $R_{max} \approx (\kappa L)^2$, in agreement with Eq. (8-11). As κL increases, the reflectivity starts to saturate around $\kappa L \sim 1$, and in the limit $\kappa L \gg 1$ it approaches the limiting value $R_{max} \to 1$.

The significance of the saturation condition $\kappa L \sim 1$ can be understood by relating κ to the attenuation of light in the Bragg grating. Coupled-mode theory predicts that the light wave's E field at resonance decreases according to

$$E(x) = E_0 e^{-\kappa x} \tag{8-18}$$

as illustrated in Fig. 8-6. After the light has propagated a distance $L_{\text{eff}} \equiv \pi/\kappa$ into the grating, its E field has been reduced by the factor $\exp(-\pi) = 0.043$. Since the parts of the grating beyond $x = L_{\text{eff}}$ do not interact significantly with the light's E field, this distance can be thought of as an effective grating length. The number of grating planes in this effective grating is

$$N_{\text{eff}} = \frac{L_{\text{eff}}}{\Lambda} = \left(\frac{\lambda_B}{\Delta n}\right)\left(\frac{2n}{\lambda_B}\right) = \frac{2n}{\Delta n} \tag{8-19}$$

In the limit $\kappa L \gg 1$, the grating's spectral width can then be found by setting $N = N_{\text{eff}}$ in Eq. (8-13), with the result

$$\frac{\Delta\lambda}{\lambda_0} = \frac{1}{N_{\text{eff}}} = \frac{\Delta n}{2n} \qquad \text{(spectral half-width, strong grating)} \tag{8-20}$$

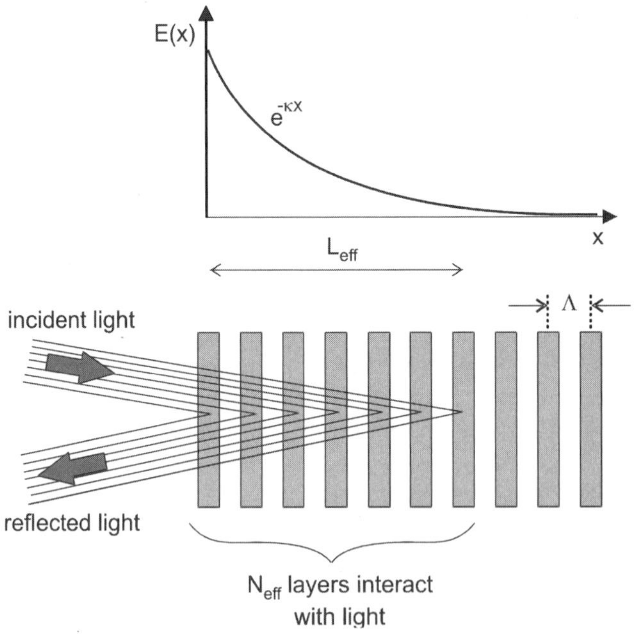

Figure 8-6 At the Bragg resonance, light is attenuated exponentially as it propagates through the grating, so that only a finite number N_{eff} of grating planes are effectively interacting with the light.

This contribution to the resonance width is present even for thin gratings where $\kappa L \ll 1$, but in that case it is smaller than the contribution $1/N = \Lambda/L$. The two limiting cases for Bragg grating linewidth can be summarized by the expressions

$$\frac{\Delta\lambda}{\lambda_0} = \begin{cases} \lambda_0/(2nL) & \text{for } \kappa L \ll 1 \\ \Delta n/(2n) & \text{for } \kappa L \gg 1 \end{cases} \tag{8-21}$$

where the Bragg condition $\lambda_0 \approx 2n\Lambda$ has been used. In the transition region $\kappa L \sim 1$, both expression above become the same order of magnitude. By varying the pump intensity, exposure time, and grating length, fiber Bragg gratings with a wide range of reflectivity bandwidths can be fabricated.

This ability to customize the spectral reflectivity for a particular application has led to the incorporation of Bragg gratings into a number of photonics devices. Perhaps the most important application of Bragg gratings has been to replace one or more of the mirrors in a laser cavity. The high reflectivity that is obtained for $\kappa L \gg 1$ leads to low lasing thresholds, and the narrow spectral width for large N provides wavelength selectivity for the laser output. Bragg gratings are integral to the operation of distributed-feedback lasers and vertical-cavity surface emitting lasers (see Chapter 11), and have become standard in fiber lasers as well (see Chapter 23).

Another application of growing importance is the use of the fiber Bragg grating as a sensor. A number of different physical parameters can be measured, such as strain, temperature change, and pressure change, but the parameter measured most directly is strain. When a fiber of length L is stretched by some applied force, the elongation δL is generally proportional to L, with the ratio $\delta L/L$ defined as the *strain*. Since $\delta L/L$ is usually quite small, the "unit" *microstrain* is often used, which corresponds to a fractional extension $\delta L/L = 10^{-6}$. If a fiber containing a Bragg grating is uniformly stretched, the grating spacing Λ is increased by the same fractional amount as the length L. Since the Bragg wavelength is $\lambda_B = 2n\Lambda$, it also increases by this same fractional amount.

EXAMPLE 8-1

A Bragg grating in silica fiber ($n = 1$) has a length of 7.5 mm, and is designed to reflect light of free-space wavelength 1500 nm in first order. (a) Determine the shift $\delta\lambda_B$ in wavelength of the Bragg peak for an elongation of 1 microstrain. (b) Determine the Bragg resonance half-width assuming $\kappa L \ll 1$. (c) If it is possible to measure a shift in λ_B equal to 5% of the half-width, what is the minimum strain that can be measured in this fiber?

Solution: (a) For $\delta L/L = 1 \times 10^{-6}$, the fractional change in λ_B is also 10^{-6}, so

$$\delta\lambda_B = 10^{-6}\,\lambda_B = (10^{-6})(1.5 \times 10^{-6}\text{ m}) = 1.5\text{ pm}$$

where 1 pm $= 10^{-12}$ m. The sensitivity to strain can then be expressed as 1.5 pm/microstrain.

(b) From Eq. (8-21) in the limit $\kappa L \ll 1$,

$$\Delta\lambda = \frac{\lambda_B^2}{2nL} = \frac{(1.5 \times 10^{-6}\text{ m})^2}{2(1.5)(7.5 \times 10^{-3}\text{ m})} = 1.0 \times 10^{-10}\text{ m} = 0.1\text{ nm}$$

(c) The minimum fractional shift in wavelength that can be measured is

$$\frac{\delta\lambda}{\lambda} = (0.05)\frac{\Delta\lambda}{\lambda} = (0.05)\frac{10^{-10}\text{ m}}{1.5 \times 10^{-6}\text{ m}} = 3.3 \times 10^{-6}$$

which corresponds to a sensitivity of 3.3 microstrain.

Fiber grating sensors have a number of advantages over conventional electrical strain gauges. They are robust, compact, and linear over a wide range of strains. They are also inherently "calibrated," and since wavelength is measured rather than optical power, they are not affected by power fluctuations in the light used to probe the grating. Perhaps the greatest advantage is that a number of different fiber sensors can be cascaded together in series along a fiber, so that the probe light passes through them all sequentially. If each grating has a slightly different center Bragg wavelength, then the wavelength shift for each one can be determined in a single spectral measurement with a broadband light source. The set of all such wavelength shifts gives information about how the strain is distributed along the length of the fiber; it is in effect a *distributed fiber sensor.* Such an arrangement can be used, for example, to probe the development of strains in buildings, bridges, and other structures so that structural defects can be remedied prior to catastrophic failure. This type of application has come to be termed *smart-structure technology.*

An alternative type of distributed fiber sensor is possible, in which the reflections from the different gratings are separated in the time domain rather than in wavelength. We have already seen one example similar to this—the optical time-domain reflectometer (OTDR)—which gives information on the distribution of losses due to Rayleigh scattering and splices throughout a fiber link. If there are Bragg gratings distributed throughout the fiber as well, the OTDR can separate out the reflections from the different gratings.

Photonic Band Gap

One of the important characteristics of a Bragg grating is the exponential attenuation of light for wavelengths close to λ_B. For a sufficiently long grating, this leads to nearly 100% reflection of the light, with essentially none transmitted. According to coupled-mode theory, this exponential attenuation occurs only for wavelengths λ_0 in the range $\lambda_B - \Delta\lambda < \lambda_0 < \lambda_B + \Delta\lambda$, with $\Delta\lambda$ given by Eq. (8-21) in the limit $\kappa L \gg 1$. Outside this wavelength range, the solutions to the wave equation are oscillatory rather than exponentially damped, and the effect of the Bragg resonance there is to alter the wave velocity, not the amplitude. The range of wavelengths for which light is attenuated is referred to as the *stop band,* since light is "stopped" from propagating in this region. A similar situation arises in the propagation of infrared light through a crystal, where a stop band (the *reststrahlen band*) arises from the strong interaction between the light and the vibrational modes of the lattice.

The existence of a stop band in the Bragg grating implies that there is a "gap" in the spectrum of allowed propagation frequencies; that is, there is a certain range of frequencies for which there are no propagating electromagnetic modes. Since the energy of a photon is $\hbar\omega$, this is equivalent to saying that there is a gap in the allowed energy states in the system. The idea of an energy gap may be familiar if you have studied the optical properties of solids. We will see in Chapter 10 that energy bands and gaps between those bands are a fundamental feature of the energy spectrum of electrons in a solid. These gaps arise from the interaction of the electron with the periodic potential presented by the crystalline lattice. In much the same way, the gap in the photon energy spectrum arises from the interaction of the photon with a periodic dielectric "lattice." Because of this close analogy, a frequency gap in the photon spectrum has come to be known as a *photonic band gap.*

When a material has a photonic band gap, the dispersion relation $\omega(k)$ is altered for frequencies near the gap. Figure 8-7a shows the dispersion curve for a perfectly homogeneous material of refractive index n. For simplicity, we will assume here that n does not vary with frequency (no material dispersion), so the phase and group velocities are both equal to c/n [see Eqs. (2-5)–(2-7)]. If now a dielectric lattice with spatial period Λ is introduced into the medium, then Bragg scattering occurs at wavelengths satisfying the Bragg condition $m\lambda_0 = 2n\Lambda$. Here, m is an integer specifying the diffraction order, and λ_0 is the free-space wavelength. The corresponding wavelength in the medium is $\lambda = \lambda_0/n$, with wave vector magnitude

$$k_B = \frac{2\pi}{\lambda_0/n} = (2\pi n)\left(\frac{m}{2n\Lambda}\right) = m\frac{\pi}{\Lambda} \tag{8-22}$$

At the values $k = \pi/\Lambda, 2\pi/\Lambda, \ldots$, the wave becomes nonpropagating, which means that the group velocity $d\omega/dk$ must go to zero there. The dispersion curve, therefore, bends as shown in Fig. 8-7b, in such a way that frequency gaps open up. For frequencies below a gap, v_g decreases with increasing ω, which is similar to the effect of "normal" material dispersion. For frequencies above a gap, v_g increases with increasing ω, which is similar to "anomalous" material dispersion. This photonic band gap dispersion adds to whatever material dispersion may also be present.

One result of the dispersion shown in Fig. 8-7b is that there are two values of the light frequency at $k = \pi/\Lambda$. This seems strange at first, because for a propagating wave one

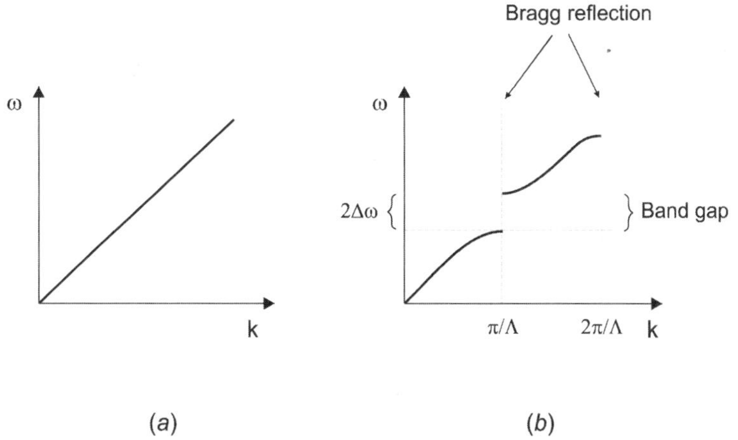

Figure 8-7 Dispersion relation $\omega(k)$ for (a) homogeneous material and (b) photonic band gap material. Bragg reflection opens up frequency gaps in the dispersion curve.

would expect a unique frequency for every possible wavelength. However, at the Bragg condition $k = \pi/\Lambda$, the wave is not propagating, but is actually a standing wave created by the superposition of two equal amplitude, counterpropagating beams. The physical origin of the two frequencies can then be understood by considering two possible standing wave intensity profiles, as depicted in Fig. 8-8. If the intensity maxima are aligned with the higher-index regions, the effective refractive index n_{eff} will be higher than the average, whereas if the intensity maxima are aligned with the lower index regions, the effective refractive index n_{eff} will be lower than the average. The effective index relates the frequency and wave vector according to

$$\omega = (c/n_{\text{eff}})k = (c/n_{\text{eff}})\frac{\pi}{\Lambda} \tag{8-23}$$

so that a higher n_{eff} gives rise to a lower ω, and vice versa. The two frequencies at $k = \pi/\Lambda$ can then be interpreted as arising from these two offset intensity profiles. Qualitatively, this viewpoint suggests that a higher index difference Δn will lead to a larger frequency gap.

The width of the frequency gap can be determined quantitatively from the known width of the wavelength stop band, $2\Delta\lambda$. Using $\omega = 2\pi c/\lambda_0$, we have

$$\Delta\omega = \frac{d\omega}{d\lambda_0}\Delta\lambda = -\frac{2\pi c}{\lambda_0^2}\Delta\lambda$$

The fractional half-width is then

$$\frac{\Delta\omega}{\omega} = \frac{\Delta\lambda}{\lambda_0} = \frac{\Delta n}{2n}$$

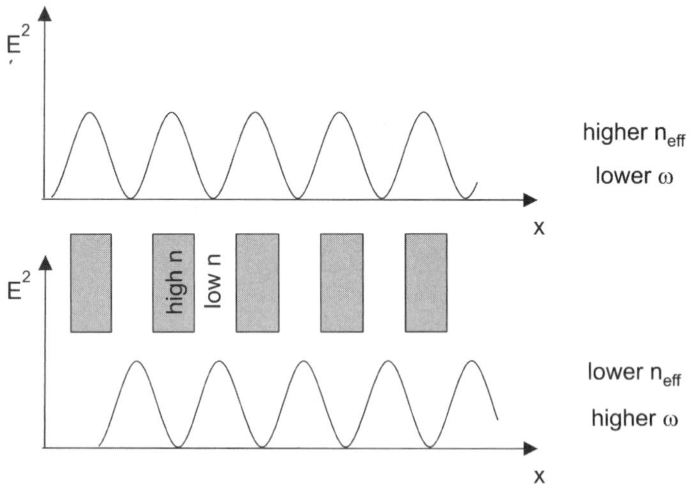

Figure 8-8 The standing wave intensity patterns that occur at Bragg resonance give a lower oscillation frequency when the intensity peaks line up with regions of higher *n*, and higher frequency when they line up with regions of lower *n*.

where the minus sign has been dropped (only the magnitude is significant) and Eq. (8-21) has been used. The ratio of gap width to center frequency is therefore

$$\frac{\text{frequency gap}}{\text{center frequency}} = \frac{2\Delta\omega}{\omega} = \frac{\Delta n}{n} \tag{8-24}$$

According to the above equation, if we want to create a material with a large photonic band gap, we need a large index difference. Although the derivation has assumed a sinusoidal index grating in one dimension, the same qualitative conclusion is true for arbitrary refractive index profiles, and also for two and three dimensions. Later in this chapter, we will see that a high Δn is especially important in the creation of 2-D and 3-D photonic band gaps.

Localized Modes

We have seen that for frequencies within the photonic band gap, the oscillations of a light wave's E field are exponentially attenuated. A grating of infinite length (a perfect photonic crystal) is, therefore, unable to support propagating light waves at these frequencies. However, if there is a discontinuity or defect in the "crystalline" structure, then it is possible to have light energy residing in a special type of optical mode that is confined to the vicinity of the defect. These are termed *localized modes,* and they play an important role in the practical application of photonic crystals.

The simplest type of defect is the edge or surface of a photonic crystal, which for the 1-D Bragg grating corresponds to the point at which the grating begins. In this case, the E field of the localized mode decays exponentially with distance from the surface, as illustrated in Fig. 8-6. This mode is localized to within a distance ~ $L_{\text{eff}} = \pi/\kappa$ of the surface. It is important to note that there is no steady-state flow of energy in the localized mode (in the language of electromagnetic theory, the Poynting vector is zero). The localized mode does store optical energy, however, and there will be a transient flow of energy in the vicinity of the mode when the stored energy changes.

Another type of defect can be formed by removing one or more of the high-index layers from an infinite Bragg grating, as illustrated in Fig. 8-9. Light cannot propagate in the re-

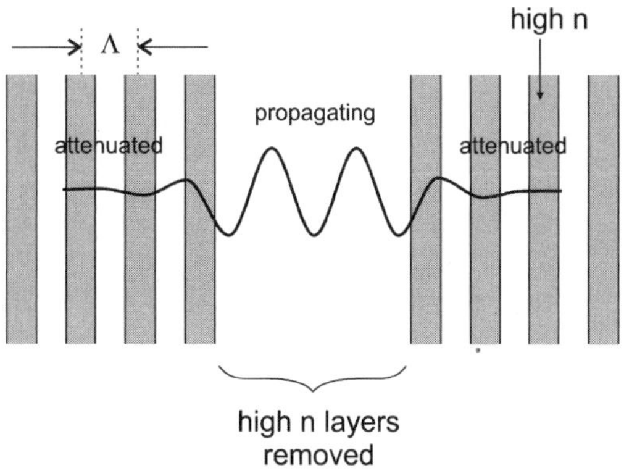

Figure 8-9 When one or more high index layers are removed from an infinite Bragg grating, the E field in the vicinity of the resulting defect is localized, with an oscillatory region surrounded by two exponentially damped regions.

gions on either side where the Bragg grating is still intact, but it can propagate in the middle where the high-index layers were removed. The result is a localized mode, with an oscillatory *E* field region surrounded by two exponentially damped *E* field regions. This localized mode can be thought of as an optical cavity formed by two semiinfinite Bragg gratings. Light initially propagating to the right is reflected from the Bragg grating on the right, and after propagating to the left for a short distance it is reflected from the Bragg grating on the left. This process continues, resulting in complete confinement of the light to the vicinity of the defect. The same general principle can be extended to propgation in two or three dimensions, and has important implications for practical devices, as we will soon see.

8-2. 2-D PHOTONIC CRYSTALS

The concepts developed for the 1-D photonic crystal can now be used to develop a qualitative understanding of 2-D photonic crystals. An example of such a structure is the array of dielectric rods shown in Fig. 8-10a, which is periodic in both the *x* and *y* directions.

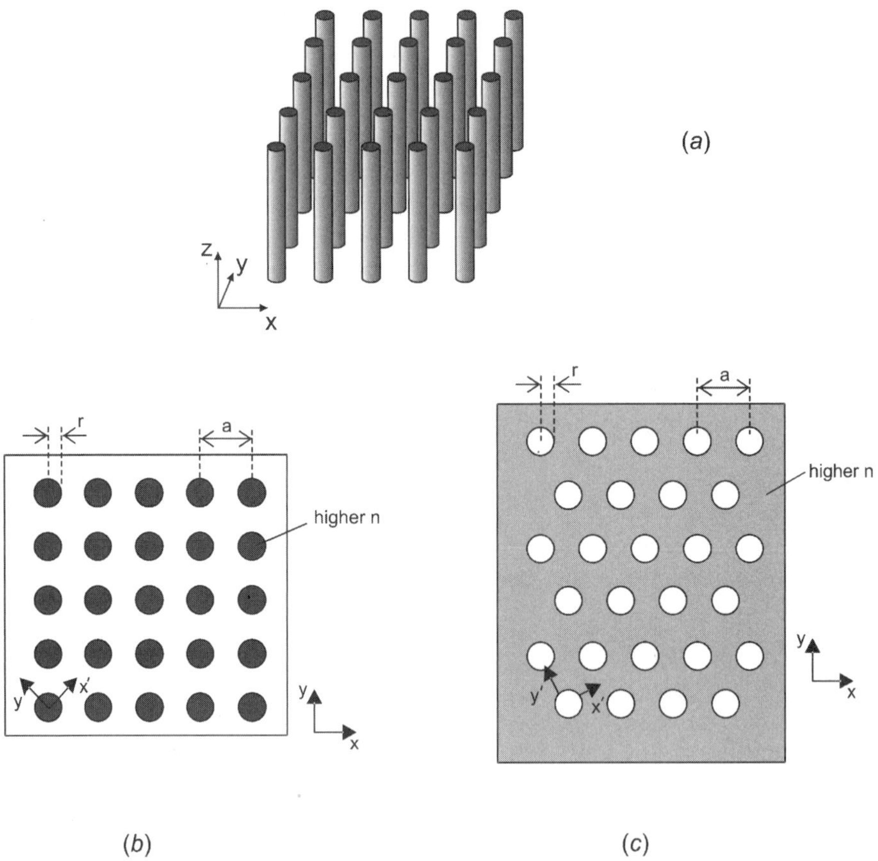

Figure 8-10 (a) A periodic array of parallel dielectric rods forms a 2-D photonic crystal. (b) Cross-sectional view of square lattice, showing distinct symmetry directions *x* and *x*′. (c) Triangular lattice of air holes in solid substrate, with *r* the hole radius and *a* the spacing between hole centers. Two symmetry directions *x* and *x*′ are shown.

Historically, the understanding of 2-D photonic crystals has developed along two parallel lines. In one approach, the light is assumed to propagate mostly in the x–y plane, perpendicular to the rods. In the other approach, light is assumed to propagate mostly along the z axis, parallel to the rods. These two limiting cases lead to rather different types of devices and applications, and we consider them separately in the following.

Planar Geometry

First, consider light propagating in the x–y plane perpendicular to the dielectric rods of Fig. 8-10a. We will assume initially that the rods are arranged in a square lattice with spacing a, as shown in the cross-sectional view of Fig. 8-10b. Because of the periodicity in the $+x$ direction, light propagating along that axis encounters a spatially varying refractive index, and is scattered as in a 1-D Bragg grating. When the propagation constant satisfies the Bragg condition, the light will be strongly reflected, and gaps will open up in the dispersion relation $\omega(k_x)$, as depicted in the left panel of Fig. 8-11a. This is similar to the gap that develops in the 1-D Bragg grating dispersion curve of Fig. 8-7, at the Bragg condition given by Eq. (8-22). In accordance with Eq. (8-24), we would expect the band gap to increase with increasing refractive index difference between the rods and surrounding medium.

These analogies with the 1-D Bragg grating are helpful in developing a qualitative understanding of 2-D photonic crystals. However, there are important differences in the two cases. For example, whereas the Bragg grating is periodic in only one direction, the 2-D photonic crystal is periodic in more than one direction. The position and width of the band gaps will in general be different for different directions, and they may or may not overlap. Another difference is that polarization becomes important for 2-D photonic crystals. Light propagating in the x direction can have an E field along y (with B along z), or an E

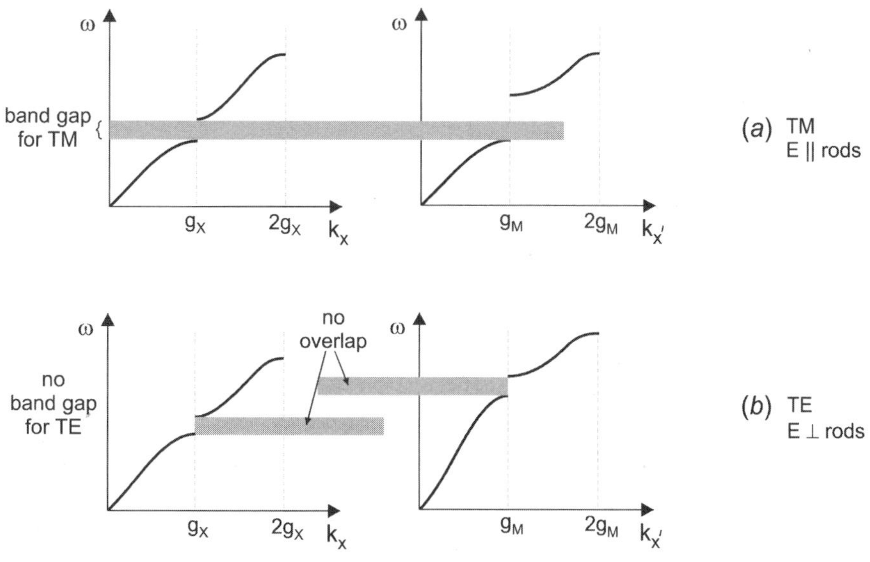

Figure 8-11 Illustration of dispersion curves for (a) TM polarized and (b) TE polarized light propagating along symmetry axes x and x' in a square lattice of dielectric rods embedded in air. The position and width of the band gaps correspond to rods with dielectric constant $\varepsilon_r \approx 9$, and $r/a \approx 0.2$.

field along z (with B along $-y$).* The former is termed *TE polarization* (transverse electric), since the E field is perpendicular to the rods, whereas the latter is termed *TM polarization* (transverse magnetic), since the B field is perpendicular to the rods. The dispersion curves and corresponding band gaps will be different for the two polarizations.

The dispersion for different propagation directions and polarizations is illustrated in Fig. 8-11 for the square lattice of side a. There are two distinct directions, labeled x and x' in Fig. 8-10b, which have different periodicities. There is also periodicity in the y direction, but it is the same as that in the x direction, and is therefore not distinct. The separation between neighboring planes in the x direction is a, while the separation between neighboring planes in the x' direction is $a/\sqrt{2}$. Bragg reflection, therefore, occurs at $k_x = mg_X$ in the x direction and at $k_x' = mg_M$ in the x' direction, where $g_X = \pi/a$, $g_M = \sqrt{2}(\pi/a)$, and m is an integer. For TM polarization, the bandgaps in the x and x' directions partially overlap, which results in a range of frequencies for which no light can propagate in any direction (in the x–y plane). We say in this case that there is a *complete photonic band gap* for TM polarized light.

For TE polarized light, the band gaps for the x and x' directions are narrower, and do not overlap for the assumed relative dielectric constant $\varepsilon_r = 8.9$. There is thus no complete photonic band gap for this polarization. The physical origin of this difference between TM and TE polarization is easily understood by referring to Fig. 8-12, which shows the field distribution and electric field orientation with respect to the rods. We discussed previously (see Fig. 8-8) the interpretation of the 1-D band gap as arising from the relative alignment of the high-index material with the intensity peaks in the standing waves that occur near Bragg resonance. The same argument can be made for the array of 2-D rods, except that now the effective index n_{eff} depends on polarization as well. The difference in n_{eff} for the two standing waves is largest in the case of TM polarization, because the E field is mostly in the rods for one standing wave, and mostly outside of the rods for the other standing wave. For TE polarization, there is less difference, because the E field in both standing waves passes through a certain amount of the low-index material. We thus expect in general that TE waves will have a smaller band gap than TM waves.

To confine light in two dimensions, we need a range of frequencies for which light cannot propagate for any direction and any polarization. This will occur only when the band gaps for different directions and polarizations all overlap in some frequency range. Since the band gap increases with increasing refractive index difference $\Delta n = n_{\text{rod}} - n_{\text{air}}$, the existence of a complete photonic band gap becomes more likely for higher Δn. In the case of a square lattice of dielectric rods, it is found that at sufficiently high Δn there are indeed complete photonic band gaps for both TE and TM polarization. However, the band gap for TE polarization does not overlap the band gap for TM polarization, so there is no frequency range for which light of any polarization is blocked. A square lattice of dielectric rods is, therefore, not the best structure for a photonic crystal.

A better structure consists of a triangular lattice of air holes embedded in a solid dielectric substrate. Figure 8-10c shows a cross section of the geometry, with r and a the radius and center-to-center spacing of the holes, respectively. A complete photonic band gap is found to occur in such a structure when the relative dielectric constant of the substrate is $\varepsilon_r > 7.2$, which corresponds to a refractive index $n > 2.7$.† For practical applications, we would like the refractive index of the substrate to be considerably larger than

*Light with arbitrary polarization can be written as a linear combination of these two fundamental polarization directions.

†The relation between dielectric constant and refractive index is $n = \sqrt{\varepsilon_r}$.

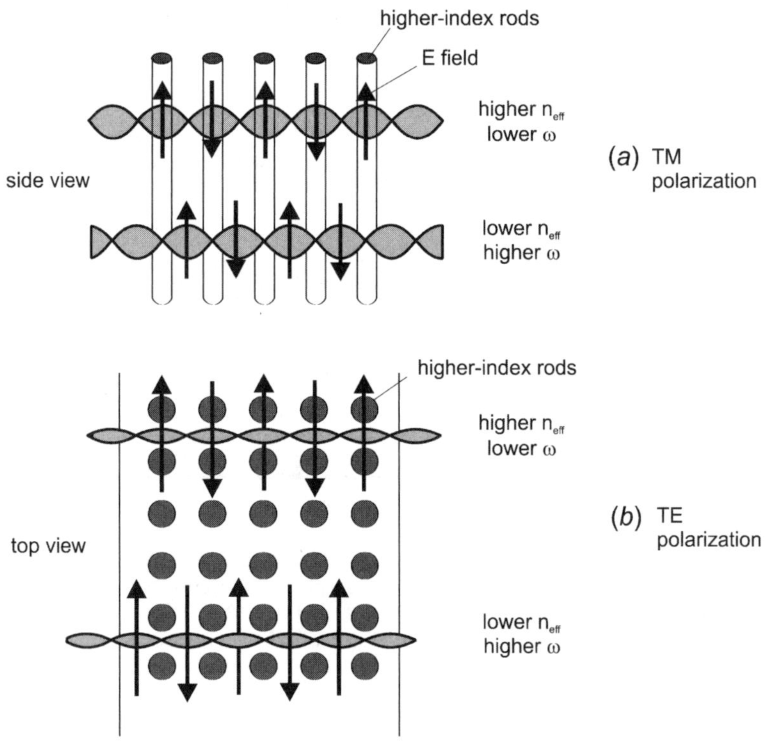

Figure 8-12 (a) Side view of square lattice of dielectric rods, showing possible alignments of standing waves and rods for TM polarization. (b) Top view of the same, for TE polarization.

this, so that the band gap is reasonably large compared to the center frequency of the gap. If we use the semiconductor GaAs, for example, which has $\varepsilon_r \simeq 13$ and $n \simeq 3.6$, then the band gap extends over a frequency range that is $\approx$ 19% of the mid-gap frequency. The magnitude of the gap also depends on the ratio r/a, and is maximized in this case when $r/a \approx 0.48$.

The photonic band gap for the triangular lattice of air holes can be represented in terms of the dispersion graph $\omega(k)$, as shown in Fig. 8-13. This graph represents the same type of information presented in Fig. 8-11, but in a more compact fashion. Instead of separate $\omega(k)$ plots for the different directions x and x', there is a single plot showing $\omega(k)$ as k varies from zero (Γ) to the maximum value in the x direction (K), and also from zero to the maximum value in the x' direction (M). Another difference is that the higher-order bands (beyond the first-order Bragg condition) have been translated along the k axis to be more easily visualized. The results for both polarizations are presented, with TE polarization (E in x–y plane) shown by the solid curves, and TM polarization (E along z) shown by the dotted curves. It is clear from this graph that there is a range of frequencies for which no light can propagate, regardless of direction or polarization. There is a complete photonic band gap for both polarizations.

Light with a frequency within the photonic band gap will be prevented from propagating, but only if the photonic crystal is perfectly uniform. Defects in the photonic crystal structure will partially relax this prohibition, as we saw in the case of the 1-D Bragg grating. For example, if one or more rows of holes is removed, a linear defect is created, and

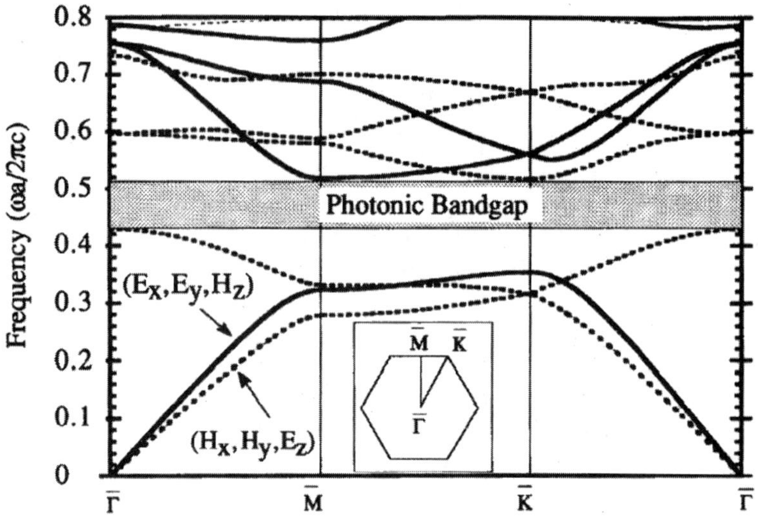

Figure 8-13 Compact representation of dispersion $\omega(k)$ for triangular lattice of air holes in solid dielectric. Vertical axis is frequency ω normalized to $c(2\pi/a)$, and horizontal axis gives k along different symmetry directions. The substrate has dielectric constant $\varepsilon_r = 13$, and the hole radius is $r = 0.48\,a$. There is a complete photonic band gap for both TE (solid line) and TM (dotted line) polarizations (after Meade et al. 1992).

local modes are allowed in the vicinity of this defect. Light is prevented from leaving the vicinity of the defect, because the E field is evanescent (exponentially decaying) in the surrounding 2-D photonic crystal. The only direction in which light can propagate is along the path of the removed holes. The line of missing holes thus acts much like an optical waveguide, confining light to a certain path in the x–y plane.

This optical waveguiding effect has considerable potential for applications in integrated optics. Practical devices would not extend infinitely in the z direction, however, but would have the geometry of a slab waveguide, such as that shown in Fig. 8-14. Confine-

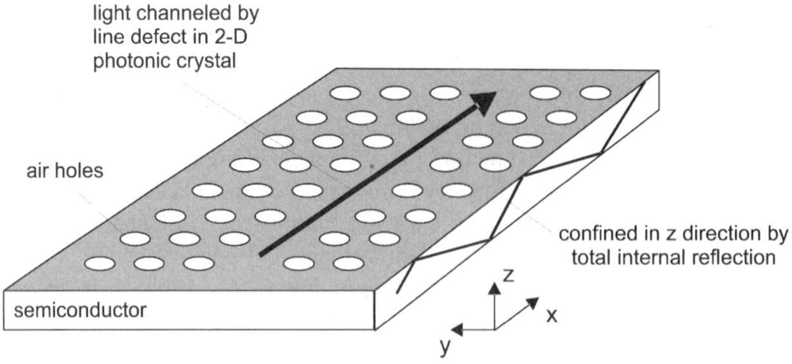

Figure 8-14 In a photonic crystal slab waveguide, light is confined to a row of missing air holes by two different mechanisms. In the plane of the waveguide, confinement is due to the photonic band gap in the surrounding 2-D photonic crystal, whereas in the perpendicular direction, confinement is by conventional total internal reflection.

ment to the slab is provided by total internal reflection (TIR), as discussed in Chapter 3. These structures are not true 2-D photonic crystals because of their finite extent in the z direction. Furthermore, the light propagation is not strictly in the x–y plane, but has a small component along z. Nonetheless, the essential qualitative features of a 2-D photonic crystal still do apply, and light can be guided through such structures with very little loss.

The confinement of light to the vicinity of a line defect in a 2-D photonic crystal slab waveguide is illustrated in Fig. 8-15. This shows the calculated E field distribution for a triangular lattice of air holes in InP, which has three rows of air holes missing to form a line defect. At the position of the arrow, the direction of the line defect changes abruptly by 60°, and from there the line of missing holes continues again in a straight line to the edge of the photonic crystal region. Before the bend, the wavefronts are uniform, characteristic of a low-order waveguide mode. After the bend, the wavefronts become irregular due to scattering of light into a mixture of higher-order modes. However, light is still confined to the line of missing holes, and exits the photonic crystal region with high efficiency.

This ability to transmit light efficiently around sharp bends is a key advantage of photonic crystal waveguides, as compared to traditional waveguides based on TIR. We saw in Section 5-3 that bending losses in a traditional optical waveguide increase as the bend radius is reduced, so the losses would be very high around a sharp bend. In contrast, a photonic crystal waveguide suffers little loss around even a sharp bend, because light is totally reflected from the photonic crystal boundary for any angle of incidence. This manner of channeling light through a photonic crystal structure represents a new paradigm for integrated optics, with considerable potential for planar waveguide devices.

Fiber Geometry

The opposite limiting case for a 2-D photonics crystal is that in which the light is propagating mostly in the z direction, with only a small component perpendicular to the rods or air holes in the structure. This is the appropriate limit for propagation in an optical fiber, and we will see that the photonic band gap can be used to confine light to the core of an

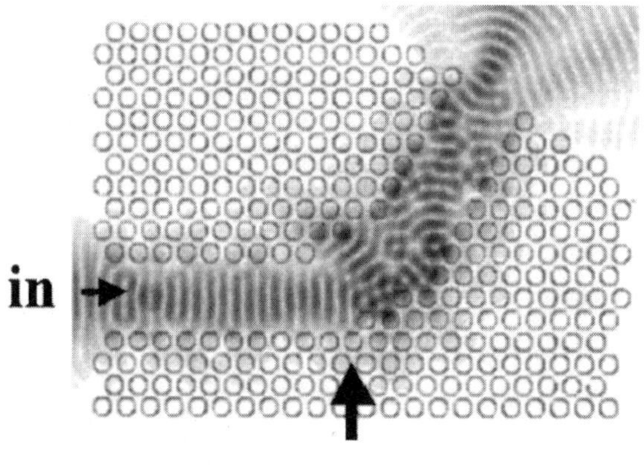

Figure 8-15 Calculated E field distribution for 1500 nm light propagating along a line defect with a sharp 60° bend. The photonic crystal consists of a triangular lattice of air holes, separated by 450 nm, in an InP substrate. The hole radius is such that about half the volume of the structure is air, and the effective refractive index for the waveguide mode is $n_{eff} = 3.21$ (after Talneau et al. 2002).

optical fiber in much the same way that it confines light in a planar waveguide structure. However, light in such a fiber can also be confined by total internal reflection (TIR), just as in a conventional fiber, with the photonic crystal acting to create an appropriately lower refractive index. This TIR principle was the first to be exploited, and we begin our discussion of the fiber geometry in photonic crystals with this type of application.

Guiding by Effective Index

Figure 8-16 shows an electron microscope image of an early *photonic crystal fiber,* consisting of a triangular array of air holes parallel to the fiber axis. Fibers such as this can be made by stacking silica rods and tubes into a mm-scale preform, and then drawing the preform into a fiber by heating and pulling in a conventional fiber draw tower. Surface tension forces tend to keep the holes properly aligned as the preform is pulled out into a μm-scale fiber. Because the fiber has a periodic array of air holes running down its length, it is also referred to as "holey fiber."

At the center of the fiber is a single missing air hole, which constitutes a defect in the 2-D photonic crystal structure. It is this defect that allows confinement of light by TIR. The effective refractive index for light in the vicinity of the defect is approximately that of the glass material itself, which is $\simeq$ 1.45 for pure silica. For light that propagates into the surrounding photonic crystal structure, however, the effective index is some appropriate average between that of silica and that of air. Since air has $n = 1$, this average index is lower than that of pure silica. The fiber, therefore, contains a high-index "core" region, surrounded by a low-index "cladding" region. This is exactly what is needed for confinement of light to the core by TIR, as discussed in Chapters 3 and 4.

One practical aspect of this scheme is that only one glass type is needed for the fiber. Normally, either the core must be doped with atoms that raise the index (such as Ge), or the cladding must be doped with atoms which lower the index (such as F). In the photonic crystal fiber, no doping of the glass is required, although one can think of the cladding region as being "doped" with air holes.

Photonic crystal fibers can guide light in either a single transverse mode (single-mode), or in a combination of higher order modes (multimode), just as in a conventional

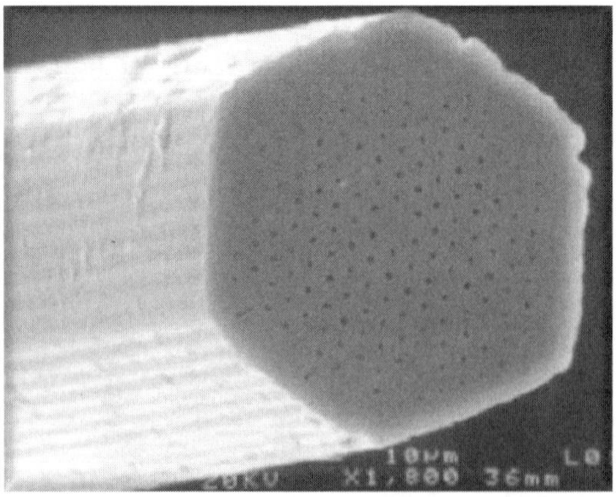

Figure 8-16 Electron microscope image of an early photonic crystal fiber, showing the triangular lattice of air holes with one hole missing at the center. The hole spacing is Λ = 2.3 μm, and the fiber is $\approx$ 40 μm across (after Birks et al. 1997).

fiber. We saw in Chapter 4 that single-mode guidance of light occurs in conventional step-index fiber when $V < 2.405$, where the V parameter is given by Eq. (4-9). For holey fiber of the type shown in Fig. 8-16, it is found that single-mode guidance occurs for the similar condition $V_{PCF} < \pi$, where

$$V_{PCF} \equiv \frac{2\pi\Lambda}{\lambda}\sqrt{n_c^2 - n_{cl}^2(\lambda)} \tag{8-25}$$

is the V parameter as defined for photonic crystal fiber. The definitions for V are the same in the two cases, except that the hole spacing Λ is used for holey fiber, rather than the core radius a. In this equation, λ is the free-space wavelength, n_c is the "core" index (essentially that of silica glass), and $n_{cl}(\lambda)$ is an effective cladding index, as discussed above.

Although light confinement by TIR proceeds in much the same way for the two types of fibers, there is one important aspect in which photonic crystal fiber behaves quite differently. The effective cladding index is not a constant, characteristic of one particular type of glass, but instead varies strongly with wavelength. This can be understood by considering that n_{cl} is an average over the glass and air hole parts of the cladding structure, weighted by the strength of the light field distribution in each region. At shorter wavelengths, light is better able to be confined to the glass between the air hole regions, where the index is higher. At longer wavelengths, however, diffraction prevents the light from being well confined to the glass regions, and the light field spreads out to "sample" more of the air hole regions, thereby lowering the effective index. The net result is that the refractive index of the cladding increases strongly with decreasing wavelength, and approaches that of the core in the short-wavelength limit. This amounts to a strong dispersion in the cladding structure (see Chapter 6), which is in addition to the material dispersion of the glass.

The variation of $n_{cl}(\lambda)$ with wavelength has important implications for the single-mode condition $V_{PCF} < \pi$. As λ decreases, the factor $2\pi\Lambda/\lambda$ in Eq. (8-25) causes V_{PCF} to increase, and in a conventional fiber this would eventually cause the fiber to go from single-mode to multimode. In a photonic crystal fiber, however, this increase is offset by a decrease in the factor $\sqrt{n_c^2 - n_{cl}^2}$, since n_{cl} increases and approaches n_c as λ decreases. The fiber, therefore, has the remarkable property of being single-mode over a much wider wavelength range than conventional fibers, and such fibers have been termed *endlessly single-mode.*

In order for a fiber to be endlessly single-mode, it is necessary that the glass region between the edges of neighboring air holes be sufficiently wide; if it is too narrow, diffraction will prevent the light field from being confined there, and n_{cl} will not become sufficiently close to n_c to keep $V_{PCF} < \pi$. This means that there is a maximum ratio of hole diameter d to hole spacing Λ for endlessly single-mode operation. Above this value of d/Λ, the fiber will still be single-mode for sufficiently long wavelength. However, as the wavelength is decreased, the fiber will become multimode when the condition $V_{PCF} = \pi$ is reached.

There are thus two boundaries between single-mode and multimode behavior: one as d/Λ increases and another as λ decreases. This can be summarized graphically in a kind of "phase diagram,"* as shown in Fig. 8-17. On one side of the "phase boundary" line, the fiber is single-mode, and on the other it is multimode. For the triangular lattice of air

*This terminology comes from pressure/temperature diagrams that show the boundary between liquid and solid phases of matter, for example.

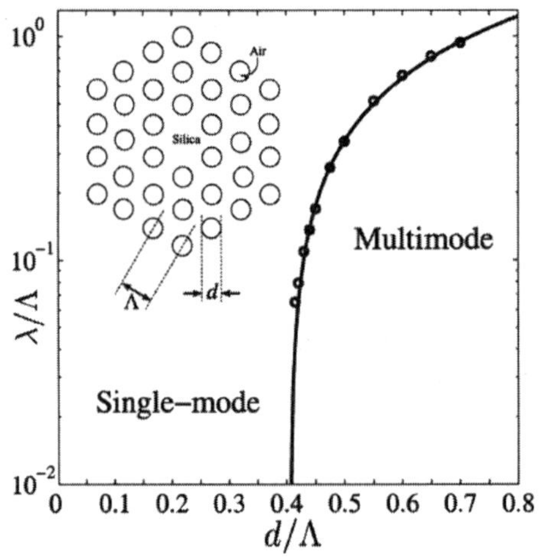

Figure 8-17 The boundary between single-mode and multimode behavior can be represented on a plot of free-space wavelength λ versus air hole diameter *d*, both normalized to the hole spacing Λ. This fiber is single-mode for any wavelength when $d/\Lambda < 0.4$ (after Mortensen et al. 2003).

holes with one hole missing, the fiber is always single-mode when $d/\Lambda < 0.4$. For larger d/Λ, the fiber will be single-mode for large enough λ/Λ, but the required wavelength becomes longer as d/Λ increases. It should be noted that the quantitative results of Fig. 8-17 apply only when the core consists of a single missing air hole. When there are three or seven missing air holes, the endlessly single-mode condition becomes $d/\Lambda < 0.25$ and $d/\Lambda < 0.15$, respectively.

An important feature of the endlessly single-mode condition is that it depends only on the ratio d/Λ, and not on d or Λ separately. The fiber can, therefore, be scaled up in size, and it will remain single-mode as long as $d/\Lambda < 0.4$ (for a single missing air hole). This is in distinct contrast to a conventional step-index fiber, which becomes multimode when the core radius *a* exceeds a certain value.

This scaling property of the photonic crystal fiber has important applications in devices such as high-power fiber lasers (see Chapter 23). The optical power achievable in such devices is limited ultimately by various nonlinear optical processes and by optical damage, both of which depend on the optical intensity (power per unit area). Increasing the effective core area of the fiber, therefore, increases the maximum optical power that can be generated before nonlinear processes become significant or optical damage occurs. In principle, the core size (and hence maximum power) can be made arbitrarily large by increasing both Λ and d in the same proportion. However, the fiber becomes very sensitive to small bends and inhomogeneities when $\Lambda > 10\ \lambda$, which puts a practical limit on the achievable output power. Output powers of several kilowatts in a single transverse mode are predicted to be possible using this scheme.

In the example of the fiber laser just discussed, nonlinear optical effects are detrimental, limiting the optical power that can be generated at the desired wavelength. In other applications, however, nonlinear effects may be desirable. For example, we may wish to convert light at one optical frequency into light at another frequency, a process called frequency conversion (see Chapter 9). The efficiency of such a conversion process increases with increasing light intensity, which means that the light should be confined to a core area that is as small as possible.

In a conventional step-index fiber, if the core radius a is made too small, the optical mode is not well confined to the core. This can be seen from Eq. (4-18)—if a (and hence V) becomes very small, then $w \gg a$. In a photonic crystal fiber, on the other hand, the optical mode can be well confined even to a very small core. The reason for this difference is that the cladding index of the photonic crystal fiber can be made close to 1 by making the cladding consist mostly of air (large d/Λ ratio). This makes NA $= \sqrt{n_c^2 - n_{cl}^2}$ large, which in turn keeps V_{PCF} reasonably large even for very small Λ. Although Eq. (4-18) does not apply quantitatively to the modes of a photonic crystal fiber, it remains true qualitatively that the mode is well confined to the core when $V_{PCF} > 1$.

A small core size not only allows higher optical intensity, but also changes the dispersion properties of the fiber (see Chapter 6). The zero-dispersion point in such fibers can be shifted from the usual value of 1300 nm to a wavelength as short as 800 nm. These altered dispersion characteristics, in combination with the high intensity that is possible, serve as the basis for a whole new class of nonlinear optical applications.

Guiding by Photonic Band Gap

When the photonic crystal fiber guides light by total internal reflection, the function of the periodic air holes is simply to provide an effective refractive index in the cladding that is lower than that in the core. However, we saw earlier in this chapter that a 2-D photonic crystal structure can do more than this—it can prevent light from propagating in the cladding at all, regardless of the index of the core, provided that the cladding structure has a complete photonic band gap. In the case of a planar-waveguide photonic crystal, it was found that a complete photonic band-gap in the most favorable crystal type (triangular lattice of air holes) only occurs when the substrate material has a dielectric constant $\varepsilon_r > 7$, which corresponds to a refractive index $n > 2.65$. Since silica glass has $n \leqslant 1.5$, it might seem that this scheme of photonic band gap guiding will not work for silica-based holey fiber. However, the condition $\varepsilon_r > 7$ applies only for light propagating perpendicular to the air holes. In an optical fiber, just the opposite limit occurs, since the direction of propagation is mostly parallel to the holes. It turns out that in this case, a complete photonic band gap can indeed be obtained for the proper geometry of air holes in silica glass fiber.

Figure 8-18 shows a cross-sectional view of a typical fiber that has a photonic band gap in the cladding. The size of the air holes has expanded to take up most of the volume, leaving thin webs of silica between them in a honeycomb structure. The core is hollow, an important and distinguishing characteristic of this type of fiber. In conventional fiber, a hollow core could not guide light efficiently, because TIR requires that the refractive index of the core be higher than that of the cladding. When the guiding of light is due to a photonic band gap in the cladding, however, there is no need for the core index to be higher than 1.

We have seen in this chapter that a photonic band gap occurs over some range of light frequencies ω, which depends on the index difference Δn and the particular photonic crystal structure. The gap frequency also depends on the direction of light propagation, and in an optical fiber the direction of light propagation is related to the axial wave vector β, as depicted in Fig. 8-19a. For a given ω, modes with larger β are directed more nearly down the fiber axis, whereas those with smaller β have a larger transverse component. A photonic band gap will, therefore, occur over some range of the parameters ω and β.

The values of ω and β for which a photonic band gap occurs can be represented on a plot of ω/c vs. β, as depicted in Fig. 8-19b. In the cladding, where the effective refractive

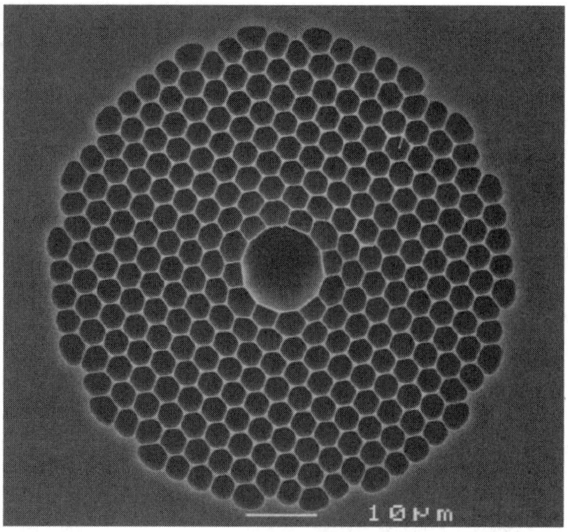

Figure 8-18 Electron micrograph of photonic crystal fiber with hollow core. The core diameter is ≈10 μm (after Couny et al. 2005).

index is n_{cl}, the magnitude of the wave vector is $n_{cl}\,\omega/c$. Propagating modes are then allowed in the cladding when

$$\beta < n_{cl}\frac{\omega}{c} \qquad \text{(propagating mode allowed in cladding)} \qquad (8\text{-}26)$$

and in the core when

$$\beta < \frac{\omega}{c} \qquad \text{(propagating mode allowed in core)} \qquad (8\text{-}27)$$

Both of the above conditions amount to requiring that the component of the wave vector be less than its magnitude. The boundaries between propagating and nonpropagating modes are then straight lines of slope $1/n_{cl}$ for modes in the cladding, and slope 1 for modes in the air.

The regions in which light cannot propagate are shaded in the graph, and they consist of two types. Any mode with $\beta > n_{cl}\omega/c$ is automatically prohibited, because the axial component of the wave vector would be larger that its magnitude. The other shaded regions correspond to the photonic band gaps, where light is prohibited from propagating in the cladding (but not in the core). These photonic band gap regions have the shape of narrow "fingers," which gives rise to the name *finger plot* for such a graph. Light will be guided in the hollow core for values of ω/c and β that are within these finger regions, and also above the slope = 1 boundary line.

Hollow-core fibers have some important advantages over conventional solid-core fibers. The E field of the light field mode is distributed mostly in the hollow core, rather than in the glass, and this makes the fiber propagation characteristics much less dependent on the properties of the glass. The loss coefficient, for example, which in conventional fiber is limited by absorption and Rayleigh scattering in the fiber core, could potentially be much lower in hollow-core fiber. This would allow light to propagate

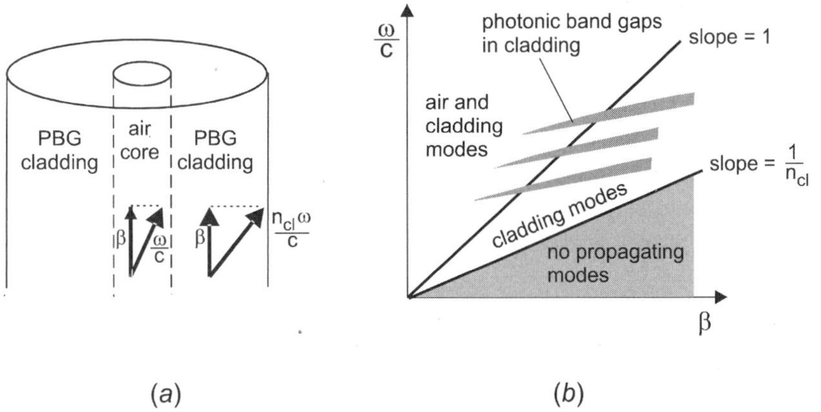

Figure 8-19 (a) For a propagating wave, the wave vector has a projection β along the fiber axis. (b) The "finger plot" shows the values of ω and β (shaded regions) for which there is a complete photonic band gap in the cladding.

further in a telecommunications fiber before amplification is required. Dispersion could also be reduced, allowing very short optical pulses to propagate without significant spreading in time. Another advantage of hollow-core fibers is the ability to transmit much higher optical powers before damage occurs. Optical damage occurs when the E field in the glass material exceeds a limiting value, and the high damage threshold of hollow-core fiber is a consequence of the fact that the light is propagating mostly in air, rather than in the glass.

Some novel applications are made possible by injecting various gases, liquids, or small particles into the hollow core of the fiber. Light guided by the fiber interacts with these materials over a long path length and at high optical intensity, which is ideal for nonlinear optical effects. For example, new optical frequencies can be generated by stimulated Raman scattering or harmonic generation (see Chapter 9) when the proper gas is introduced into the core, and this frequency conversion is efficient at much lower optical power than would normally be required. Another application is the guiding of atoms or small particles down the core of the fiber using optical trapping. Optical trapping takes advantage of the natural tendency of objects to drift toward a region of very high optical intensity, such as occurs at the core of an optical fiber. In this way, atoms or small particles can be kept near the fiber axis, and prevented from striking and sticking to the glass material at the core boundary. The ability to move small particles around in a controlled way using optical fibers has considerable potential for a variety of applications.

8-3. 3-D PHOTONIC CRYSTALS

To have complete control over the path that light can take through a material, we need a photonic crystal that is periodic in all three dimensions. The concept of the 3-D photonic crystal was introduced independently by E. Yablonovitch and S. John in 1987. The original motivation for Yablonovitch was to control the wavelength distribution of light spontaneously emitted by matter, so as to improve the efficiency of semiconductor laser devices. For John, the interest was in studying the fundamental mechanisms by which light could be "localized," or confined to a small region of space. Both of these applications would require a material with a complete photonic band gap over some frequency range, so that light of any polarization or direction would be unable to propagate.

It was not clear at first whether any such material could actually be constructed. An early candidate was the close-packed array of dielectric spheres depicted in Fig. 8-20a. The crystalline symmetry for this arrangement is *face-centered cubic,* or fcc, which means that the unit cell is cubic, with dielectric sphere "atoms" at each corner and face of the cube. Unfortunately, theoretical calculations showed that this structure does not have a complete photonic band gap, regardless of the refractive index of the spheres. However, further calculations by K. Ho and coworkers showed that this structure can be made to have a complete gap, provided that the single dielectric sphere at each lattice location is replaced by a pair of spheres (a dimer). For the proper spacing between spheres in the dimer, this configuration is identical to what would be obtained if single dielectric spheres were placed at the lattice points of a diamond lattice. The resulting structure, depicted in Fig. 8-20d, is actually a special case of the more general configuration in which asymmetrical dielectric objects are placed at the lattice points of an fcc lattice. It is found that a complete photonic band gap can occur in such a structure, provided that the dielectric objects have a sufficiently high refractive index ($n > 1.87$) and asymmetry. Although these more general structures are not true diamond lattice configurations, they are closely related, and are typically referred to as "diamond-type structures."

It would certainly be nice if dielectric spheres would spontaneously organize themselves into a diamond lattice, thereby creating the desired photonic crystal. Nature is not so inclined, however, and a collection of dielectric spheres tends instead to assemble into an fcc lattice, which is the structure of the naturally occuring gem opal (see Fig. 8-20a). Although the opal structure itself does not have a complete photonic band gap, it is possible to derive from this structure a material that does have a band gap. The situation is similar to the 2-D case, in which it was found that an array of dielectric rods has no complete photonic band gap, but the complementary array of air holes in a solid dielectric does

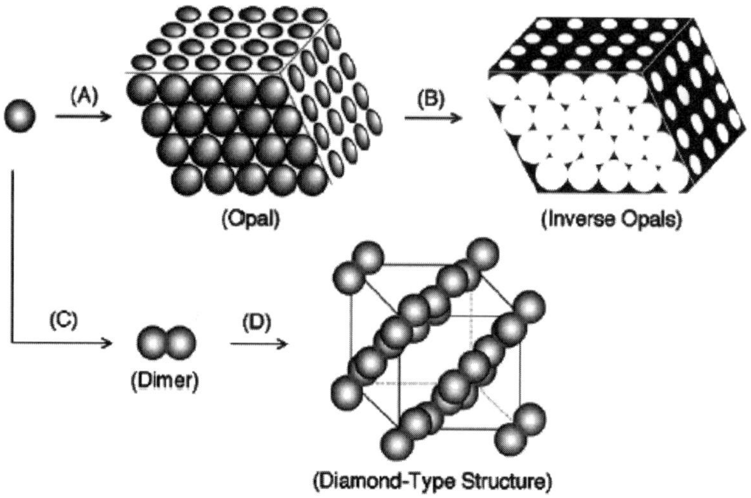

Figure 8-20 (a) Spheres of SiO_2 in suspension naturally assemble into the fcc close-packing structure of opal. (b) Interchanging the silica and nonsilica regions yields the inverse opal structure. (c) A dimer consists of two associated silica spheres. (d) Replacing each sphere in the opal structure with a dimer yields the diamond-type structure. (After Xia et al. 2001.)

have a band gap. In the 3-D case, it is the so-called *inverse opal* structure, depicted in Fig. 8-20b, that has a complete photonic band gap. It is derived from opal by replacing the dielectric spheres by air, and the space between spheres by dielectric. For this structure, it is found that a refractive index of $n > 2.8$ is needed for a complete photonic band gap. Compared with a diamond-type structure, therefore, the inverse opal structure requires a higher index for a complete band gap, and for a given index ($n = 3.6$, say, for GaAs), the frequency width of the gap will be smaller.

There are thus two basic approaches to creating a 3-D structure with a complete photonic band gap. The first approach is to modify materials in a deliberate and controlled way so that a diamond-type structure is obtained. This is called the "top-down" approach,

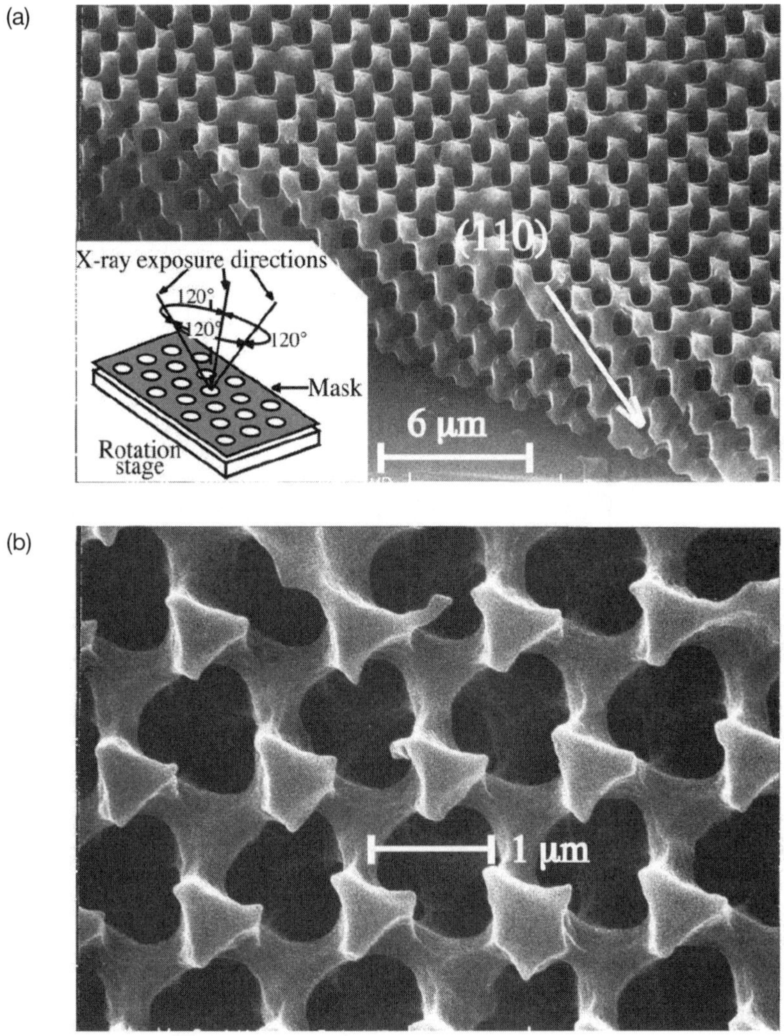

Figure 8-21 Yablonovite is formed by drilling holes into a dielectric at three precise angles, and results in a diamond-like structure. The holes can be created by exposure to X-rays through a mask (after Cuisin et al. 2000).

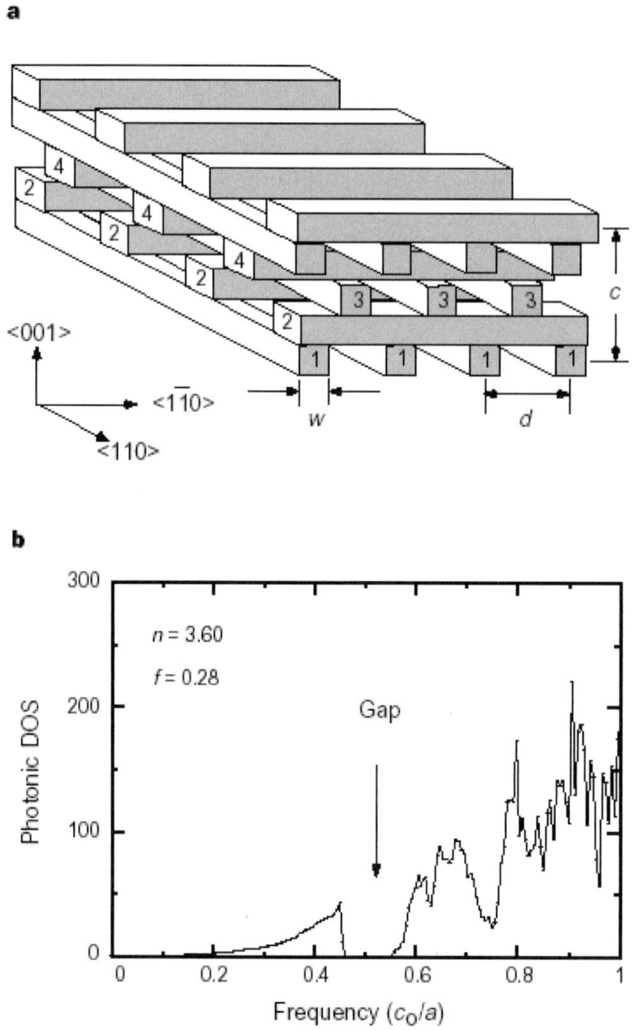

Figure 8-22 (a) The "Lincoln log" structure has a diamond-type lattice symmetry, and exhibits a 3-D photonic band gap. (b) Calculated density of states when the dielectric bars have a refractive index n = 3.6 (that of GaAs) and fill 28% of the space. Reproduced by permission of Nature Publishing Group (after Lin et al. 1998). Courtesy Sandia National Laboratories.

and has the advantage that a wider band gap can be achieved for a given refractive index. The second approach is to allow the constituents to self-assemble in an opal structure, and to then use this as a template in forming an inverse opal structure. This is called the "bottom-up" approach, and is more readily scaled up to mass-production levels than the first method.

Both approaches for constructing a 3-D photonic bandgap material have been used with some success. The earliest experimental demonstration of a complete photonic band gap (Yablonovitch, 1991) is an example of the top-down approach. The structure was created by drilling holes into a dielectric substrate at precise angles, as indicated in

Fig. 8-21. The intersection of the holes in the material forms a diamond-like pattern, and this structure (now known as *yablonovite*) was found to have a complete photonic band gap. In the initial experiments, the hole spacing was on a millimeter scale, and the band gap was in the microwave frequency region. To give a band gap for optical frequencies, the hole spacing needs to be scaled down to the μm range. Figure 8-21 shows a more recent realization of this type of structure, using X-ray exposure through a mask to create the holes.

Another example of the top-down approach is the "Lincoln log" or "woodpile" structure depicted in Fig. 8-22a. This sequence of crisscrossing dielectric bars can be shown to have the symmetry of a diamond-like structure. The computed *density of states* (number of propagating modes per unit frequency interval) is shown in Fig. 8-22b, and is characterized by a robust frequency gap of $\leqslant$ 20% of the center frequency. This structure can be fabricated on the μm scale with conventional lithographic techniques, using a sequence of steps involving deposition of layers followed by selective etching. In this way, it has been possible to create structures that have a complete photonic band gap at the important 1.3 μm telecommunications wavelength.

Although these top-down approaches have been successful in creating materials with a robust photonic band gap, it is not clear if they can be easily scaled up for mass production. More promising in this regard are the bottom-up or self-assembly approaches. These take advantage of the natural tendency of spherical colloids (such as monodispersed SiO_2 spheres with diameters ranging from 100 to 1000 nm) to self-assemble in the fcc structure. The inverse opal structure can be obtained by first injecting a high-index material into the spaces between the spheres, and then removing the original SiO_2 material through selective etching. The process is somewhat akin to fossilization, with the original opal material (the "living organism") serving as a template for creation of the desired inverse opal material (the "fossil"). By using silicon as the high-index material, complete photonic band gaps with $\Delta\omega/\omega = 5\%$ have been generated in this way. Although these band gaps are not as wide as those possible in the diamond-type structures, the simplicity of the method and the potential for scaling up to mass production make these bottom-up processes promising for future photonic applications.

PROBLEMS

8.1 The sum in Eq. (8-8) was determined graphically by adding vectors in the complex plane. An alternative is to use the mathematical identity

$$1 + x + x^2 + \ldots + x^{N-1} = \frac{1 - x^N}{1 - x}$$

taking $x = \exp(-i\delta)$. Show that this approach leads to a reflectivity

$$R = \left(\frac{\Delta n}{n}\right)^2 \frac{\sin^2(N\delta/2)}{\sin^2(\delta/2)}$$

8.2 Using the result from Problem 8.1, show that the reflectivity at Bragg resonance is given by Eq. (8-9). Also use this to show that the wavelength interval from the Bragg resonance to the first reflection zero on either side is given by Eq. (8-13).

8.3 Use the result from Problem 8.1 to determine the full width at half maximum

(FWHM) of the center peak in the grating reflectivity spectrum. Show that this is approximately equal to the width $\Delta\lambda$ given in Eq. (8-13).

8.4 The properties of a step index Bragg grating formed by uniformly spaced layers of GaAs and $Al_xGa_{1-x}As$ are evaluated by illuminating it with broadband light and measuring the spectrum of reflected light. A Bragg peak with maximum reflectivity of 5% is found at 900 nm, and the first zero reflection near the Bragg peak is located at 960 nm. GaAs is known to have $n = 3.6$, but the refractive index of the $Al_xGa_{1-x}As$ layers is unknown. Determine (a) the spacing between layers, (b) the number of layers, and (c) the refractive index of the $Al_xGa_{1-x}As$ layers.

8.5 Using the geometry of Fig. 8.5, show that the Bragg grating spacing in the fiber core is given by Eq. (8-14).

8.6 A fiber Bragg grating of length $L = 1$ cm is to be written into silica fiber ($n = 1.5$) so that it reflects light of free-space wavelength 1500 nm. It is desired that the peak reflectivity be 10%. (a) Determine the spectral width and quality factor Q of the Bragg resonance. (b) What Δn is needed to achieve the desired reflectivity?

8.7 In Problem 8.6, what grating length is required to make the peak reflectivity 98%? Assume the same value of Δn. For this case, calculate both contributions to the resonance width given in Eq. (8-21). If the two contributions are the same order of magnitude, the total width can be obtained by adding the contributions together "in quadrature" (i.e., add the squares and take the square root).

8.8 A fiber Bragg grating is designed to measure a strain of 2 microstrain. It will use a Bragg wavelength peak around 975 nm, and the instrumentation can detect a shift of the peak as small as 10% of the resonance half-width $\Delta\lambda$. Determine the grating length required, assuming weak reflection ($\kappa L \ll 1$).

8.9 A fiber Bragg grating with $\lambda_B = 1300$ nm is written into the core of a silica fiber. The fiber used is highly doped with GeO_2, and has $n = 1.5$ and $\Delta n = 2.5 \times 10^{-4}$. How far will light of wavelength 1300 nm propagate through the fiber grating before the light wave's E field is reduced to 1% of its initial value? At this point, how many grating periods has the light passed through?

Chapter 9

Nonlinear Optics

In the preceding chapters we have assumed that light interacts with matter in a linear fashion. A linear interaction is illustrated in Fig. 9-1a, which shows two light waves of different frequency intersecting in a material medium. The presence of wave 2 has no affect on wave 1, and vice versa. The waves are uncoupled, and propagate independently. In contrast, a nonlinear interaction is characterized by a coupling of two or more intersecting waves, as illustrated in Fig. 9-1b. In this case, each wave can modify the properties of the other, changing, for example, the other wave's amplitude or phase. One of the waves can also modify its own amplitude or phase, a phenomenon termed *self-action.* Still another possibility in a nonlinear interaction is the generation of new waves with a frequency different from those of the incident beams. This *frequency conversion* process does not occur in a linear medium.

In a perfect vacuum, there is no coupling between two light waves because Maxwell's equations (which govern the propagation of an electromagnetic wave) are linear in the electric and magnetic field variables. The superposition principle then applies, which states that the sum of two solutions is itself another solution to the equations. Coupling between two light waves is only allowed when light propagates in a material medium, and it is an indirect type of process. One wave changes the properties of the medium in some way, and then the second wave is affected by the changed properties of the medium. The degree of coupling between two light waves, therefore, depends on how strongly the light wave interacts with the medium.

The interaction between light and matter is normally quite weak. The order of magnitude of this interaction can be estimated by comparing the strength of the light wave's electric field E_ℓ to the electric field E_a in the atoms of the material. Light from the sun, for example, has a typical field $E_\ell \approx 600$ V/m, whereas typical atomic fields are $E_a \sim 10^{11}$ V/m (see Problem 9.1). Since $E_\ell \ll E_a$, the light wave deviates the electrons in the material only slightly from their normal positions, which means that the light–matter interaction is weak. It is actually a good thing that nonlinear optical effects are usually negligible, because linear behavior is necessary for the image-forming property of lenses. To form a proper image, light from each point of an object must propagate to the image plane in a manner that does not depend on the presence or absence of light from other points on the object, which is simultaneously passing through the lens. Nonlinear interactions would cause a distorted image.

Nonlinear effects are expected to become important only for very high optical intensity I, where the field E_ℓ (given by Eq. 2-9) is large. It is therefore not surprising that the developement of lasers, which are capable of very high intensities, has spurred on progress in understanding and applying nonlinear optical phenomena. Indeed, the first important experimental demonstration of nonlinear optics, that of second-harmonic generation by Franken and coworkers in 1961, occured just after the demonstration of the first laser in

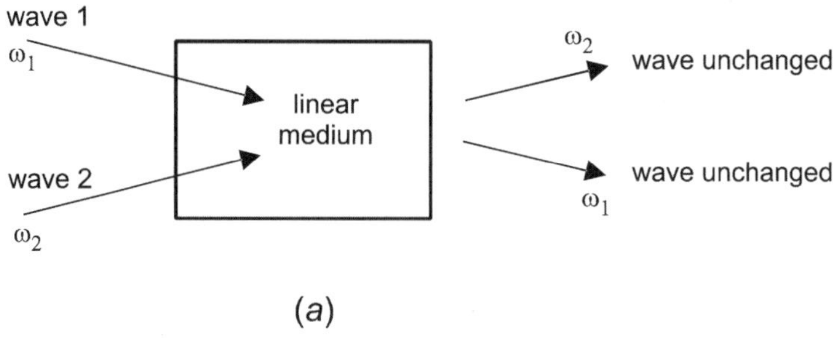

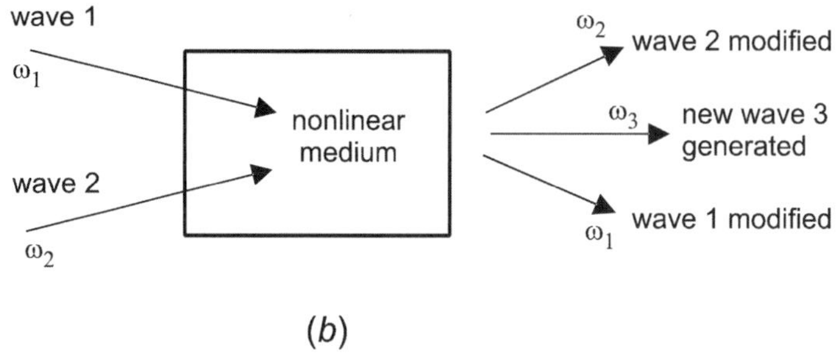

Figure 9-1 (a) In a linear medium, two waves pass through the same region of space without interacting. (b) In a nonlinear medium, two waves that overlap spatially may each modify the properties of the other, and create additional waves with different frequencies.

1960. Since that time, laser physics and nonlinear optics have matured into two rich and multifaceted subdisciplines, each evolving in synergy with the other. Today, they are inextricably linked, lasers being needed to study nonlinear effects, and nonlinear effects being needed for the operation of many lasers.

In this introductory survey, space permits us to sample just a few of the many diverse aspects of nonlinear optics. After first reviewing the mechanisms that give rise to nonlinear effects, we next consider those phenomena in which new frequencies are generated, and then those in which the frequency remains unchanged, but some other property (such as phase) is modified. Finally, we consider the electrooptic effects, in which the optical properties of a material are modified by a static electric field. More complete treatments of these and other nonlinear optical phenomena will be found in the Bibliography.

9-1. FUNDAMENTAL MECHANISMS

In this section, we describe some of the more important mechanisms that can give rise to nonlinear optical phenomena. It should be kept in mind that not all of these can occur in every material, and also that two or more may both play a role in a particular nonlinear phenomenon. The various mechanisms are each characterized by a response time, which governs how fast the medium can change its properties in response to the incident light. This becomes relevant for optical switching applications, in which the medium response time limits the time response of the switch.

Electron Cloud Distortion

Perhaps the most universal nonlinear optical mechanism is that which arises from a distortion of the electron clouds around the atoms in the material. This is illustrated in Fig. 9-2a, which shows the effect of the light wave's electric field **E** (we now drop the subscript ℓ for simplicity) on the charge distribution within the atom. The effect of **E** is to cause a charge separation, in which the positively charged nucleus is displaced in the direction of **E**, whereas the negatively charged electron is displaced in the opposite direction. This results in an *electric dipole* **p** for each atom, which is in the same direction as **E**. If there are N atoms per unit volume, then there is a *polarization density* $\mathbf{P} = N\mathbf{p}$ induced in the medium.

The induced dipoles are forced to oscillate at the frequency of the incident light, as in a driven harmonic oscillator, with the charges in the dipoles accelerating periodically in the direction of **E**. According to a fundamental principle of electricity and magnetism, any accelerated charge radiates in a direction perpendicular to the acceleration vector. The oscillating dipoles therefore radiate light with the same frequency and direction as the incident

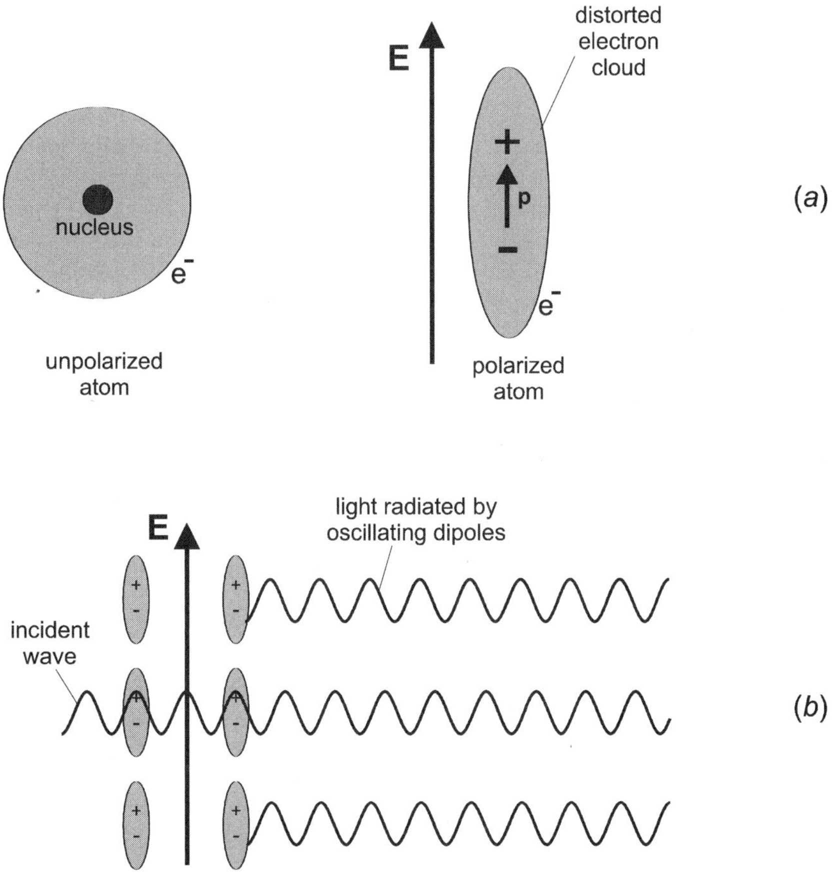

Figure 9-2 (a) An electric field distorts the electron cloud in an atom, creating an electric dipole moment **p**. (b) The oscillating dipoles driven by the light wave's E field radiate additional waves that interfere with the original wave, modifying its propagation speed.

light, as illustrated in Fig. 9-2b. The reradiated light is shifted in phase from the incident light, just as it is in any driven, damped harmonic oscillator. When this phase-shifted light from the oscillating dipoles is added to the original light wave, the result is a single sinusoidal wave of the same frequency, which moves through the medium with a phase velocity different (usually less) than c, the speed of light in vacuum. The phase velocity is given by $v_p = c/n$ (Eq. 2-6), where n is the index of refraction.

According to this physical picture of the origin of a material's index of refraction, a larger degree of polarization would be expected to lead to a larger value of n. This is borne out by an analysis of Maxwell's equation in a material medium, in which it is found that

$$n = \sqrt{\varepsilon_r} = \sqrt{\varepsilon/\varepsilon_0} = \sqrt{1+\chi} \tag{9-1}$$

where ε and ε_0 are the permittivity of the medium and of free space, respectively, ε_r is the relative permittivity or *dielectric constant* (actually not a constant, but a function of frequency), and χ is the *electric susceptibility,* given by

$$\chi \equiv \frac{P_x}{\varepsilon_0 E_x} \qquad \text{(electric susceptibility)} \tag{9-2}$$

Although the susceptibility is actually a tensor quantity (P_x may be related to E_y or E_z, etc.), for simplicity we will treat it here as a scalar.

The origin of optical nonlinearity can be seen in Fig. 9-3a, which shows the potential energy of an electron as it is displaced from equilibrium by the electric field. When the E field (and hence the displacement x) is small, the restoring force on the electron varies linearly with x, and the potential energy varies quadratically with x. In this regime, the polarization P_x is linear with the field E_x, as shown by the dotted line in Fig. 9-3b. For sufficiently large x, however, the restoring force becomes smaller, due to the varying $1/r^2$ Coulomb force. The potential energy curve flattens out for large x, making it easier to polarize the atom. The polarization, therefore, exhibits a nonlinear variation with E_x, as depicted by the solid line in Fig. 9-3b. The susceptibility χ defined in Eq. (9-2) is now no longer a constant, but instead increases with increasing field.

Since the deviation from linearity is small for atom–field interactions, it is possible to expand χ in powers of E_x, keeping only a few low-order terms. It is customary to do this by writing the polarization as

$$P_x = \varepsilon_0(\chi_1 E_x + \chi_2 E_x^2 + \chi_3 E_x^3) \tag{9-3}$$

where χ_1 is the usual linear susceptibility, and χ_2 and χ_3 are the second- and third-order susceptibilities, respectively. Terms of higher order than E_x^3 are usually negligible. This is not the most general starting point for nonlinear behavior, because it assumes a local relation between P and E; that is, the value of P at one location and time depends on the value of E at that same location and time. However, most nonlinear optical processes can be understood using this equation, and in the rest of this chapter we consider a number of such examples.

One important general result can be deduced right away from Eq. (9-3). Consider a material medium that has inversion symmetry; that is, if x is replaced by $-x$, y by $-y$, and z by $-z$, there is no physically distinguishable change in the material. This would be the property of a perfect cubic lattice, for example. Say that a field E_x is applied in the $+x$ direction,

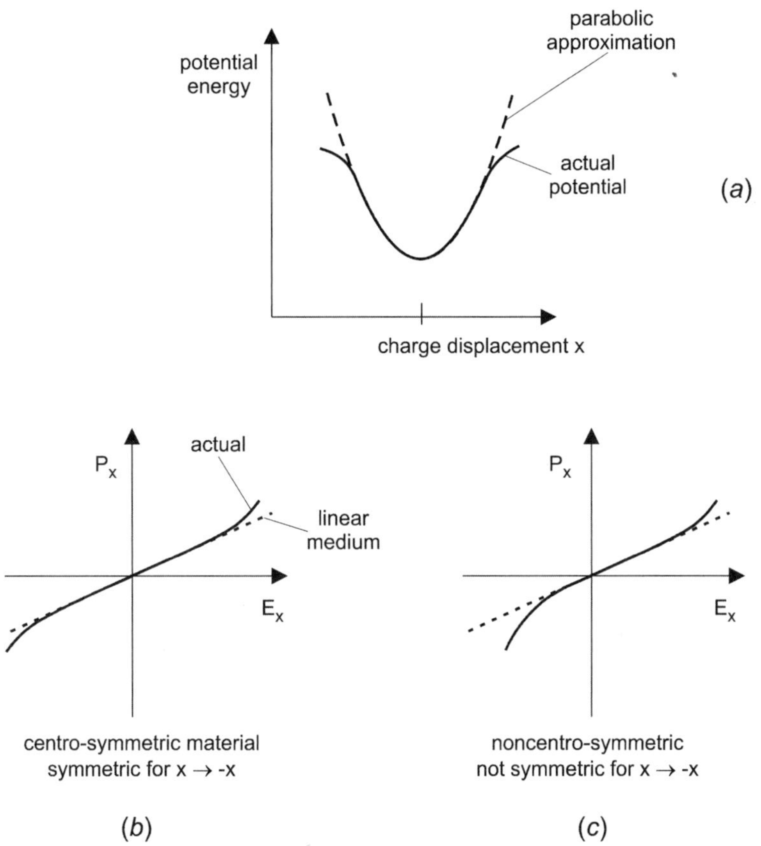

Figure 9-3 For a small electric field, in which the displacement x of the electron cloud is small, an atom's potential energy varies quadratically with x and the polarization density is linear with E_x. In larger fields, there are deviations from this linear behavior. In materials that lack inversion symmetry, the polarization magnitude for applied field E is not the same as that for applied field –E.

with a corresponding P_x in the $+x$ direction. Upon inverting the lattice, there can be no change in the physical polarization or field, because the new lattice is physically the same. However, since the x axis has been reversed, the signs of both E_x and P_x are reversed. That is, $E_x \rightarrow -E_x$ and $P_x \rightarrow -P_x$ upon the inversion. This means that P_x must be an odd function of E_x, containing terms E_x, E_x^3, and so on, but not E_x^2. If P_x contained a term involving E_x^2, it would not reverse sign when E_x reversed sign. We therefore come to the important conclusion that for *centrosymmetric* materials (those with inversion symmetry), χ_2 must be zero. For χ_2 to be nonzero, the material must lack inversion symmetry, and have an asymmetrical P_x versus E_x relation, as illustrated in Fig. 9-3c. This restriction will play a key role in selecting the proper medium for second-order nonlinear optical processes.

One advantage of electron cloud distortion as the nonlinear mechanism is its speed. The electron distribution in an atom can change on a femtosecond (fs, 10^{-15} s) time scale, which corresponds to the semiclassical "orbital" time of an electron around the nucleus. This means that the nonlinear output can respond to changes in the input on a fs time scale, much faster than electronic circuit elements. There is the potential, then, for extremely fast optical switches or modulators.

Other Nonlinear Mechanisms

Although distortion of the electron cloud is the most universal type of optical nonlinearity, other mechanisms become important in certain materials and applications. We overview them briefly here, along with their relevant time scales.

Molecular Orientation

This nonlinearity occurs for asymmetrical molecules that are free to rotate in a liquid. In the absence of an applied field, thermal agitation will cause the molecules to settle into a random distribution of orientations, as depicted in Fig. 9-4a. When a strong electric field is applied, it tends to orient the molecules, due to the interaction of the E field with the induced dipoles. This alignment, shown in Fig. 9-4b, creates a macroscopic asymmetry in the material that changes its optical properties. A light wave passing through the modified medium would experience a different refractive index for different orientations of its **E** field vector. Generally, the polarizability of the molecules (and hence the refractive index) is higher when **E** is parallel to the long axis of the molecules. The result is a field-induced *birefringence,* in which the refractive index depends on the direction of polarization of the light wave. Changes in n due to this type of process occur on a picosecond (ps, 10^{-12} s) time scale, a typical time scale for molecular rotations.

Electrostriction

This nonlinear mechanism is quite common, and like molecular orientation, it arises from the tendency of an induced dipole to lower its potential energy in an applied electric field. Instead of accomplishing this by rotation, however, electrostriction involves the translation of atoms or molecules into a region of higher optical intensity. The potential energy of a dipole moment **p** in a field **E** is given by $U = -\mathbf{p} \cdot \mathbf{E}$, as depicted in Fig. 9-5a. This energy would be lower if the atom were to move from a region of lower intensity (where E is smaller) to a region of higher intensity (where E is larger). Since systems naturally tend to relax to a state of lower energy, atoms in an intense optical beam experience a force di-

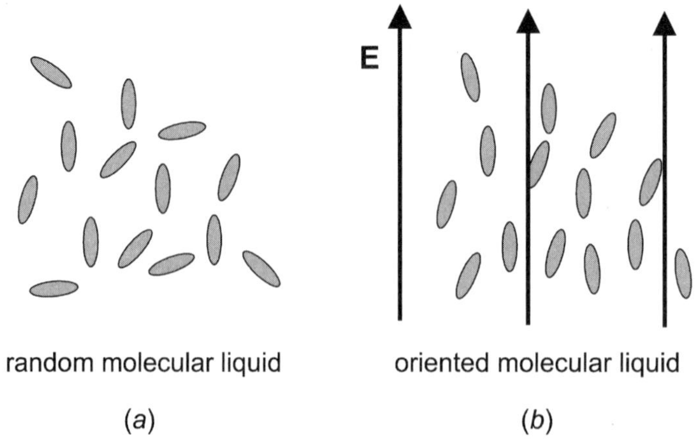

Figure 9-4 A strong electric field can orient asymmetrical molecules in a liquid, producing a nonlinear response.

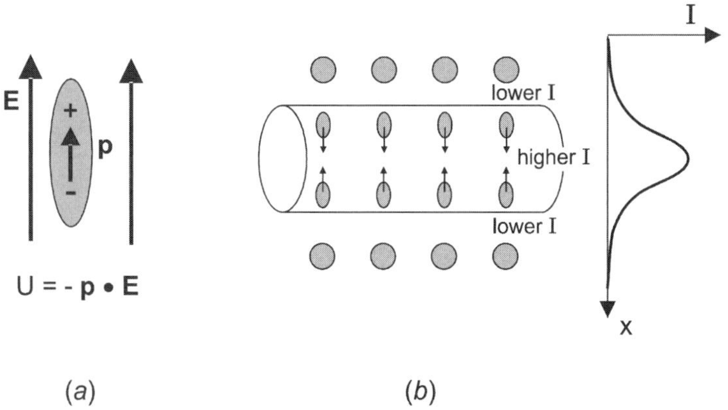

Figure 9-5 (a) A dipole's energy is lower when it is in a larger *E* field. (b) Atoms tend to move to the center of an optical beam, where the *E* field is highest, a phenomenon known as electrostriction.

rected toward regions of higher optical intensity, as illustrated in Fig. 9-5b. This is the phenomenon of *electrostriction,* and it occurs quite generally in solids, liquids, and gases.

To the extent that atoms in a material are able to move, electrostriction causes an increase in mass density in the high-intensity region of an optical beam. This increases the index of refraction, since the polarization density **P** (and hence χ and n) increases with increasing density. The change of refractive index affects not only the original light wave, but also any other light wave passing through the same region. In a solid, atoms are more constrained than in a liquid or gas, but they can move a little bit, creating strain in the lattice when they do so. The increased lattice energy from this strain counterbalances the decreased potential energy from electrostriction, and this determines how far the atoms will move. The response time for electrostriction is limited by the transit time of an acoustic wave across the width of the optical beam. For optical fibers with core diameters of a few μm, this is on the order of 1 ns, much slower than molecular orientation (ps) or electron cloud distortion (fs), but still fast enough to play a role in many nonlinear optical phenomena.

Resonant Absorption

In the nonlinear mechanisms discussed so far, there is no actual change in the population of energy states in the medium. In these situations, the incident photon energy $h\nu$ does not match up with any difference ΔE in the energy levels of the medium, and the atom–field interaction is said to be *nonresonant.* A different type of nonlinearity can occur when there is resonance between $h\nu$ and ΔE, as illustrated in Fig. 9-6. Incident photons are absorbed by the material with a probability that is proportional to the number of atoms in the ground state (level 1). If there are many photons incident simultaneously (high optical intensity), a large fraction of the atoms will be promoted from the ground state to the excited state (level 2). The probability of a photon being absorbed is then reduced, as shown in Fig. 9-6b, due to the reduced number of atoms in the ground state. The absorption probability therefore decreases with increasing optical intensity, an effect known as *optical bleaching.* We will see in Chapter 22 how this can be used to advantage in generating short laser pulses.

In certain materials it is possible for the absorption probability to increase with increasing optical intensity, rather than decrease. Fig. 9-7a shows how this can occur when

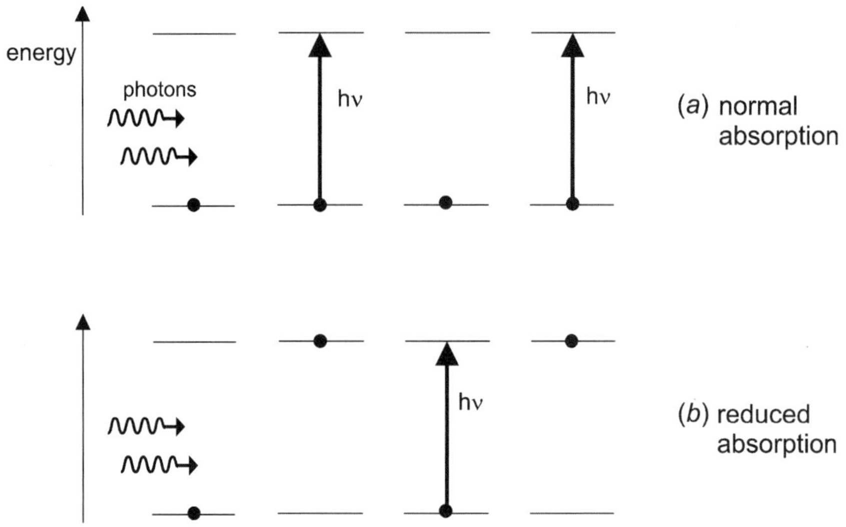

Figure 9-6 (a) At low intensity, most atoms are in the ground state 1 and are available for absorbing a photon. (b) At high intensity, a significant fraction of atoms (50% in this example) are in the excited state 2 and are unavailable for absorbing a photon. This is optical bleaching, an intensity-dependent absorption.

there are three levels interacting with the photons. After the atom has been raised in energy to level 2 by absorption of a photon, it can subsequently be raised to a higher level 3 by the absorption of an additional photon. This process is termed *stepwise absorption,* and occurs with a probability that is nonlinear in the incident intensity. If the absorption probability for the 2 → 3 transition is greater than the absorption probability for the 1 → 2 transition, the result will be increased absorption at high optical intensity. This can serve

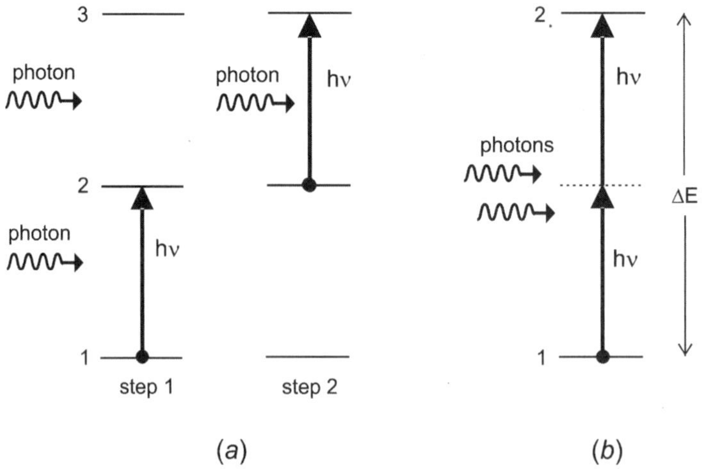

Figure 9-7 (a) In a stepwise absorption process, an atom is first raised to level 2 by absorbing a photon, and then raised to level 3 by absorbing a second photon. (b) In two-photon absorption, the two photons are absorbed simultaneously, without excitation of any real intermediate level. Both of these processes can be used to increase the absorption at high intensity.

as the basis for an *optical limiter,* which functions to protect people or equipment from undesired high-power laser pulses.

Another mechanism which increases the absorption probability at high intensity is that of *multiphoton absorption,* illustrated in Fig. 9-7b. In this process, two or more photons are absorbed simultaneously, raising the atom's energy from level 1 to level 2. By conservation of energy, $2h\nu = \Delta E$ for the two-photon absorption illustrated, where $h\nu$ is the energy of one photon and ΔE is the energy difference between levels 1 and 2. This process is similar to stepwise absorption in that it occurs with a probability that is nonlinear in the optical intensity—I^2 for two-photon absorption and I^m for m-photon absorption. It differs from stepwise absorption, however, in that there is no real intermediate state between the initial and final levels. Because of the lack of any one-photon resonance, multiphoton absorption is in general much weaker than stepwise absorption, and is important only at much higher intensities. It can also serve as the basis for an optical limiter.

The response time for optical bleaching or optical limiting depends on how long the atom remains in the excited state 2, after the incident light is switched off. This is known as the *excited state lifetime,* and varies from milliseconds to nanoseconds, depending on the material.

Inelastic Scattering

Another type of nonlinear mechanism arises from the Raman and Brillouin scattering processes discussed earlier in Chapter 5. These involve a type of resonance, since the difference between incident and scattered photon frequencies must match up with a vibrational frequency of the material. The interaction can be considered nonlinear in the sense that there is coupling between the electronic and vibrational motions in the molecule or solid, which generates new frequencies of light. However, under conditions in which the vibrational amplitude is due mostly to thermal agitation, the probability for scattering to occur is independent of the light intensity. In this sense, the scattering is a linear phenomenon, with the scattered intensity proportional to the incident intensity. The scattering process becomes nonlinear only when the vibrational amplitude becomes large and dependent on the light intensity. This leads to stimulated scattering and optical gain, as we will see in Chapter 24. The response time for this nonlinearity corresponds to the period of vibration in the medium, typically on the ps time scale.

Thermal Effects

The last nonlinear mechanism we will discuss is rather indirect, but it turns out to be quite important in many applications. When an intense optical beam (from a laser, for example) passes through a material, some fraction of the beam's energy is usually absorbed and converted into heat. This results in a temperature rise, which in turn causes a thermal expansion. Thermal expansion makes the material less dense (fewer atoms per unit volume), and this lowers the index of refraction. Each of these processes are themselves linear, so the net result is that the index n decreases linearly with the beam intensity I.

The optical response is effectively nonlinear because n is not a constant, but rather depends on I. To see what effect this has, consider the propagation of a nearly collimated optical beam through a partially absorbing medium, as illustrated in Fig. 9-8. The optical intensity is highest at the beam center, so the decrease in n will be greatest there. Since an optical wave travels faster in a medium with lower n ($v_p = c/n$), the middle part of the

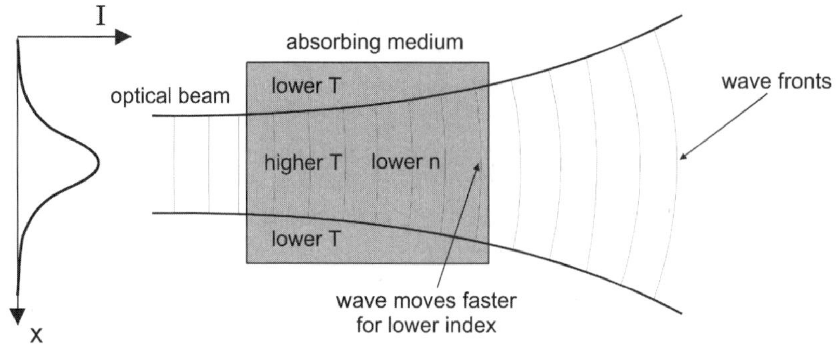

Figure 9-8 Heating of a material by an intense optical beam lowers the refractive index near the beam center. This creates a thermal lens that causes a collimated beam to diverge.

wave front will move faster than the edges, making the wave front "bow out" as it propagates. The beam therefore diverges, just as it would if it had passed through a negative (diverging) lens. This phenomenon is known as thermal blooming or *thermal lensing,* and is an issue that must be accounted for in designing high-power solid-state lasers. The time response of this nonlinearity is much slower than the other mechanisms, since temperature changes depend on the thermal diffusion of heat. Typical time scales are on the order of milliseconds or even slower.

9-2. FREQUENCY CONVERSION

The mixing of two different frequencies to obtain a third frequency is common in electrical circuits. This is done when demodulating a radio signal and "tuning in" to different channels, for example. Certain electronic circuit elements are (or can be made to be) highly nonlinear, and this makes such a mixing process fairly routine. In contrast, the relatively weak nonlinear interaction of optical waves requires that special techniques and materials be used for the optical generation of new frequencies. In this section, we consider several important examples, showing how they are all related to the nonlinear susceptibility.

Second Harmonic Generation

Consider an optical wave with frequency ω propagating inside a material medium.* At a fixed position within the material, the light wave's electric field has a time dependence given by

$$E(t) = A \cos \omega t \tag{9-4}$$

where A is the electric field amplitude, and the phase has been set to zero at $t = 0$. This time-varying electric field creates a time-varying polarization density according to Eq. (9-

*In the rest of this chapter, we will leave off the word "angular" and simply refer to ω as the "frequency." Also, we drop the subscript x on E and P for simplicity, keeping in mind that they still refer to components of a vector along an axis.

3). Assuming that the medium lacks inversion symmetry, so that $\chi_2 \neq 0$, the two lowest-order terms in Eq. (9-3) give

$$P(t) = \varepsilon_0 \chi_1 A \cos \omega t + \varepsilon_0 \chi_2 A^2 \cos^2 \omega t \tag{9-5}$$

for the time-dependent polarization. Using the identity $\cos^2 \theta = (1 + \cos 2\theta)/2$, this can be written in the form

$$P(t) = P_0 + P_\omega \cos \omega t + P_{2\omega} \cos 2\omega t \tag{9-6}$$

where

$$\begin{aligned} P_0 &= \varepsilon_0 \chi_2 A^2/2 \\ P_\omega &= \varepsilon_0 \chi_1 A \\ P_{2\omega} &= \varepsilon_0 \chi_2 A^2/2 \end{aligned} \tag{9-7}$$

According to Eq. (9-6), the induced polarization varies with time in three distinct ways. The first term P_0 corresponds to *optical rectification,* in which a static (dc) polarization is produced in response to the rapidly varying electric field of the light wave. Although this can actually be observed by placing the material between the plates of a capacitor, it is seldom used in practice. The second term, $P_\omega \cos \omega t$, causes the atoms to radiate light at frequency ω, and this results in the linear refractive index as discussed in the previous section.

It is the third term, $P_{2\omega} \cos 2\omega t$, that is of particular interest here, since the dipoles oscillating at frequency 2ω will radiate light at that same frequency 2ω. This radiated light is at a frequency twice that of the incident light, and the phenomenon is, therefore, referred to as *second harmonic generation* (SHG) or *frequency doubling.* Its most important application is in generating new frequencies of laser light, especially in the ultraviolet, where laser operation is more difficult. Continuously tunable laser light is often desirable, and frequency doubling provides a convenient way to extend the tuning range of a near-infrared or visible laser into the blue or ultraviolet regions.

Since second-harmonic generation requires that $\chi_2 \neq 0$, it only occurs in materials that lack inversion symmetry. This eliminates many common materials such as optical glass, which is isotropic and therefore has inversion symmetry. Only crystals of a particular type of symmetry are suitable for frequency doubling. Table 9-1 lists the χ_2 values for a few representative materials, along with their index of refraction and the range

Table 9-1 Second-order nonlinear susceptibility χ_2 for selected crystals

Material	Transparency range (μm)	Linear index[a] at ω n_o/n_e	Linear index[a] at 2ω n_o/n_e	χ_2 (10^{-12} m/V)[b,c]
KDP	0.18–1.45	1.495/1.460	1.512/1.471	0.86
$LiNbO_3$	0.4–5.5	2.234/2.155	2.325/2.233	12
$AgGaS_2$	0.5–13	2.316/2.347	2.383/2.341	40
$CdGeAs_2$	2.4–18	3.505/3.591	3.530/3.621	470

[a]Fundamental at 1.064 μm for KDP and $LiNbO_3$, 10.6 μm for $AgGaS_2$ and $CdGeAs_2$.
[b]For 1.064 → 0.532 μm SHG (KDP, $LiNbO_3$) and 10 → 5 μm SHG ($AgGaS_2$, $CdGcAs_2$).
[c]The parameter $d = \chi_2/2$ is an often used alternative definition.

of wavelengths for which they are highly transparent. The numbers given are effective values that apply to SHG at a particular wavelength. They are intended to illustrate the order of magnitude of the parameters, and will be somewhat different for other wavelengths or applications.

It is seen from Table 9-1 that materials with a higher index of refraction also tend to have a higher χ_2, and that their range of transparency is shifted to longer wavelengths. This correlation can be understood in terms of the bandgap in the material (see Chapter 10). A smaller bandgap leads to a stronger photon–material interaction, which increases both n and χ_2. Generally, both n and χ_2 increase with increasing frequency (decreasing wavelength) as the photon energy approaches the bandgap energy. This is referred to as *dispersion,* and was discussed in Chapter 6 to explain the spreading of a light pulse in time. It was seen to be a detrimental process, limiting the rate at which data can be sent down a fiber.

We will see now that dispersion plays a similarly detrimental role in SHG. Consider how the E fields from the two waves at ω and 2ω vary with position z as they propagate together through the material. Recalling from Eq. (6-3) that the propagation constant for a wave of frequency ω is $k = n\omega/c$, we can write the two propagation constants k_ω and $k_{2\omega}$ as

$$k_\omega = \frac{2\pi}{\lambda_\omega} = n_\omega \frac{\omega}{c} \tag{9-8a}$$

$$k_{2\omega} = \frac{2\pi}{\lambda_{2\omega}} = n_{2\omega} \frac{2\omega}{c} \tag{9-8b}$$

where λ_ω and $\lambda_{2\omega}$ are the wavelengths in the medium at the two frequencies ω and 2ω, and n_ω and $n_{2\omega}$ are the corresponding indices of refraction. If $n_{2\omega} = n_\omega$, then Eqs. (9-8) give $k_{2\omega} = 2k_\omega$ and $\lambda_{2\omega} = \lambda_\omega/2$. If this were the case, the field maxima for the fundamental wave would occur at the same positions z as the field maxima for the second harmonic wave, as illustrated in Fig. 9-9a. Since the fundamental wave continues to create newly radiated light as it propagates, this would ensure that the second harmonic light radiated by atoms at one position z would be in phase with the light radiated by atoms at a different z. The condition $n_\omega = n_{2\omega}$, then, would result in constructive interference of the different radiated waves, and efficient SHG.

Unfortunately, dispersion generally results in the condition $n_{2\omega} > n_\omega$, so that $k_{2\omega} > 2k_\omega$ and $\lambda_{2\omega} < \lambda_\omega/2$. This situation is depicted in Fig. 9-9b, which shows that the second harmonic waves from different atoms will now be out of phase. For atoms that are not too far apart, however, the phases of the second harmonic waves will still be sufficiently in phase to give efficient SHG. The efficiency will start to decrease when the radiation from different atoms is out of phase by more than 180°. This occurs for a propagation distance L_c, given by

$$L_c(k_{2\omega} - 2k_\omega) \equiv L_c\Delta k = \pi \tag{9-9}$$

where $\Delta k \equiv k_{2\omega} - 2k_\omega$ is the wave vector mismatch. Using Eqs. (9-8), the wave vector mismatch can be written as

$$\Delta k = (n_{2\omega} - n_\omega) \frac{2\omega}{c} \qquad \text{(wavevector mismatch)} \tag{9-10}$$

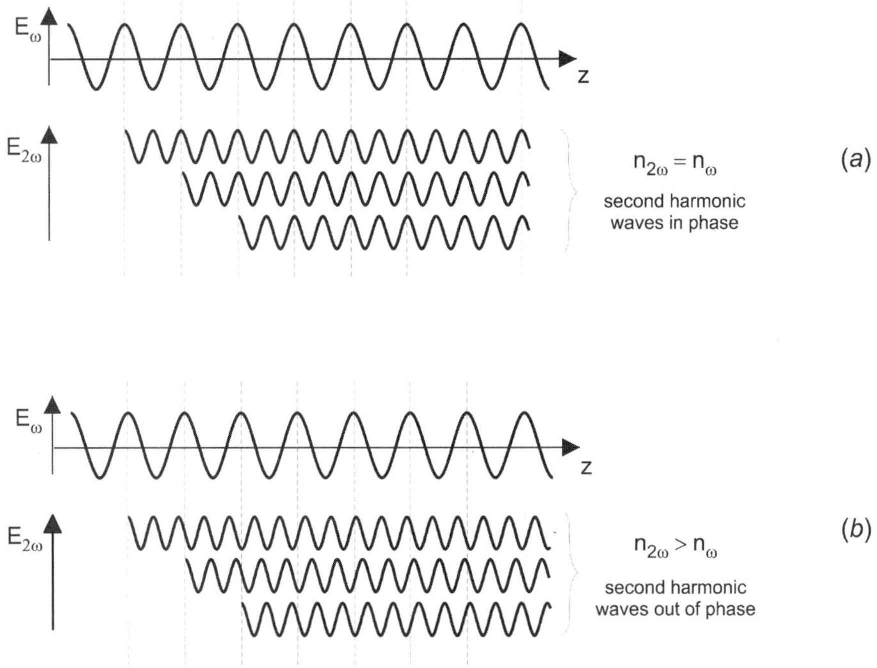

Figure 9-9 (a) The upper wave is the fundamental and the lower waves are the second harmonic waves radiated by different atoms in the material. Constructive interference occurs when the refractive index is the same at the two frequencies. (b) When the two indices are different, the waves get out of phase and add by destructive interference.

The optimum SHG conversion is obtained by passing the fundamental wave through a crystal of length equal to L_c. In most cases, this length is quite small, as illustrated by the following example.

EXAMPLE 9-1

Light of free-space wavelength 1064 nm from a Nd:YAG laser is passed through a lithium niobate ($LiNbO_3$) cystal for frequency doubling to 532 nm. The index of refraction at 1064 nm is 2.234, and at 532 nm is 2.325. Determine the optimum crystal length for SHG.

Solution: The optimum length is

$$L_c = \frac{\pi}{\Delta k} = \frac{\pi c}{2\omega \Delta n} = \frac{\lambda_0}{4\Delta n}$$

where λ_0 is the free space wavelength of the fundamental wave and $\Delta n \equiv n_{2\omega} - n_\omega$. This evaluates to

$$L_c = \frac{1064 \times 10^{-9}\text{ m}}{4(2.325 - 2.234)} = 2.92\ \mu\text{m}$$

The small optimum lengths indicated by the preceding example lead to very inefficient conversion of the fundamental frequency into the second harmonic. To increase the efficiency, some type of *phase matching* is usually employed, as explained below.

Phase Matching

For crystals with certain symmetries, the refractive index varies not only with wavelength, but also with the direction of polarization of the light wave. This *birefringence* provides a way of adjusting the indices to be equal at ω and 2ω. Consider a uniaxial crystal, which has refractive index n^o when the light wave's E field is in the xy plane (an "ordinary" wave), and index n^e when the E field is along the z axis (an "extraordinary" wave). Figure 9-10a shows a light wave propagating through such a crystal, with its **k** vector in the yz plane, making an angle θ with the z axis. When the wave's E field is along the x axis, the refractive index is the ordinary one n^o. When it is polarized perpendular to this, however, the E field has components along both y and z, and the refractive index is intermediate between n^o and n^e. The effective refractive index for this "extraordinary" wave will then depend on the angle θ.

It is this variation of index with propagation angle that provides a means for phase matching. Figure 9-10b shows the variation of refractive index with frequency for a pure ordinary wave, a pure extraordinary wave, and a mixture of the two at some angle θ. If the fundamental wave is polarized along x and the second harmonic wave polarized perpendicular to this, then phase matching will occur when θ is adjusted so that $n^o_\omega = n_{2\omega}(\theta)$. It is clear from Fig. 9-10b that for phase matching to work, the birefringence $(n^o - n^e)$ must be greater than the dispersion $(n^o_{2\omega} - n^o_\omega)$. Crystals commonly used for frequency doubling in the visible and near-IR regions are lithium niobate ($LiNbO_3$) and KDP.

The astute reader may object that this scheme should not work, because the fundamental and second harmonic waves are polarized perpendicular to each other, and as such should not interact at all. Although this would be true in an isotropic material, where **P** is in the same direction as **E**, it is not the case in the anisotropic materials we are discussing. Here, the tensor nature of χ allows a field in the x direction to give rise to polarization in any other direction, and this permits the coupling of the two waves.

An alternative approach to phase matching is that of *quasi-phase matching,* illustrated in Fig. 9-11. In this method, no attempt is made to match the refractive indices of the fundamental and second harmonic waves, and they get out of phase after propagating a distance L_c. However, the direction of the crystalline symmetry axis is now made to alternate spatially with a period equal to the coherence length L_c. This reverses the sign of the nonlinear susceptibility in a periodic fashion, and resynchronizes the fundamental and second harmonic waves when they have gotten out of phase.

The most common material used for quasi-phase matching is lithium niobate ($LiNbO_3$), which is *ferroelectric* (possessing a permanent electric dipole moment). The alternating crystalline symmetry is created by poling, a process in which a high voltage is applied for a short time to patterned electrodes deposited on the surface. The resulting structure is known as *periodically poled lithium niobate,* or PPLN (pronounced "piplin"), and is commercially available. It can be designed to work with wavelengths that are diffucult to phase match using conventional angle tuning.

Three-Wave Mixing

Second harmonic generation is actually a special case of the more general phenomenon of *three-wave mixing.* In this process, two photons interact to give rise to a third photon,

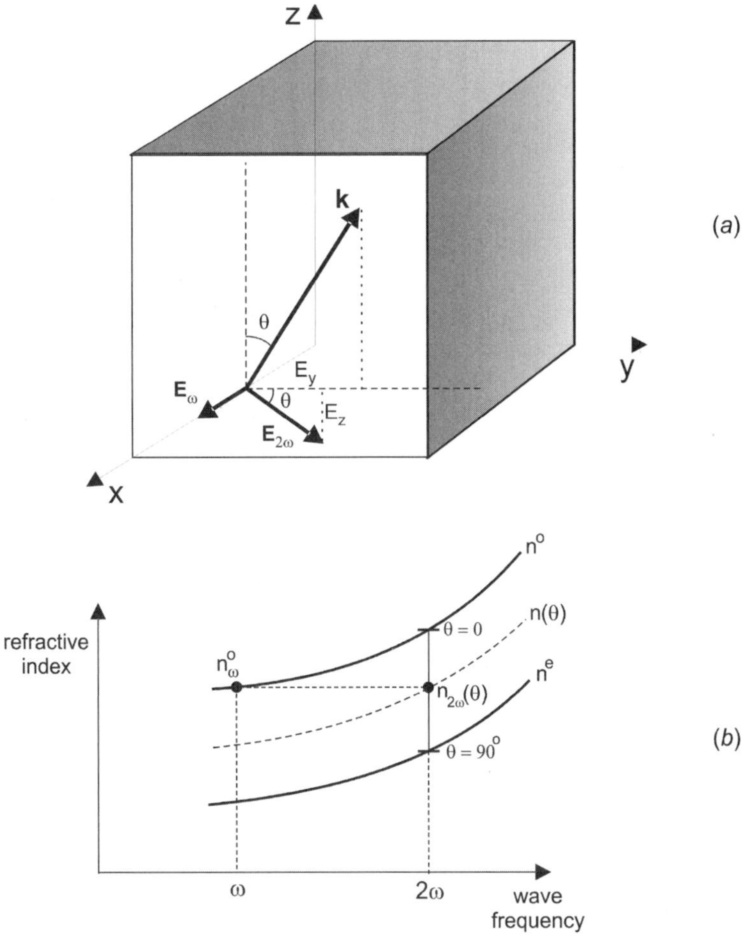

Figure 9-10 (a) For phase matching, the fundamental and second harmonic waves propagate in the same direction with wave vector **k**, but with different polarizations (fundamental along *x*, second harmonic perpendular to this). (b) The refractive index varies with frequency for both waves, but varies with θ only for the second harmonic. At the proper value of θ, the refractive indices for the two waves are the same.

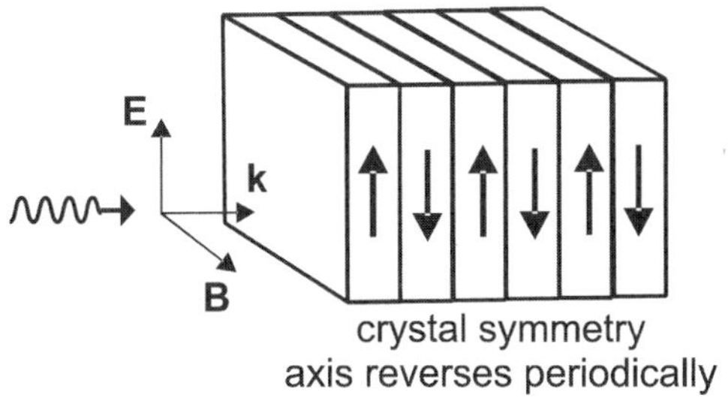

Figure 9-11 In quasi-phase matching, a nonlinear material is polled with a high voltage to produce a periodic variation in the crystalline symmetry. This periodically resets the phase relation between the fundamental and second harmonic waves so that on average they remain approximately in phase.

which may be of different frequency than either of the first two. To see how this works, consider two light waves with frequencies ω_1 and ω_2 that add together at a given point in a material to give a total electric field

$$E(t) = A_1 \cos \omega_1 t + A_2 \cos \omega_2 t \tag{9-11}$$

where A_1 and A_2 are the amplitudes of the two waves. Assuming that the material has a nonzero χ_2, the time-dependent polarization can be evaluated using Eq. (9-3), keeping terms up to order E^2. Writing out the terms and expanding E^2 gives

$$\begin{aligned}\frac{1}{\varepsilon_0} P(t) &= \chi_1[A_1 \cos \omega_1 t + A_2 \cos \omega_2 t] \\ &\quad + \chi_2[A_1^2 \cos^2 \omega_1 t + A_2^2 \cos^2 \omega_2 t + 2A_1A_2(\cos \omega_1 t)(\cos \omega_2 t)]\end{aligned} \tag{9-12}$$

The terms involving $\cos \omega_1 t$, $\cos \omega_2 t$, $\cos^2 \omega_1 t$, and $\cos^2 \omega_2 t$ give the refractive index, optical rectification, and frequency doubling for the two frequencies ω_1 and ω_2, just as in Eqs. (9-5)–(9-7). The new term here is the cross term $(\cos \omega_1 t)(\cos \omega_2 t)$, which can be expanded as

$$2(\cos \omega_1 t)(\cos \omega_2 t) = \cos(\omega_1 + \omega_2)t + \cos(\omega_1 - \omega_2)t \tag{9-13}$$

using the trig identity $2 \cos \theta_1 \cos \theta_2 = \cos(\theta_1 + \theta_1) + \cos(\theta_1 - \theta_1)$.

There are now components of the polarization density oscillating at the sum frequency $\omega_1 + \omega_2$ and the difference frequency $\omega_1 - \omega_2$, and these oscillations give rise to newly created waves at a third frequency $\omega_3 = \omega_1 \pm \omega_2$. When $\omega_3 = \omega_1 + \omega_2$, the process is termed *upconversion* (the frequency is shifted upward), and when $\omega_3 = \omega_1 - \omega_2$, the process is termed *downconversion* (the frequency is shifted downward). These are depicted in Fig. 9-12.

Although either upconversion or downconversion is possible, only one of these will normally be efficient in any given situation, due to the need for phase matching. The general condition for phase matching can be understood most easily from the quantum point

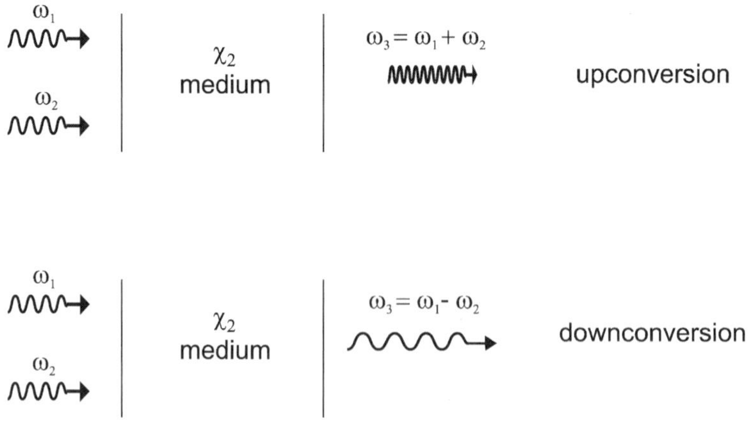

Figure 9-12 Two photons of frequencies ω_1 and ω_2 incident on a χ_2 medium can create a new photon at either frequency $\omega_3 = \omega_1 + \omega_2$ (upconversion) or $\omega_3 = \omega_1 - \omega_2$ (downconversion).

of view by considering momentum conservation for the interacting photons. Since a photon's momentum is $\mathbf{p} = \hbar\mathbf{k}$, the condition for upconversion becomes

$$\hbar\mathbf{k}_1 + \hbar\mathbf{k}_2 = \hbar\mathbf{k}_3 \qquad \text{(three-photon phase matching)} \tag{9-14}$$

where the magnitude of $\mathbf{k}$ is $k_i = n_i\omega_i/c$ for the ith photon. If the photons are all collinear (same direction) then this reduces to the scalar equation

$$n_1\omega_1 + n_2\omega_2 = n_3\omega_3 \tag{9-15}$$

If there is no dispersion ($n_1 = n_2 = n_3$), this is automatically satisfied by energy conservation, which in the quantum view states that the sum of all photon energies is the same before and after the interaction. For three photons this is

$$\hbar\omega_1 + \hbar\omega_2 = \hbar\omega_3 \qquad \text{(three-photon energy conservation)} \tag{9-16}$$

Dispersion is generally present, however, and this necessitates the phase matching methods discussed in the previous section for efficient operation.

Frequency doubling can be thought of as a special case of upconversion, where $\omega_1 = \omega_2$. In this case energy conservation becomes $2\hbar\omega_1 = \hbar\omega_3$, or simply $2\omega_1 = \omega_3$ (the frequency is doubled). Phase matching then reduces to $2n_1\omega_1 = n_3\omega_3$, or simply $n_1 = n_3$.

Another important application of three-wave mixing is the *optical parametric oscillator* (OPO), depicted in Fig. 9-13. The OPO consists of a χ_2 medium placed between two mirrors, with pump light from a laser incident on the medium through one of the mirrors. Inside the medium, a pump photon at frequency ω_P is split into a "signal" photon at ω_S and an "idler" photon at ω_I. The distinction between signal and idler waves is arbitrary, and simply separates the two generated wavelengths into one that is of interest (signal), and one that is extraneous (idler).

The interaction can be considered to be a downconversion process, with ω_P and ω_I mixing together to generate the difference frequency $\omega_S = \omega_P - \omega_I$. If there is no light initially in the crystal at frequency ω_I, it is the zero-point fluctuations of the idler's E field

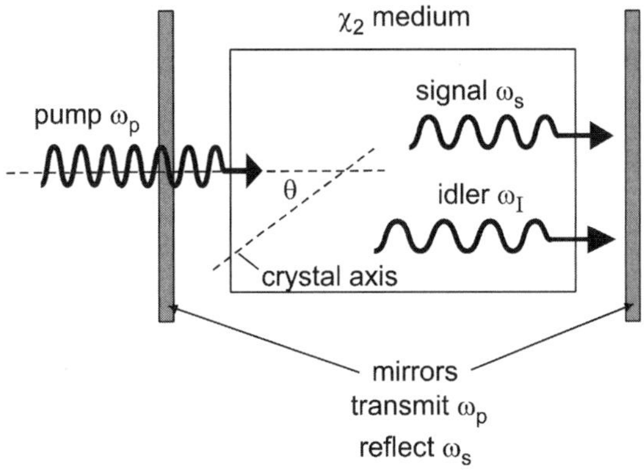

Figure 9-13 In an optical parametric oscillator (OPO), pump photons are converted into lower frequency signal and idler photons by a nonlinear χ_2 medium. Phase matching enables lasing at a frequency ω_S, which depends on the angle θ.

that mix with the pump to produce light at ω_S. This process is termed *parametric fluorescence,* and is purely quantum mechanical in nature. Once the idler wave has been established, however, then the addition of light to the signal wave can be understood semiclassically, as previously discussed.

The buildup of energy in the signal wave depends not only on the pump and idler power, but also on the power of the signal wave itself. This is the phenomenon of *stimulated emission,* which will be discussed in Chapter 18. The mirrors in the OPO allow the signal wave to propagate back and forth through the medium many times, all the while increasing in intensity. This can result in laser oscillation at frequency ω_S if the phase matching condition of Eq. (9-14) is satisfied. By varying the crystal angle θ, one can change the frequency at which phase matching occurs, and this provides a way of continuously tuning the laser light. Tunability over a wide frequency range is a key advantage of the OPO over other types of laser sources.

Four-Wave Mixing

Second harmonic generation and three-wave mixing can only occur in a χ_2 medium, which lacks a center of inversion symmetry. For isotropic materials such as glass, the lowest-order nonlinearity is due to χ_3, and the time-dependent polarization becomes

$$P(t) = \varepsilon_0[\chi_1 E(t) + \chi_3 E^3(t)] \qquad \text{(centrosymmetric medium)} \tag{9-17}$$

Consider the simplest case of a single wave of frequency ω incident on the material, with a time-dependent E field given by

$$E(t) = A \cos \omega t \tag{9-18}$$

Substituting this expression into Eq. (9-17), and using the trig identity $\cos^3 \theta = (3 \cos \theta + \cos 3\theta)/4$, we have

$$P(t) = P_0 + P_\omega \cos \omega t + P_{2\omega} \cos 2\omega t + P_{3\omega} \cos 3\omega t \tag{9-19}$$

where

$$\begin{aligned} P_0 &= 0 \\ P_\omega &= \varepsilon_0 A[\chi_1 + (3/4)\chi_3 A^2] \\ P_{2\omega} &= 0 \\ P_{3\omega} &= \tfrac{1}{4}\,\varepsilon_0 \chi_3 A^3 \end{aligned} \tag{9-20}$$

Since $P_0 = P_{2\omega} = 0$, there is no optical rectification or SHG in the χ_3 medium. However, there is a term oscillating at frequency 3ω, which will generate additional light at a frequency three times that of the incident light. This is *third harmonic generation* or frequency tripling, and can be used to produce even shorter wavelengths than SHG. However, the efficiency is low, and phase matching is more difficult than in SHG. In practice, shorter wavelengths are usually obtained by using SHG in successive steps (doubling the doubled frequency, etc.).

Third harmonic generation is a special case of the more general phenomenon known as *four-wave mixing*. In this process, three waves of frequencies ω_1, ω_2, and ω_3 mix together to generate additional waves at various sum and difference frequencies such as $\omega_1 + \omega_2 - \omega_3$, $\omega_1 - \omega_2 + \omega_3$, and so on. Many third-order nonlinear processes that generate new frequencies can be described in this way, including anti-Stokes Raman scattering (see Problem. 9.2). A full treatment of these is given in the more advanced texts listed in the Bibliography.

9-3. NONLINEAR REFRACTIVE INDEX

We saw in the previous section that a wave of frequency ω incident on a χ_3 medium will produce polarization oscillations at 3ω. Of equal (or perhaps greater) importance, however, are the additional polarization oscillations at frequency ω that are produced by the χ_3 nonlinearity. According to Eq. (9-20), the linear susceptibility χ_1 is replaced by an effective susceptibility

$$\chi_1' = \chi_1 + \tfrac{3}{4}\chi_3 A^2 \qquad \text{(effective } \chi_1\text{, optical Kerr effect)} \qquad (9\text{-}21)$$

which is a change $\Delta\chi_1 = \frac{3}{4}\chi_3 A^2$ from the usual linear susceptibility χ_1. Since it is the susceptibility χ_1 which determines the refractive index, this change in χ_1 results in a change in the refractive index. Taking the differential of Eq. (9-1) gives

$$\Delta n = \frac{1}{2\sqrt{1+\chi_1}}\Delta\chi_1 = \frac{\Delta\chi_1}{2n} = \frac{3}{8n}\chi_3 A^2 \qquad (9\text{-}22)$$

We see that the refractive index change is proportional to the square of the electric field amplitude, and that the magnitude of this change is governed by χ_3.

It is common (and useful) to relate the change in refractive index to the light intensity I rather than to the electric field. Using Eq. (2-9), the square of the electric field amplitude is

$$A^2 = \frac{2I}{cn\varepsilon_0} \qquad (9\text{-}23)$$

so the refractive index change becomes

$$\Delta n = \frac{3\chi_3 I}{4n^2 c\varepsilon_0} \qquad (9\text{-}24)$$

The refractive index can then be written in the simple form

$$n \equiv n_0 + n_2 I \qquad (9\text{-}25)$$

where n_0 is the refractive index at low intensity (the usual index of refraction), and

$$n_2 = \frac{3\chi_3}{4n^2 c\varepsilon_0} \qquad \text{(nonlinear refractive index)} \qquad (9\text{-}26)$$

is the *nonlinear refractive index.* If I is expressed in MKS units of W/m^2, then n_2 will be in units of m^2/W, so that $n_2 I$ is dimensionless.* This variation of refractive index with light intensity is sometimes referred to as the *optical Kerr effect.*

Table 9-2 lists typical values of n_2 for a few representative materials. In general, those that contain heavier elements such as lead (Pb) and tantalum (Ta) have a higher n_2 because the outer electrons are more weakly bound to the nucleus, and the atoms are therefore more easily polarized. In a polymer, electrons are relatively free to move along the direction of the carbon chain, resulting in high polarizability in that direction. The n_2 for molecular liquids such as CS_2 is particularly high due to the contribution from molecular orientation. CS_2 is often used as a standard for determining absolute values of n_2 from relative measurements.

An intensity-dependent refractive index has important consequences for the propagation of light through a material. The propagation constant now varies with intensity according to

$$k = \frac{n\omega}{c} = \frac{n_0\omega}{c} + \frac{n_2 I\omega}{c} = k_0 + \Delta k \tag{9-27}$$

where $k_0 \equiv n_0\omega/c$ is the low-intensity value, and $\Delta k \equiv n_2 I\omega/c$ is the change in k at high intensity. When the light propagates a distance L in the material, the intensity causes the phase to shift by

$$\Delta\phi = \Delta kL = \frac{n_2 I\omega}{c}L = \frac{2\pi n_2 IL}{\lambda_0} \tag{9-28}$$

where $\lambda_0 = 2\pi c/\omega$ is the free-space wavelength of the light. This intensity-dependent shift of phase is termed *self-phase modulation.* It is a type of self-action phenomenon in which the wave acts on itself, via the material medium. We previously encountered another type of self-action effect when discussing optical bleaching and optical limiting (Section 9-1). In that case, it was an absorption that varied with intensity, rather than a phase.

Self-phase modulation has a number of consequences, some of which lead to useful applications. In the following, we consider a few important examples.

Optical Switching

One potential application of self-phase modulation is in optical switching. Figure 9-14 shows one configuration that will accomplish this, the *Mach–Zehnder interferometer.* In this device, an input light wave is split by a 50% reflecting beam splitter and directed along two different paths. One path contains a nonlinear χ_3 medium and the other does not. The two light waves are then combined by a second beam splitter into a single wave, which is the output of the device.

If the light is coherent (see Chapter 15), the electric fields of the waves from the two paths will add together constructively or destructively, depending on the relative phase of the two waves when they recombine. If the E fields of the two waves are both maximum at the same time, the waves add constructively and the output is maximum. If one wave is shifted by π radians, however, then a maximum of one wave coincides with a minimum of the other wave, resulting in destructive interference and zero output. One source of

*Some authors use the alternative definition $n = n_0 + n_2 A^2/2$, in which case the units of n_2 are m^2/V^2.

Table 9-2 Nonlinear refractive indices for selected crystals

Material	λ (nm)	n_2 (10^{-20} m^2/W)
Pure silica	1300	2.4
Ge–silica	1300	2.6
Water	500–1000	4
Lead silicate glass	1000	30–70
Ta_2O_5	800	72
PPV polymer	880	80
As_2S_3	1320	170
CS_2	1000	310
GaAs	1000	3000

such a phase shift is any difference in path length for the two waves, as they propagate from the first beam splitter to the second one. A path-length change for one of the waves of only $\lambda_0/2$ will cause the output to go from maximum to minimum (or vice versa). Indeed, this sensitivity to path-length difference is the fundamental advantage of interferometers for measuring small displacements.

The other source of phase shift between the two waves is a change of refractive index along one of the paths. The light propagating along the lower path through the χ_3 medium has its phase shifted by an amount given in Eq. (9-28), whereas light propagating along the upper path does not. The device output therefore goes from maximum to minimum when $\Delta\phi = \pi$, and this occurs at an optical intensity I_π given by

$$\pi = \frac{n_2 I_\pi 2\pi L}{\lambda_0}$$

or

$$I_\pi = \frac{\lambda_0}{2n_2 L} \qquad \text{(switching intensity)} \qquad (9\text{-}29)$$

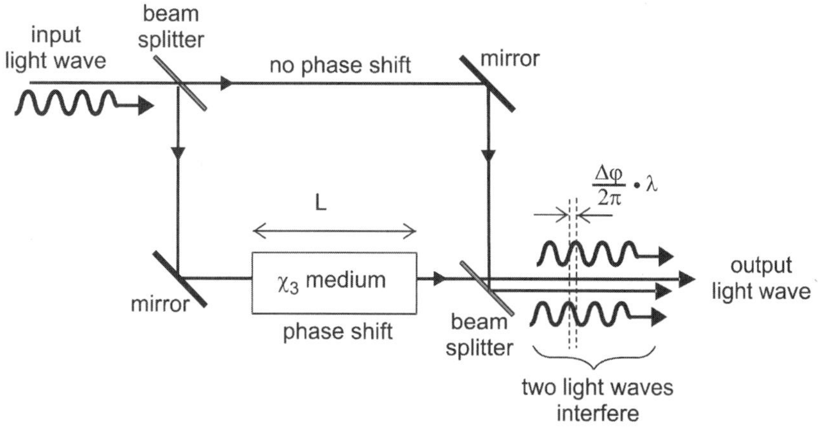

Figure 9-14 Self-phase modulation in one arm of a Mach–Zehnder interferometer can result in switching of the optical beam if the intensity is high enough to cause a 180° phase shift.

where L here is the length of the nonlinear medium in the direction of propagation. This method of switching is attractive because it is all optical (the light "switches itself") and very fast (fs time scale). However, very high intensities are needed, as seen in the following example.

EXAMPLE 9-2

A planar Ta_2O_5 rib waveguide has transverse dimensions 1×3 μm and a length of 1 cm. Determine the optical power needed for all-optical switching of 800 nm light.

Solution: The optical power is $P = IA$, where A is the cross-sectional area of the waveguide. Therefore,

$$P_\pi = \frac{\lambda_0 A}{2n_2 L} = \frac{(8 \times 10^{-7}\ \text{m})(3 \times 10^{-12}\ \text{m}^2)}{2(7.2 \times 10^{-19}\ \text{m}^2/\text{W})(10^{-12}\ \text{m})} = 167\ \text{W}$$

Optical powers this high that are cw (continuous-wave) would not be practical in an integrated optical circuit. However, if the light is in the form of pulses, the peak intensity can be high while still maintaining a sufficiently low average power to be compatible with integrated optics.

Pulse Chirping and Temporal Solitons

If the intensity of a light wave is constant in time (a cw beam), self-phase modulation has the effect of simply shifting the phase by a constant value. If the intensity varies in time, however, as it would in an optical pulse, the changing intensity causes a time-dependent phase shift. To see what effect this has, consider a pulse of frequency ω_0 that propagates in the $+z$ direction and enters a χ_3 medium of thickness L, as shown in Fig. 9-15. We will take the propagating wave in the medium to be of the form

$$E(z, t) = A \cos(\omega_0 t - kz) \qquad \text{(9-30)}$$

where

$$k = (n_0 + n_2 I)k_0 \qquad \text{(9-31)}$$

and $k_0 = \omega_0/c$. Taking $z = 0$ to be at the left edge of the medium (labeled point #1), the electric field there varies in time as

$$E_1(t) = A \cos \omega_0 t \qquad \text{(9-32)}$$

At $z = L$ (point #2), where the wave leaves the medium, the field can be evaluated using Eqs. (9-30)–(9-31) to be

$$E_2(t) = A \cos [\phi(t)] \qquad \text{(9-33)}$$

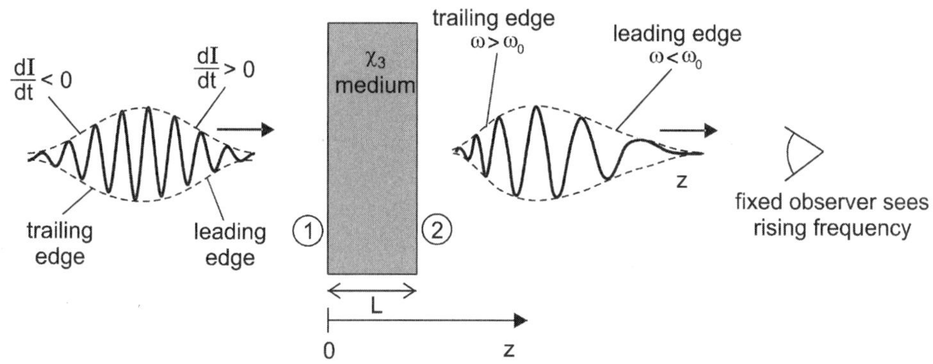

Figure 9-15 In a nonlinear medium, the instantaneous frequency in a pulse is lowered when the intensity is increasing in time, and raised when the intensity is decreasing. This gives a frequency chirp to the pulse, in which the leading edge is red-shifted and the trailing edge is blue-shifted.

where

$$\phi(t) = \omega_0 t - [n_0 + n_2 I(t)]k_0 L \tag{9-34}$$

is the time-dependent phase of the wave.

Writing the time dependence of $E_2(t)$ in terms of a time-dependent phase is useful because, from a fundamental point of view, the frequency is defined as the rate at which the phase is changing.* The optical frequency is therefore

$$\omega \equiv \frac{d\phi}{dt} = \omega_0 - n_2 k_0 L \frac{dI}{dt} \qquad \text{(instantaneous frequency)} \tag{9-35}$$

If the intensity is constant in time ($dI/dt = 0$), then $\omega = \omega_0$, and the frequency is simply the coefficient of t in the cosine function of Eq. (9-30). This is the usual notion of frequency for a sinusoidal wave. If the intensity is changing in time, however, then the frequency of the wave differs from ω_0. Furthermore, if dI/dt is itself changing in time, then the frequency ω changes in time. We should therefore interpret Eq. (9-35) as giving the instantaneous frequency of the wave.

Consider now how this changing frequency affects the pulse of Fig. 9-15 as it passes through the nonlinear medium. When the leading edge of the pulse enters the medium, $dI/dt > 0$, so that $\omega < \omega_0$. When the trailing edge of the pulse is passing through the medium, however, $dI/dt < 0$, and in this case $\omega > \omega_0$. The center of the pulse is unshifted in frequency, since $dI/dt = 0$ there. We can summarize this by saying that the leading edge is "red-shifted" (shifted to lower frequency, or longer wavelength), while the trailing edge is "blue-shifted" (shifted to higher frequency, or shorter wavelength). The spatial variation of the pulse's E field after it has passed through the medium is depicted in Fig. 9-15.

An observer at a fixed position z who monitored this pulse as it went by would first see the leading edge, with its lowered frequency, and then the trailing edge, with its increased frequency. The result would be a measured frequency that changes in time, a phenomonon termed *frequency chirp.* Since the frequency increases in time in this example,

*Remember that in this chapter, "frequency" means angular frequency ω.

it is called a positive chirp. Positive chirp results from our assumption that $n_2 > 0$, which is usually the case. Negative n_2 can occur at frequencies near a strong material resonance, in which case there would be negative chirp (frequency decreasing in time).

This separation of a pulse into low- and high-frequency regions is reminiscent of the phenomenon of material dispersion that was discussed in Chapter 6. We found there that in the normal dispersion regime ($\lambda_0 < 1300$ nm in silica glass), the material dispersion parameter D_m in Eq. (6-10) is negative. This means that longer wavelengths travel faster and are shifted toward the leading edge of the pulse, whereas shorter wavelengths travel slower and are shifted back toward the trailing edge. Dispersion, therefore, results in the same spatial separation of frequencies that we observe in the nonlinear pulse of Fig. 9-15. We can then think of the nonlinear medium as creating an intensity-dependent dispersion, which adds to the actual material dispersion to broaden a pulse even further.

For wavelengths longer than 1300 nm in silica glass, the material dispersion is anomalous—longer wavelengths travel slower and tend to move toward the trailing edge of the pulse. The nonlinear index n_2 remains positive for these longer wavelengths, however, and the nonlinearity continues to force longer wavelengths toward the leading edge. At some critical value of intensity, these two tendencies will cancel, and there will be no movement of different frequencies to different parts of the pulse. The resulting pulse preserves its shape and frequency distribution as it propagates, and is known as an *optical soliton.*

Optical solitons have considerable potential in fiber optic communications because they eliminate one of the primary limits on span length—that of dispersion. It might be thought that solitons would be rather fragile entities, breaking up when the light intensity becomes less than the required value. However, they turn out to be actually rather robust, since the critical quantity for maintaining the soliton is not the intensity alone, but rather the product of peak electric field magnitude and pulse duration. As the pulse loses energy due to attenuation, it broadens slightly in time, but remains a soliton. Periodic replenishment of these energy losses in an optical amplifier keeps the pulse energy and corresponding pulse duration in the desired range.

Pulse Compression

Frequency chirping can be used to reduce the duration of an optical pulse, using the scheme shown in Fig. 9-16. In this procedure, known as *pulse compression,* an input pulse is first broadened and chirped by a χ_3 medium. This broadened pulse is then sent to a pair of reflective diffraction gratings that are oriented parallel to each other. Because of the geometry of the gratings and the diffraction grating condition (Eq. 2-28), longer wavelengths must travel a greater distance through the grating pair combination than shorter wavelengths (see Problem 9.3). The longer wavelengths therefore become relatively more delayed in time, which has the effect of making the longer wavelength components (initially near the leading edge) move toward the trailing edge of the pulse. As a result, the duration of the pulse can be considerably reduced. A pair of prisms can be arranged to give the same effect.

Using pulses from a Ti:sapphire laser, compressed pulse durations under 5 fs have been achieved at wavelengths ~ 780 nm. This corresponds to about two cycles of oscillation of the E field during the pulse, a number small enough to call into question its description as a wave. The spectral width of such a pulse* is $\Delta\nu \sim 1/\Delta t = 2 \times 10^{14}\ \text{s}^{-1}$, which

*See Appendix B on the Fourier transform.

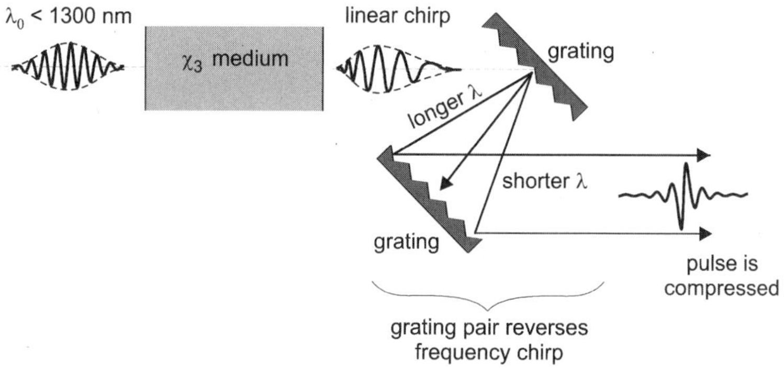

Figure 9-16 A chirped pulse can be compressed by a pair of diffraction gratings, due to the different path lengths for longer and shorter wavelengths.

corresponds to a wavelength spread $\Delta\lambda \sim 400$ nm. Pulses this short have such an ill-defined wavelength that they are of limited utility in optical communications. However, they are very useful for probing rapid physical and chemical processes that occur in photonic materials.

Self-Focusing and Spatial Solitons

In the previous examples we have seen how a nonlinear refractive index can lead to changes in the time dependence of an optical pulse. We consider now another consequence of the nonlinear index, in which the spatial profile of a propagating beam is modified. The basic idea is illustrated in Fig. 9-17, which shows a beam of finite lateral extent incident on a nonlinear χ_3 medium. The wavefronts propagate through the medium with phase velocity

$$v_p = \frac{c}{n} = \frac{c}{n_0 + n_2 I} \tag{9-36}$$

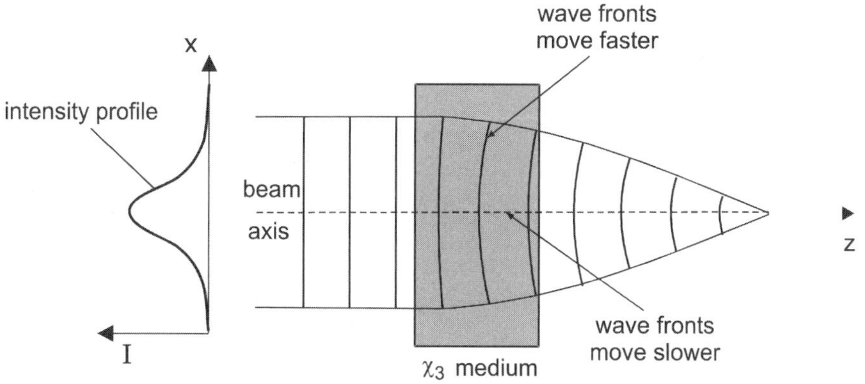

Figure 9-17 The variation of optical intensity across a beam's transverse profile causes a spatially varying refractive index in a nonlinear χ_3 medium, and this acts to focus the light.

where Eq. (9-25) has been used. Since portions of the wave front farther from the beam axis have a smaller intensity I than portions of the wave front near the axis, they move relatively faster there, distorting the wave front as shown. The curvature of the wave front makes the beam converge to a point, just as if a lens had been placed in the path of the beam. Because of its similarity to the focusing of light by a lens, this phenomenon is referred to as *self-focusing*.

Self-focusing has some undesirable effects on the propagation of very high power laser pulses through a material. As the beam becomes narrower due to self-focusing, the intensity increases ($I = P/A$ with P constant and A decreasing). This leads to even stronger self-focusing, which in turn narrows the beam more rapidly still. These two effects feed on each other and soon result in beam breakup and optical damage. This becomes an issue, for example, in proposed schemes that use high-power lasers to initiate nuclear fusion. The problem can be minimized by employing laser glasses with a low n_2, and by spreading the beam over a large cross-sectional area A to limit the intensity.

Under certain conditions, the collapse of a beam by self-focusing can be controlled and used to advantage. This is possible because of diffraction, which is the natural tendency of optical beams to diverge (see Chapter 2, Section 2-3). The smaller the beam diameter becomes, the more it tries to spread out by diffraction. If the beam intensity has the right spatial profile, then at some value of beam diameter the two effects of self-focusing and diffraction can exactly cancel one another. This leads to a beam of constant diameter, as depicted in Fig. 9-18, which is known as a *spatial soliton*. It is the spatial counterpart to

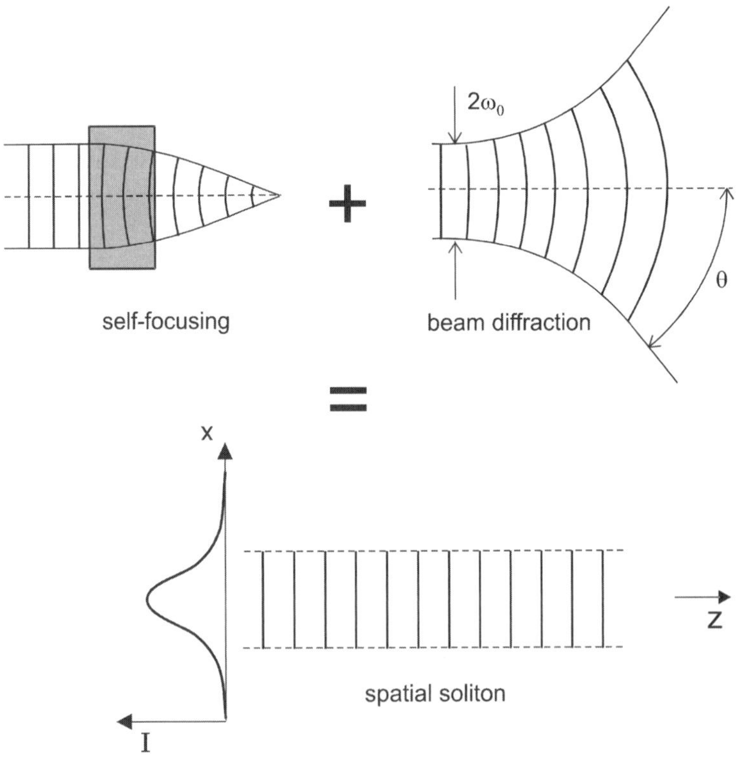

Figure 9-18 The combination of self-focusing and diffraction can lead to a spatial soliton, in which the beam's transverse profile remains constant with position.

the temporal soliton discussed earlier, and has potential for applications in all-optical switching and signal processing.

Another application that takes advantage of self-focusing is the optical switching process depicted in Fig. 9-19. An aperture is placed in the path of a diverging beam, which allows only a small fraction of the light to be transmitted when the optical power is low. When the optical power is increased, self-focusing narrows the divergence of the beam, so more light is transmitted through the aperture. As a result, the fraction of light transmitted by the aperture varies with beam power in the manner shown in Fig. 9-19b. This power-dependent transmission efficiency is similar to what would be obtained in the optical bleaching process discussed earlier. Since it depends on the optical Kerr effect, this type of arrangement is referred to as a *Kerr lens shutter.* We will see in Chapter 22 how it can be used to create very short laser pulses.

9-4. ELECTROOPTIC EFFECTS

In the preceeding section, we saw how a light wave of high intensity propagating through a nonlinear medium can change the refractive index of that medium. We turn now to a related phenomenon, in which a material's refractive index is changed by applying a static (dc) electric field. This is known as the *electrooptic effect,* and can be considered as a special case of multiwave interaction in which one of the waves has zero frequency. Depending on the material, the refractive index can vary linearly with field (*Pockels effect*) or quadratically with field (*electrooptic Kerr effect*). In general, the refractive index varies with applied field E_0 as

$$n = n_0 + aE_0 + bE_0^2 \tag{9-37}$$

In the following, we show how the coefficients a and b are related to the nonlinear susceptibilities χ_2 and χ_3, and discuss some typical applications.

Pockels Effect

Consider a light wave of frequency ω incident on a nonlinear medium as shown in Fig. 9-20. The light wave is propagating in the $+x$ direction, and enters the medium of thickness

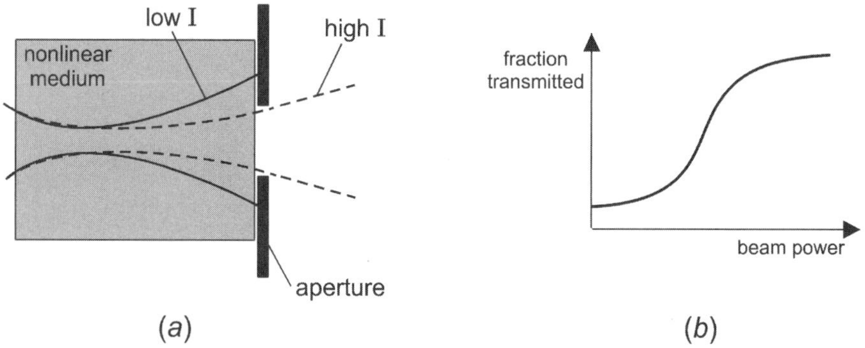

Figure 9-19 In a Kerr lens shutter, the fraction of light transmitted through a fixed aperture increases at high beam power due to self-focusing.

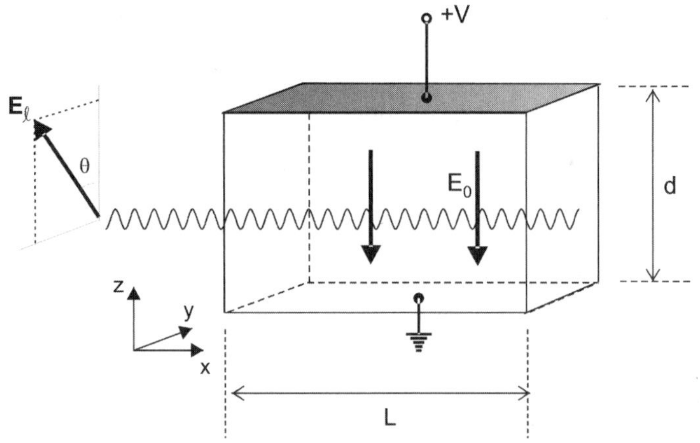

Figure 9-20 Light wave polarized in the *y–z* plane, propagating along the *x* axis. The change in refractive index is greatest for $\theta = 0$ and smallest for $\theta = 90°$.

d and length L. A voltage V is applied across the distance d, creating a static electric field of magnitude $E_0 = V/d$. The incident light can be polarized at any angle θ, but for now we assume that it is polarized along the z axis ($\theta = 0$), in the same direction as the static field. The total field at a point inside the medium can then be written as

$$E(t) = E_0 + A \cos \omega t \tag{9-38}$$

where A is the amplitude of the light wave's E field.

The time-dependent electric field creates a time-dependent polarization in the medium, given by Eq. (9-3). In a material that lacks inversion symmetry, $\chi_2 \neq 0$, and the polarization to lowest order is

$$P = \varepsilon_0\chi_1(E_0 + A \cos \omega t) + \varepsilon_0\chi_2(E_0 + A \cos \omega t)^2 \tag{9-39}$$

Writing this in the form of Eq. (9-6), we find that the oscillation in polarization at frequency ω has an amplitude

$$\begin{aligned} P_\omega &= \varepsilon_0\chi_1 A + \varepsilon_0 2\chi_2 E_0 A \\ &= \varepsilon_0\chi_1' A \end{aligned} \tag{9-40}$$

where the effective susceptibility χ_1' here is

$$\chi_1' = \chi_1 + 2\chi_2 E_0 \qquad \text{(effective } \chi_1\text{, Pockels effect)} \tag{9-41}$$

The effect of the static E field is therefore to change the susceptibility by an amount $\Delta\chi_1 = 2\chi_2 E_0$. Following the steps of Eq. (9-22), this changes the refractive index by

$$\Delta n = \frac{1}{n}\chi_2 E_0 \tag{9-42}$$

The coefficient of E in Eq. (9-37) is therefore $a = \chi_2/n$. It is customary to write the refractive index change in the Pockels effect as

$$\Delta n \equiv \tfrac{1}{2} n^3 r E_0 \qquad \text{(index change, Pockels effect)} \qquad (9\text{-}43)$$

where r is the *Pockels coefficient.* Comparing Eqs. (9-42) and (9-43), we see that $r = (2/n^4)\,\chi_2$.

The analysis leading up to Eq. (9-42) is somewhat simplified, since it assumes that the susceptibility χ_2 is a scalar. If this were the case, then an applied static field in the z direction would only interact with light waves polarized with their E field along z. However, χ_2 is actually a tensor quantity, which means that in general the applied static field changes the refractive index for light polarized along either x, y, or z. Furthermore, the refractive index changes differently for each polarization, so that there is a field-induced birefringence in the medium.

These polarization-dependent effects are well illustrated by lithium niobate, an important electrooptic material. It is a uniaxial crystal, with one axis of symmetry (usually taken as the z axis). The refractive index has the value n_e when the light wave's E field is parallel to the z axis, and the value n_o when the light wave's E field is in the xy plane.* Assume first that the applied static field is along z. If the light wave's E field is also along z, then the appropriate Pockels coefficient is $r_{33} = 30.8 \times 10^{-12}$ m/V (30.8 pm/V), and the field-induced change in index is $\Delta n = n_e^3 r_{33} E_0/2$. If the light wave's E field is in the xy plane, however, then the appropriate Pockels coefficient is $r_{13} = 8.6$ pm/V, and $\Delta n = n_o^3 r_{13} E_0/2$.

Other Pockels coefficients describe the index changes for a static field in the xy plane. For example, if light is propagating along the z axis, there would normally be no birefringence, since the index is the same (n_o) for polarization along either x or y. If a static field is applied in the xy plane, however, it reduces the index for one polarization by $n_o^3 r_{22} E_0/2$ and increases it by the same amount for the other. This results in an index difference between opposite polarizations (birefringence) of $n_o^3 r_{22} E_0$, where $r_{22} = 3.4$ pm/V for lithium niobate. Certain applications take advantage of this induced birefringence, while others simply utilize the change in index for one polarization. The value of r to use in Eq. (9-43) depends on the material as well as on the crystal orientation and the type of application. Typical values for r are in the range $10^{-12} \rightarrow 10^{-10}$ m/V (1–100 pm/V).

Figure 9-21 shows an application of the Pockels effect that allows electrically controlled modulation or switching of a light wave in a planar waveguide. Light entering from the left is split at the Y junction into two paths, one of them (the upper path) passing through a static electric field region. The field in this region is controlled by the voltage applied across a pair of parallel electrodes. Light from the two paths then comes together in the second Y junction, and the two component waves combine to give a single output wave.

This arrangement constitutes an integrated-optic version of the Mach–Zehnder interferometer, and the device operates in much the same way as the optical Kerr effect switch discussed on page 142. In both cases, the output changes from zero to maximum when the light propagating along one of the paths experiences a phase shift of π radians. The difference is that here the phase shift arises from an applied static field, rather than from the intensity of the light itself. From the point of view of the incident light wave, the Pockels effect modulator exhibits a linear optical response, in the sense that the optical output

*Subscripts are used here on n_e and n_o, rather than the superscripts n^e and n^o used previously.

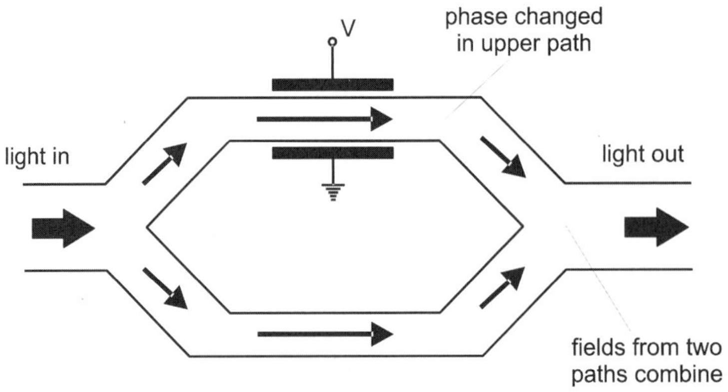

Figure 9-21 A Mach–Zehnder interferometer is formed from lithium niobate planar waveguides. A phase shift in one path changes the amplitude of the output wave, which enables electrical modulation of the light wave.

varies linearly with the optical input. It is nonlinear only in the sense that the modulator's output is changed by applying a voltage. This optically linear behavior is an advantage in parallel processing of multiple signals, because each signal can then be processed independently without being affected by the presence of other signals. Electrooptic modulators such as this are now commonly used for fast modulation and switching in high-speed optical communications.

EXAMPLE 9-3

An integrated-optical waveguide modulator uses $LiNbO_3$ (lithium niobate) in a Mach–Zehnder configuration. The crystal is oriented so that light propagates along the x axis, with the applied field and the light wave's field both along the z axis. The waveguide width (and spacing of electrodes) is 20 μm, and the length of the electrode region is 1 cm. Find the voltage V_π necessary to switch the output from low to high for light of free-space wavelength 1500 nm.

Solution: The condition for switching is $\Delta kL = \pi$, or

$$\frac{2\pi}{\lambda_0}\Delta nL = \pi$$

The index change here depends on Pockels coefficient r_{33}, so

$$\frac{2\pi}{\lambda_0}\left(\frac{1}{2}n_e^3 r_{33}\frac{V_\pi}{d}\right)L = \pi$$

or

$$V_\pi = \frac{d\lambda_0}{n_e^3 r_{33} L}$$

Using the value $n_e = 2.139$ at $\lambda_0 = 1500$ nm,

$$V_\pi = \frac{(20 \times 10^{-6}\ \text{m})(1.5 \times 10^{-6}\ \text{m})}{(2.139)^3(30.8 \times 10^{-12}\ \text{m/V})(10^{-2}\ \text{m})} = 9.9\ \text{V}$$

This result shows that relatively modest voltages are needed for electrooptic modulation in a planar waveguide configuration.

Another method for modulating a light wave's amplitude is illustrated in Fig. 9-22. An electrooptic crystal with electrodes attached (a *Pockels cell*) is placed between two crossed polarizers, and oriented with the z axis along the direction of light propagation. If no voltage is applied between the electrodes, the refractive index of a uniaxial crystal is the same for any polarization of the light, and this polarization is preserved as it passes through the crystal. Since the polarizers are oriented with their transmission axes perpendicular to each other, no light is transmitted by the second polarizer.

When voltage is applied, the electric field induces a birefringence in the crystal such that the indices of refraction along the x and y axes are different. If the incident light's E field were initially oriented along either x or y, then it would remain oriented in this direction, the phase velocity being determined by the corresponding refractive index. However, the first polarizer is arranged so that the incident E field is along the axis x', which is at 45° from the x and y axes. To see how the light propagates in this case, we must decompose the incident $\mathbf{E}_1$ vector into components along the crystal axes x and y, and treat the propagation of each of these components separately.

Just before entering the crystal, the field has components

$$E_{1x} = (A/\sqrt{2}) \cos \omega t$$
$$E_{1y} = (A/\sqrt{2}) \cos \omega t \qquad (9\text{-}44)$$

where the incident field magnitude is taken to vary in time as $E_1(t) = A \cos \omega t$. The two components propagate with different phase velocities, and accumulate a phase difference

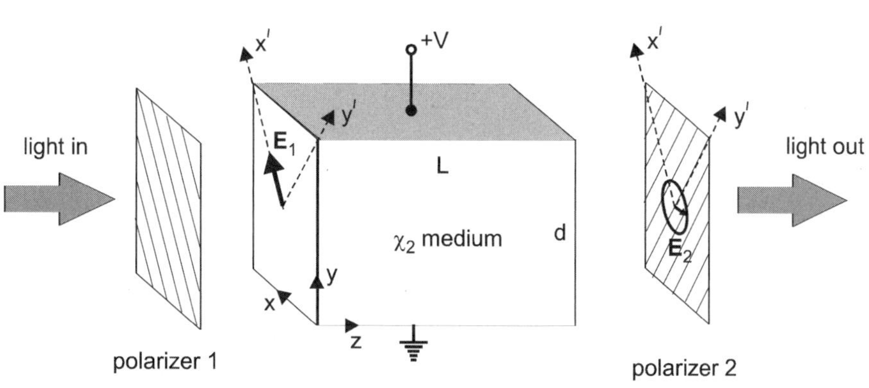

Figure 9-22 In a Pockels cell, the applied voltage changes the light's polarization from linear to elliptical. When it is placed between crossed polarizers, this can be used to modulate the transmitted light intensity.

$$\begin{aligned}\Delta\phi &= \frac{2\pi}{\lambda_0}\Delta nL \\ &= \frac{2\pi}{\lambda_0}\left(n_o^3 r_{22}\frac{V}{d}\right)L\end{aligned} \tag{9-45}$$

after propagating a distance L through the crystal. The field $\mathbf{E}_2$ of the light wave exiting the crystal therefore has x and y components

$$\begin{aligned}E_{2x} &= (A/\sqrt{2})\cos\omega t \\ E_{2y} &= (A/\sqrt{2})\cos(\omega t + \Delta\phi)\end{aligned} \tag{9-46}$$

where we have suppressed the phase change that the two components experience in common. A light wave with the time dependence of Eq. (9-46) is termed *elliptically polarized* because the direction of the field $\mathbf{E}_2$ rotates in time, the tip of $\mathbf{E}_2$ tracing out an ellipse in space. When $\Delta\phi = \pi/2$, the light is said to be *circularly polarized.*

It is this change in the light's polarization from linear to elliptical that allows some light to get through the second polarizer. The elliptically polarized field has some component along the transmission axis y' of polarizer 2, whereas the initial field was linearly polarized along x', and had no component along y'. The component of $\mathbf{E}_2$ along y' can be evaluated as

$$\begin{aligned}E_{2y'} &= \frac{1}{\sqrt{2}}E_{2y} - \frac{1}{\sqrt{2}}E_{2x} \\ &= \frac{A}{2}[\cos(\omega t + \Delta\phi) - \cos\omega t] \\ &= -A\sin\left(\omega t + \frac{\Delta\phi}{2}\right)\sin\left(\frac{\Delta\phi}{2}\right)\end{aligned} \tag{9-47}$$

where the trigonometric identity $\cos\alpha - \cos\beta = -2\sin(\alpha + \beta)/2\sin(\alpha - \beta)/2$ has been used. This represents a sinusoidal oscillation of frequency ω and amplitude $A' = A\sin(\Delta\phi/2)$. Since the light intensity is proportional to E^2, the fraction of incident light which is transmitted by the second polarizor is

$$\frac{I_{out}}{I_{in}} = \left(\frac{A'}{A}\right)^2 = \sin^2\left(\frac{\Delta\phi}{2}\right) \tag{9-48}$$

According to Eq. (9-48), the fraction of light transmitted increases as the phase shift $\Delta\phi$ increases from zero. When $\Delta\phi = \pi$, the transmitted intensity becomes a maximum, and then it decreases when $\Delta\phi$ is increased further. The condition $\Delta\phi = \pi$ is equivalent to a simple change of sign in Eq. (9-46), with $E_{2y} = -(A/\sqrt{2})\cos\omega t$. This corresponds to linear polarization for the light field $\mathbf{E}_2$, but with a polarization axis 90° different than that of the incident field $\mathbf{E}_1$. Under these conditions, the induced birefringence has rotated the plane of polarization in just the right way so that all the light is transmitted by the second polarizer.

The voltage V_π required to induce a π phase shift is found from Eq. (9-45) to be

$$V_\pi = \frac{\lambda_0 d}{2n_o^3 r_{22}L} \qquad \text{(switching voltage, Pockels cell)} \tag{9-50}$$

and the transmission characteristics for the Pockels cell modulator can be written in the form

$$I_{out} = I_{in} \sin^2\left(\frac{\pi}{2}\frac{V}{V_\pi}\right) \quad \text{(Pockels cell transmission)} \tag{9-50}$$

The Pockels cell can be used in two different ways to modulate a light signal. In the first application, a small-signal modulating voltage $V_m(t)$ is added to a dc bias voltage V_0 to give a total voltage $V(t) = V_0 + V_m(t)$. If $V_m(t)$ is sufficiently small, the change in optical output will be linearly proportional to $V_m(t)$, the desired relation for analog signal processing. In the second type of application, the Pockels cell operates in a switching mode, where $V(t)$ changes in steps of V_π. This would be appropriate for digital applications, for example, in which the "signal" is either on or off. As we will see in Chapter 22, the switching mode is also useful in generating short laser pulses.

Kerr Electrooptic Effect

The Pockels effect is absent in materials having inversion symmetry, such as liquids and glasses, because $\chi_2 = 0$. In these materials, there is still an electrooptic effect, but it is a higher-order process that depends on χ_3. To see how it arises, we write the total field in the material as the sum of a static applied field E_0 and the light wave's field $A \cos \omega t$, as in Eq. (9-38). Substituting this into Eq. (9-17) for the time-dependent polarization, it is straightforward to show (see Problem 9.4) that the component of polarization oscillating at frequency ω has amplitude

$$P_\omega = \varepsilon_0 \chi_1' A \tag{9-51}$$

where

$$\chi_1' = \chi_1 + 3\chi_3 E_0^2 \quad \text{(effective } \chi_1\text{, Kerr electrooptic effect)} \tag{9-52}$$

The applied field E_0 therefore changes the susceptibility by $\Delta\chi_1 = 3\chi_3 E_0^2$. Following the steps of Eq. (9-22), we find a corresponding index change

$$\Delta n = \frac{3\chi_3}{2n} E_0^2 \tag{9-53}$$

This quadratic variation of refractive index with applied static E field is termed the *Kerr electrooptic effect.** Comparing Eq. (9-53) with the definition $\Delta n = bE_0^2$ from Eq. (9-37), we identify $b = 3\chi_3/(2n)$. It is conventional to write the change of index as $\Delta n = K\lambda_0 E_0^2$, where K is the *Kerr electrooptic coefficient*. However, it is b and not K that is most nearly independent of wavelength, and so b is the more useful parameter when comparing different materials. Table 9-3 shows a few typical values of b in selected liquids and glasses. Glasses have a larger b when they contain heavy metal components that are highly polarizable, and liquids have a larger b when they contain molecules with a permanent dipole moment.

*This should be distinguished from the optical Kerr effect discussed earlier. Unfortunately, both of these are sometimes referred to as the "Kerr effect."

Table 9-3 Kerr electrooptic properties for some materials

Material	$b = K\lambda$ $(10^{-20}\ m^2/V^2)$
Nitrobenzene	206
Water	1.85
Glass	10^{-2}–1

A device in which a material is made birefringent via the Kerr electrooptic effect is called a *Kerr cell,* and it can serve the same function as a Pockels cell. The advantage of the Kerr cell is that it can use a variety of isotropic materials such as liquids and glasses. The disadvantage is that the response is not linear with applied voltage, and that very high voltages are needed. Most electrooptic modulators today utilize Pockels cells rather than Kerr cells.

PROBLEMS

9.1 Estimate the rms value of the electric field in sunlight having intensity 1 kW/m^2, using Eq. (2-9). ($E_{rms} = E_{peak}/\sqrt{2}$). Compare this with a typical atomic electric field obtained from Coulomb's law at a distance 0.1 nm from a single proton charge.

9.2 Two light waves with frequencies ω_L and ω_S are incident on a nonlinear χ_3 material, with $\omega_L > \omega_S$. If the combined E field in the crystal is taken as $E(t) = A_L \cos \omega_L t + A_S \cos \omega_S t$, use Eq. (9-17) to show that there is a generated lightwave of frequency $\omega_{AS} = \omega_L + (\omega_L - \omega_S)$. This can be used to describe anti-Stokes Raman scattering.

9.3 Use Eq. (2-28) to show that longer wavelengths take a longer path through the grating pair in Fig. 9-16.

9.4 Show that Eqs. (9-51) and (9-52) follow from substituting the field $E(t) = E_0 + A \cos \omega t$ into Eq. (9-17).

9.5 Both the Kerr electrooptic coefficient K and the nonlinear refractive index n_2 are related to the third-order susceptibility χ_3. Obtain a relation between b and n_2, and use this to compare the values given for water in Tables 9-2 and 9-3. Explain any difference by considering the time response of the different contributions to χ_3 (such as electronic, molecular rotations, etc.).

9.6 A He–Ne laser beam of power 1 mW is focused to a beam diameter of 0.2 mm in a $LiNbO_3$ crystal. Use Eq. (2-9) to calculate the peak E field in the crystal (use the ordinary index n_o), and use this value of E to calculate P_ω and $P_{2\omega}$. How do they compare?

9.7 For the previous problem, calculate the ratio $P_{3\omega}/P_\omega$, and compare this with the ratio $P_{2\omega}/P_\omega$. Assume that the value of n_2 for $LiNbO_3$ is the same as that for Ta_2O_5.

9.8 In our description of second harmonic generation, we considered only the conversion of a strong wave at ω into a new wave at 2ω. However, this new wave at 2ω can subsequently interact with the original wave at ω, generating still additional waves. Explore this phenomenon by considering a total E field in the material given by

$$E(t) = A_\omega \cos \omega t + A_{2\omega} \cos(2\omega t + \delta)$$

Show that this leads to additional polarization oscillations at ω and 3ω, and that the new wave at ω can interfere constructively or destructively with the original wave at ω, depending on the phase angle δ.

9.9 In writing the power series expansion for polarization given in Eq. (9-3), it is assumed that $\chi_2 E \ll 1$. This assumption may break down, however, for sufficiently high optical intensity. (a) For $LiNbO_3$ illuminated with 1 μm light, determine the optical intensity at which $\chi_2 E = 0.01$. (b) If the beam is focused to a diameter of 2 μm in the material, what optical power does this represent? (c) If the light source consists of pulses of duration 10 ns, what is the corresponding energy per pulse?

9.10 A crystal of $CdGeAs_2$ is used for second harmonic generation of 10.6 μm light. Determine the wave vector mismatch Δk between the waves at ω and 2ω if the light propagates along the z axis. From this, determine the crystal length at which the efficiency of SHG reaches a maximum.

9.11 Light at 1.064 μm is converted to 0.532 μm by SHG in a KDP crystal. Phase matching is achieved using the scheme illustrated in Fig. 9-10. As the angle θ is varied by tilting the crystal, the index $n_\omega = n_\omega^o$ remains constant, whereas the index at 2ω is given by

$$\left(\frac{1}{n_{2\omega}(\theta)}\right)^2 = \left(\frac{\cos\theta}{n_{2\omega}^o}\right)^2 + \left(\frac{\sin\theta}{n_{2\omega}^e}\right)^2$$

Use this equation, along with the data in Table 9-1, to determine the proper tilting angle θ for phase matching.

9.12 Two optical beams with wavelengths 800 and 650 nm interact in a χ_2 material, generating a frequency-upconverted output beam. Determine the wavelength of the new beam.

9.13 The arrangement of Fig. 9-14 for optical switching is used, with a long optical fiber as the χ_3 medium. The pulses to be switched have a peak power of 100 mW at wavelength 1300 nm. Assume that the light uniformly fills the 8 μm diameter core of the fiber. (a) Assuming that Ge–silica fiber is used, determine the fiber length required to switch the beam. (b) Repeat the calculation if As_2S_3 fiber is used instead of silica.

9.14 Light with wavelength 1064 nm is switched using the Pockels effect, as shown in Fig. 9-21. The lithium niobate crystal is oriented with the z axis aligned with the applied static field, and the light is polarized with its E field perpendicular to the static field. The electrode spacing and length are 15 μm and 0.75 cm, respectively. Determine the voltage required to switch the optical output from zero to maximum.

9.15 A Pockels cell is configured as in Fig. 9-22 to modulate the intensity of a 500 mW Nd:YAG laser beam. Using lithium niobate with the z axis along the direction of propagation, it is found that the transmitted power goes from zero to maximum when the applied voltage V goes from zero to 1.4 kV. (a) Determine the power of the transmitted laser beam when $V = 300$ V. (b) If the applied voltage is modulated according to $V(t) = 300 + 10\cos(5 \times 10^3 t)$, determine the minimum and maximum transmitted power of the transmitted laser beam. (c) Characterize the degree of distortion in the output waveform by calculating the fractional difference between the positive-going and negative-going amplitudes of the output power oscillations. (d) Determine the L/d ratio for this Pockels cell modulator.

9.16 We saw in Section 9-3 that an intense optical beam changes the index of refraction seen by that same beam, a phenomenon known as self-phase modulation. It is also possible for one intense beam to change the refractive index seen by a second, weaker beam, and this is termed *cross-phase modulation.* Consider a strong beam of frequency ω_1 and a weak beam of frequency ω_2 copropagating in a χ_3 medium, with the total electric field at a point in the crystal given by

$$E(t) = A_1 \cos \omega_1 t + A_2 \cos \omega_2 t$$

Show that the change of index seen by the weaker beam is

$$\Delta n = \frac{3\chi_3 I_1}{2n^2 c\varepsilon_0}$$

which is twice the value for self-phase modulation given in Eq. (9-24).

Chapter 10

Review of Semiconductor Physics

The previous chapters of this book have been concerned with how light behaves as it propagates, either in free space or in a waveguide geometry. In general, the propagation of light can be understood by considering it to be a wave. We turn now to a study of devices in which light is generated or detected. The generation or detection of light involves the interaction of light with matter, and it is here that the particle nature of light (quantum viewpoint) becomes relevant. Quantum mechanics is also important for understanding the properties of matter such as semiconductors, which are key materials for use in light emitters and detectors. In this chapter, we briefly review those aspects of semiconductors that are essential for an understanding of light generation and detection.

10-1. UNIFORM SEMICONDUCTOR

We begin by ignoring the boundaries between different materials and consider a semiconductor that is infinite in extent, with properties that do not vary from place to place within the material (i.e., a uniform material). The simplest treatment of semiconductor physics considers only the allowed energies of electrons in the material. A more refined treatment includes the effects of electron momentum. Both viewpoints are useful in understanding the optical properties of semiconductors.

Energy Bands

One of the fundamental principles of quantum mechanics is that the energy of a system cannot take on arbitrary values but is quantized (that is, can only take on discrete values). The electrons in a free atom (such as hydrogen with one electron or oxygen with eight) must reside in one of these "sharp" energy levels, with the restriction that not more than two (one with spin up, one with spin down) can be in any one level. This restriction is known as the *Pauli exclusion principle.* Electrons can make a transition from one energy level to another unocuppied level by absorbing or emitting a photon. In such a process, the photon energy $h\nu$ must equal the difference in energy of the two levels to satisfy energy conservation. The study of atomic absorption and emission spectra (such as the familiar Balmer series in hydrogen) played an important role in the development of quantum mechanics.

When two atoms come together to form a molecule, there is an interaction between the electrons and nuclei of the two atoms that causes the energy levels to change. This process is illustrated in Fig. 10-1a, which shows how the energy of an atomic energy level varies as the interatomic separation R decreases. At large R, there are two energy levels, one for each atom, which have the same energy. The levels are then said to be *degenerate.* As R decreases, the two levels split, one going to lower energy (the "bonding" level) and

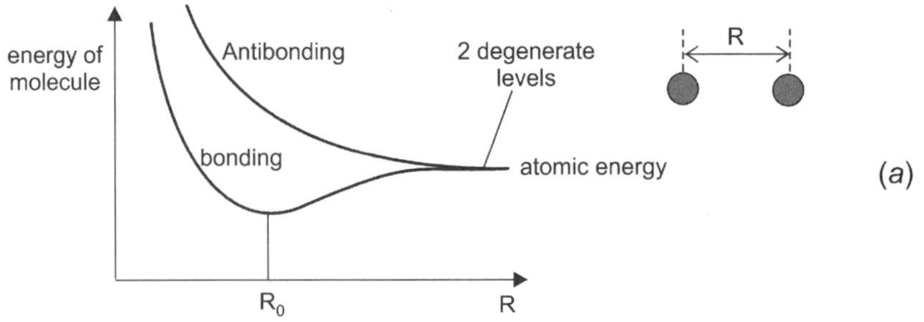

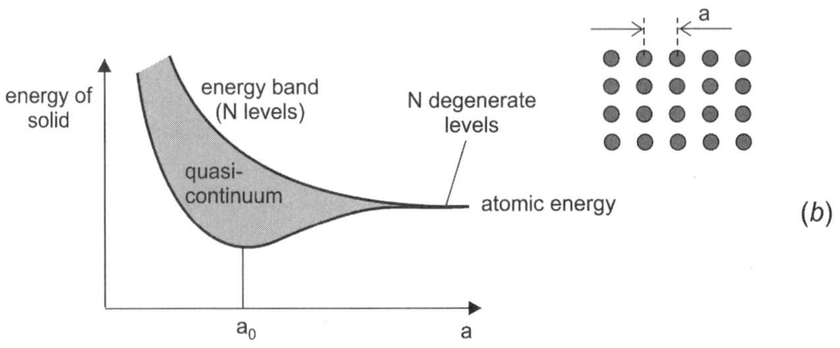

Figure 10-1 (a) Splitting of degenerate atomic energy levels as atoms come together to form a molecule. (b) Splitting of many degenerate atomic energy levels as atoms come together to form a solid, giving rise to a quasicontinuum of levels, or energy band.

the other going to higher energy (the "antibonding" level). These two levels are not associated with either atom individually, but are rather associated with the combined two-atom system. As R continues to decrease, eventually the bonding level also increases in energy, due to Coulomb repulsion by the two nuclei. Since systems have a natural tendency to settle into the lowest energy configuration, the equilibrium separation of the atoms in the molecule will be R_0 as shown.

The energy levels in a solid can be understood in a similar way. Instead of two degenerate levels at large separation, however, there are some 10^{23} degenerate levels corresponding to the ~ 10^{23} atoms in a macroscopic solid of centimeter dimensions. As the interatomic separation a decreases in the solid, these levels split to form a quasicontinuous band of allowed electron energies, as shown in Fig. 10-1b. Although in principle the individual levels in the band are still discrete, the small spacing between them and the natural width of each one leads to a range of electron energies that for all practical purposes is continuous. The equilibrium separation a_0 will correspond to the configuration in which the combined energy of all electrons is minimized.

Fig. 10-1 depicts the splitting of a single degenerate energy level into a band. Each atom in the solid has other energy levels as well, which are similarly split into bands. These bands may be completely filled with electrons, partially filled, or empty. In semiconductors, the highest-energy filled band is termed the *valence band,* and the band above this (which is empty at zero temperature) is termed the *conduction band.* In between these is the *bandgap,* as illustrated in Fig. 10-2. The energy separation between the top of the valence band and the bottom of the conduction band is known as the *bandgap energy* E_g, and is typically ~ 1 eV. Table 10-1 gives values of E_g for a few important semiconductor materials.

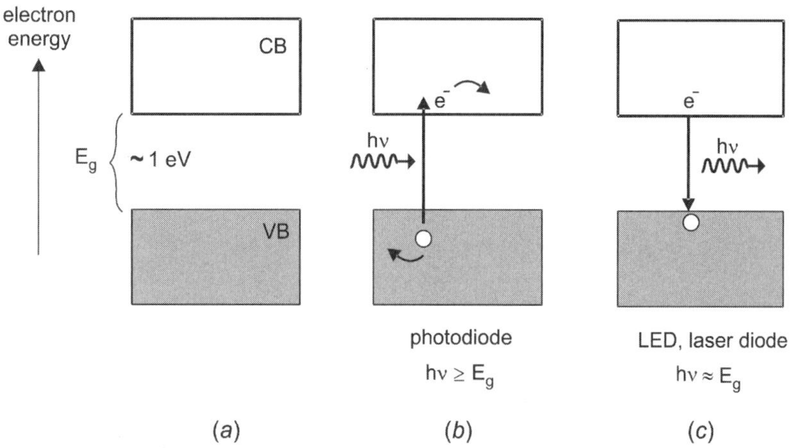

Figure 10-2 (a) The valence band (VB) is filled with electrons at low temperature, and the conduction band (CB) is empty. (b) In a photodiode detector, a photon of energy $h\nu$ is absorbed, creating an electron–hole pair. (c) In an LED or laser diode, an electron–hole pair recombines, creating a photon.

Using the band picture of solids, it is easy to obtain a simple understanding of how light is absorbed by a semiconductor. Consider light with photon energy $h\nu$ incident on a semiconductor of bandgap energy E_g, as shown in Fig. 10-2b. If $h\nu \geq E_g$, the photon has sufficient energy to promote an electron from the valence band to the conduction band. When this occurs, there is not only an extra electron in the conduction band, but also a deficit of one electron in the valence band. This deficit of one electon is termed a *hole,* and acts in many ways like a positive particle. The hole can be visualized as a particle that represents the absence of an electron, in the same way that a bubble in water acts like a particle. The net result of this process is the *absorption* of a photon (the photon disappears), and the creation of an *electron–hole (e–h) pair.*

The excess energy $h\nu - E_g$ in an absorption process gives the electron and hole some kinetic energy in the conduction and valence bands, which is quickly lost by inelastic collisions. As the electron loses kinetic energy, it settles to the bottom of the conduction

Table 10-1 Bandgap energy E_g and relative dielectric constant ε_r for selected semiconductors.

Material	E_g (eV)	Type of gap	ε_r
Si	1.12	indirect	11.9
Ge	0.66	indirect	16
GaAs	1.42	direct	13.2
AlAs	2.15	indirect	10.1
$Al_{0.3}Ga_{0.7}As$[a]	1.85	direct	10.2
InAs	0.33	direct	15.2
$In_{0.53}Ga_{0.47}As$[b]	0.74	direct	12.5
InP	1.35	direct	12.6
GaP	2.27	indirect	11.1
GaN	3.44	direct	10.4/9.5[c]

[a]Becomes indirect for more than 35% Al.
[b]This composition lattice matched to InP.
[c]Birefringent.

band, whereas the hole settles to the top of the valence band. To understand the behavior of the hole, it should be kept in mind that diagrams such as Fig. 10-2 depict the energy of an electron, with increasing energy upward. The energy of a hole then increases downward, since to move the hole downward requires electrons to move up in energy. One can, therefore, think of holes as naturally "floating" to the top of the valence band, just as a bubble naturally rises in a liquid.

Once the electron and hole (collectively called *charge carriers*) are created by absorption of a photon, they are free to move in the conduction and valence bands, respectively. If there is an electric field in the semiconductor, the charge carriers will move in response to this field and give rise to a *photocurrent,* which can be measured in an external circuit. This is the basis for photodiode detectors, to be discussed in Chapter 14. It is clear from this discussion that if $h\nu < E_g$, the photon cannot be absorbed, since there are no available energy levels for the electron in the bandgap. This leads to the important feature of photodiode detectors that there is a minimum photon energy for photodetection, $h\nu_{min} \approx E_g$. Since $\nu = c/\lambda$, this condition can be written as a maximum wavelength for photodetection, $\lambda_{max} \approx hc/E_g$.

EXAMPLE 10-1

Compare the long wavelength detection limit for a Si photodetector with that of an $In_{.53}Ga_{.47}As$ detector. Which is suitable for a fiber optic communications system in the 1.5 μm band?

Solution: Using the values from Table 10-1, for Si,

$$\lambda_{max} = \frac{hc}{E_g} = \frac{(6.63 \times 10^{-34}\ \mathrm{J \cdot s})(3 \times 10^8\ \mathrm{m/s})}{(1.12\ \mathrm{eV})(1.6 \times 10^{-19}\ \mathrm{J/eV})} = 1110\ \mathrm{nm}$$

and for $In_{.53}Ga_{.47}As$,

$$\lambda_{max} = \frac{(6.63 \times 10^{-34}\ \mathrm{J \cdot s})(3 \times 10^8\ \mathrm{m/s})}{(0.74\ \mathrm{eV})\ (1.6 \times 10^{-19}\ \mathrm{J/eV})} = 1675\ \mathrm{nm}$$

The $In_{.53}Ga_{.47}As$ detector will be suitable for 1500 nm, but the Si detector will not. This particular mixture of 53% In and 47% Ga is often chosen so that the atomic spacing matches that of InP, a commonly used substrate.

The energy band picture can also be used to understand the emission of light by a semiconductor, which is the basis for operation of the LED (light emitting diode) and laser diode. If electrons are somehow promoted into the conduction band, these electrons can recombine with holes in the valence band, as indicated in Fig. 10-2c. Since the electrons and holes will settle to the band edge before recombining, the energy of the photon generated is $h\nu = hc/\lambda \approx E_g$. For example, GaAs has a bandgap of 1.42 eV, so the corresponding emission wavelength is $\approx$ 870 nm. This material is historically important for photonics, since it was used as a light source in the earliest optical communications systems.

Energy and Momentum

The picture presented in the previous section based on energy alone is not complete, because the electron has momentum as well as energy. According to quantum mechanics, the momentum p of a particle is associated with a wavelength known as the *de Broglie wavelength,* given by

$$\lambda = \frac{h}{p} \qquad \text{(de Broglie wavelength)} \tag{10-1}$$

where h is Planck's constant. It is convenient to define a wave vector magnitude $k = 2\pi/\lambda$ for the electron, just as for light, so the electron's momentum can be expressed as

$$p = \frac{h}{\lambda} = \left(\frac{h}{2\pi}\right)\left(\frac{2\pi}{\lambda}\right) = \hbar k \tag{10-2}$$

where $\hbar \equiv h/2\pi$. The kinetic energy of a free electron can then be written

$$E(k) = \frac{p^2}{2m} = \frac{\hbar^2 k^2}{2m} \tag{10-3}$$

which is shown graphically in Fig. 10-3a. The wave nature of the electron is described mathematically by the *wave function* $\psi(x, t)$, which for a freely propagating electron is a plane wave of the form

$$\psi(x, t) = Ae^{i(kx-\omega t)} \tag{10-4}$$

where A is a normalization constant and $\omega = E/\hbar$. The wave function gives a complete description of our knowledge about the electron's behavior. Most importantly, the square of the wave function $|\psi|^2$ gives the relative probability that the electron will be found at a particular value of x at time t when a measurement is made. For the plane wave of Eq. (10-4), $|\psi|^2$ is independent of x, which means that the particle could be anywhere. This is in accord with the *Heisenburg uncertainty principle* (see Appendix B), since a momentum that is precisely known leads to a position that is completely unknown.

For the electron to be localized, it is necessary to add together plane waves having slightly different values of k and ω. The velocity of the electron then corresponds to the group velocity of the wave packet, as given by Eq. (2-7). Written in terms of energy this becomes

$$v = \frac{d\omega}{dk} = \frac{1}{\hbar}\frac{dE}{dk} \tag{10-5}$$

It is left as an exercise (Problem 10.1) to show that this velocity is consistent with the momentum in Eq. (10-2).

When the electron is in a solid, it is no longer perfectly free to propagate, because of interactions with the atoms in the solid. Thinking of the electron as a propagating wave with de Broglie wavelength λ, the situation is analogous to that of a light wave scattering off a periodic array of refractive index "bumps." As we saw in Chapter 8, this gives rise to efficient reflection of the wave at the Bragg condition:

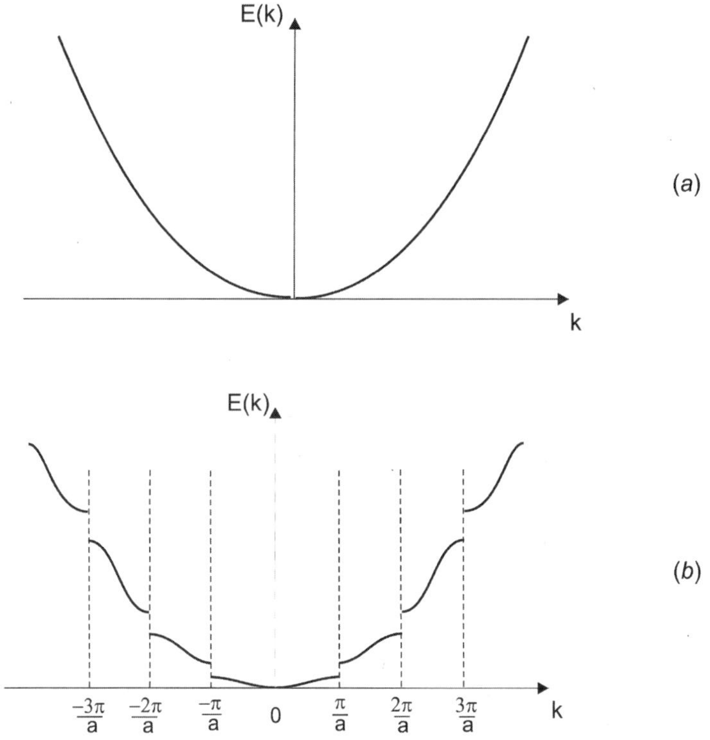

Figure 10-3 (a) For a free electron, *E*(*k*) is parabolic, and the electron can have any energy. (b) In a solid, the *E*(*k*) is distorted around multiples of π/a due to Bragg scattering of the electron's wave function.

$$2a = n\lambda \qquad \text{(Bragg condition)} \tag{10-6}$$

where a is the spacing between scattering centers (in this case the interatomic spacing) and n is an integer. The electrons in the solid will, therefore, be highly reflected (i.e., they cannot propagate) at discreet values of k given by

$$k_n = \frac{2\pi}{\lambda_n} = n\frac{\pi}{a} \tag{10-7}$$

At the values k_n where efficient Bragg scattering occurs, the electron's wave function does not propagate, but rather takes the form of a stationary standing wave. This corresponds to an electron that does not move, having a group velocity of zero. But a zero group velocity for the electron implies that the slope dE/dk of the $E(k)$ curve is zero, according to Eq. (10-5). We therefore expect the $E(k)$ curve to take the form shown in Fig. 10-3b. The curve deviates from the free-electron parabola of Fig. 10-3a only when k is close to a multiple of π/a, the condition for Bragg reflection.

The deviation of the $E(k)$ curve near $k \approx n(\pi/a)$ has the important consequence that the energy of the electron can no longer take on any arbitrary value. There are now forbidden energy regions, which correspond to the bandgaps discussed previously. The allowed energy regions are a series of bands, most easily visualized in the *reduced zone scheme* shown in Fig. 10-4. In this scheme, each part of the $E(k)$ curve is translated by an integer

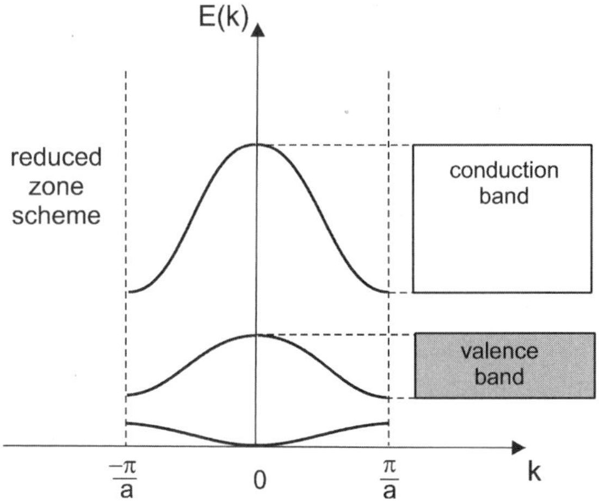

Figure 10-4 In the reduced zone scheme, the allowed energy regions of the *E*(*k*) curves correspond to the energy bands shown in Fig. 10-2.

multiple of $2\pi/a$ so that the entire curve is within the range $-\pi/a \leq k \leq \pi/a$. Such translations are allowed due to the periodicity of the lattice. Each point on the $E(k)$ curves of Fig. 10-4 corresponds to a particular quantum state for the electron, which may be occupied by an electron, or unoccupied. In the valence band at $T = 0$ K, all the available states are occupied, and in the conduction band all states are vacant.

If an electron is promoted into one of the states of the conduction band, with a vacancy (hole) in one of the states of the valence band, the electron and hole can recombine radiatively to generate a photon. This was discussed previously in terms of energy conservation, but now we add the restriction that momentum be conserved in the recombination process. Consider a transition in which the k of the electron changes by Δk as illustrated in Fig. 10-5. The electron momentum is $\hbar k$, so the change in momentum

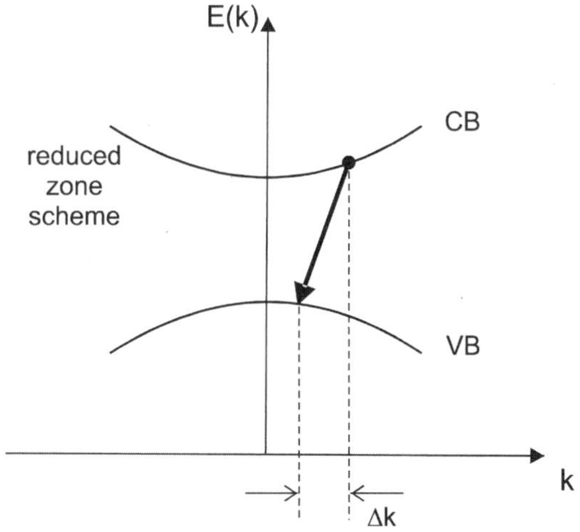

Figure 10-5 In a direct radiative transition, the *k* of the electron must change by an amount Δ*k* that balances the momentum of the emitted photon.

is $\hbar\Delta k$. To conserve momentum, the photon that is produced must have this same momentum, or

$$p_{\text{photon}} = \frac{h}{\lambda} = \hbar\left(\frac{2\pi}{\lambda}\right) = \hbar\Delta k$$
$$\frac{2\pi}{\lambda} = \Delta k \tag{10-8}$$

where λ here is the photon wavelength. To get a sense of how large the Δk must be for a typical photon of wavelength ~ 1 μm, it can be compared with the maximum range of k values for each band, which is $2\pi/a$. Taking a typical atomic spacing of $a = 1 \times 10^{-10}$ m,

$$\frac{\Delta k}{2\pi/a} = \frac{a}{\lambda} \sim \frac{10^{-10}\text{ m}}{10^{-6}\text{ m}} = 10^{-4}$$

which means that on the scale of a diagram such as Fig. 10-4, the transition must be essentially vertical. If the electron and hole initially have very different values of k, they must first each settle to the bottom and top of their respective bands, before recombining in a vertical transition.

The band structure depicted in Fig. 10-5 is said to have a *direct gap,* since the bottom of the conduction band is directly over the top of the valence band, allowing a vertical radiative transition. Some materials instead have an *indirect gap,* illustrated in Fig. 10-6, in which the conduction and valence bands are offset. In this case a vertical transition is not allowed, because the states in the valence band directly below the electron in the conduction band are already filled. The transition must, therefore, be indirect, with the electron's momentum change $\hbar\Delta k$ much larger than that of the photon. Since this does not satisfy momentum conservation, radiative decay of the electron and hole is largely suppressed. It

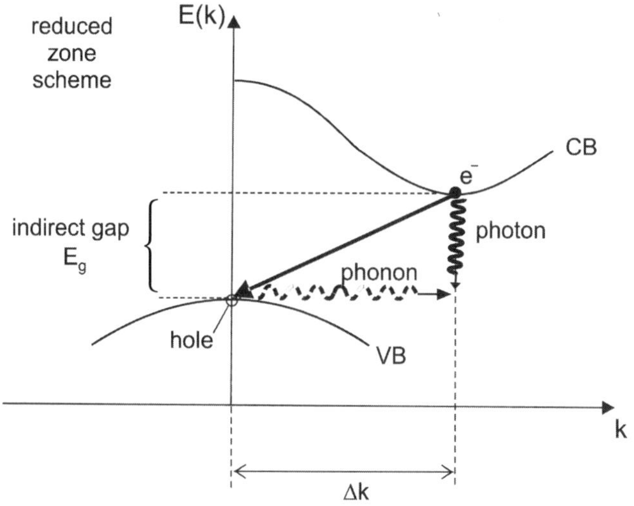

Figure 10-6 In an indirect radiative transition, the Δk of the electron is too large to balance the photon's momentum. A phonon simultaneously emitted or absorbed can allow momentum to be conserved.

can be weakly allowed, however, by the simultaneous emission or absorption of *phonons* (the quanta of atomic vibrations in the solid) which serve to conserve momentum. Since the probability of occurance for a quantum process decreases when the number of objects involved increases, the efficiency of these "indirect transitions" is quite low, typically some four to six orders of magnitude smaller than direct (vertical) transitions.

It is unfortunate that the elemental semiconductors silicon and germanium, so important and well developed for the electronics industry, happen to be indirect-gap materials. This makes them unsuitable as light emitters, at least in the traditional crystalline form. Research on nanostructured silicon has shown an improved efficiency due to the breakdown of translational symmetry and consequent relaxation of the momentum requirement. However, this technology is not yet mature, and it remains true that an efficient light emitter requires a direct-gap material.

Many binary semiconductors such as GaAs have a direct gap, as indicated in Table 10-1, whereas others such as AlAs have an indirect gap. If GaAs and AlAs are mixed together to form the ternary alloy $Al_xGa_{1-x}As$, the material has a direct gap for $x < 0.35$ and an indirect gap for $x > 0.35$. Throughout the direct-gap range, the bandgap energy varies with x as

$$E_g = 1.424 + 1.427x + 0.041x^2 \text{ eV} \qquad (Al_xGa_{1-x}As \text{ bandgap}) \qquad (10\text{-}9)$$

so the emission wavelength of the material can be chosen by selecting different compositions. The alloy $Al_xGa_{1-x}As$ has the added advantage that its lattice constant (separation between atoms) is nearly independent of x, which allows layers with different bandgaps to be grown on top of each other without a lattice mismatch. Not all ternary semiconductors behave so nicely, however. For example, $In_xGa_{1-x}As$ is lattice-matched to the substrate InP only for $x = 0.53$, resulting in the bandgap energy $E_g = 0.74$ eV. To obtain a different bandgap energy, while at the same time keeping the lattice constant matched to InP, an additional element can be added to form the quaternary alloy $In_xGa_{1-x}As_yP_{1-y}$. If proper combinations of x and y are chosen, this material can have an emission wavelength anywhere in the range $920 < \lambda < 1650$ nm, with the same lattice constant as InP. InGaAsP has proved to be quite useful for fiber optic communications, since its wavelength range includes the important second and third telecommunications windows around 1.3 and 1.5 μm.

Although indirect-gap materials such as Si make poor light emitters, they are actually good materials (and commonly used) for photodetectors. To understand why, compare the emission process shown in Fig. 10-6 with the absorption process shown in Fig. 10-7. In either case, a fundamental requirement (the Pauli exclusion principle) is that the final state for the electron be unoccupied (i.e., a hole must be there). For the emission process, the final state for the electron is in the valence band, and since any holes in the valence band will be near the top, vertical transitions are not allowed. For absorption, however, the final state for the electron is in the conduction band, which is nearly empty (i.e., full of holes). In this case, vertical transitions for a wide range of photon energies are allowed, restricted only by energy conservation.

Radiative Efficiency

If the only way that an electron and hole could recombine were radiatively (i.e., by emitting a photon), one might imagine that even indirect-gap materials might emit efficiently,

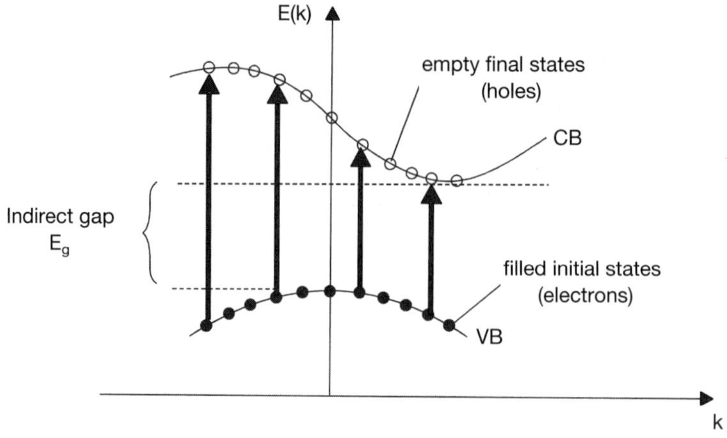

Figure 10-7 Vertical absorption transitions can readily occur even in an indirect-gap material.

because all the energy put into the material to create e–h pairs would eventually come out as light. There are other processes, however, by which the charge carriers can lose their energy without producing a photon. These *nonradiative decay* processes compete with radiative recombination and limit the radiative efficiency.

One nonradiative mechanism, illustrated in Fig. 10-8a, is the transfer of the electron's energy to a trap state located in the band gap. Trap states can be associated with impurities or defects in the volume of the semiconductor, or with defects (unsatisfied "dangling" bonds) found at the surface. Surface losses can be quite significant, although they can be reduced by proper treatment (passivation) of the surface. A typical passivation process is to grow an oxide layer on the surface, which serves to satisfy the bonding requirements of atoms at the surface.

Another nonradiative mechanism is *Auger recombination,* illustrated in Fig. 10-8b. In this process, an electron and hole recombine, but the recombination energy does not go into creating a photon, but is rather transferred as kinetic energy to another electron in the conduction band. This kinetic energy is quickly dissipated as heat by collisional process-

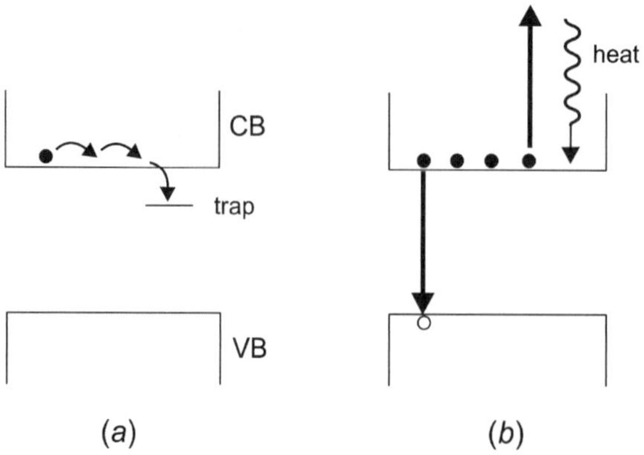

Figure 10-8 An electron can decay nonradiatively by (a) energy transfer to a trap state, or (b) the Auger process.

es, so the net result of the Auger process is conversion of the e–h energy into heat. Since this process requires that there be excess electrons in the conduction band, it becomes especially important for laser action, in which a large number of charge carriers are injected into a small recombination region.

A quantitative measure of the radiative efficiency can be obtained by considering the rate at which radiative and nonradiative decay processes occur. An electron can decay radiatively only when a hole is in close proximity. The probability per unit time that the electron will recombine with a hole is, therefore, proportional to the number of holes that are sufficiently close, which in turn is proportional to the number of holes per unit volume, denoted by p. We can therefore write the *radiative transition rate* for the electron as

$$\frac{\text{probability of recombination}}{\text{unit time}} \equiv W_r = B_r p \tag{10-10}$$

where B_r is a constant of proportionality. Typical values for B_r are 10^{-11} to 10^{-9} cm^3/s for direct transitions and 10^{-15} to 10^{-13} cm^3/s for indirect transitions. For example, GaAs has $B_r \approx 7 \times 10^{-10}$ cm^3/s. At sufficiently high hole concentration this relation breaks down, because the transition rate becomes limited not by the availability of a nearby hole, but rather by the intrinsic rate at which an electron can recombine with a hole. The transition rate, therefore, saturates at some maximum value, which in the case of GaAs is ~ 3×10^9 s^{-1}.

The radiative rate below saturation can also be written in terms of electron density. In a laser diode, an equal number of holes and electrons are injected into a recombination region, so $p \approx n$, where n is the number of free electrons per unit volume. The radiative rate for an electron can therefore also be written $W_r = B_r n$.

Nonradiative decay of an electron to a trap requires that a trap state be in close proximity to the electron but does not involve any other charge carriers. The probability per unit time for this process, designated A_{nr}, is therefore proportional to the number of trap states per unit volume, but independent of the density of charge carriers n and p. An electron decaying by the Auger process, on the other hand, requires not only a hole for recombination, but also another electron in the conduction band. The probability of encountering a hole is $\propto p$, and the probability of encountering an electron is $\propto n$, so the joint probability is $\propto pn \approx n^2$ (since $p \approx n$ for laser action). We then have

$$W_{\text{Auger}} = C_A n^2 \tag{10-11}$$

where the proportionality constant C_A is the *Auger constant.*

There are then three different processes by which the electron can decay out of the conduction band: radiative decay, nonradiative decay to traps, and Auger relaxation. Since the probabilities for independent processes add, the total probability that an electron decays per unit time is given by

$$W_{\text{total}} = A_{nr} + B_r n + C_A n^2 \tag{10-12}$$

where Eqs. (10-10) and (10-11) have been used, along with the condition $p \approx n$ for laser action. The radiative efficiency η_i is defined as the fraction of all decays that are radiative. This is equal to the radiative probability divided by the total probability, or

$$\eta_i = \frac{W_r}{W_{\text{total}}} = \frac{B_r n}{A_{nr} + B_r n + C_A n^2} \quad \text{(radiative efficiency)} \tag{10-13}$$

It is seen from Eq. (10-13) that η_i increases with carrier density n, up to a point. At sufficiently high n, however, the Auger process degrades the efficiency, and this is one contribution to the limits on output power in semiconductor lasers.

EXAMPLE 10-2

Compare the radiative efficiencies for GaAs (E_g = 1.42 eV) and $In_{.53}Ga_{.47}As$ (E_g = 0.74 eV), for the same electron density of $n = 5 \times 10^{18}$ cm^{-3}. Take Auger constants of 5×10^{-30} cm^6/s and 1×10^{-28} cm^6/s, and B_r values of 7.2×10^{-10} and 4×10^{-11} cm^3/s for GaAs and InGaAs, respectively.

Solution: For GaAs, the radiative and Auger rates are

$$W_r = B_r n = \left(7.2 \times 10^{-10}\ \frac{\text{cm}^3}{\text{s}}\right)(5 \times 10^{18}\ \text{cm}^{-3}) = 3.6 \times 10^9\ \text{s}^{-1}$$

$$W_{\text{Auger}} = C_A n^2 = \left(5 \times 10^{-30}\ \frac{\text{cm}^6}{\text{s}}\right)(5 \times 10^{18}\ \text{cm}^{-3})^2 = 1.25 \times 10^8\ \text{s}^{-1}$$

For $In_{.53}Ga_{.47}As$ the corresponding rates are

$$W_r = B_r n = \left(4 \times 10^{-11}\ \frac{\text{cm}^3}{\text{s}}\right)(5 \times 10^{18}\ \text{cm}^{-3}) = 2 \times 10^8\ \text{s}^{-1}$$

$$W_{\text{Auger}} = C_A n^2 = \left(1 \times 10^{-28}\ \frac{\text{cm}^6}{\text{s}}\right)(5 \times 10^{18}\ \text{cm}^{-3})^2 = 2.5 \times 10^9\ \text{s}^{-1}$$

Note that all these rates are much larger than the nonradiative decay rate to traps, $A_{nr} \sim 10^7\ \text{s}^{-1}$. Therefore, the efficiency for the two materials is

$$\text{GaAs:}\ \eta_i = \frac{36}{36 + 1.25} = 0.97$$

$$\text{InGaAs:}\ \eta_i = \frac{2}{25 + 2} = 0.074$$

This example illustrates the general trend that Auger losses are more significant in smaller-bandgap materials.

10-2. LAYERED SEMICONDUCTORS

The boundary between different semiconductor layers plays a fundamental role in the operation of many photonic devices. Junctions can occur between semiconductors of similar or different compositions, and also between semiconductors and metals. We consider here these different types of junctions and their important applications.

The p–n Junction

A pure semiconductor material such as Si is known as an *intrinsic semiconductor* because electrons in the conduction band are present intrinsically, due to thermal excitation of e–h pairs. When Si is doped with an impurity such as Al or P, there are one too few (Al) or one too many (P) electrons in the dopant atom, relative to the number needed for bonding with Si. In the case of P doping of Si, the extra electron is easily ionized from the P atom, becoming an additional electron in the conduction band. The dopant atom here is said to be a *donor,* because it donates an electron. Since the dopant P atom was initially electrically neutral, this ionization leaves behind a positively charged P^+ ion core.

For Al doping of Si, just the opposite occurs. In this case, an electron from the valence band joins the Al atom to form the negatively charged ion core Al^-, leaving an electron vacancy, or hole, in the valence band. The dopant here is said to be an *acceptor,* since it accepts an electron. Semiconductors doped with either donors or acceptors are termed *extrinsic,* since the electron and hole concentrations are determined by an extrinsic factor such as doping level. An extrinsic semiconductor with extra free electrons is referred to as *p-type,* and one with extra free holes is referred to as *n-type,* as indicated in Fig. 10-9.

The basic features of the p–n junction can be understood by imagining that the p and n type materials shown in Fig. 10-9 are grown separately and then brought together into physical contact. Although p–n junctions are not actually made this way in practice (they are grown by a successive deposition of thin layers), the conclusions arising from this simple viewpoint are still valid. When the materials come into contact, the free electrons and free holes have a natural tendency (known as *diffusion*) to move around so as to fill the entire (now larger) material uniformly. As the electrons and holes move across the boundary, they start to recombine there, creating a region near the boundary known as the *depletion region,* which is depleted of free charge carriers. One might think that this depletion region would continue to increase in width until all the free electrons and holes had recombined. However, there is a net charge density created inside the depletion region, due to the positive and negative ion cores that remain after the electons and holes recombine. This charge density creates an electric field, as indicated in Fig. 10-10, that is in a direction so as to oppose the diffusion of free holes into the n-type material, and of free electron's into the p-type material. The recombination process is thus self-limiting, and results in a depletion region of finite width.

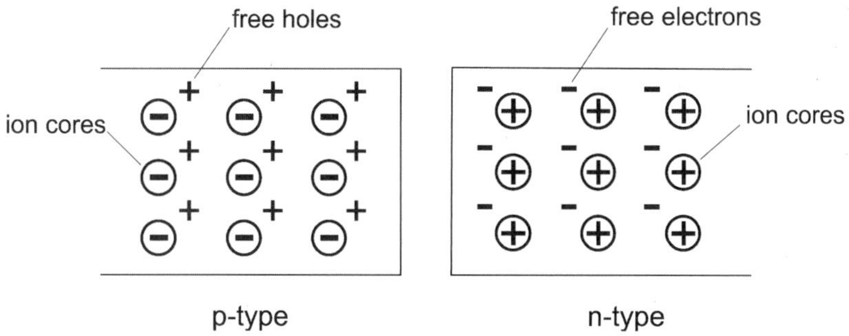

Figure 10-9 Before coming into contact, the p- and n-type materials have free holes and electrons uniformly distributed.

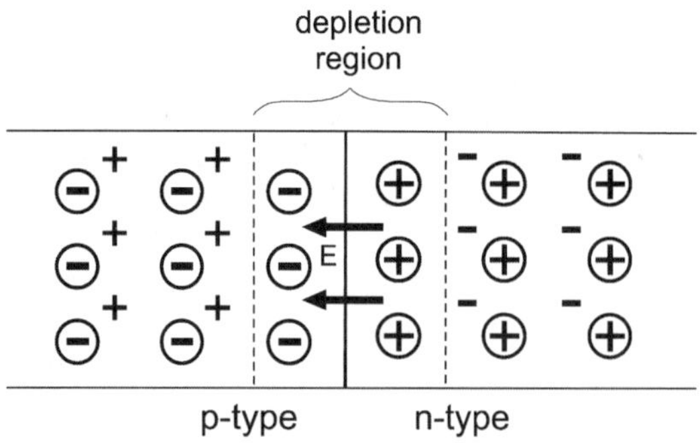

Figure 10-10 Electrons and holes recombine in the depletion region, leaving ion core charges that create an electric field.

Associated with the electric field in the depletion region is a change in electric potential across the junction, which has the effect of changing the relative energies of the conduction and valence bands on opposite sides of the junction. To see how this works quantitatively, consider a simplified model in which the charge density ρ is independent of position x on either side of the junction, located at $x = 0$. To simplify further, assume a p^+–n junction, in which the p-type material is much more heavily doped than the n-type material. The charge density in the p-type portion of the depletion region can be written $\rho_p = -eN_A$, and in the n region $\rho_n = eN_D$, where N_A and N_D are defined as the number of acceptors and donors per unit volume in the p and n regions, respectively. The charge density $\rho(x)$ then has the position dependence given in Fig. 10-11a, with d_p and d_n the width of the depletion region in the p and n regions. Since electrons and holes must recombine in pairs, the total charge on either side of the junction must be the same, or

$$|\rho_p d_p A| = |\rho_n d_n A|$$
$$eN_A d_p = eN_D d_n \tag{10-14}$$

where A is the cross-sectional area of the junction (in the y–z plane). Eq. (10-14) is equivalent to saying that the area under the curve of Fig. 10-11a must be the same on each side of $x = 0$. Since we assume $N_A \gg N_D$, then $d_p \ll d_n$, and the total junction width is $d = d_p + d_n \approx d_n$. The junction width is, therefore, determined mostly by the lightly doped material.

In a region where ρ is independent of x, it is easy to show using Gauss's law (see Problem 10.5) that the x component of the electric field varies with x as

$$E_x(x) = E_x(0) + \frac{\rho}{\varepsilon}x \tag{10-15}$$

where ε is the dielectric permitivity. In a material medium, $\varepsilon = \varepsilon_r\varepsilon_0$, where ε_0 is the dielectric permitivity of free space and ε_r is the relative dielectric constant. In Si, for example, $\varepsilon_r = 11.9$. The constant $E_x(0)$ in Eq. (10-15) can be determined by the condition that the field must go to zero outside the depletion region. This will be true because the electrical conductivity outside the depletion region is high due to the free carriers there. The

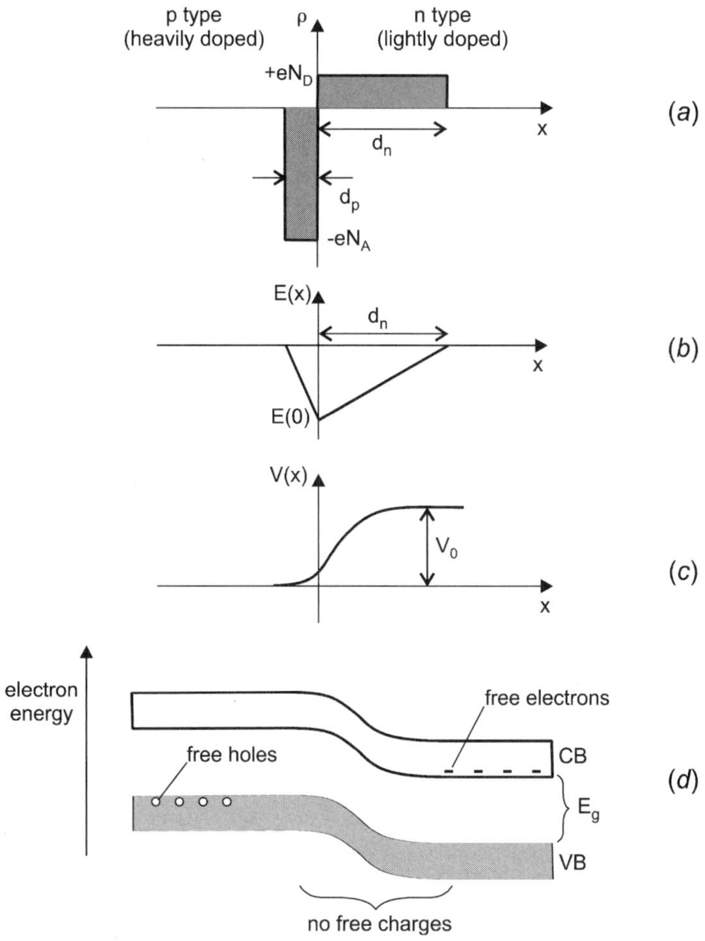

Figure 10-11 The (a) charge distribution in an ideal p–n junction, along with the corresponding (b) electric field and (c) electric potential. (d) The energy bands are offset on either side of the junction, due to the changing electric potential.

situation is similar to that of a capacitor, in which the electric field exists only between the plates of the capacitor. Requiring that $E_x = 0$ at $x = d_n$ gives

$$\begin{aligned} 0 &= E_x(0) + \frac{eN_D}{\varepsilon} d_n \\ E_x(0) &= -\frac{eN_D d_n}{\varepsilon} \end{aligned} \tag{10-16}$$

The field in the n region can then be written

$$\begin{aligned} E_x(x) &= -\frac{eN_D}{\varepsilon} d_n + \frac{eN_D}{\varepsilon} x \\ &= -\frac{eN_D}{\varepsilon}(d_n - x) \end{aligned} \tag{10-17}$$

which is shown graphically in Fig. 10-11b. Note that $E_x(x) < 0$, so the field points from the n region toward the p region, as expected. The change in electric potential is equal to the area under the $E_x(x)$ graph, or

$$
\begin{aligned}
V(x) &= -\int_0^x E_x(x')dx' \\
&= \frac{eN_D}{\varepsilon}\left[d_n x - \frac{x^2}{2}\right]
\end{aligned}
\tag{10-18}
$$

which is graphed in Fig. 10-11c. The total change in potential across the n-type part of the depletion region is then

$$
\Delta V_n \equiv V(d_n) - V(0) = \frac{eN_D}{\varepsilon}\frac{d_n^2}{2} \tag{10-19}
$$

There is also a potential change ΔV_p across the p-type part of the depletion region, but since $d_p \ll d_n$, this is negligible compared with ΔV_n. The total potential change across the junction is then $V_0 = V_p + V_n \simeq V_n$, where V_0 is known as the *built-in potential* of the junction. The built-in potential is that which exists between the two sides of the p–n junction when there is no externally applied voltage. Using the approximation $d \approx d_n$, the width of the depletion region then becomes

$$
d \simeq \sqrt{\frac{2\varepsilon V_0}{eN_D}} \qquad \text{(width of depletion region)} \tag{10-20}
$$

This expression will be useful in understanding the response times for photodetectors.

The effect of the changing electric potential $V(x)$, shown in Fig. 10-11c, is to change the potential energy of the electrons in both the valence and conduction bands. This is illustrated in Fig. 10-11d, where the band energies are plotted as a function of distance across the junction. Note that since the electron has negative charge, the electron energy increases when the electric potential decreases. This kind of diagram is quite useful in understanding how p–n junctions operate, and provides another way to see why the electrons and holes do not continue to recombine after the depletion region is formed. The electrons on the n side would need to be given additional energy to get to the p side; that is, they would have to go "uphill" over an energy barrier. Similarly, the holes would need extra energy to get into the n region (remember that the hole energy increases downward). The p–n junction thus serves to keep the electrons and holes apart.

The analysis so far has been of an "unbiased" p–n junction, that is, with no external applied voltage. When a positive voltage is applied to the n side, and a negative voltage to the p side, the junction is said to be *reverse biased.* Under these conditions, the electron energy on the two sides becomes even more different, as shown in Fig. 10-12, and the electrons and holes are prevented even more from crossing the junction. The p–n junction acts like an open circuit for an applied voltage of this polarity.

With an applied voltage of the other polarity (positive potential applied to the p side), the junction is said to be *forward biased.* In this case, the bands on the two sides of the junction come more into alignment, as illustrated in Fig. 10-13. The potential energy barrier is now low enough that electrons from the n region can diffuse into the p region, and vice versa. This process is referred to as *minority carrier injection,* because the electrons

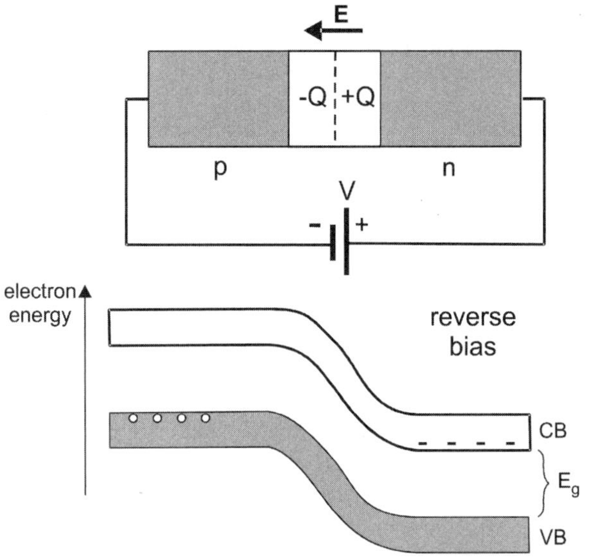

Figure 10-12 When the p–n junction is reverse biased, the bands on either side become more widely separated.

that make it into the p region are in the minority there (holes being the majority carrier in the p region). The injected electrons can now recombine with the many holes available in the p region, resulting in light emission. This is the basis for LED and laser diode operation, to be discussed further in subsequent chapters.

The current *i* that flows through the junction will be quite different for forward and reverse biases, and this gives the p–n junction its rectification property, so useful in elec-

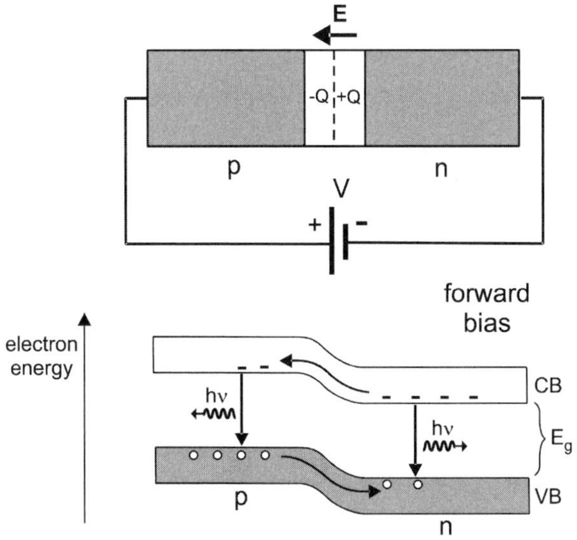

Figure 10-13 When the p–n junction is forward biased, the bands on either side come into alignment, allowing current to flow.

tronic circuits. A device containing a p–n junction for rectification is called a *diode.* The dependence of current i on applied voltage V can be written

$$i = i_0\left[\exp\left(\frac{eV}{\beta k_B T}\right) - 1\right] \quad \text{(diode equation)} \tag{10-21}$$

where k_B is Boltzmann's constant, T is the absolute temperature, and β is the *diode ideality factor*. Fig. 10-14 shows a graph of this equation, along with the sign convention for V and i. This expression can be understood by considering that the current flow through the junction is a result of electrons and holes "jumping over" an energy barrier. For zero-applied bias voltage ($V = 0$), the height E_b of this barrier is $E_b = eV_0$, where V_0 is the built-in potential. It might be supposed that if $V = 0$, then the built-in potential should be forced to zero, since V is directly applied across the p–n junction. The reason this is not the case is that there are additional voltage drops across the junctions between the metal wires and the semiconductor, and it is the sum of these voltage drops plus the built-in potential V_0 that must equal the applied voltage V. When $V = 0$, the sum of the metal–semiconductor voltage drops is exactly equal and opposite to the built-in potential. The nature of these metal–semiconductor junctions is discussed further in beginning on p. 178.

When the applied voltage $V \neq 0$, the barrier height becomes $E_b = E(V_0 - V)$, with V taken as positive for forward bias. Charge carriers can jump over the energy barrier with a probability proportional to the *Boltzmann factor* exp $(-E_b/k_BT)$, which in turn is $\propto$ exp (eV/k_BT). To ensure that $I = 0$ when $V = 0$ (a necessary condition; see Problem 10.8), the net current is $\propto$ [exp$(eV/k_BT) - 1$]. The proportionality constant is denoted as i_0.

This argument assumes that electrons and holes must jump across the entire barrier region in order to recombine; that is, they are injected minority carriers. If instead, they mostly recombine with each other in the depletion region, then they only have to "jump over" half the energy barrier, on average, and the net current is $\propto$ [exp$(eV/2k_BT) - 1$]. The diode ideality factor β, therefore, ranges from 1 to 2, depending on whether e–h recombinations occur primarily outside or inside the depletion region.

The constant i_0 is known as the *reverse saturation current,* since under a large reverse bias ($V \ll -k_BT/e$), i saturates at $\approx -i_0$. Physically, i_0 arises from thermal generation of e–h pairs in and near the depletion region, which occurs with a probability $\propto$ exp$(-E_g/k_BT)$. Therefore, $i_0 \propto$ exp$(-E_g/k_BT)$, which increases at higher temperature. The fundamental quantity is actually the reverse saturation current density $J_0 = i_0/A$, since the total number of thermally generated e–h pairs will be proptional to the junction cross-sectional area A. A typical value for a Si p–n junction at room temperature is $J_0 \approx 1.5 \times 10^{-8}$ A/cm^2.

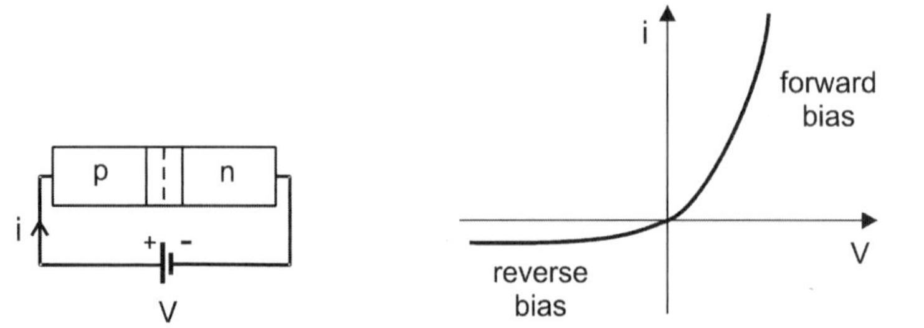

Figure 10-14 The current versus voltage relation for an ideal diode. The sign convention for voltage and current are indicated on the left.

Semiconductor Heterojunctions: The Quantum Well

The type of p–n junction discussed in the previous section is termed a *homojunction,* a boundary between two regions of the same semiconductor material, one doped with donor impurities and the other with acceptors. The energies of the valence and conduction bands shift by the same amount due to the electric potential change across the junction, and the band gap on either side of the p–n junction is the same. We now consider another type of boundary, the *heterojunction,* in which the semiconductor composition and corresponding band gap are different on either side. For example, GaAs and $Al_xGa_{1-x}As$ have different band gap energies (see Table 10-1), and are lattice-matched for any value of x. If they are grown in a layered structure, with a thin layer of GaAs sandwiched between two thick layers of $Al_xGa_{1-x}As$, the conduction and valence band energies will vary with position, as shown in Fig. 10-15. Any charge carriers that find themselves in the GaAs material will encounter a large potential energy barrier when they reach the boundary on either side, and are therefore "confined" to the GaAs layer.

When the thickness d of the GaAs layer is made sufficiently small, quantum mechanical effects become important, and the structure is called a *quantum well.* In the simplest model for the quantum well, the walls of the well can be considered infinitely high, so that there is no probability of the charge carrier getting out. This corresponds to the classic "particle in a box" problem in quantum mechanics, which can be understood by considering the electron to be a wave with deBroglie wavelength $\lambda = h/p = \hbar k$. The condition for the allowed wavelengths in the box is the same as that of a string stretched between two supports: an integer number of half-wavelengths must fit in the box, so that the vibra-

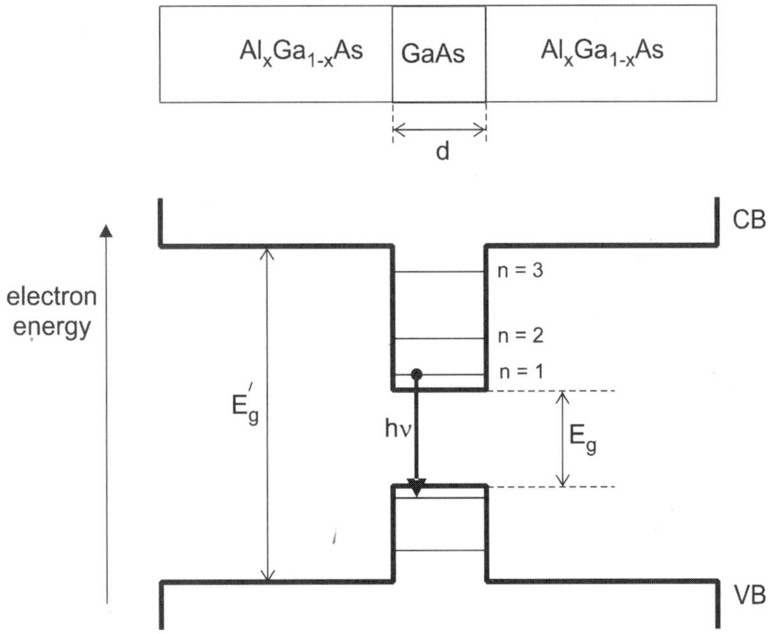

Figure 10-15 In a quantum well structure, the energy levels in the conduction and valence bands depend not only on the bandgap of the material, but also on the thickness *d* of the middle layer.

tional amplitude is zero at the ends. For a quantum well of thickness d, this condition becomes

$$n\frac{\lambda}{2} = d$$

where $n = 1, 2, 3 \ldots$. The allowed values of $k = 2\pi/\lambda$ are then

$$k_n = \frac{n\pi}{d} \tag{10-22}$$

and the corresponding allowed energies are

$$E_n = \frac{h^2 n^2}{8m^* d^2} \qquad \text{(quantum well energies)} \tag{10-23}$$

where Eq. (10-3) has been used. In a solid, the electron or hole acts like a particle having an *effective mass* m^* which is different than the mass m of a free electron, due to interactions with atoms of the solid. The energy still has a quadratic dependence on momentum, however, as expressed in Eq. (10-3). In GaAs, for example, $m_e^* \approx 0.067m$ for the electron and $m_h^* \approx 0.48\, m$ for the hole.

Equation (10-23) is only approximately correct for a real quantum well, since the potential barrier that the electron sees is not really infinite. The effect of a finite barrier height is that the energy levels are somewhat lower than predicted, and the electron has a small probability of being found outside of the well. However, for the lowest energy levels in the well, which are far from the band energy of the surrounding material, Eq. (10-23) provides a good approximation.

A most interesting and useful feature of the quantum well is that the energy difference between states in the conduction and valence bands now depends on the well width d. The photon created by a transition from the lowest conduction band state to the highest valence band state will have energy

$$h\nu = E_g + \frac{h^2}{8m_e^* d^2} + \frac{h^2}{8m_h^* d^2} \tag{10-24}$$

where E_g is the normal bandgap energy of the material in the well. This has important advantages in laser applications, since it allows the emission wavelength to be tuned by changing the quantum well thickness. Similarly, the absorption response of photodetectors can be optimized for a particular wavelength by adjusting d.

Another application of quantum well materials is in voltage-controlled modulation of light. The shift of energy levels with applied electric field is known as the *Stark effect,* and in a quantum well this shift (known as the *quantum-confined Stark effect,* or QCSE) is particularly large. Devices based on the QCSE have seen rapid development in recent years.

Metal–Semiconductor Junctions

The junctions discussed so far are those between two semiconductors with different doping levels and/or compositions. Junctions between a semiconductor and a metal are also possible; in fact, they are always necessary for connecting a semiconductor device to an external circuit. Metal–semiconductor junctions also have applications as photodetectors, and as unidirectional circuit elements (diodes). The semiconductor can be either n-type or p-type, but for simplicity we will confine our discussion here mostly to n-type semiconductors.

Energy Levels

Consider first the allowed electron energies in the two materials just before they are joined, as depicted in Fig. 10-16. In the metal, the highest energy electrons occupy a partially filled band, up to an energy known as the *Fermi level.* At zero temperature, all states below the Fermi level are occupied, and all states above are unoccupied. At finite temperature, this transition from filled to unfilled states occurs over an energy range ~ k_BT, where k_B is Boltzmann's constant and T is the absolute temperature.

In the n-type semiconductor, the valence band is filled and the conduction band is nearly empty. The free electrons that are found in the conduction band are mostly due to the doped donor atoms that have been ionized by thermal agitation at finite temperature T. If the temperature were lowered to 0 K, all these free electrons would become bound to the positively charged donor core ions, decreasing their total energy due to a lower electric potential energy. Therefore, the Fermi level in the n-type semiconductor is just below the bottom of the conduction band.

To compare the electron's energy in the metal and semiconductor, we need a common reference energy that applies to both materials. This is conventionally taken to be the *"vacuum level,"* which is the energy the electron would have if it were removed from the material and were at rest (no kinetic energy). Energy must be given to the electron to remove it (otherwise the material would spontaneously lose electrons), so the vacuum level

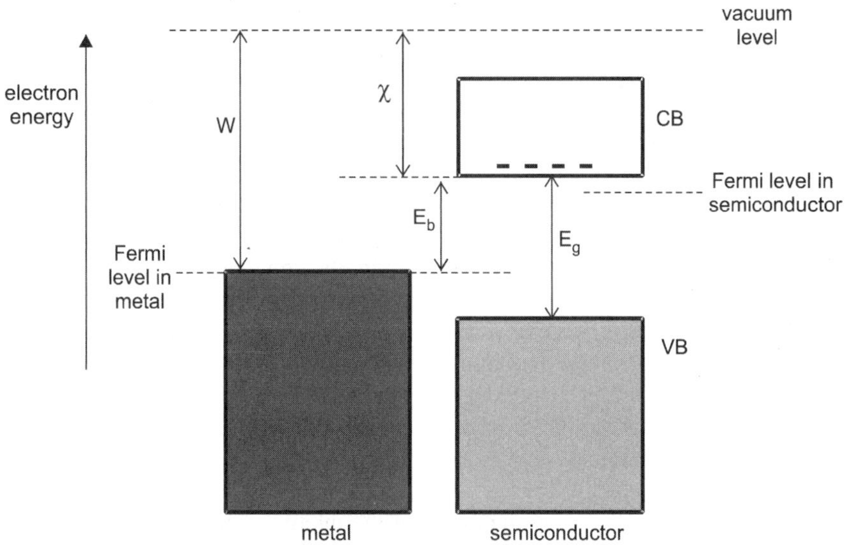

Figure 10-16 Energy of electrons in a metal and semiconductor before they are joined, using the vacuum level as a common reference. The Fermi levels are different in the two materials because they are not in contact and not in thermal equilibrium.

must be higher than the highest occupied energy level. For a metal, the minimum energy needed to remove an electron is known as the *work function,* designated as W. Since electrons have energies up to the Fermi level, the work function is the energy difference between the vacuum level and Fermi level.

For a semiconductor, the work function can be similarly defined as the energy difference between the vacuum level and Fermi level. However, since there are usually very few electrons actually at the Fermi level in a semiconductor, it is customary instead to specify χ, the *electron affinity,* which is the energy required to raise an electron from the bottom of the conduction band to the vacuum level. For most metal–semiconductor combinations, $W > \chi$, and the relative energy positions are as illustrated in Fig. 10-16.

So far, we have considered the energy levels of a separate metal and semiconductor. Imagine now moving them gradually together until their surfaces make contact. A similar type of "thought experiment" was used in our discussion of the properties of the p–n junction. For the metal–semiconductor junction, it is clear from Fig. 10-16 that electrons in the conduction band of the semiconductor have a higher energy than electrons in the metal. Electrons in the metal will not spontaneously jump over into the semiconductor, because to do so they would have to surmount a barrier energy $E_b = W - \chi$, which is usually $\gg k_BT$. On the other hand, electrons in the semiconductor can easily jump over into the metal, since they would be lowering their energy in doing so. Therefore, we conclude that just after the surfaces make contact, there will be a flow of electrons from the semiconductor to the metal.

This transfer of electrons is only momentary, however, because the charge separation that is produced creates an electric field that opposes the motion of the electrons. The generation of this built-in field, illustrated in Fig. 10-17, can be understood in the manner discussed previously for a p–n junction. Electrons leaving the semiconductor uncover positive ion cores of donor atoms, resulting in a positive space-charge density ρ to the right of the junction. The electrons accumulate on the surface of the metal, giving a negative spike to the charge density there. The electric field points from the positive charge to the negative charge, which in this case is to the left ($E_x < 0$). The force on the negatively charged electron is therefore to the right, opposite to the direction of electron flow.

The quantitative analysis of $E_x(x)$ and $V(x)$ in the semiconductor is identical with that given previously in Eqs. (10-14)–(10-20) for the n-type side of a p–n junction. The metal now takes the place of the highly doped p-type semiconductor, and the limit $d_p \to 0$ becomes a very good approximation. Inside the metal, the net charge density must be zero because of the high conductivity, and therefore $E_x = 0$ and $V(x)$= constant. The variation of electric field and potential with position are illustrated in Fig. 10-17, which should be compared with the corresponding graphs for a p–n junction given in Fig. 10-11.

As a result of the charge separation and associated electric field, there is a difference in electric potential between the metal and semiconductor, denoted by V_0. This is referred to as the built-in potential, similar to that of a p–n junction. Since the electron is negatively charged, its electric potential energy decreases as it moves from the metal surface into the semiconductor, where the electric potential is higher. Combining this electric potential energy change with the potential energies that existed without the E field (Fig. 10-16), we obtain the total potential energy curve for the electron shown in Fig. 10-18.

It is seen from Fig. 10-18 that the effect of the internal E field is to lower the Fermi level in the semiconductor, bringing it into alignment with the Fermi level in the metal. According to a general principle of statistical mechanics, a system in thermal equilibrium has a Fermi level that is uniform throughout the material. This requirement that the Fermi levels come into alignment is what determines the magnitude of the built-in potential V_0.

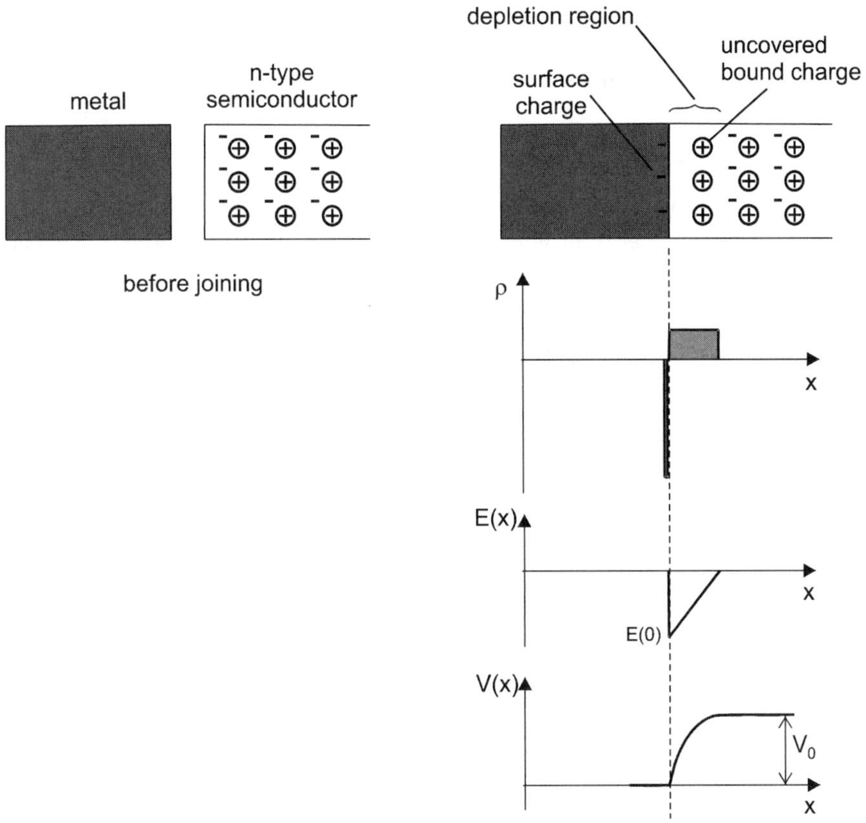

Figure 10-17 When the metal and semiconductor are brought into contact, electrons move from the semiconductor to the metal. The resulting charge separation creates an internal electric field and change of electric potential across the junction.

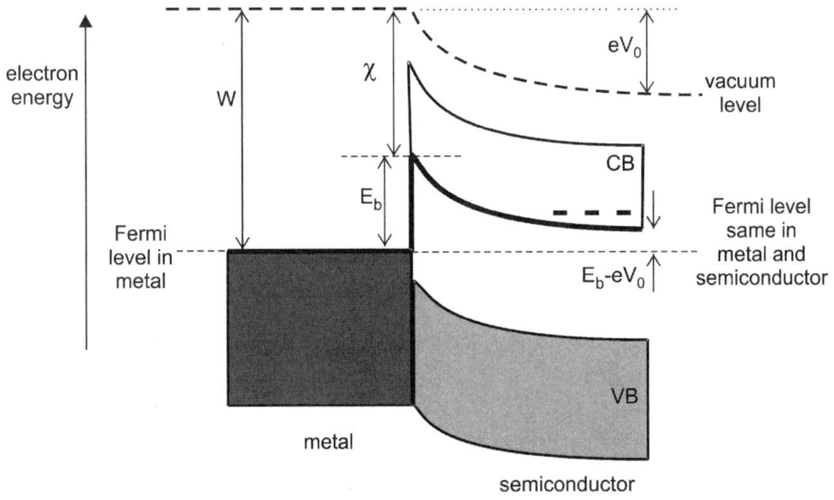

Figure 10-18 The energy bands in the semiconductor are bent due to the field in the depletion region. In thermal equilibrium, the Fermi levels in the metal and semiconductor are brought into coincidence by the built-in potential V_0.

Since the Fermi level in the semiconductor is close to the bottom of the conduction band, $eV_0 \approx E_b$. A similar analysis can be done for a p–n junction, which shows that $eV_0 \approx E_g$, where E_g is the bandgap energy of the semiconductor.

Schottky Diode

According to the energy diagram of Fig. 10-18, electrons on either side of the junction encounter a potential barrier of magnitude $E_b \approx eV_0$, which inhibits their movement across the junction. Those few that do jump across by thermal excitation do so with equal probability in either direction, so that the net current through the junction is zero in thermal equilibrium. To produce a net current, an external voltage needs to be applied, as depicted in Fig. 10-19. When positive potential is applied to the metal and negative to the semiconductor (forward bias, $V > 0$), the energy barrier for electrons jumping right to left across the junction is reduced. There is still the large energy barrier E_b for electrons jumping left to right, however, so the net result is electron flow to the left (conventional current is to the right).

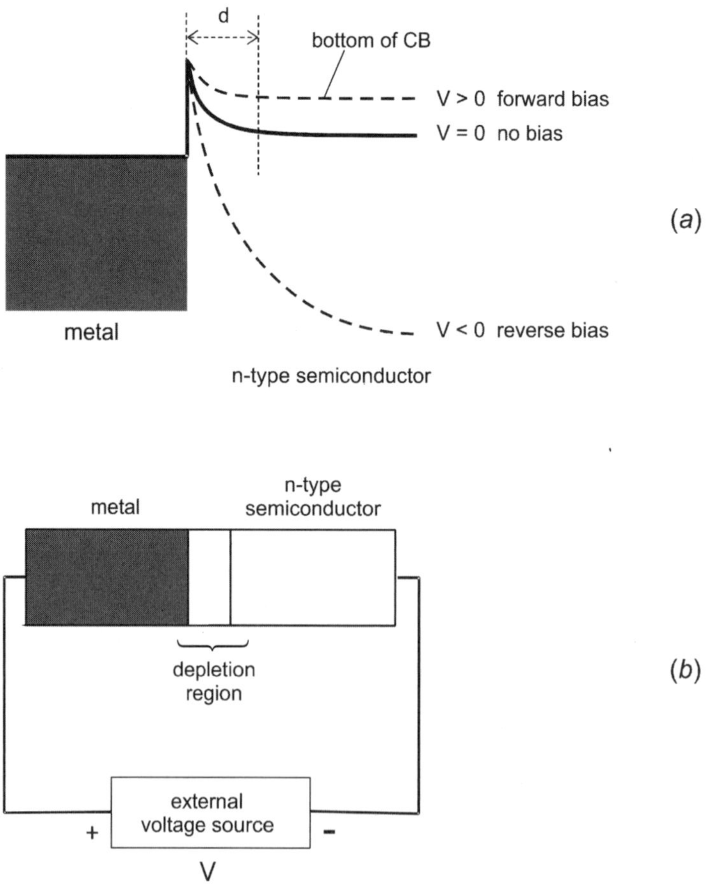

Figure 10-19 A forward bias reduces the barrier for electrons attempting to jump from the semiconductor to the metal, whereas a reverse bias increases this barrier. This results in the unidirectional properties of a diode.

For the opposite polarity of applied voltage (reverse bias, $V < 0$), the barrier for electrons jumping right to left becomes even larger than it was for zero bias. This results in a small net electron flow to the right (current to the left), which is independent of applied voltage. The metal–semiconductor junction, therefore, acts as an electrical diode, conducting current efficiently in one direction only. Such a device is termed a *Schottky diode,* and it shares the rectifying property and other behavior of a p–n junction. For example, the i–V curve has the shape given in Fig. 10-14 for the p–n junction. One difference is that the Schottky diode is a majority-carrier device—there are no minority carriers such as electrons in a p-type region or holes in an n-type region. The lack of minority carrier diffusion in the Schottky diode leads to a faster response time, which is a key advantage in applications such as high-speed electronic signal processing and high-speed photodetection (see Chapter 14).

Ohmic Contacts

When connecting a semiconductor device to the metallic wires of an external circuit, the rectifying property of a Schottky diode is generally undesirable. The ideal contact is one in which the voltage drop across the junction is small and proportional to the current. Junctions having this property are termed *ohmic contacts,* since they obey Ohm's law.

One way to achieve ohmic contacts is to increase the donor concentration in the semiconductor. According to Eq. (10-20), the junction thickness d decreases as the donor concentration N_D increases. If d becomes sufficiently small, electrons can pass through the energy barrier by the quantum mechanical process of *tunneling.* Values of $N_D > 10^{19}$ cm^{-3} are generally needed for this tunneling process to dominate.

PROBLEMS

10.1 Using the expression for $E(k)$ in Eq. (10-3), calculate a particle's velocity according to Eq. (10-5), and show that it leads to a momentum consistent with Eq. (10-2).

10.2 A slab of $Al_xGa_{1-x}As$ is illuminated with light of variable wavelength, and it is found that the transmission through the slab decreases rapidly for wavelengths shorter than 770 nm. Determine the Al fraction x in this material.

10.3 A sample of GaAs has a lifetime for nonradiative decay to traps of $\tau_{nr} = 10$ ns, where $\tau_{nr} \equiv 1/A_{nr}$. Using the value of B_r from Example 10-2, determine the hole concentration at which the radiative efficiency for an electron in the conduction band will be 20%.

10.4 Use Eq. (10-13) to determine the electron concentration that maximizes the radiative efficiency. (a) Express your answer symbolically in terms of A_{nr}, B_r, and C_A. (b) Use the data in Example 10-2 to evaluate this optimum concentration for GaAs and $In_{.53}Ga_{.47}As$. Assume $A_{nr} = 10^7$ s^{-1} for both materials. (c) Determine the radiative efficiency for the concentrations determined in part b.

10.5 Derive Eq. (10-15) by applying Gauss's law $\int \mathbf{E} \cdot d\mathbf{A} = q_{\text{enc}}/\varepsilon_0$ to a rectangular volume of length x and constant charge density ρ. Remember that the area vector $d\mathbf{A}$ points from the inside toward the outside of the volume.

10.6 A silicon diode has a p–n junction with donor density of 10^{16} Sb atoms per cm^3 on the n side, and acceptor density of 10^{14} B atoms per cm^3 on the p side. This results in a built-in potential of $V_0 = 0.56$ V. (a) Determine the junction width. (b) Determine the maximum electric field in the junction region.

10.7 (a) Give arguments to show that the capacitance of a p–n junction can be obtained using the formula for a parallel-plate capacitor with plate spacing *d*. (b) Determine the capacitance of the diode of Problem 10.6, assuming a junction area 1 mm^2.

10.8 Show that the current through a diode must be zero when the voltage across it is zero. Hint: consider the circuit of Fig. 10-14 with the voltage source replaced by a resistor.

10.9 Eq. (10-20) was derived under the assumption that $N_D \ll N_A$. Relax this assumption and derive a more general expression for the depletion width in terms of the dopant concentrations on both sides of the junction.

10.10 A GaAs quantum well is sandwiched between layers of $Al_xGa_{1-x}As$, as illustrated in Fig. 10-15. (a) Determine the well thickness *d* required to shift the usual GaAs transition wavelength from 870 nm to 830 nm. (b) What restrictions are needed on the Al fraction *x* in the surrounding layers in order to make this a practical device? Assume that the well depths are the same in the valence and conduction bands.

10.11 Determine the transition energy (in eV) from the $n = 2$ level in the conduction band to the $n = 1$ level in the valence band for a quantum well in GaAs of thickness 10 nm. Compare this with the $n = 1$ to $n = 1$ transition in the quantum well, and with the transition in bulk $Al_{0.3}Ga_{0.7}As$. Take the GaAs bandgap energy as 1.425 eV.

Chapter 11

Light Sources

The previous chapter reviewed the fundamental processes by which light is emitted or absorbed by semiconductors. We continue in this chapter by examining the principles and operating characteristics of two light-emitting devices: the LED and the laser diode.

11-1. THE LED

When a diode is forward biased, as shown in Fig. 11-1a, current flows readily through the device. The current consists of holes in the p region, and electrons in the n region, both moving toward the p–n junction. When the electrons and holes meet in the vicinity of the junction, they recombine and emit photons of energy $h\nu$. A device in which useful light is emitted in this way is termed a *light-emitting diode* or LED. We consider here some fundamental aspects of LED operation.

Biasing and Optical Power

The optical power generated in an LED is related in a simple way to the current i flowing through the device. Since each electron contributing to the current in the n region recombines with a hole at the junction, the number of recombinations per unit time is equal to the number of electrons entering the junction per unit time, i/e. The optical power P_{opt} is therefore

$$\begin{aligned} P_{\text{opt}} &= \left[\frac{\text{recombinations}}{\text{unit time}}\right]\left[\frac{\text{photons}}{\text{recombination}}\right]\left[\frac{\text{energy}}{\text{photon}}\right] \\ &= \left(\frac{i}{e}\right)\eta_i h\nu \end{aligned} \tag{11-1}$$

where η_i is the internal efficiency given by Eq. (10-13). For wavelengths in the range of interest for fiber optic communications ($0.8 < \lambda < 1.5$ μm), the quantity $h\nu/e$ evaluates to ~ 1 W/A, so the optical power in mW is roughly equal to the current in mA multiplied by η_i.

The optical power can also be related to the electrical power $P_{\text{elec}} = iV_d$ supplied to the LED. The voltage V_d applied to the diode must shift the energy bands on either side of the junction by ~ E_g for significant current to flow (see Fig. 10-13). The emitted photon energy is then $h\nu \approx E_g \sim eV_d$, and Eq. (11-1) can be written as

$$P_{\text{opt}} \sim \eta_i P_{\text{elec}} \tag{11-2}$$

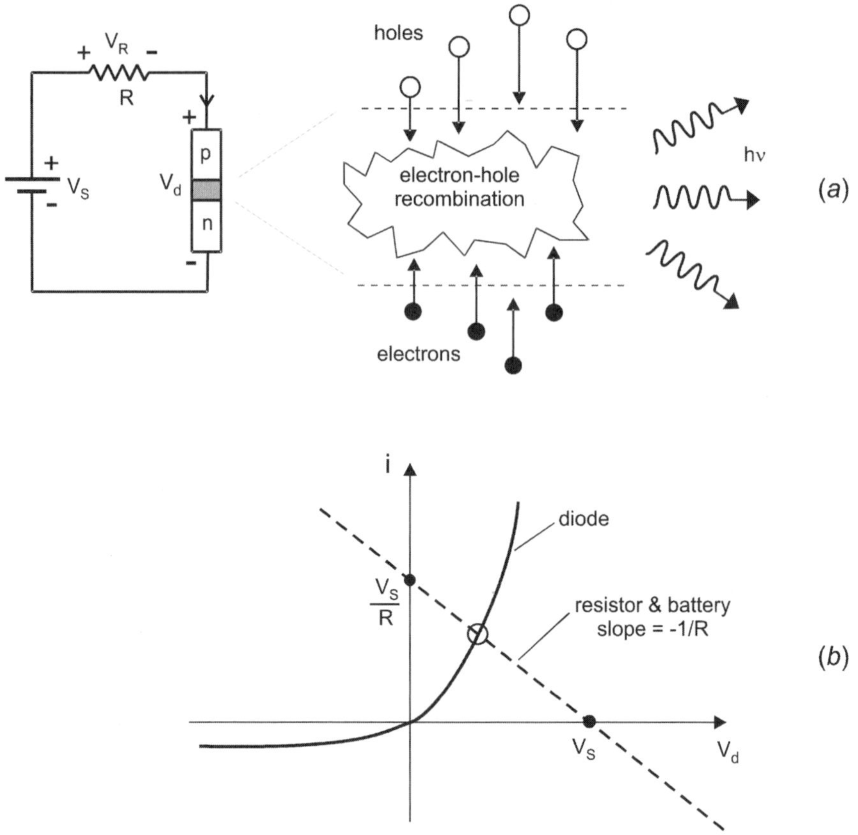

Figure 11-1 (a) A simple circuit for biasing an LED. (b) The operating point for this circuit occurs where the load line (dashed) intersects the diode curve (solid).

The electrical-to-optical conversion efficiency is thus ~ η_i. Values of η_i can be close to unity, making light generation a very efficient process in the LED.

A more accurate analysis of the light generation efficiency can be obtained by determining the actual diode voltage V_d, rather than using the approximation $V_d \sim E_g/e$. Figure 11-1a shows a simple circuit for biasing an LED, using a series combination of source voltage V_s and resistance R. Defining the polarities as shown, the voltages and current are related by $V_d = V_s - iR$, or

$$i = \frac{1}{R}(V_s - V_d) \tag{11-3}$$

This equation, known as the *load line,* gives the diode current i versus voltage V_d relation for the resistor/voltage source part of the circuit. It must be consistent with the i versus V_d relation for the diode itself, given in Eq. (10-21) and Fig. 10-14. The solution for i and V_d can be obtained by setting Eq. (11-3) equal to Eq. (10-21), with $V = V_d$. However, this results in a transcendental equation, which must be solved numerically. An alternative approach is to plot both the load line and the diode curve on the same graph, as shown in Fig. 11-1b. The intersection of the two curves then gives the operating point for the circuit.

The load line analysis provides a simple way of understanding how changes in the bias resistor effect the circuit. The load line has a slope of $-1/R$, and passes through the fixed point V_s on the V_d axis. As R increases, the slope becomes smaller and the intersection point lies lower on the diode curve, giving a smaller operating current. In the limit of infinite resistance, $I = 0$ as expected. As $R \rightarrow 0$, the current gets very large, and is ultimately limited by the resistance of the semiconductor material itself. The optical power generated is, therefore, easily adjusted by changing the bias resistance.

EXAMPLE 11-1 A GaAs LED is designed to generate 15 mW of light in a series circuit with a 3 V battery and load resistor. Assume that the emission wavelength is 860 nm, $T = 293$ K (room temperature), $\eta_i = 0.80$, the reverse saturation current density is 1×10^{-8} A/cm^2, and the junction area is (1 mm)^2. Determine the required load resistance.

Solution: Using Eq. (11-1), the required current is

$$i = \frac{e\lambda P_{\text{opt}}}{\eta_i hc} = \frac{(1.6 \times 10^{-19})(8.6 \times 10^{-7})(1.5 \times 10^{-2})}{(0.8)(6.63 \times 10^{-34})(3 \times 10^{8})} = 0.0130 \text{ A}$$

Using $i_0 = (10^{-8}$ A/cm^2) $(10^{-2}$ cm$^2) = 10^{-10}$ A, Eq. (10-21) gives the diode voltage,

$$V_d \simeq \frac{k_B T}{e} \ln\left(\frac{i}{i_0}\right) = \frac{(1.38 \times 10^{-23})(293)}{1.6 \times 10^{-19}} \ln\left(\frac{0.0130}{10^{-10}}\right) = 0.473 \text{ V}$$

Note that this assumes $\beta = 1$. For $\beta = 2$, V_d will be twice this, or 0.965 V. The resistance is then found from Eq. (11-3),

$$R = \frac{V_s - V_d}{i} = \frac{3 - 0.473}{0.013} = 194\Omega$$

Time and Frequency Response

In certain applications, it is required that the LED be turned on and off quickly. This is especially true in optical communications, where the rate at which data can be sent depends (among other things) on the time response of the light source. A simple circuit that modulates the LED output is shown in Fig. 11-2. The transistor serves as a switch, offering a low-resistance path between emitter and collector only when the input voltage V_{in} is sufficiently high. The speed with which the LED output responds to the input voltage depends not only on the LED itself, but also on the transistor and associated circuitry. We will focus here only on the effect of the LED on the time response.

One limit to the time response that is always present to some degree is due to capacitance. The capacitance C of the p–n junction, in combination with the load resistance R, gives a response time equal to the *time constant RC*. We will find in Chapter 14 that this is often significant for photodiode detectors, where R can be quite large. In the case of LEDs, R is typically small, and the RC time constant does not usually dominate the time response.

More important for the LED time response is the electron lifetime in the conduction

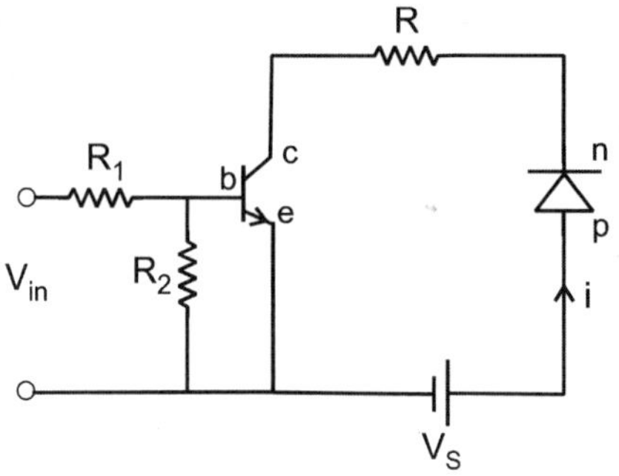

Figure 11-2 The current through the LED can be modulated by inserting a transistor into the circuit.

band, τ. If the current i in Fig. 11-2 is switched on and off instantaneously by the transistor, the light emitted by the LED will not follow the switched current exactly, but will exhibit a delayed response as depicted in Fig. 11-3. This can be understood by considering that light is emitted whenever there are e–h pairs available for recombination. The population of e–h pairs cannot change instantaneously, however, but rather increases and decreases with the characteristic time τ.

To analyze the time response quantitatively, we define the number of excess electrons in the recombination region as $\mathcal{N}$. This is related to the electron density n by $\mathcal{N} \equiv nAd$, where A is the junction area and d is the recombination region thickness. The electron population $\mathcal{N}$ satisfies the *rate equation*

$$\frac{d\mathcal{N}(t)}{dt} = \frac{i(t)}{e} - \mathcal{N}(t)W_{tot} \tag{11-4}$$

where W_{tot} is the probability of decay for a single electron per unit time, given by Eq. (10-12). This equation basically says that the net change in $\mathcal{N}$ per unit time is equal to the

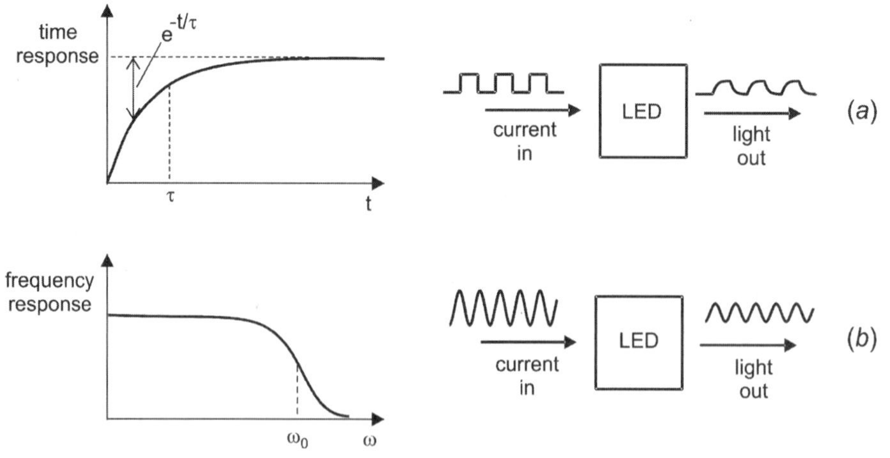

Figure 11-3 The response of the LED can be described by (a) a relaxation time that rounds off edges in the time waveform, or (b) a frequency bandwidth, above which the amplitude of a sinusoidal waveform is attenuated.

number added by injected current minus the number taken away by e–h recombinations or other losses. The solution for $i(t) = 0$ is given by

$$\begin{aligned} \mathcal{N}(t) &= \mathcal{N}(0)e^{-W_{\text{tot}}t} \\ &= \mathcal{N}(0)e^{-t/\tau} \end{aligned} \tag{11-5}$$

which is easily verified by substituting into Eq. (11-4). The *electron lifetime* τ has been defined here as $\tau \equiv 1/W_{\text{tot}}$. According to Eq. (11-5), when the current is abruptly cut off, the electron population decreases exponentially with a time constant τ. Since the light generated by the LED is $\propto \mathcal{N}$, it will decay exponentially in the same way.

When $i(t)$ is switched from zero back to a constant value, the solution for $\mathcal{N}(t)$ is an exponential rise with the same time constant τ (see Problem 11.2). The situation is analogous (indeed mathematically identical) to the charging and discharging of a capacitor in an RC electrical circuit. The result is a rounding of the leading and falling edges of light pulses from the LED, as illustrated in Fig. 11-3.

To produce light pulses that accurately follow the input current, a small value of τ is desirable. If the electron decay is primarily radiative ($\eta_i \approx 1$), then Eq. (10-10) gives

$$\frac{1}{\tau} = W_r = B_r p \tag{11-6}$$

where p is the number of holes per unit volume in the recombination region. At low injection-current levels, most of the holes available for recombination come from the dopant acceptor atoms in the p-type material, with concentration N_A. In this case $p \approx N_A$, and

$$\tau \approx \frac{1}{B_r N_A} \qquad \text{(electron radiative lifetime)} \tag{11-7}$$

For the fastest response, N_A should be as large as possible, up to a certain limit. One reason for the limit comes from the creation of nonradiative trap sites that decrease the radiative efficiency η_i [see Eq. (10-13)]. For dopants such as Ge, C, and Be in GaAs, η_i starts to decrease for N_A in the range 10^{18} to 10^{19} cm^{-3}. Increasing N_A above this range makes the LED response faster, but at the expense of decreased output efficiency.

Another reason for the limit is that the radiative recombination rate W_r does not increase indefinitely with N_A, but rather saturates at a maximum rate characteristic of the material. For electrons in GaAs, the maximum radiative rate is $\approx 3.3 \times 10^9$ s^{-1}, with a corresponding minimum response time $\tau_{min} \approx 0.3$ ns. Using $B_r = 7.2 \times 10^{-10}$ cm^3/s, this minimum response time would be reached at an acceptor concentration of $\sim 5 \times 10^{18}$ cm^{-3}. In practice, response times as short as 0.1 ns have been measured in heavily (7×10^{19} cm^{-3}) Be-doped GaAs. It is thought that the improved time response is due to Auger processes contributing to the decay rate.

The speed with which an LED responds to switching is often expressed in terms of modulation bandwidth rather than response time. The basic idea is illustrated in Fig. 11-3b, which shows a sinusoidal current driving the LED at modulation frequency $f = \omega/2\pi$. Here ω is the angular frequency of the modulation. If the modulation amplitude is not too large, the optical output will also vary sinusoidally at the same frequency. Over some range of frequencies, the response (ratio of output to input amplitudes) is approximately constant, independent of ω. At some value $\omega = \omega_0$, the response decreases to half its low-frequency value, and the corresponding $f_0 = \omega_0/2\pi$ is known as the modulation bandwidth.

To see how the bandwidth f_0 is related to the response time τ, we will solve Eq. (11-4) for a sinusoidally modulated current of the form

$$i(t) = i_0 e^{j\omega t} \tag{11-9}$$

where i_0 is real and $j \equiv \sqrt{-1}$. In this analysis, it is assumed that when there is a complex quantity, the real part of the expression corresponds to the actual physical quantity. For example, the actual current is $i(t) = \Re[i_0 \exp(j\omega t)] = i_0 \cos(\omega t)$. The electron population $\mathcal{N}$ is expected to oscillate at frequency ω with some amplitude A, so we write

$$\mathcal{N}(t) = Ae^{j\omega t} \tag{11-9}$$

The amplitude A will in general be complex, to account for a phase difference between $i(t)$ and $\mathcal{N}(t)$. Substituting Eqs. (11-8) and (11-9) into Eq. (11-4) gives

$$j\omega A = \frac{i_0}{e} - \frac{A}{\tau}$$

or

$$A = \frac{i_0 \tau / e}{1 + j\omega\tau} \tag{11-10}$$

The numerator of Eq. (11-10) is real, and corresponds to the low-frequency limit for the amplitude A. The denominator is complex, and can be written as

$$1 + j\omega\tau = \sqrt{1 + (\omega\tau)^2}\, e^{j\phi}$$

where $\tan\phi = \omega\tau$. Eq. (11-9) can then be expressed as

$$\mathcal{N}(t) = \frac{i_0 \tau / e}{\sqrt{1 + (\omega\tau)^2}}\, e^{j(\omega t - \phi)} \tag{11-11}$$

Since the optical power generated is $\propto \mathcal{N}(t)$, it will have a modulated amplitude P_{mod} which depends on frequency according to

$$P_{\text{mod}}(\omega) = \frac{P_{\text{mod}}(0)}{\sqrt{1 + (\omega\tau)^2}} \quad \text{(LED frequency response)} \tag{11-12}$$

There are two commonly used definitions for the bandwidth of the LED. The 3 dB electrical bandwidth is conventionally defined as the frequency at which the electrical power is reduced by a factor of two. Since the power in an electrical circuit is proportional to V^2 or i^2, the 3 dB bandwidth occurs at the frequency at which i^2 is reduced by a factor of two. If the LED is thought of as an optoelectronic circuit element that generates optical power in proportion to current, it makes sense to define the 3 dB *electrical bandwidth* $f_e = \omega_e/2\pi$ by

$$\left[\frac{P_{\text{mod}}(\omega_e)}{P_{\text{mod}}(0)}\right]^2 = \frac{1}{1 + (\omega_e\tau)^2} = \frac{1}{2}$$

Solving for f_e gives $\omega_e\tau = 1$ or

$$f_e = \frac{1}{2\pi\tau} \qquad \text{(3 dB electrical bandwidth)} \tag{11-13}$$

Using a typical minimum lifetime of $\tau_{\min} \approx 0.3$ ns, this equation gives a maximum modulation bandwidth for the LED of 530 MHz.

The other definition for bandwidth recognizes that the light output of the LED is a power, and so the 3 dB point should occur at the frequency at which this optical power is reduced by a factor of two. Defining the *optical bandwidth* as $f_o = \omega_o/2\pi$, we have

$$\frac{P_{\text{mod}}(\omega_o)}{P_{\text{mod}}(0)} = \frac{1}{\sqrt{1 + (\omega_o\tau)^2}} = \frac{1}{2}$$

which yields $\omega_o\tau = \sqrt{3}$, or

$$f_o = \frac{\sqrt{3}}{2\pi\tau} \qquad \text{(3 dB optical bandwidth)} \tag{11-14}$$

The relation between response time and bandwidth can also be understood in terms of the uncertainty relation (see Appendix B). According to this viewpoint, the time waveform can be reconstructed by a linear superposition of many sinusoidal waves with a distribution of frequencies ω. When only frequencies in a range $\Delta\omega$ are allowed in the reconstruction, the edges of any pulse will arrive with an uncertainty in time given by by $\Delta t \sim 1/\Delta\omega$. This is consistent with Eqs. (11-13) and (11-13) if we associate $\Delta\omega$ with the bandwidth and Δt with the response time τ.

Emission Efficiency

In Section 10-1, we considered the efficiency with which e–h pairs recombine to give light. This is the internal efficiency, and can be quite high ($\eta_i \sim 1$). However, photons that are generated still have to make it out of the LED to count as useful output power. The fraction of generated photons that escape from the LED is known as the *external efficiency, η_{ext}*, and is generally much lower than η_i.

There are two principal causes of reduced η_{ext}. The first is reabsorption of emitted photons by the semiconductor material, shown in Fig. 11-4a. Photons generated by e–h recombination have an energy $h\nu \simeq E_g$ that is large enough to promote an electron back up across the bandgap energy E_g in an absorption process. When this happens, the photon is lost. One solution to this problem is to make the active region very thin, and sandwich it between two buffer layers with a larger band gap, as in Fig. 11-4b. The photon energy $h\nu$, which is determined by the band gap of the thin active region, is now too small to cause absorption across the larger band gap in the buffer layers. This will be discussed further in connection with diode lasers.

The other principal reason for reduced external efficiency in LEDs is total internal reflection at the semiconductor-air interface. Figure 11-5 illustrates this problem by showing light emitted inside the semiconductor, propagating at different angles θ to the surface normal. For angles greater than the critical angle ($\theta > \theta_c$), the light is totally reflected back into the semiconductor and does not "escape" from the LED. The critical angle depends

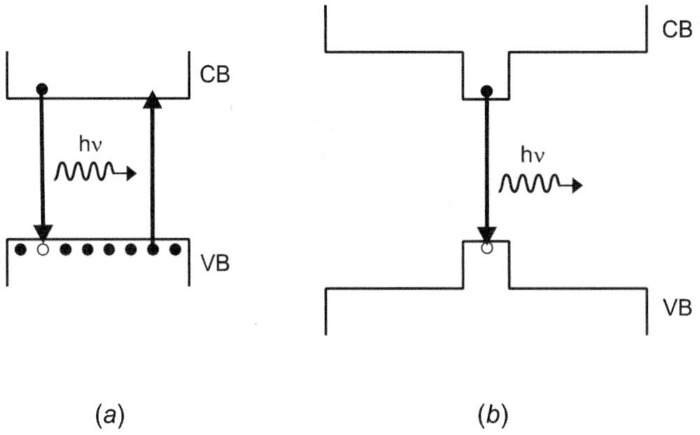

Figure 11-4 (a) Emitted light is reabsorbed by the semiconductor from which it originated. (b) In a layered structure, this reabsorption is reduced.

on the indices of refraction as $\sin\theta_c = n_2/n_1$, from Eq. (2-18). For example, a GaAs–air interface has $n_1 = 3.6$ and $n_2 = 1$, giving $\theta_c \simeq 16°$. Therefore, of all the light that is emitted inside the GaAs material, only the fraction emitted within a cone of half-angle 16° escapes to become useful output.

An estimate of this fraction that escapes can be made using the concept of the solid angle (see Appendix A). Assuming that light is emitted uniformly into all 4π steradians inside the material (isotropic emission), the fraction emitted into the solid angle Ω of the cone is

$$\eta_{\text{ext}} = \frac{\Omega}{4\pi} T \qquad \text{(external efficiency)} \qquad (11\text{-}15)$$

where T accounts for Fresnel reflection losses at the interface. A useful approximation here is $\Omega \leqslant \pi\,\theta_c^2$, valid when $\theta_c \ll 1$, with θ_c in radians. It is also a good approximation to use the normal incidence reflectivity of Eq. (2-14) in determining T when $\theta_c \ll 1$.

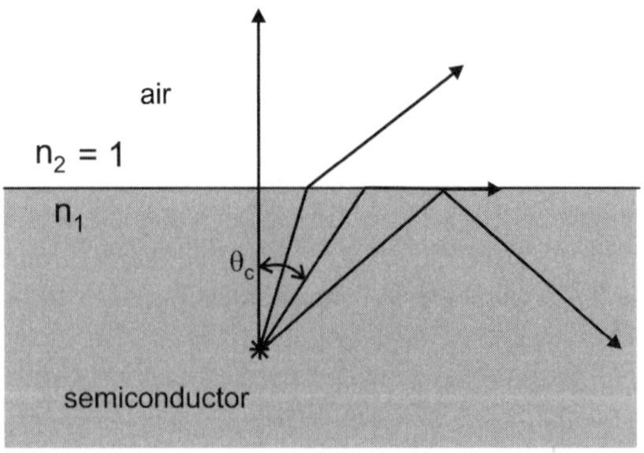

Figure 11-5 Light emitted inside the semiconductor suffers total internal reflection when the angle of incidence on the surface exceeds the critical angle θ_c.

EXAMPLE 11-2

Estimate the external efficiency for a GaAs LED emitting into air.

Solution: Light is emitted into a cone of half-angle 16° = 0.28 rad. The solid angle is

$$\Omega = \pi\theta_c^2 = \pi(0.28)^2 = 0.246 \text{ sr}$$

and the Fresnel transmission is

$$T = 1 - \left(\frac{n_1 - n_2}{n_1 + n_2}\right)^2 = 1 - \left(\frac{3.6 - 1}{3.6 + 1}\right)^2 = 0.68$$

The external efficiency is then

$$\eta_{\text{ext}} = \frac{0.246}{4\pi}(0.68) = 0.013$$

The above example shows that the external efficiency of an LED is naturally quite low. There are, however, ways to improve this efficiency for certain applications. When the goal is simply to get photons out into the air, as in an indicator light, the semiconductor can be encapsulated in a plastic dome structure, as shown in Fig. 11-6. The critical angle for the semiconductor–plastic interface is larger than that of a semiconductor–air interface, giving a larger Ω and η_{ext}. In this design, the Fresnel losses at the dome–air interface are also minimized because all emitted rays pass through at normal incidence. A higher refractive index material for the dome, such as another semiconductor, would give an even greater improvement in η_{ext}. However, this represents a manufacturing challenge, and takes away one of the key advantages of LEDs: they are inexpensive.

For coupling LED light into an optical fiber, a "Burrus" type geometry is often used. As illustrated in Fig. 11-7, the end of the fiber is brought into close proximity to the emission region by etching a well in one side, and fixing the fiber with an index-matching epoxy. The

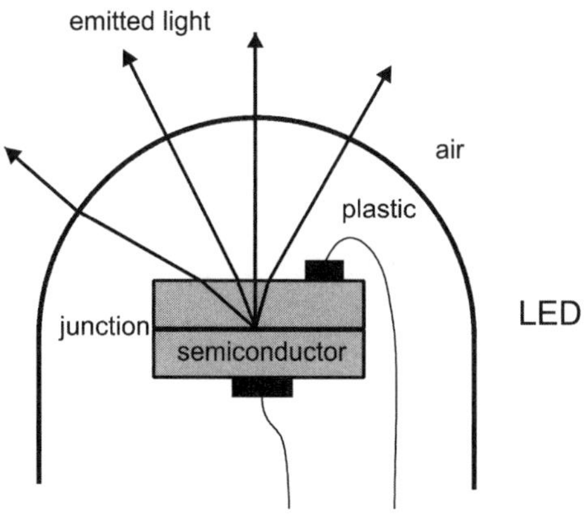

Figure 11-6 Total internal reflection losses are reduced by encapsulating the semiconductor in a dome-shaped structure with high refractive index.

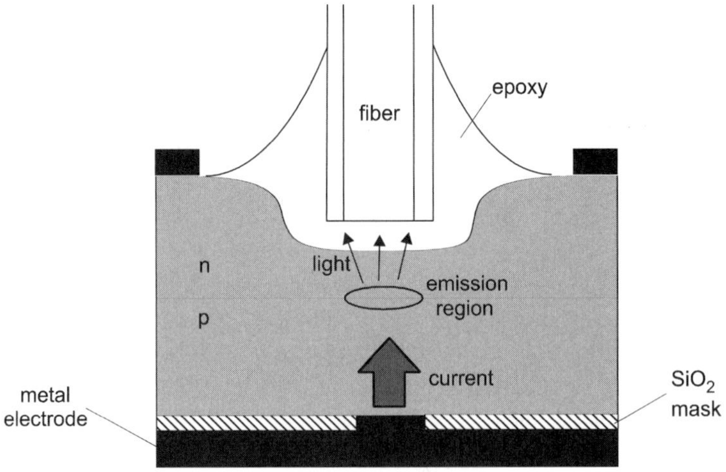

Figure 11-7 In the Burrus geometry, emission from the surface of the p–n junction is collected by a fiber attached to the diode.

emission region is restricted to a small part of the p–n junction by an insulating SiO_2 mask that confines the current flow laterally. If there is good optical contact between the fiber and semiconductor, the critical angle is determined by the refractive indices of the semiconductor and glass fiber. Using Eq. (11-15), η_{ext} is then $\simeq 0.04$, which is still fairly low but higher than the value $\eta_{ext} \simeq 0.01$ for emission into air. However, only a small fraction of this light is coupled into guided modes of the fiber, as shown in Problem 11.7.

An LED in the Burrus geometry is termed a *surface emitter,* since the light emission is perpendicular to the p–n junction surface. Alternatively, the LED can be operated as an *edge emitter,* as illustrated in Fig. 11-8. The emission region here is a thin active layer, sandwiched between two layers of lower refractive index. This structure forms a planar waveguide (see Chapter 3), and allows light in the various modes to propagate with little loss. When light is generated inside the active layer, some of it is trapped by the waveguide, eventually exiting through the edge of the material. If the refractive index differ-

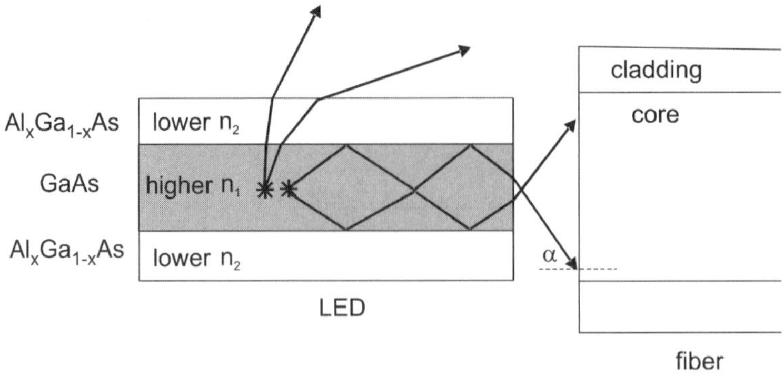

Figure 11-8 In an edge-emitting LED, the structure creates a planar waveguide that traps some of the light generated in the active layer. This light is then emitted from the edge with a narrow angular distribution.

ence between the semiconductor layers is not very large, a fairly small fraction of the light generated will be trapped by the waveguide (see Problem 11.8), resulting in a low output power. However, the light that is emitted will have a narrow angular distribution when it leaves the LED, and will be efficiently coupled into an optical fiber. This is especially beneficial for fibers with a low NA, which accept light only over a narrow range of angles. The efficiency of coupling light into a fiber is further discussed in Chapter 12.

11-2. THE LASER DIODE

In an LED, light is generated by electrons and holes as they recombine radiatively, in a process known as *spontaneous emission*. The electron decay rate is given by Eq. (10-10), and depends on the number of holes per unit volume but not on the light intensity. In a *laser diode,* on the other hand, light is generated by a different process known as *stimulated emission.* In this process, first proposed by Albert Einstein in 1917, the probability that a photon is generated depends on the number of photons already present, that is, on the light intensity. The result is an amplification of the light, with additional photons being produced by those already created. This amplification can be made self-sustaining by adding reflective elements to the ends of the device. As the light makes multiple passes through the semiconductor, it is increasingly amplified until laser light is produced.

There are many interesting facets of laser physics to explore. In this section, we give a first overview of laser characteristics, focusing especially on the contrast between LEDs and laser diodes. We also describe a number of specific types of semiconductor lasers that have been developed, along with their applications. A more detailed accounting of laser principles and operation will be given in Chapters 15–23.

Properties of Lasers

Threshold

One fundamental aspect of laser operation is that of threshold: lasing does not occur until a minimum amount of power is injected into the material. This behavior is illustrated in Fig. 11-9, which compares the power output versus drive current characteristics for a laser diode and LED. Although the laser diode does emit light spontaneously below threshold, just like an LED, this spontaneous emission is much weaker than the laser light emitted above threshold. Above threshold, the power output P_{out} increases approximately linearly with drive current i according to

$$P_{\text{out}} = \beta_s(i - i_{th}) \tag{11-16}$$

where i_{th} is the threshold current. The slope of the curve is β_s, which relates the increment in output power ΔP_{out} to the increment in drive current Δi according to $\beta_s = \Delta P_{\text{out}}/\Delta i$.

To derive an estimate for β_s, we must understand how the additional photons are generated above threshold. A fascinating feature of stimulated emission is that the additional photons generated above threshold are duplicates, or "optical clones," of those that already exist. This means that if light is already propagating in the modes of an edge-emitting diode structure (Fig. 11-8), the additional photons generated will be emitted into those same modes. Loosely speaking, stimulated emission has the effect of channeling additional photons into a particular path, that of the waveguide modes. Since the extraction

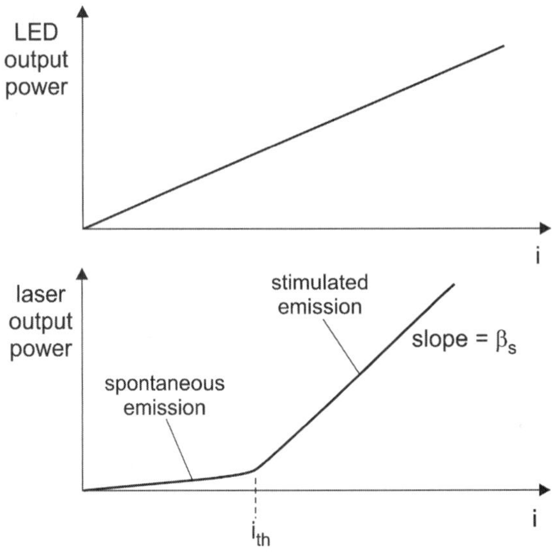

Figure 11-9 The output power versus drive current for a laser diode exhibits a threshold behavior, whereas that for an LED does not.

efficiency of these guided modes can be quite high, the extra optical power output ΔP_{out} is approximately the same as the extra optical power generated, ΔP_{opt}. Using Eq. (11-1), we then have

$$\beta_s \simeq \frac{\Delta P_{\text{out}}}{\Delta i} \simeq \frac{\Delta P_{\text{opt}}}{\Delta i} \simeq \eta_i \frac{h\nu}{e} \tag{11-17}$$

which for wavelengths ~ 1 μm and $\eta_i = 1$ evaluates to $\beta_s \sim 1$ mW/mA. For example, if $i_{th} = 40$ mA and $i = 140$ mA, the output power would be $P_{\text{out}} \simeq 100$ mW. Note that although Eq. (11-1) predicts a similarly high conversion of current into light for an LED, most of this light does not escape the LED as useful output.

The higher output power of a laser diode compared to an LED is one of its practical advantages. A downside is that the laser diode does not respond linearly to i over the entire range. Therefore, if linear modulation of the output is desired, special circuitry is needed to keep the laser diode biased well above threshold.

Directionality

A property of lasers that is quite useful for many applications is the highly directional nature of the emitted light. For a nonlaser source such as an LED, the angular distribution of emitted light often obeys the relation

$$P_{\text{out}}(\theta) \propto \cos\theta \qquad \text{(LED, Lambert's law)} \tag{11-18}$$

This is known as *Lambert's law,* and will be discussed further in Chapter 12. In contrast, the angular distribution of light from a laser can be modeled approximately as

$$P_{\text{out}}(\theta) \propto \cos^n\theta \qquad \text{(laser, approximate)} \tag{11-19}$$

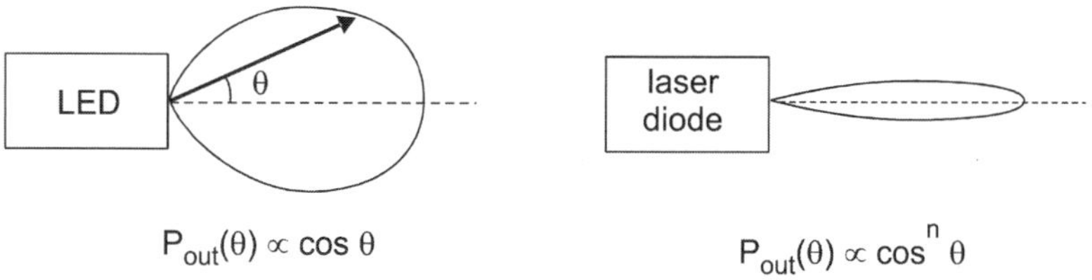

Figure 11-10 The angular distribution pattern is much more directional for a laser than for an LED.

where n is a large number. This becomes highly directional for large n, as shown in Fig. 11-10. It should be noted that Eq. (11-19) is not derived from fundamental principles, but is rather just a convenient form for modeling the laser output.

The physical origin of the laser's directionality lies in the optical cloning process that occurs during stimulated emission. Each photon is added to the beam with the same phase, so that the entire optical wavefront is oscillating in a synchronized fashion. In such a situation, the angular spreading of light is given by the diffraction condition $\theta \sim \lambda/D$ of Eq. (2-25). When light is emitted from the edge of a laser diode, as in Fig. 11-11a, the effective aperture dimension D can be taken to be the active region thickness d. The angular spread of light in a direction perpendicular to the layer will then be

$$\Delta\theta_\perp \sim \frac{\lambda}{d} \qquad \text{(half-width perpendicular to layer)} \qquad (11\text{-}20)$$

where λ is the wavelength after leaving the diode, and $\Delta\theta_\perp$ is the half-width of the distribution. There will also be a spreading of the beam parallel to the layer, given by

$$\Delta\theta_\parallel \sim \frac{\lambda}{w} \qquad \text{(half-width parallel to layer)} \qquad (11\text{-}21)$$

where w is the width of the active region parallel to the junction. Since $d < w$ for most laser diodes, the angular distribution is asymmetrical, with greater spreading in the direction perpendicular to the layers. This creates complications for coupling laser diode light into symmetrical elements like optical fibers, although special aspherical lenses can help to circularize the beam. The problem of coupling laser light into optical fibers is treated further in Chapters 12 and 17.

EXAMPLE 11-3

A GaAs laser diode emitting at 830 nm (free-space wavelength) has an angular width of 18° (full width at half maximum, or FWHM) perpendicular to the plane of the junction. Determine (a) the thickness of the active region, and (b) the value of n in the $\cos^n\theta$ angular dependence perpendicular to the junction.

Solution: (a) The half-width is $\Delta\theta_\perp = 9° = 0.157$ rad. The active layer thickness is then

$$d \simeq \frac{0.830\ \mu\text{m}}{0.157} = 5.3\ \mu\text{m}$$

(b) We require that

$$\frac{1}{2} = \cos^n(9°)$$

which can be solved for n by taking the log of both sides:

$$\ln(0.5) = n \ln(\cos 9°)$$

or

$$n = \frac{\ln(0.5)}{\ln(\cos 9°)} \simeq 56$$

Spectral Purity

In addition to having a narrow angular distribution, laser light also has a narrow spectral distribution, as illustrated in Fig. 11-12a. Typical spectral linewidths for a diode laser are ~ 2 nm, in contrast to typical widths of ~ 50 nm for an LED. This is another consequence of the "optical cloning" process. The photons generated by stimulated emission all have the same phase, and add constructively to produce a larger amplitude wave with well-defined frequency and phase. This property of laser light is known as *coherence,* and will be discussed in detail in Chapter 15.

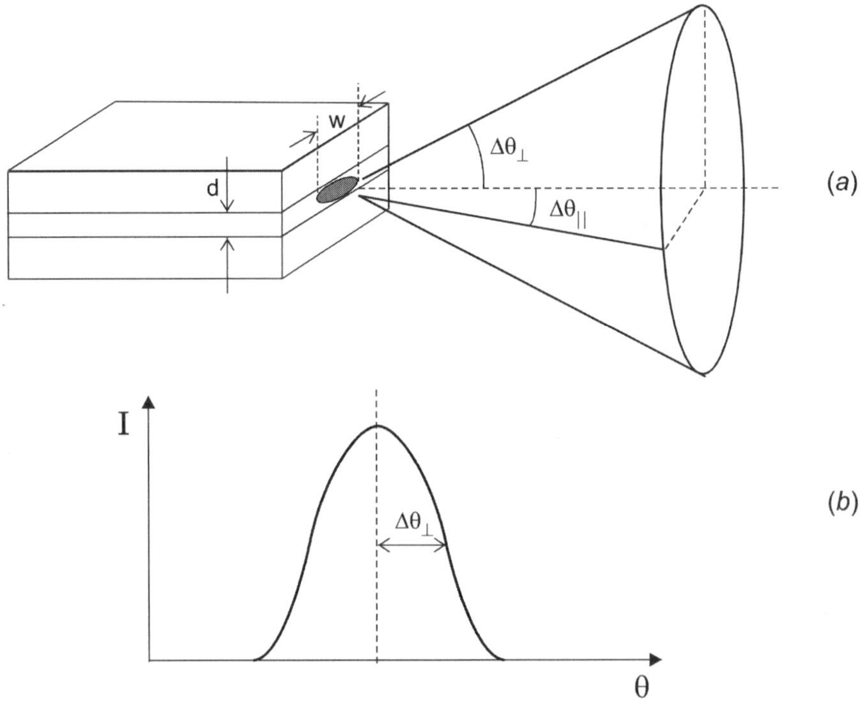

Figure 11-11 For laser light, the half-width $\Delta\theta$ is inversely related to the emitting region's dimensions, and is widest in the direction perpendicular to the plane of the junction.

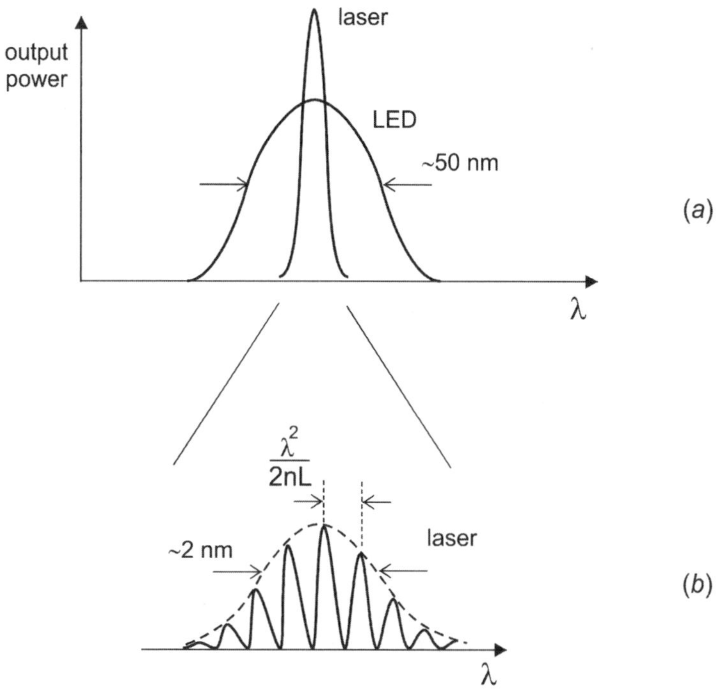

Figure 11-12 (a) Light from a laser diode has a narrower spectral distribution than that from an LED. (b) Laser light has a mode structure, with peaks separated in wavelength by $\lambda^2/(2nL)$.

At sufficiently high resolution, the laser diode spectrum is seen to have a mode structure, as depicted in Fig. 11-12b. A detailed discussion of the nature and origin of these modes is presented in Chapter 16. The spacing of the modes in frequency is $c/(2nL)$, where n is the refractive index and L is the laser cavity length. The corresponding spacing in wavelength is $\lambda^2/(2nL)$, with λ the free-space wavelength. If light is emitted in several modes (multimode operation), the effective linewidth is the width of the distribution of light in the various modes, as indicated in Fig. 11-12b. If only one mode is allowed to oscillate (single-mode operation), the linewidth becomes that of a single mode, and can be in the megahertz range in frequency. Methods for obtaining single-mode operation are discussed later in this chapter.

The narrow linewidth of a laser source is an advantage in certain applications. For example, the spreading in time of a light pulse in fiber optic communications is proportional to the spectral linewidth (see Chapter 6), and is therefore much less of a problem for a laser diode source than for an LED. A narrow spectral width is also desirable for wavelength division multiplexing (WDM), in which information can be sent simultaneously on a number of closely spaced frequency channels in an optical communications system. The narrower the linewidth, the more channels can fit into the finite available bandwidth. Clearly, single-mode lasers are preferred for this application.

Response Time

We found earlier that the response time of an LED is limited by the spontaneous lifetime τ of an electron in the conduction band. In a laser diode, the response time can be faster than this. As illustrated in Fig. 11-13, stimulated emission provides an additional way for the electron to decay out of the conduction band. The total decay rate W_{tot} expressed in Eq. (10-12) can be generalized to include stimulated emission as well as the spontaneous processes. When this is done, the resulting lifetime $\tau = 1/W_{tot}$ becomes shorter, indicating an improved time response. The time response of the laser diode is best at high excitation, where the stimulated emission rate is large. Response times in the range 15–30 ps can be obtained in this way.

The modulation of laser diode output power by varying the drive current is termed *gain switching.* While this is quite practical (and commonly used) at lower modulation rates, there is a complication at the highest rates. The optical frequency of the laser output is found to vary with time at the beginning of a pulse, a phenomenon known as *frequency chirping.* This is undesirable because it can create cross talk in optical communications systems that utilize closely spaced frequency channels (dense wavelength division multiplexing, or DWDM). An alternative approach that avoids frequency chirping is to run the laser diode in continuous-wave (cw) mode, and switch the beam on and off using an external modulator (see Chapter 9). Still another method is that of mode locking, discussed in Chapter 22. Mode-locked lasers can produce pulse widths in the femtosecond (fs) regime—much faster than is possible with gain switching.

Types of Semiconductor Lasers

Laser diodes have many applications, ranging from optical communications and optical sensors to high-power optical pumping of other lasers. In all these, it is desirable to have the conversion of electrical to optical power as high as possible. For certain applications, other attributes are desirable as well, such as narrow spectral linewidth, symmetrical angular distribution, or reliability in manufacturing. In the following, we take a look at how the different types of semiconductor lasers have evolved to meet these different needs.

Double Heterostructure Laser

The simplest semiconductor laser is the *homojunction* laser, consisting of a single junction between n- and p-type semiconductors of the same material. As electrons and holes are injected across the junction, they form a gain region in which light can be amplified. The width of this gain region is determined by how far from the junction the electrons and

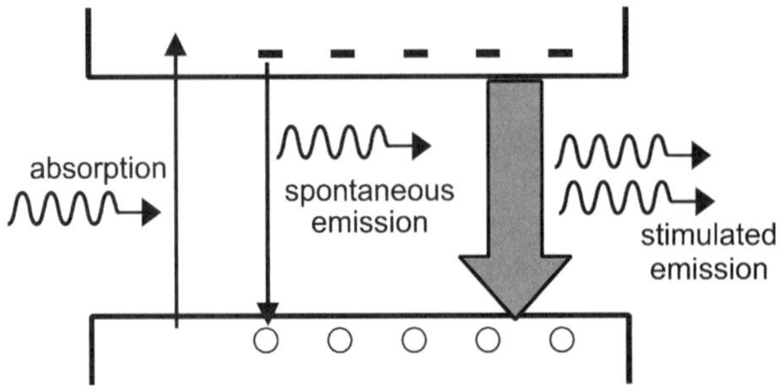

Figure 11-13 In a laser diode, light interacts with the semiconductor via absorption (upward arrow), spontaneous emission (downward arrow), and stimulated emission (downward thick arrow).

holes diffuse before recombining, a typical length being ~ 1 μm. Light propagating parallel to the junction will be amplified more than light propagating in other directions, so the light distribution tends to follow the junction, a phenomenon known as *gain guiding.* The depletion region provides some index guiding as well, because the free electrons and holes in the n and p regions lower the refractive index there, making the depletion region a weak optical waveguide. Both gain guiding and index guiding must compete, however, with the natural tendency of light to spread out by diffraction. The result is a weakly guided light distribution, as illustrated in Fig. 11-14, with a width much larger than that of the gain region.

Because of the poor overlap of the light distribution with the gain region, the gain per unit length is low in the homojunction laser, and the part of the lightwave that is outside the gain region suffers absorption rather than gain. The threshold current densities for lasing are therefore quite high, ~ 10^5 A/cm^2, and highly temperature dependent, features that are undesirable.

The *double heterostructure* (DH) laser overcomes these limitations. As shown in Fig. 11-15, this structure involves junctions between different semiconductor materials, known as *heterojunctions.* It also involves two junctions, hence the "double" heterostructure. In the example shown, one junction is between n and lightly doped p, and the other is between this lightly doped p and a more heavily doped p (p^+). The energy bands in the different sections shift due to the bandgap differences between materials, as well as the p–n junction potential. With the proper choice of composition and doping in each of the three sections, the energy bands in the three sections can be arranged as in Fig. 11-16, with electrons and holes overlaping only in the middle section. This results in a well-defined gain region, the width of which can be readily controlled by changing the thickness of the middle region.

The DH laser has a number of important advantages over a homojunction laser. Optical confinement is much stronger and independent of current and temperature, because it arises from index guiding in the planar waveguide formed by the higher-index GaAs and the lower-index $Al_xGa_{1-x}As$. An additional advantage is that the tail of the lightwave distribution extending into the lower-index material will not be absorbed there, due to the higher bandgap energy of that material. As a rule, higher band gap materials have lower refractive index, so this will be generally true for all such heterostructure devices. A special advantage of the GaAs-$Al_xGa_{1-x}As$ heterostructure is that the lattice constants are nearly the same for any x, so that layers can be grown on top of each other without strain-induced defects.

The threshold current for a DH laser can be estimated by considering the simple model shown in Fig. 11-17. Electrons and holes enter an active region of thickness d, where they

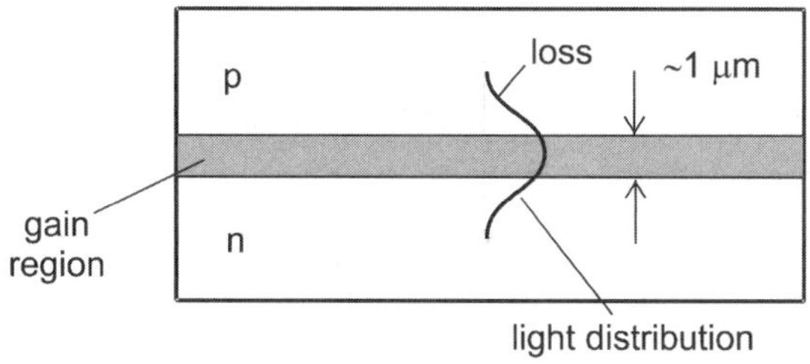

Figure 11-14 In a homojunction laser, the light distribution is not well guided, and suffers absorption losses outside the gain region.

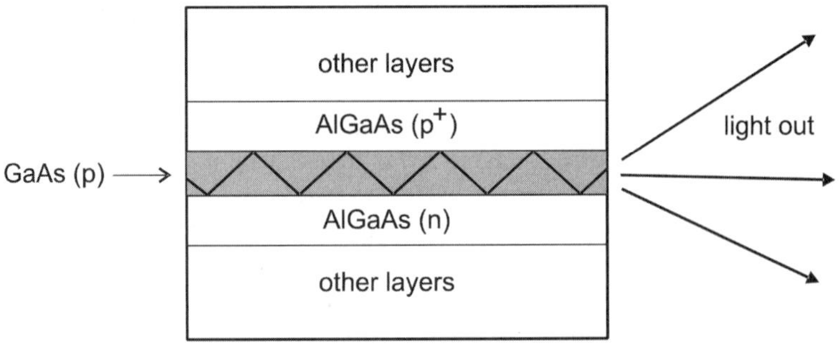

Figure 11-15 The light distribution is more confined in a double heterostructure (DH) laser, due to index guiding in the waveguide structure.

recombine and emit light. Under steady-state conditions, the number $\mathcal{N}$ of electrons in the active region is found by setting $d\mathcal{N}/dt = 0$ in Eq. (11-4). This gives

$$\frac{i}{e} = n(Lwd)\frac{1}{\tau} \tag{11-22}$$

where n is the number of electrons per unit volume in the active region, and the junction cross-sectional area is $A = Lw$. Assuming primarily radiative decay, the lifetime is obtained from Eq. (11-6), giving

$$i = eB_r n^2 Lwd \tag{11-23}$$

where we have also taken $p \simeq n$. This assumption is justified under high excitation, where equal numbers of electrons and holes are injected into the gain region from op-

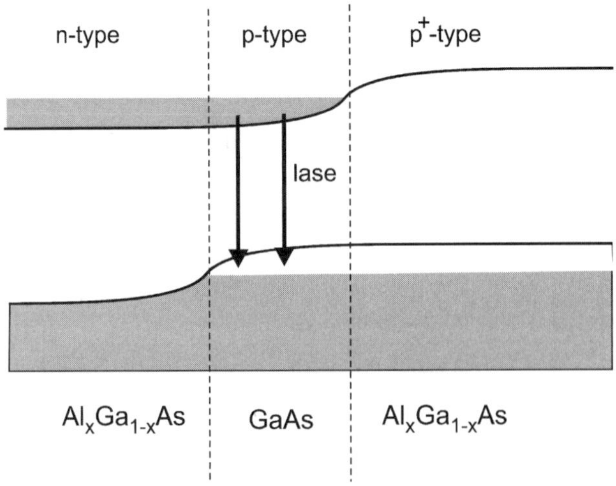

Figure 11-16 In a DH laser, the energy bands are shifted so as to permit overlap of free electrons and holes in a well-defined active region.

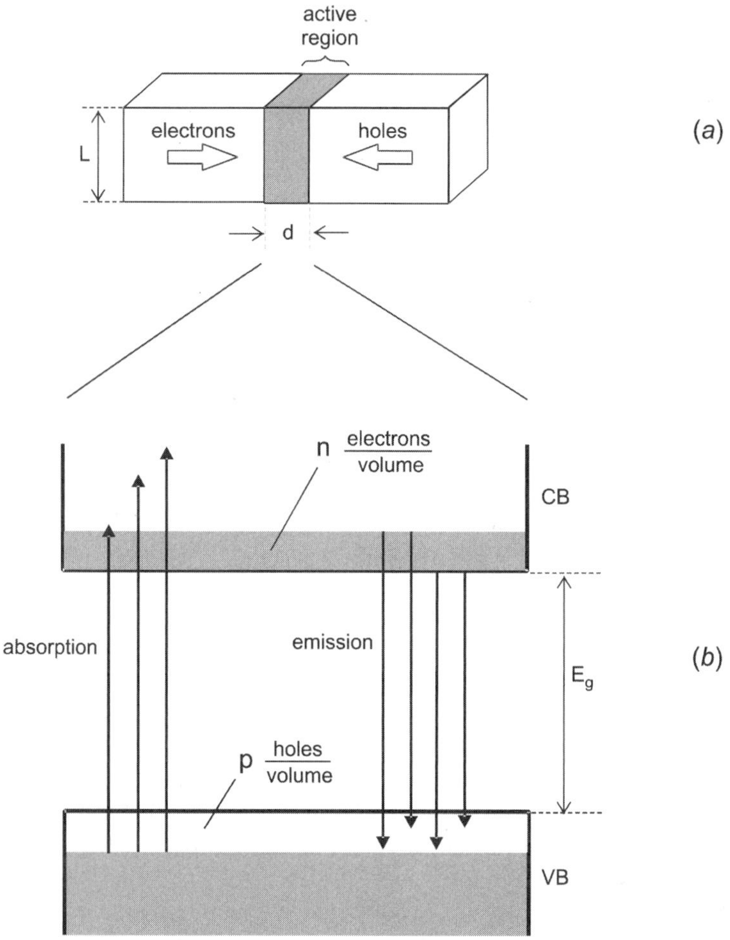

Figure 11-17 (a) Electrons and holes enter a recombination region of thickness *d* from opposite sides. (b) Band filled by the injected carriers separates the average energies of absorbed and emitted photons, resulting in net gain for certain photon wavelengths.

posite sides, decaying in pairs by e–h recombination. The current density at threshold is then

$$J_{th} = \frac{i_{th}}{Lw} = eB_r n_{th}^2 d \tag{11-24}$$

where n_{th} is the electron density required to achieve lasing.

One may well wonder why there needs to be a certain minimum electron density for light amplification and lasing; after all, light is emitted whenever there are any electrons and holes in the gain region. The most fundamental reason is that there are absorption as well as emission processes in a semiconductor (see Fig. 10-2). Whether a propagating light wave will be amplified or attenuated will then depend on which of these two processes predominates. Under weak excitation conditions, the absorption process wins out and lasing does not occur.

To see how a higher electron density improves the balance between emission and absorption, consider Fig. 11-17b, which shows various absorption and emission transitions that are possible when the conduction band contains n electrons per unit volume. As electrons are added to the conduction band, they must go into unfilled energy states, according to the Fermi exculsion principle. They therefore fill the conduction band from the bottom upward, a phenomenon known as *band filling.* The boundary between filled and empty states is not perfectly sharp, but rather is spread out over an energy range $\sim k_B T$.

An optical transition, either absorption or emission, must take an electron from a filled state to an empty state. For absorption, the photon energy must therefore be $h\nu \geq E_g + \Delta E_{\text{fill}}$, where ΔE_{fill} is the combined filling energy in the conduction and valence bands. For emission, on the other hand, photons in the range $E_g < h\nu < E_g + \Delta E_{\text{fill}}$ are possible. The result is that the average energies for absorption and emission become more different as n increases. Above some minimum value of n, the probability of stimulated emission becomes greater than the probability of absorption, and net gain is achieved.

When the probabilities for stimulated emission and absorption just become equal, there is no net change in the intensity of propagating light. The material is then said to be transparent, and the value of n for which this occurs is the *transparency density,* n_{tr}. For GaAs at room temperature, $n_{tr} \sim 10^{18}$ cm^{-3}. In practice, n needs to be somewhat higher than n_{tr} to achieve lasing, because there must be a finite positive gain to balance the other losses that are always present. The nature of these losses and their effect on laser performance will be discussed in detail in Chapter 20. For the present purpose, we make the approximation $n_{th} \approx n_{tr}$, which is adequate for making rough estimations.

It is clear from Eq. (11-24) that the threshold current for lasing is reduced when d is made smaller. If d is made too small, however, diffraction causes a large fraction of the lightwave distribution to extend outside of the inner guided region. Since it is only the portion of the lightwave inside the waveguide that gets amplified, this results in weaker amplification and requires a higher current for lasing. The balance between these two effects occurs around $d \sim 0.1$ μm for a GaAs DH laser.

It is not just the current density J that should be minimized at threshold, but also the actual current i. According to Eq. (11-23), this means that the cross-sectional area Lw of the active region should be minimized. This can be done by using the *stripe geometry,* illustrated in Fig. 11-18, in which the width w of the active waveguide is reduced. In addition to lowering the threshold current, this also serves to stabilize the transverse mode distribution of the laser output, preventing irregularities in the light versus current response. It is not generally feasible to significantly reduce the other dimension L, because then light propagating a distance $2L$ in a round-trip of two passes through the gain medium would no longer be sufficiently amplified to balance the round-trip losses. Typically, $L \sim 1$ mm for a DH laser. When the active region is surrounded on all sides by other conducting or insulating materials, the laser is said to have a *buried heterostructure.*

EXAMPLE 11-4

A GaAs DH laser has an active region of thickness 0.1 μm, width 8 μm, and length 1 mm. Determine (a) the threshold current density, and (b) the threshold current.

Solution: (a) The threshold current density can be estimated from Eq. (11-24) as

$$J_{th} \simeq (1.6 \times 10^{-19}\ \text{C})\left(7 \times 10^{-10}\ \frac{\text{cm}^3}{\text{s}}\right)(10^{18}\ \text{cm}^{-3})^2(1 \times 10^{-5}\ \text{cm}) = 1.1 \times 10^3\ \frac{\text{A}}{\text{cm}^2}$$

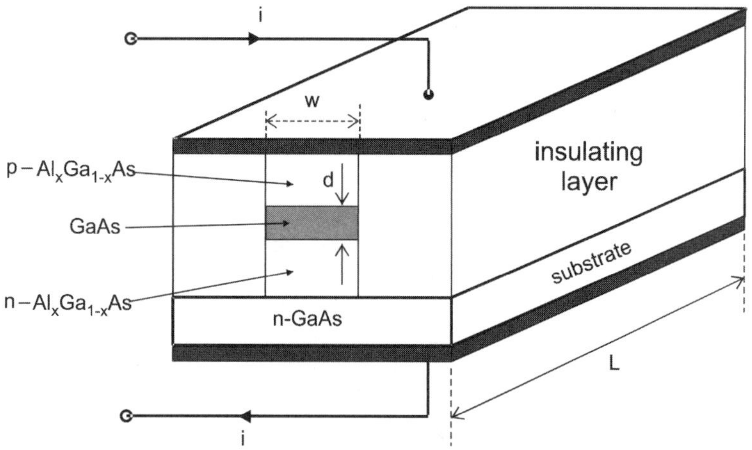

Figure 11-18 In the stripe geometry laser, the waveguide width w is reduced for improved mode stability. Shown is the buried heterostructure configuration, with the active GaAs region surrounded by other materials.

This calculated value is in good agreement with experimental current thresholds for DH lasers, which are ~ 10^3 A/cm^2.

(b) The cross-sectional area for current flow is

$$A = Lw = (10^{-1}\text{ cm})(8 \times 10^{-4}\text{ cm}) = 8 \times 10^{-5}\text{ cm}^2$$

so the threshold current is

$$i_{th} = \left(1.1 \times 10^3 \frac{\text{A}}{\text{cm}^2}\right)(8 \times 10^{-5}\text{ cm}^2) = 0.088\text{ A} = 88\text{ mA}$$

The typical numbers presented above are for room temperature. It is found that the threshold is a fairly strong function of temperature, varying as $i_{th} \propto \exp(T/T_0)$, where T_0 is the *characteristic temperature.* This can be understood qualitatively by remembering that the boundary between filled and unfilled states is spread out over an energy ~ k_BT. At lower temperatures, the electron and hole distributions are more sharply defined, resulting in less spectral overlap between absorption and emission. Less band filling is then needed to achieve transparency, so that n_{tr} and J_{th} are reduced. Similarly, n_{tr} and J_{th} increase at higher temperatures. A typical value for a GaAs DH laser is $T_0 \approx 100$ K, so that i_{th} doubles when T is raised ~ 70°C above room temperature.

Quantum Well Laser

We saw for the DH laser that an active region thickness d smaller than ~ 0.1 μm is not beneficial, because the lightwave mode cannot stay confined to such a small dimension. However, if d is reduced still further to the ~ 10 nm range, the picture changes. A quantum well is formed (see p. 187), and the motion of the electron (or hole) perpendicular to the layer is constrained by the well, resulting in discrete energies as shown in Fig. 10-15.

The electrons and holes can still move freely parallel to the layers, however, making the movement of charge carriers effectively two-dimensional.

The probability for electron–hole recombination, as for any quantum mechanical transition, depends on the number of unoccupied states having energies in the immediate vicinity of the transition energy, a quantity known as the *density of states.* This density of states is enhanced in the quantum well compared with a bulk semiconductor, because the energy levels are more well defined. Therefore, the gain is correspondingly higher in a quantum well, and this partially offsets the weaker optical confinement that arises from the extremely small values of d.

To improve the optical confinement in a quantum well laser, separate *optical confinement layers* (OCL) may be added on either side of the quantum well, as shown in Fig. 11-19. These layers have an energy gap and index intermediate between that of the quantum well and the surrounding layers and this forms an optical waveguide for confining the optical mode. Lasers incorporating such structures are termed *separate confinement heterostructure* (SCH) lasers.

Another way to increase the effective optical confinement is to simply add more quantum wells, as shown in Fig. 11-19b. This arrangement is termed a *multiple quantum well* (MQW) structure, in contrast to the *single quantum well* (SQW) of Fig. 11-19a. In the MQW, e–h pairs in each individual well interact with the same optical mode, increasing the net gain. This comes, however, at the expense of an increased threshold current, be-

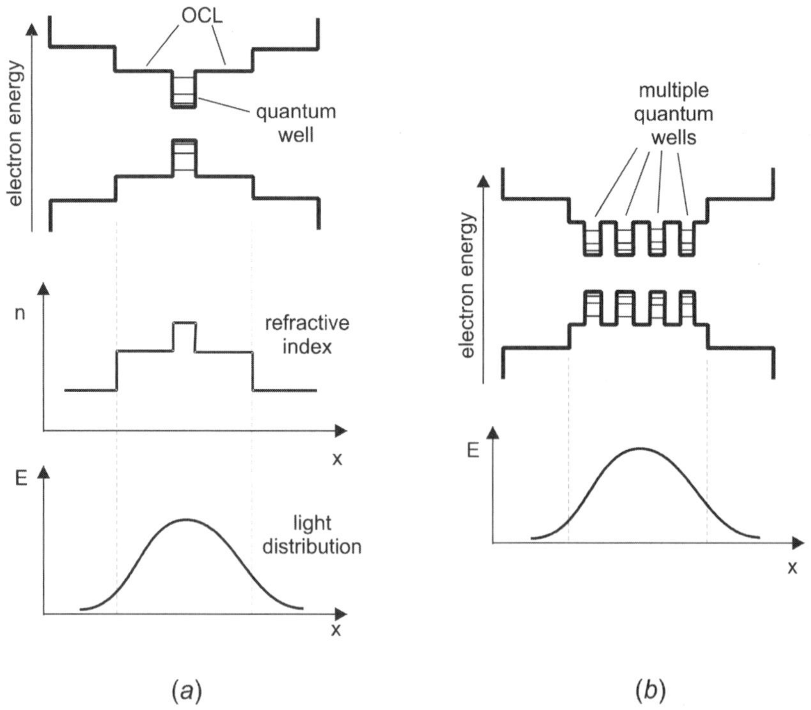

Figure 11-19 (a) In a separate confinement heterostructure laser, the light wave is guided by optical confinement layers (OCLs), whereas the electrons and holes are confined inside the quantum well. (b) In a multiple quantum well (MQW) device, electrons and holes in each quantum well interact with the same light field.

cause the n for each individual well must be brought to the transparency value n_{tr}. One benefit of the MQW structure is that more of the injected carriers go into the wells, and fewer stay in the OCLs. This results in higher efficiency and a faster time response.

Quantum well lasers have a number of advantages over traditional DH lasers. For example, the lasing wavelength can be adjusted simply by selecting the value of d, according to Eq. (10-24). This provides considerable flexibility in manufacturing laser diodes for particular applications. The properties of the laser can also be fine-tuned by selecting materials for the layers that are mismatched in lattice constant. This gives rise to lattice strain, which in thicker layers would result in defects, leading to losses and device failure. One of the remarkable discoveries about quantum wells during their development was that strain is no longer a problem, but can actually make the devices work better.

Another advantage of quantum well lasers is the very small current thresholds made possible by the small values of d. Current density thresholds of ~ 50 A/cm^2 have been obtained in strained InGaAs SQW lasers. The threshold also tends to increase less strongly with temperature in a quantum well laser, with characteristic temperatures > 250 K. This weaker temperature dependence arises from the more well-defined energies of the quantum well levels, with less blurring of the filled–unfilled energy boundary. The QW laser has so many advantages that it has largely supplanted the DH laser for everyday applications.

Single-Frequency Laser

As discussed earlier, the spectral linewidth of a free-running laser diode is ~ 2 nm, compared with ~ 50 nm for a typical LED. Although this is a significant improvement, it is not sufficient for certain applications. For example, the commonly used C-band for telecommunications spans the range $1530 < \lambda < 1560$, giving an available bandwidth of 30 nm. With 2 nm per channel, this would allow only ~ 15 channels to propagate without interference. To increase the number of channels so that each fiber is used most efficiently, it is necessary to decrease the laser linewidth.

The method commonly used to narrow the linewidth of a semiconductor laser utilizes a Bragg grating built into the structure of the device. This grating, indicated schematically by the undulating line in Fig. 11-20, can be formed by continuously varying the thickness of one of the layers in the structure. The evanescent field of the optical waveguide mode interacts with this periodic modulation, causing some of the light to be scattered in the backward direction. When light that is scattered from different undulation peaks adds together constructively with the same phase, the lightwave is nearly entirely reflected, just

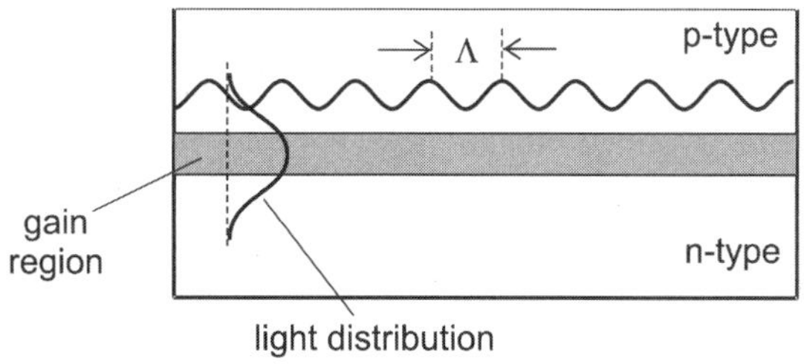

Figure 11-20 In a distributed feedback (DFB) laser, single-frequency operation is obtained by Bragg reflection from corrugations near the gain region.

as in a fiber Bragg grating (see Chapter 8). If the spacing between undulations is Λ, this occurs for wavelengths satisfying the Bragg condition

$$\Lambda = m\frac{(\lambda/n)}{2} \tag{11-25}$$

where λ is the free-space wavelength, n is the refractive index of the semiconductor, and m is an integer. Light within some range $\Delta\lambda$ around this center wavelength λ will be efficiently reflected, as discussed in Chapter. 8. When $\Delta\lambda$ is less than the mode spacing $\lambda^2/(2nL)$, single-mode operation is obtained.

EXAMPLE 11-5

A GaAs laser has a cavity length of 0.8 mm, and operates at 860 nm. The refractive index of GaAs is 3.6. Determine the modulation period for a Bragg reflector in this laser, and the maximum allowable width of the grating reflection spectrum.

Solution: For first-order diffraction, $m = 1$, so

$$\Lambda = \frac{860\text{ nm}}{2\,(3.6)} = 119\text{ nm}$$

The patterning of a semiconductor surface on this length scale is feasible, but it presents a manufacturing challenge. The width of the reflection spectrum must be smaller than

$$\Delta\lambda_{max} = \frac{(860\text{ nm})^2}{2(3.6)(8 \times 10^5\text{ nm})} = 0.128\text{ nm}$$

The structure described in Fig. 11-20 is referred to as a *distributed feedback* (DFB) laser, because the reflections from the grating (the feedback) are distributed along the entire gain region. An alternative structure is the so-called *distributed Bragg reflector* (DBR) laser, depicted in Fig. 11-21. This is similar in concept, except that the gain and feedback regions are kept separate, with the two Bragg grating regions acting effectively as mirrors. The reflection from these mirrors does not occur at any one point, but is distributed over the grating instead.

Single-frequency diode lasers have advantages in addition to the narrower linewidth. They are in general less sensitive to changes in temperature, and have a more linear output power versus current relation. Both of these characteristics are due to the lack of mode hopping, a phenomenon in which small perturbations in current, temperature, or other environmental factors cause the laser light energy to jump from one mode to another. Mode hopping is eliminated in a single-frequency diode laser by the stabilizing influence of the Bragg grating.

Vertical Cavity Surface-Emitting Laser

All the diode laser types discussed so far are edge emitters, in which light is amplified while propagating parallel to the layers. Of course, light can also be emitted perpendicular

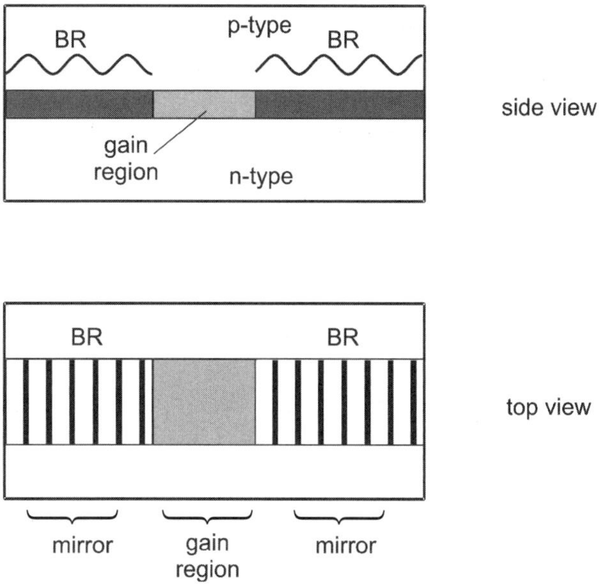

Figure 11-21 In a distributed Bragg grating (DBG) laser, single-frequency operation is obtained by reflection from Bragg gratings that are separate from the gain region, and which act like mirrors.

to the layers, as in a surface-emitting LED. The problem with this scheme for a laser is that the amplification of light in one pass through the thin active layer is very small, insufficient to overcome the reflectivity losses of ordinary mirrors. For this reason, the development of surface-emitting lasers initially lagged behind that of edge emitters.

What made surface emitters finally successful was the use of Bragg grating reflectors both above and below the active layer, as illustrated in Fig. 11-22. Each Bragg reflector consists of alternating layers of semiconductors with different band gaps, such as GaAs and $Al_xGa_{1-x}As$. By adjusting x and the layer thicknesses, the reflectivity can be made very high ($R > 0.995$). The loss per bounce off the mirrors is then < 0.5%, which is small enough that the amplification per pass through the active layer (~ 1%) produces a net round-trip gain. A device operating in this way is termed a *vertical cavity surface-emitting laser,* or VCSEL (pronounced "vick-sel"). The active layer of the VCSEL often consists of a quantum well.

There are a number of features of VCSELs that make them ideal for certain applications. The output beam is inherently symmetrical, with emission into a diffraction-limited cone of half-angle typically ~ 7–10°. This makes for more efficient coupling into optical fibers, and in general facilitates manipulation of the beam. The small width of the VCSEL reduces capacitive effects, which enables high-speed modulation (> 10 GHz). It also reduces the threshold current (< 1 mA), and results in high efficiency (> 50%). The small cavity length L naturally promotes single-frequency operation, since the mode spacing $\lambda^2/(2nL)$ is greater than the width of the gain spectrum.

Some of the biggest advantages of VCSELs over edge emitters come from practical considerations rather than fundamental device properties. These advantages arise from the manufacturing method, in which many individual VCSELs are deposited simultaneously on a single semiconductor wafer. This leads to efficiency and cost savings in production, with the lasers individually testable in situ. Another benefit is that 2-D arrays of

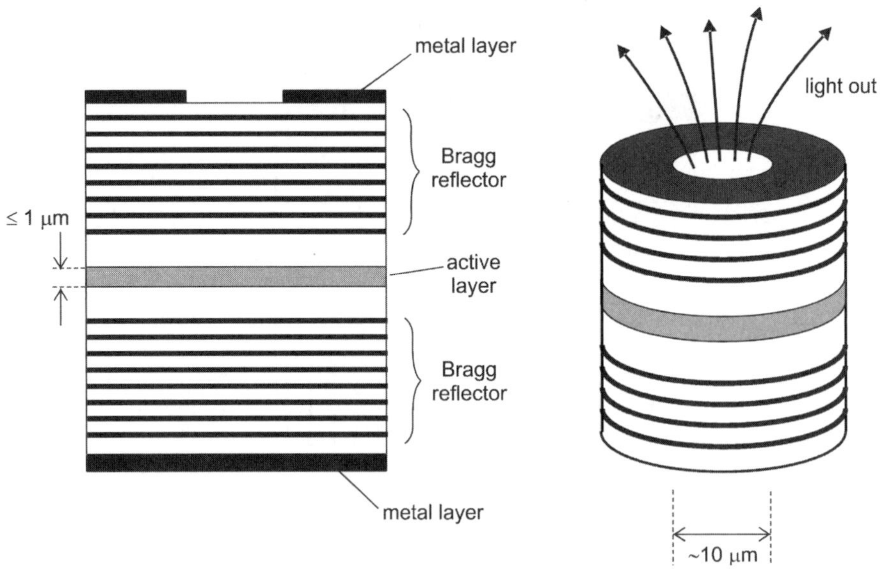

Figure 11-22 In a vertical cavity surface-emitting laser (VCSEL), the laser cavity is perpendicular to the active layer, with feedback provided by Bragg reflectors.

VCSEL's can be naturally created in this way, each laser being individually controllable. The laser wavelengths in the array can be designed to be the same, or made different by careful control of the individual cavity lengths L. Laser diode arrays have applications in optical switching, optical processing, and interconnects between different integrated optical circuits.

Quantum Cascade Laser

A common feature of the light sources discussed so far is the recombination of electrons and holes to produce light. In this process, the electron falls from the conduction band to the valence band, a so-called *interband transition.* Light can also be emitted when an electron goes from one energy level in a quantum well (QW) to another level in the same QW. This *intraband transition* provides the basis for an entirely different kind of semiconductor laser, in which the wavelength is determined solely by the spacing of the QW energy levels in the conduction band. Since there is no e–h recombination involved, the wavelength does not depend on the band gap of the semiconductor.

Figure 11-23 illustrates one realization of such a laser, known as the *quantum cascade laser* (QCL). It consists of alternate layers of higher and lower band gap material, forming a series of closely spaced QWs. A voltage is applied across the device, which causes the electron's potential energy to vary linearly with position as shown. There is no p–n junction, so in a circuit the device behaves much like a resistor. It is termed a *unipolar* circuit element, because its operation depends on only one type of charge carrier (i.e., the electron). Conventional laser diodes, in contrast, are *bipolar* circuit elements.

As electrons flow through the QCL, they lose potential energy due to their motion opposite to the electric field direction. However, they lose this energy not continuously, but in steps, as they jump from one QW to the next. Electrons in the lowest energy level of a given QW see a potential barrier on both sides, and according to classical me-

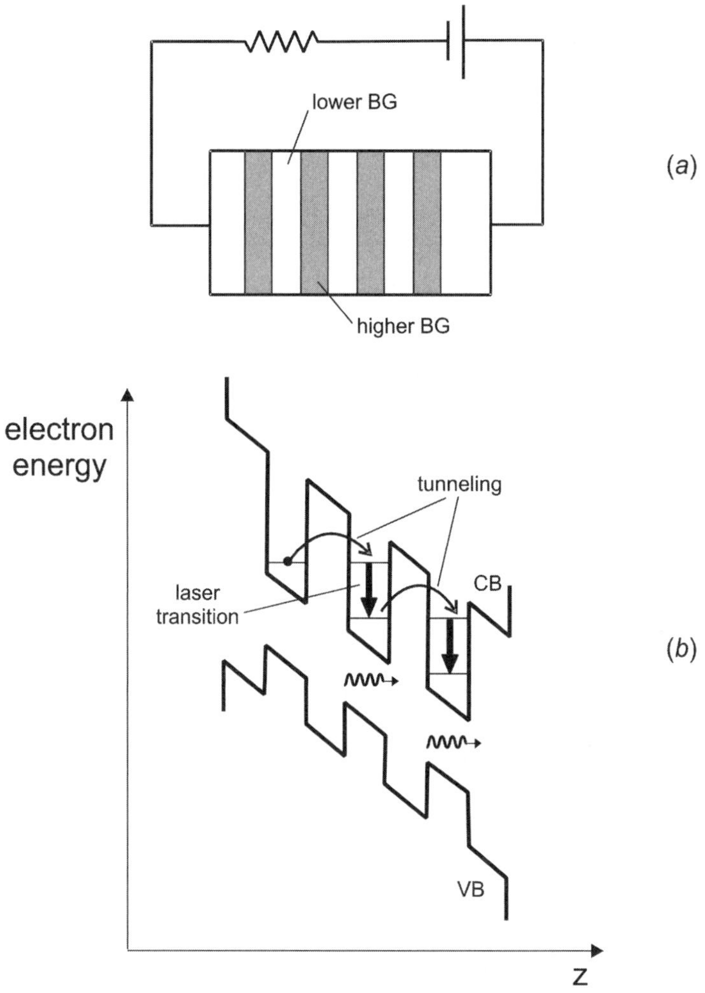

Figure 11-23 (a) A quantum cascade laser (QCL) is formed by alternating layers with higher and lower band gaps. (b) In a QCL, an electron tunnels between adjacent quantum wells, and drops from one QW level to another, emitting a photon. The process is repeated some 20–25 times for a single electron (only three wells are shown for simplicity).

chanics they would be trapped there indefinitely. For example, a marble rolling in a bowl does not spontaneously jump out of the bowl. But in quantum mechanics, where the electron is described by a wave, there is a finite probability that it will jump over the potential barrier, a process known as *tunneling.* The probability for tunneling is higher when the barrier is thinner and lower, and in a QCL this probability can be designed to be high.

The principle of QCL operation is then as follows. An electron tunnels from one QW to a different energy level in an adjacent QW. It then falls to a lower energy level in this QW, emitting a photon of energy $h\nu$. From that lower level, it again tunnels through a barrier into the next QW, where it falls to a lower energy level, emitting another photon. This process is repeated many times (typically 20–25) during the transit of a single electron across the device, so that a single electron gives rise to many individual photons. The situation is similar to that of water cascading down steps in a stream, which inspired the name "quantum cascade" laser.

The most important application of the QCL is as a light source in the mid-IR wavelength range (4–12 μm). For a conventional diode laser, this would require working with materials with a very small band-gap energy, which is undesirable due to the high levels of thermally generated electrons and holes. Since the photon energy in the QCL is independent of the band gap, wide band-gap materials can be used in the layers for better performance. The lasing wavelength can be selected simply by choosing the proper QW width and spacing. In fact, it is possible to operate the QCL laser at multiple wavelengths simultaneously, by varying the QW energy levels across the device.

The concept for the QCL was originally proposed by Kazarinov and Suris in 1971, but practical devices were not developed until the work of Capasso and Faist at Bell Labs in 1994. Commercial devices are now available, operating in pulsed mode at room temperature in the wavelength ranges 5–6 μm and 10–11 μm. These wavelength ranges are important for applications such as remote sensing and trace detection of contaminants, since many molecules have characteristic absorption and emission features in those spectral regions.

PROBLEMS

11.1 A GaAs LED that emits at 860 nm is connected as shown in Fig. 11-1, and driven with a current of 20 mA from a supply voltage of 9 V. (a) Calculate the voltage drop across the diode, assuming $\beta = 2$ and $i_0 = 150$ pA. (b) Determine the load resistor needed. (c) If $\eta_i = 0.85$, determine the optical power generated. (d) Compute the overall electrical to optical conversion efficiency (optical power divided by electrical power supplied by battery).

11.2 Solve Eq. (11-4) for the electron population $\mathcal{N}(t)$ when the current is switched from zero to a constant value. That is, take $i(t) = 0$ for $t < 0$, and $i(t) = i_c$ for $t \geq 0$. Assume $\mathcal{N}(0) = 0$.

11.3 The 3 dB electrical bandwidth of an LED is 80 MHz. (a) If the output power at low modulation frequency is 10 mW, what is the output power at 250 MHz? (b) Determine the carrier lifetime, assuming that this is what limits the bandwidth. (c) Determine the corresponding 3 dB optical bandwidth.

11.4 An LED circuit has an optical 3 dB bandwidth of 750 kHz. Determine the modulation frequency at which the LED output is reduced to 20% of the low-frequency value.

11.5 A GaAs LED has the dome configuration shown in Fig. 11-6. In designing the LED, three different types of plastic are considered, with index $n_2 = 1.5$, 1.9, or 2.9. (a) For each choice of n_2, calculate the fraction of light transmitted through both interfaces (i.e., consider only Fresnel reflection losses here, and assume a close to normal angle of incidence). (b) For each choice of n_2, calculate the fraction of all light generated within the GaAs material that is within the solid angle for transmission through the first interface (i.e., that is not internally reflected). (c) Combine the results of parts a and b above to determine the net external efficiency of the LED for each value of n_2. (d) Organize your results above into a table, showing the different contributions to the external efficiency for the assumed values of

n_2. What is the best choice for n_2 in terms of maximizing efficiency? How do the results compare with a GaAs LED that has no plastic cap?

11.6 The GaAs material of an LED is in direct contact with a glass medium on one side. Calculate the fraction of all light generated inside the GaAs that escapes into the glass medium (at any angle in the glass). For simplicity, use the normal incidence relation Eq. (2-14) in calculating the Fresnel loss. Take $n = 1.5$ for glass.

11.7 In the Burrus geometry of Fig. 11-7, calculate the fraction of all light emitted inside the GaAs material that is coupled into guided modes of the fiber. The fiber has core index 1.5 and NA = 0.25. Assume that the distance from the emitting region to the fiber end is much less than the fiber core diameter, and also assume that the GaAs material is in direct contact with the fiber core (neglect the effect of the epoxy filler between). How does this compare with the results of Problem 11.6?

11.8 A planar waveguide consists of a layer of index n_1 sandwiched between two layers of lower-index n_2, as depicted in Fig. 11-8. Light is emitted from the interior of the inner layer, and some of this light is trapped in the waveguide by TIR. Determine the fraction of all emitted light that becomes trapped in the waveguide. Give your answer in terms of the relative index difference $\Delta = (n_1 - n_2)/n_1$ for small Δ. Also compute this fraction for a GaAs/$Al_{0.3}Ga_{0.7}As$ waveguide structure with $n_1 = 3.6$ and $n_2 = 3.4$. (Note: Not all of this trapped light comes out as useful output in an edge-emitting LED, because some of it is going in the backwards or sideways directions.)

11.9 A laser diode has a threshold current of 10 mA and an output power of 18 mW when driven with a current of 30 mA. (a) Determine β_s for this laser. (b) Calculate the laser output power if $i = 40$ mA. (c) The drive current is now modulated according to $i(t) = 15 + A \cos \omega t$. Sketch the time-dependent optical power $P(t)$ for $A = 10$ mA, $A = 5$ mA, and $A = 2$ mA.

11.10 For the laser diode in Problem 11.9, assume that the forward voltage drop across the laser diode is 2.5 volts. (a) What is the slope efficiency of the laser diode, defined as the change in output power divided by change in pumping power above threshold? (b) What is the electrical to optical conversion efficiency at an operating current of 20 mA (defined as the output power divided by the pump power)?

11.11 The angular distribution from a diode laser has half-widths of 15° and 4° in the vertical and horizontal directions, respectively. (a) If the laser operates at 808 nm, what do these angular spreads tell you about the dimensions of the laser's active region and its orientation? (b) If the angular distributions are approximated by a $\cos^n(\theta)$ function, determine the values of n for the two directions.

11.12 (a) A He–Ne laser has a cavity length of 15 cm. If light is emitted by the gas atoms over a frequency range of 1.7 GHz, how many modes can lase simultaneously? Take $n = 1$. (b) A GaAs edge-emitting laser diode has a cavity length of 1 mm and a center operating wavelength of 850 nm. If light is emitted over a wavelength range of 5 nm, how many modes can lase simultaneously?

11.13 The threshold current for a GaAs DH stripe laser is 80 mA at room temperature (20° C). Determine the threshold current at 150° C.

11.14 A GaAs DH laser operates at 850 nm, and has an active region of thickness 500 nm, stripe width 8 μm, and length 300 μm. (a) Estimate the current threshold at room temperature. (b) What would the current threshold be if the laser were operated at liquid nitrogen temperature (77 K)? (c) Determine the longitudinal mode spacing in picometers.

11.15 Determine the modulation period for the Bragg reflector in an InGaAsP DFB laser diode that operates at 1500 nm. Take the refractive index of InGaAsP to be 3.35 at this wavelength.

11.16 A quantum cascade laser is to operate on a transition between the two lowest quantum well levels in the conduction band of a GaAs/$Al_xGa_{1-x}As$ structure. (a) If an operating wavelength of 8 μm is desired, what must be the GaAs layer thickness? (b) For this layer thickness, what is the minimum value of x in the $Al_xGa_{1-x}As$ composition such that the energy of the second quantum well level is no higher than the well depth (so that the level is still "in the well")? Assume that the well depths are the same in the conduction and valence bands.

Chapter 12

Light Source to Waveguide Coupling

We have seen in previous chapters how light propagates in a waveguide, and how the light can be generated by an LED or laser diode. A related issue is the efficiency with which light can be coupled from a light source into a waveguide. In this chapter we examine this efficiency for sources with different degrees of directionality. It will be seen that the coupling efficiency is greater for more directional sources.

12-1. POINT SOURCE

Consider first isotropic emission, in which light appears to emanate from a point. We would like to calculate the fraction of emitted power that is collected by an optical fiber, as illustrated in Fig. 12-1. If a point source emitting a total power P_s is embedded in a uniform medium of refractive index n_0 outside the fiber, it will radiate light equally into all 4π steradians of solid angle, with an emitted power per unit solid angle of $P_s/4\pi$. As discussed in Chapter 4, however, only those rays making an angle $\alpha < \alpha_{max}$ with the fiber axis will be trapped by the fiber, where $\sin \alpha_{max} = \text{NA}/n_0$ and NA is the numerical aperture of the fiber [Eq. (4-2)]. From Appendix A Eq. (A-2), the solid angle within which light will be trapped is

$$\begin{aligned} \Omega &= 2\pi(1 - \cos \alpha_{max}) \\ &= 2\pi\left(1 - \sqrt{1 - \sin^2 \alpha_{max}}\right) \\ &= 2\pi\left(1 - \sqrt{1 - (\text{NA}/n_0)^2}\right) \\ &\approx 2\pi \frac{1}{2}\left(\frac{\text{NA}}{n_0}\right)^2 = \pi \frac{\text{NA}^2}{n_0^2} \end{aligned} \tag{12-1}$$

where in the last step the expansion $(1 + x)^n \approx 1 + nx$ has been used, along with the approximation (good for most optical fibers) $NA^2 \ll 1$.

Since the light is distributed uniformly with angle, the power P_{in} coupled into the fiber is given by

$$P_{in} = \left(\frac{\Omega}{4\pi}\right) P_s \tag{12-2}$$

which leads to the coupling efficiency

$$\eta_c \equiv \frac{P_{in}}{P_s} = \frac{\Omega}{4\pi} = \frac{1}{4}\frac{\text{NA}^2}{n_0^2} \quad \text{(point source coupling efficiency)} \tag{12-3}$$

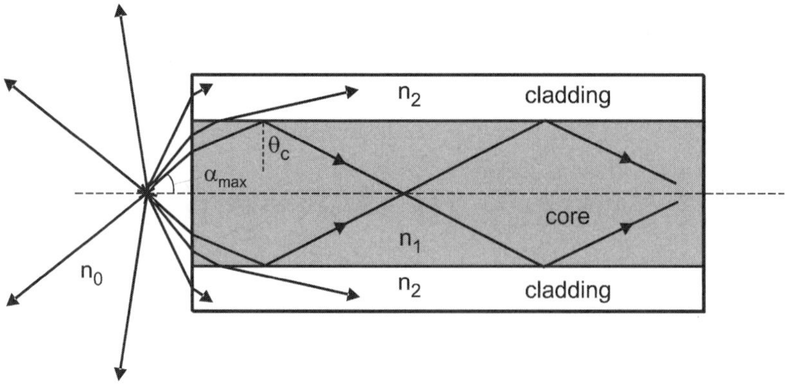

Figure 12-1 Light rays from a point source making a sufficiently small angle α with the fiber axis are coupled into the fiber core (shaded area). The solid angle of this cone compared with 4π sr gives the coupling efficiency.

This result shows that higher numerical aperture fibers accept a larger fraction of the emitted light from a point source. It should be noted that we have assumed the point source to be very close to the end of the fiber. When it is too far away, the solid angle for collection is limited by the core diameter rather than the NA (see Problem 12.1).

12-2. LAMBERTIAN SOURCE

Many practical light sources such as the LED are best treated as *extended sources,* in which light is emitted over some surface area A_s. The emitting surface is characterized by a *brightness* (see Appendix A), defined as the power emitted per unit solid angle, per unit surface area. This brightness is found in many cases to vary with direction as

$$B(\theta) = B(0)\cos\theta \qquad \text{(Lambert's law)} \tag{12-4}$$

where θ is the angle to the surface normal. A light-emitting surface with an angular distribution given by Eq. (12-4) is known as a *Lambertian source.*

The origin of the cos θ dependence can be understood by referring to Fig. 12-2, which shows light being generated inside a high refractive index material and emitted into air. Inside the material, light is emitted isotropically, with an equal amount of light radiated into each differential solid angle $d\Omega = 2\pi \sin\alpha\, d\alpha$ [see Eq. (A-1)]. Because of refraction at the boundary, however, the corresponding solid angle on the air side, $d\Omega = 2\pi\sin\theta\, d\theta$, becomes proportionately larger as θ increases. Since the same optical power is spread out over a larger solid angle as θ increases, the brightness decreases. This argument can be made quantitative by using Snell's law to relate the two solid angles (see Problem 12.2). The resulting brightness (including Fresnel reflection losses at the boundary) is

$$B(\theta) \simeq \frac{P_s}{\pi A_s n(n+1)^2}\cos\theta \tag{12-5}$$

where P_s is the total power emitted inside the material, and A_s is the emission surface area. It is assumed in deriving this that the index of refraction n is large ($n > 2.5$), which is valid for most semiconductors.

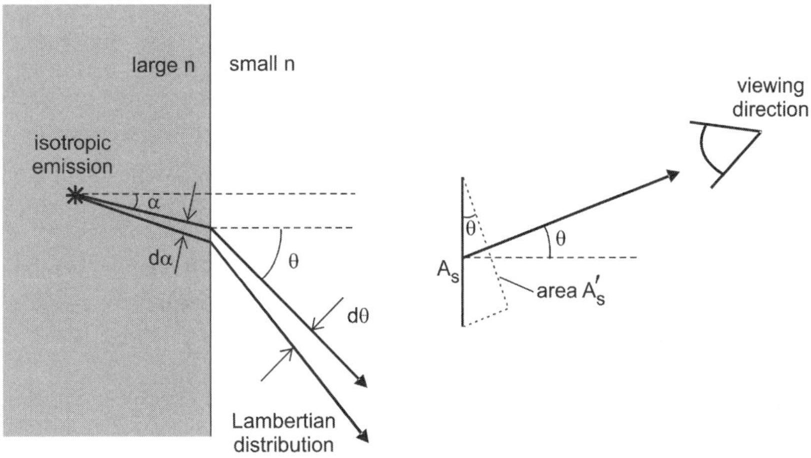

Figure 12-2 (a) Light emitted isotropically inside a high-index medium becomes distributed according to Lambert's law after refraction at a boundary. (b) When viewed off-axis, light from an emitting surface appears to come from the projected area A_s'.

An alternative view of Lambert's law is to define an "apparent brightness" B_{eff} as the power per solid angle emitted in a certain direction, divided by the projected (apparent) area of the surface when viewed from that direction. From Fig. 12-2b, this projected area is $A_s' = A_s \cos\theta$, so the apparent brightness is

$$B_{\text{eff}} = \frac{\Delta P}{A_s'\Delta\Omega} = \frac{\Delta P}{A_s\Delta\Omega\cos\theta} = \frac{B(\theta)}{\cos\theta} \tag{12-6}$$

If B_{eff} is independent of θ, then $B(\theta) \propto \cos\theta$, which is Lambert's law. We can say, then, that a Lambertian surface is one for which the apparent brightness is independent of viewing angle.

To determine the coupling efficiency into a fiber, we first calculate the total power P_0 emitted from the Lambertian surface. Integrating Eq. (12-4) over one hemisphere, we have

$$\begin{aligned} P_0 &= A_s \int_0^{\pi/2} B(\theta)\, d\Omega \\ &= A_s \int_0^{\pi/2} [B(0)\cos\theta]\,[2\pi\sin\theta\, d\theta] \\ &= 2\pi A_s B(0) \int_0^{\pi/2} \cos\theta\sin\theta\, d\theta \end{aligned} \tag{12-7}$$

Using the substitution $u = \sin\theta$, $du = \cos\theta\, d\theta$, this can be evaluated as

$$\begin{aligned} P_0 &= 2\pi A_s B(0) \int_0^1 u\, du \\ &= 2\pi A_s B(0)\, \tfrac{1}{2} u^2 \big|_0^1 \\ &= \pi A_s B(0) \end{aligned} \tag{12-8}$$

Combining this with Eq. (12-5) for $\theta = 0$ gives

$$P_0 \simeq \frac{P_s}{n(n+1)^2} \tag{12-9}$$

The ratio P_0/P_s corresponds to the LED external efficiency η_{ext} defined in Eq. (11-15). For GaAs ($n = 3.6$), Eq. (12-9) gives $\eta_{ext} \simeq 0.013$. This should be compared with the value for η_{ext} obtained earlier in Example 11-2.

The power P_{in} coupled into the fiber is calculated in the same way as the total power, except that the upper limit on θ is α_{max} rather than $\pi/2$. The upper limit on u in Eq. (12-8) is then $\sin \alpha_{max}$ rather than 1, giving

$$\begin{aligned} P_{in} &= \pi A_s B(0) \sin^2 \alpha_{max} \\ &= P_0 \sin^2 \alpha_{max} \end{aligned} \tag{12-10}$$

where Eq. (12-8) has been used. The coupling efficiency is then

$$\eta_c = \frac{P_{in}}{P_0} = \sin^2 \alpha_{max} = \left(\frac{\text{NA}}{n_0}\right)^2 \quad \text{(Lambertian source coupling efficiency)} \tag{12-11}$$

This efficiency is the same as for a point source, except that it is four times higher. The difference comes from the more directional emission of the Lambertian source, with a greater fraction of the emitted light lying within the angular acceptance range of the fiber.

For an extended source like an LED, the coupling efficiency also depends on the size of the emitting area compared with the core area of the optical fiber. As seen in Fig. 12-3, light that is emitted outside of the core area will not be accepted by the fiber, regardless of emission angle. Only the fraction $\pi a^2/A_s$ will be accepted by the fiber if $\pi a^2 < A_s$. In this case, the coupling efficiency will be

$$\eta_c = \frac{\pi a^2}{A_s}\left(\frac{\text{NA}}{n_0}\right)^2 \quad \text{(coupling efficiency when } \pi a^2 < A_s\text{)} \tag{12-12}$$

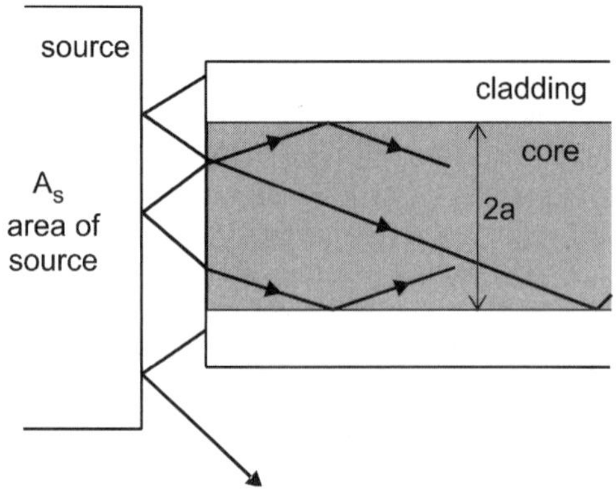

Figure 12-3 Light striking the fiber outside the core area is not coupled, regardless of angle. This decreases the coupling efficiency when the emission area A_s is greater than the fiber core area πa^2.

When $\pi a^2 > A_s$, the coupling efficiency depends only on emission angle, and Eq. (12-11) applies. These formulae for η_c assume negligible spacing between source and fiber. It should be noted that edge-emitting LEDs have a smaller emission area, making them more efficient for coupling into an optical fiber.

EXAMPLE 12-1

A GaAs LED with a square emitting area of side 0.2 mm emits light through air into an optical fiber with core index 1.5, fractional index difference $\Delta = 0.008$, and core diameter 50 μm. Determine (a) the coupling efficiency, and (b) total efficiency with which light generated inside the LED material is coupled into the fiber.

Solution: (a) The numerical aperture is given by Eq. (4-5) as

$$\text{NA} \simeq n\sqrt{2\Delta} = (1.5)\sqrt{2(0.008)} = 0.190$$

and the ratio of areas is

$$\frac{\pi a^2}{A_s} = \frac{\pi(25 \times 10^{-6})^2}{(2 \times 10^{-4})^2} = 0.049$$

Eq. (12-12) then gives

$$\eta_c = (0.049)(0.19)^2 = 1.77 \times 10^{-3}$$

(b) Using Eq. (12-9),

$$\frac{P_0}{P_s} = \frac{1}{n(n+1)^2} = \frac{1}{3.6(4.6)^2} = 0.013$$

The total efficiency is then

$$\eta_{\text{tot}} = \frac{P_{\text{in}}}{P_s} = (0.013)(1.77 \times 10^{-3}) = 2.32 \times 10^{-5}$$

This illustrates how inefficient the LED actually is in getting light generated within the material into an optical fiber. The laser, as we will see, does a much better job.

12-3. LASER SOURCE

In the previous section, we saw that light obeying Lambert's law is coupled more efficiently into an optical fiber than light from a point source, due to increased directionality in the angular distribution. We would expect laser light to be even more efficiently coupled, due to its high degree of directionality (see Fig. 11-10). To make a quantitative estimate of this improved coupling efficiency, the angular variation of a laser's brightness is taken to be

$$B(\theta) = B(0)\cos^m\theta \tag{12-13}$$

where m is large, and not necessarily an integer. Although this is not the actual angular distribution for a laser, it can be a close approximation when the proper value of m is chosen, and it makes the calculations simpler.

As for the Lambertian emitter, we first calculate the total power emitted,

$$\begin{aligned} P_0 &= A_s \int_0^{\pi/2} B(0) \cos^m\theta \, 2\pi \sin\theta \, d\theta \\ &= 2\,\pi A_s B(0) \int_0^{\pi/2} \cos^m\theta \sin\theta \, d\theta \end{aligned} \tag{12-14}$$

which can be evaluated using the substitution $u = \cos\theta$, $du = -\sin\theta\, d\theta$ as

$$\begin{aligned} P_0 &= 2\pi A_s B(0) \int_0^1 u^m \, du \\ &= 2\pi A_s B(0) \left[\frac{u^{m+1}}{m+1} \right]_0^1 \\ &= \frac{2\pi}{m+1} A_s B(0) \end{aligned} \tag{12-15}$$

The power coupled into the fiber is determined in the same way, except that the upper limit on θ is α_{max}, which leads to a lower limit on u of

$$u_{min} = \cos\alpha_{max} = \sqrt{1 - \sin^2\alpha_{max}} = \sqrt{1 - (\mathrm{NA}/n_0)^2}$$

The power into the fiber is then

$$\begin{aligned} P_{in} &= \frac{2\pi}{m+1} A_s B(0)\,(1 - u_{min}^{m+1}) \\ &= \frac{2\pi}{m+1} A_s B(0) \left[1 - \left(1 - \frac{\mathrm{NA}^2}{n_0^2} \right)^{(m+1)/2} \right] \end{aligned} \tag{12-16}$$

which corresponds to a coupling efficiency

$$\eta_c = \frac{P_{in}}{P_0} = 1 - \left(1 - \frac{\mathrm{NA}^2}{n_0^2} \right)^{(m+1)/2} \tag{12-17}$$

For small NA and m not too large, the binomial expansion $(1 + x)^n \approx 1 + nx$ can be used to obtain the simplified result

$$\eta_c \simeq \frac{m+1}{2} \left(\frac{\mathrm{NA}}{n_0} \right)^2 \qquad \text{(laser coupling efficiency)} \tag{12-18}$$

This last expression must be used with some caution, since it is obviously not true for arbitrarily large m (see Problem 12.7). However, it illustrates the improvement in coupling efficiency for a laser source as m increases and the beam becomes more directional.

EXAMPLE 12-2

A laser beam has an angular distribution with a full width at half maximum (FWHM) of 20°. Determine the value of m for this distribution, and calculate the coupling efficiency into a large core area optical fiber with NA = 0.22.

Solution: The half width is 10°, so m is found from

$$\tfrac{1}{2} = \cos^m 10°$$

Taking the natural log of both sides yields

$$\ln(0.5) = m \ln(\cos 10°)$$

so

$$m = \frac{\ln(0.5)}{\ln(\cos 10°)} = 45.3$$

Putting this into the approximate expression Eq. (12-18) gives the result

$$\eta_c \simeq \frac{46.3}{2}(0.22)^2 = 1.12$$

which is clearly not exact since η_c must be < 1. To get an accurate result here, it is neccesary to use Eq. (12-17):

$$\eta_c = 1 - [1 - (0.22)^2]^{46.3/2} = 0.683$$

This efficiency is, as expected, much higher than that for an LED.

In the preceding example we neglected to account for the overlap of the laser beam area with the fiber core area. This is often a good approximation because, as we shall see in Chapter 15, a laser has (or can be made to have) a very small effective emitting area. The coupling efficiency can, therefore, be improved by using a lens to make the beam even more directional. There are special considerations for coupling into a single-mode fiber, which are discussed in Section 17-3.

PROBLEMS

12.1 A point light source is located on the fiber axis a distance d_s from the end of a multimode fiber of core radius a and numerical aperture NA. Determine the value of d_s at which the coupling efficiency starts to depend on a. Derive an approximate expression for η_c when d_s is large.

12.2 Use Snell's law to relate the angles α and θ in Fig. 12-2, and use this to derive Eq. (12-5).

12.3 A large-core, multimode step- index fiber of radius a is excited by a surface-emitting LED with area $A_s < \pi a^2$. The fiber has NA = 0.2 and loss coefficient 4 dB/km. The LEDs total output power is 5 mW. Compute the power propagating in the fiber core at 1 m, 1 km, and 10 km. Repeat this problem if the fiber has instead NA = 0.5 and a loss of 20 dB/km.

12.4 A source has a half-power emission angle of 32.8° as measured from the normal to the emitting surface. Compute the coupling efficiency into a multimode SI fiber having NA = 0.2.

12.5 For a Lambertian emitter, calculate the angle with respect to the surface normal at which the emitted intensity is (a) 50% of the peak intensity, (b) 20% of the peak intensity, and (c) 5% of the peak intensity. (d) What is the full width at half maximum (FWHM) of the Lambertian radiation pattern?

12.6 In Example 12-1, the light from the LED was assumed to be first emitted into the air, and then coupled from the air into the fiber. Consider the situation in which the GaAs material of the LED is in direct optical contact with the core of the fiber. Assume that light is emitted isotropically within the GaAs material, and determine the fraction of generated light that ends up being coupled into the guided modes of the fiber. Using all the same parameters as in that example, determine what effect this has on η_{tot}.

12.7 Derive Eq. (12-18) from Eq. (12-17). Why can't Eq. (12-18) be valid for arbitrarily large m? What is the maximum value of η_c for which the approximate expression in Eq. (12-18) differs from the exact expression in Eq. (12-17) by no more than 10%?

12.8 Two multimode fibers with the same core diameter but different numerical aperture are butt-coupled together with no lateral, angular, or longitudinal offsets. Light is transmitted from fiber 1 (numerical aperture NA_1) into fiber 2 (numerical aperture NA_2). Show that the coupling efficiency is $\eta_c = (NA_2/NA_1)^2$ if $NA_2 < NA_1$, and $\eta_c = 1$ if $NA_1 < NA_2$. Assume that the angular distribution of light emitted from the fibers is Lambertian up to the maximum angle, and zero past this angle.

12.9 It might seem that a lens could be used to increase the efficiency with which light from an LED can be coupled into a fiber. Determine whether this is possible by considering a lens of focal length f inserted between an LED and fiber that are separated by a distance d. The LED has surface area A_s and brightness $B_s(\theta)$, and the fiber has core radius a and numerical aperture NA. Assume the lens collects most of the light from the LED.

12.10 Consider a surface that emits light with a "flat-top" brightness distribution, such that $B(\theta) = B_0$ for $\theta < \theta_0$ (with B_0 a constant), and $B(\theta) = 0$ for $\theta > \theta_0$. Determine the efficiency with which light from this surface is coupled into a multimode fiber having core radius a, core index n_1, and numerical aperture NA. Write your results in general form, and also in an approximate form valid when $NA \ll 1$ and $\theta_0 \ll 1$.

Chapter 13

Optical Detectors

In Chapter 11 we saw how an electric current can give rise to light emission in semiconductor devices such as LEDs and laser diodes. Equally important for photonics applications is the counterpart to this, in which light is detected and converted into an electrical signal. Optical detectors may be classified as either thermal or photon detectors, depending on how the electrical signal is generated. In a thermal detector, the optical power is absorbed by a sensor element, causing a rise in the element's temperature which is then converted into a voltage. In a photon detector, the light absorbed in the detector material directly creates charge carriers, which give rise to a photocurrent and signal voltage.

Each of these two detector types has advantages and disadvantages. Thermal detectors tend to be slow and not very sensitive, but they generally detect light over a very wide wavelength range. Photon detectors have essentially the opposite properties, being faster and more sensitive, but with a more restricted wavelength range. In this chapter, we consider the fundamental operating principles of thermal and photon detectors, and show how they give rise to these complementary properties.

13-1. THERMAL DETECTORS

There are several ways to convert a temperature rise into an electrical signal, any of which could be used in a thermal detector of light. We will describe two such methods commonly used in light detectors, based on the thermoelectric effect and the pyroelectric effect. Before looking at these specific devices, however, we consider in general how the flow of heat limits the time response and sensitivity in thermal detectors.

Time Response

The time response of a thermal detector can be determined using the simplified model shown in Fig. 13-1. Light is incident on a sensor element of mass m, thickness d, and area A, which is connected by some support structure to a large object (the heat sink). The temperature of the heat sink is assumed to remain at the constant value T_0, whereas the temperature of the sensor element increases by an amount ϕ due to heating by the absorbed light. There will then be a temperature gradient in the support structure that results in a flow of heat energy from the sensor element to the sink. For small ϕ, the heat leaving the sensor element per unit time will be given by $G\phi$, where G is the thermal conductance of the support structure. If the support structure is made of thermally insulating material, and has a small cross-sectional area for conducting heat, G can be made very small. We say in this case that the mass m is thermally well insulated from its surroundings.

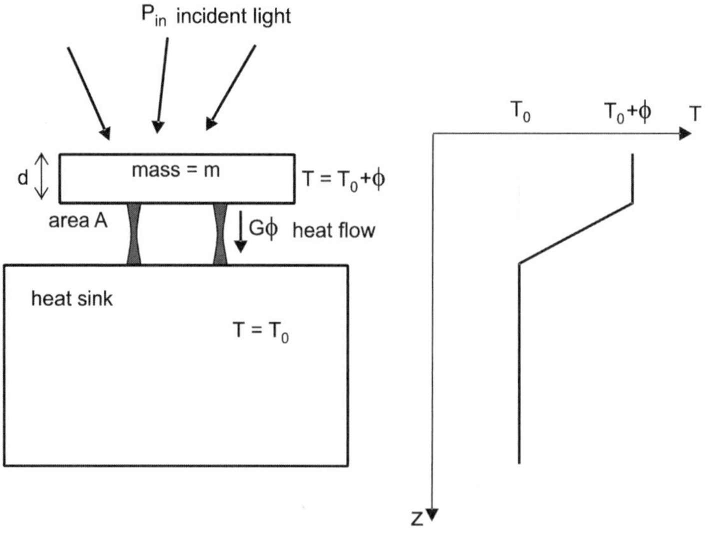

Figure 13-1 Model for temperature rise in a thermal detector. Heat flows along the temperature gradient from the sensor element to the heat sink, in proportion to the temperature difference ϕ.

The net change in the sensor element's heat energy in a time Δt is then given by

$$\begin{bmatrix}\text{increase in}\\ \text{heat energy}\end{bmatrix} = \begin{bmatrix}\text{light energy}\\ \text{absorbed}\end{bmatrix} - \begin{bmatrix}\text{heat lost}\\ \text{by conduction}\end{bmatrix} \tag{13-1}$$

$$mC\,\Delta\phi = (P_{\text{in}} - G\phi)\,\Delta t$$

where C is the specific heat (heat capacity per unit mass) of the sensor material, and $\Delta\phi$ is the change in sensor element temperature in time Δt. Dividing by Δt and letting $\Delta t \to 0$, this becomes

$$\frac{d\phi}{dt} + \frac{G}{mC}\phi = \frac{P_{\text{in}}}{mC} \tag{13-2}$$

which is a simple linear first-order differential equation, similar in form to Eq. (11-4). Like any such equation, it has exponential time-dependent solutions when the driving term (P_{in}/mC in this case) is constant in time.

The solution to Eq. (13-2) is depicted graphically in Fig. 13-2, for an incident light power that switches from zero to a constant value P_{in}. It can be expressed analytically as

$$\phi(t) = \phi_{\max}(1 - e^{-t/\tau}) \tag{13-3}$$

where the response time τ and maximum temperature rise $\phi_{\max}$ are defined by

$$\tau = \frac{mC}{G}$$
$$\phi_{\max} = \frac{P_{\text{in}}}{G} \tag{13-4}$$

This solution can be verified by direct substitution of Eq. (13-3) into Eq. (13-2).

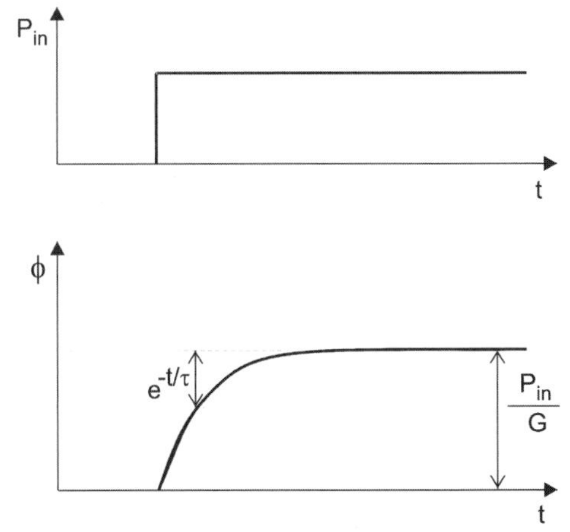

Figure 13-2 The temperature in a thermal detector rises exponentially when the incident optical power has a step-function time dependence.

It can be seen from Eq. (13-4) that the choice of G in the detector design involves a trade-off between speed and sensitivity. If G is small, the sensitivity is good, because there is a large temperature change ϕ_{max} for a given incident power, and this makes it easier to measure small powers. However, τ will then be large, which means that it takes a long time for the detector to respond to a change in input power, that is, it has a slow time response. The detector can be made to respond faster by increasing G, but this will lower the sensitivity. The value of G can be adjusted by changing the geometry of the support structure, so as to optimize the performance for a particular application.

The other variable under design control is the mass of the sensor element. Since m only appears in the expression for τ (not ϕ_{max}), a smaller mass gives an improved time response without degrading the sensitivity. The mass of the sensor element can be reduced by simply making it smaller in size. However, if the sensor element is made small in all dimensions, it will not intercept as much of the incident light, and P_{in} is then effectively reduced. The optimum geometry for the sensor element is, therefore, that of a thin disk, which allows the sensor element to intercept as much light as possible, while still having a small mass. Detectors used as laser power meters are often designed in this way. The disk is coated with a black layer that absorbs light equally over a wide wavelength range.

Thermoelectric Detector

One method for converting the temperature rise of the sensor element into an electrical signal uses a *thermocouple,* which is formed by the junction of two dissimilar metals. The electric potential is found to be different on the two sides of the junction, by an amount that varies with temperature, a phenomenon known as the *thermoelectric effect.* A practical way to incorporate thermocouples into a thermal detector is shown in Fig. 13-3. One thermocouple formed by wires A and B is attached to the sensor element, and a second, identical thermocouple between wire types B and A is attached to a point in the detector housing that is held at fixed temperature. Because the relative order of the materials is opposite for the two junctions (A → B for the first, B → A for the second), the potential differences have opposite polarity, and cancel when the two junctions are at the same tem-

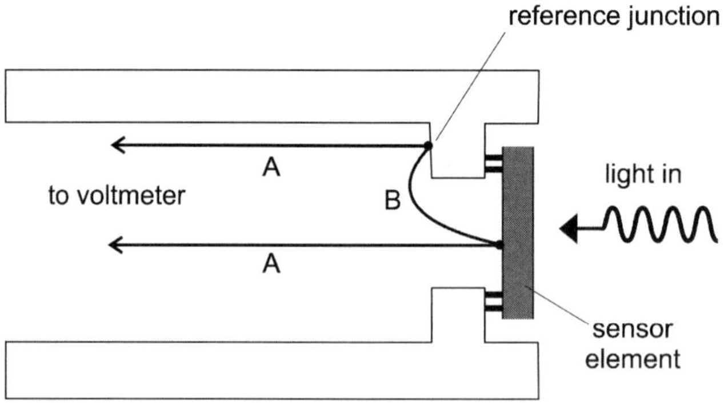

Figure 13-3 In a thermoelectric-type detector, the voltage difference across a series combination of two thermocouple junctions measures the temperature increase of the sensor element.

perature. When the sensor temperature increases relative to the reference, there will be a voltage between the two wire leads of type A, which can be read by a voltmeter or other electronic circuit element. Typical thermocouple sensitivities around room temperature are ~ 40 μV/°C. To increase this sensitivity, thermocouples can be configured in series so that the voltages across the individual thermocouples add. Such a device is termed a *thermopile.* Detectors based on this principle are commonly used for measuring absolute optical power for continuous-wave (CW) lasers.

Pyroelectric Detector

Another method for detecting the temperature rise of the sensor element uses special crystalline materials known as *ferroelectrics.* Inside a ferroelectric crystal, there is a spontaneous displacement of charge, creating electric dipoles inside the material. At the edge of the material, there are unbalanced charges, as depicted in Fig. 13-4, which effectively act as bound surface charges. If the ferroelectric is sandwiched between two parallel metallic plates, these bound charges induce free charges $+Q$ and $-Q$ on the plates.

The net dipole moment in the ferroelectric is found to decrease with increasing temperature, as shown in Fig. 13-5, going to zero above a critical temperature T_c. This behavior is analogous to that of a ferromagnet with a temperature-dependent spontaneous magnetization. In both cases, the decrease in net dipole moment is due to the competition between a naturally ordered state and thermally induced disorder.

When a change in temperature causes the net polarization to change, the induced charge on the plates changes, and this results in a flow of current through an external circuit connected to the plates. If the current flows through a load resistor R_L, a voltage $V_{sig} = iR_L$ is developed across the resistor, which can be measured. This conversion of temperature changes into an electric current or voltage is known as the *pyroelectric effect,* and is the basis for operation of the pyroelectric detector.

It is important to note that current only flows when the temperature is changing, since $\Delta Q = 0$ when the temperature (and hence the polarization) is constant. Therefore, when the light power being detected is constant in time, there is no change in temperature and no signal from the detector. This insensitivity to a constant background light level can be a useful feature for certain applications. For example, in thermal IR imaging, the display screen will show only changes in the image, providing a greater contrast for observing

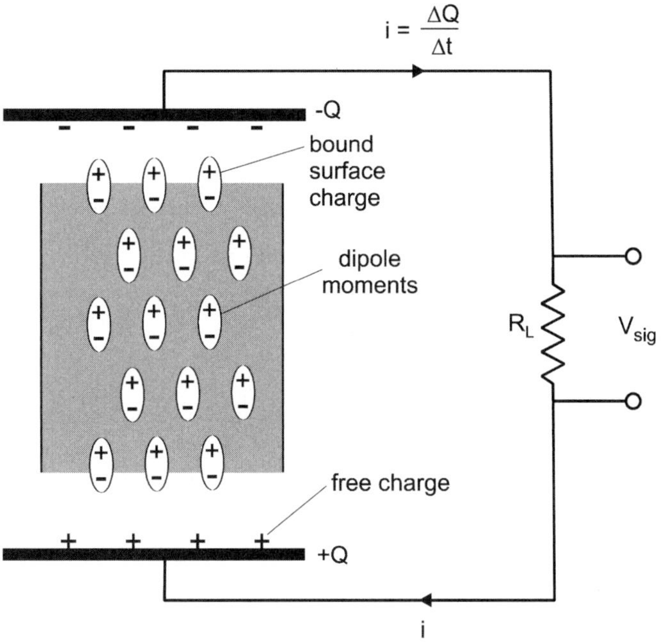

Figure 13-4 A ferroelectric material has spontaneously aligned electric dipole moments, which induces a charge on nearby metallic electrodes. When the induced charge changes in time, a current is generated through an external circuit.

small changes in a scene. When the detector is used in an intrusion or fire alarm, it naturally suppresses a slowly varying ambient light level, and responds only to the target-induced light signal that is changing in time. This type of detector is also well suited for use in power meters for pulsed lasers.

Typical ferroelectric materials used for pyroelectric detectors include $LiNbO_3$ (lithium niobate) and $LiTaO_3$ (lithium tantalate). One surface is coated with a thin black film, and incident light is absorbed in this film after passing through a transparent electrode. The sensitivity of the pyroelectric detector decreases at lower temperatures, where the slope of Fig. 13-5 is smaller. Therefore, unlike many other types of detectors, there is no advan-

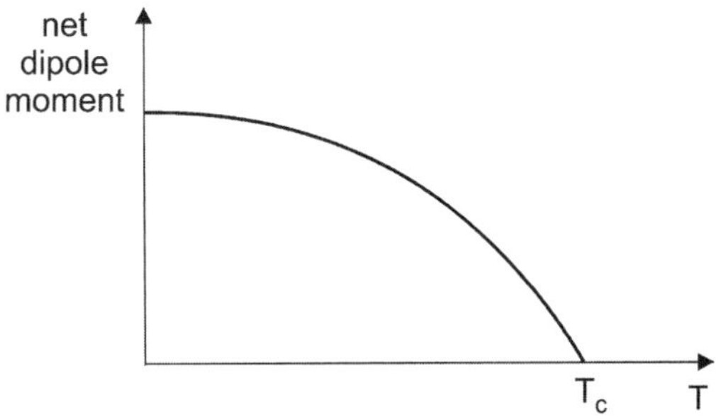

Figure 13-5 Below some critical temperature T_c, a ferroelectric material has a spontaneous electric dipole moment that increases with decreasing temperature.

tage to cooling the detector. Although the sensitivity becomes quite large near T_c, there is also increased detector noise due to fluctuations near the critical temperature. This is generally not a problem unless T_c is close to room temperature.

The pyroelectric detector is best suited for measuring time varying light levels, but it can also be used to measure light that is constant in intensity. To do this, the light can be passed through a rotating slotted blade that alternately passes and blocks the beam. The beam intensity then has an artificially induced time dependence, to which the pyroelectric detector will respond.

13-2. PHOTON DETECTORS

In contrast to the thermal-type detectors discussed in the previous section, photon detectors operate by the direct conversion of photons into charge carriers (electrons and/or holes). In this section, we discuss several types of photon detectors based on *photoemission* (ejection of an electron by an absorbed photon) and *photoconductivity* (change in electrical conductivity due to an absorbed photon). The other important type of photon detector, the photodiode, will be discussed in Chapter 14.

Photoelectric Effect

The basic principle of light detection using photoemission can be understood by considering the photoelectric effect, a phenomenon discovered toward the end of the 19th century. Negative charge was found to be emitted by a clean metallic surface when illuminated with ultraviolet light, and the kinetic energy of the emitted electrons was found to depend not on the optical power, but rather on the frequency of the light wave.

In 1905, Einstein proposed an elegant explanation of this effect, which has had a profound and lasting influence on our conceptual understanding of light. In this view, light consists of discrete energy packets called photons, and in an absorption process the entire photon energy is given to an electron in the material. The energy of each photon is $h\nu$, where h is Planck's constant and ν is the frequency [see Eq. (2-1)]. The kinetic energy of the photoejected electron can then be determined by referring to the energy diagram of Fig. 13-6. Inside a metal, the electrons have a distribution of energies up to some maximum value, referred to as the Fermi level. This Fermi level is lower than the electrons'

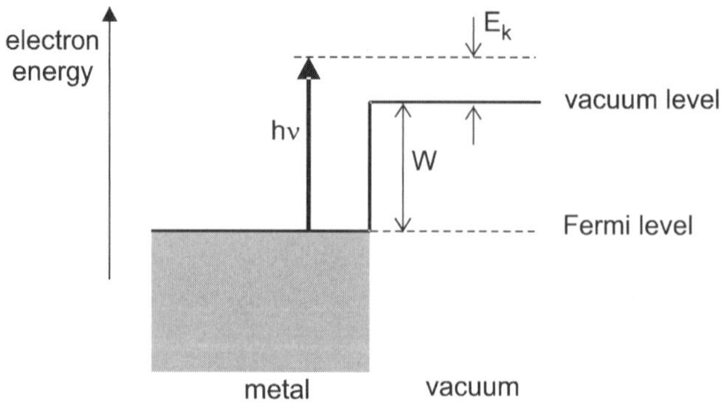

Figure 13-6 In the photoelectric effect, photons with energy greater than the work function W are ejected from a metal with maximum kinetic energy $h\nu - W$.

energy in a vacuum outside the material, the difference being known as the work function W (see section on metal–semiconductor junctions, p. 178). For an electron to escape from the metal, it must be given a minimum extra energy W, so the photon energies that will result in photoemission are

$$h\nu > W \qquad \text{(photon energy for photoemission in metal)} \tag{13-5}$$

For photon energies greater than the work function, the electron will be ejected with kinetic energies $E_k \leq h\nu - W$. Kinetic energies less than the maximum occur because some of the electrons absorbing a photon of energy $h\nu$ have an initial energy less than the Fermi level.

Although the condition $h\nu > W$ ensures that photoemission is energetically possible, it does not mean that all electrons absorbing a photon will actually be ejected. Unless the electron is initially close to the surface, it must travel a certain distance through the metal to reach the boundary. Along the way, it will suffer inelastic collisions with the many other electrons in the metal, which tend to decrease its kinetic energy. In practice, the fraction of electrons that make it out is quite small in an elemental metal, typically $\sim 10^{-3}$. These metals also have an inconveniently large work function ($W > 2$ eV), and are therefore not suitable for detection of near-IR wavelengths. For these reasons, elemental metals are seldom used as photodetector materials.

Metallic alloys are much more suited for photoemission-type detectors. The alloy compositions of interest are actually semiconductors, with the energy level structure shown in Fig. 13-7. The energy difference between the bottom of the conduction band and the vacuum level is termed the electron affinity, and denoted by χ. This represents the energy required to eject an electron initially in the conduction band. Since most of the electrons are initially in the valence band, the photon energies that will result in photoemission are

$$h\nu > E_g + \chi \qquad \text{(photon energy for photoemission in semiconductor)} \tag{13-6}$$

where E_g is the band-gap energy.

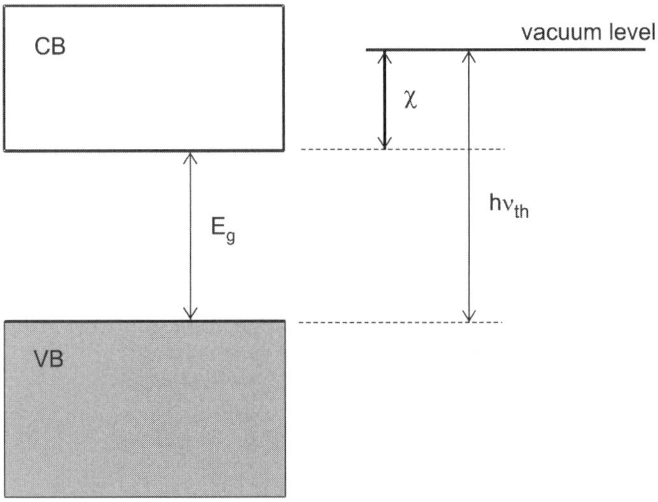

Figure 13-7 In a semiconductor, photoemission takes place when the photon energy exceeds the sum of the band-gap energy and electron affinity χ.

The escape efficiency for semiconductors is much higher than for metals, because there are very few free electrons in the conduction band to cause collisions with the ejected electron. There will still be some collisions with lattice vibrations (phonons), however, so the efficiency is less than unity. Efficiencies in the 10–20% range are typical at room temperature.

Metallic alloys also have the advantage that the threshold energy $E_g + \chi$ can be much lower than for any elemental metal. For example, the widely used composition Na_2KSb:Ce (containing small amounts of Ce) has a good response out to a wavelength of 800 nm, which corresponds to a theshold photon energy $h\nu_{th} = E_g + \chi \approx 1.55$ eV. Metallic alloys used in photodetectors often contain atoms from group I of the periodic table, because their outermost electrons have smaller binding energies, leading to a smaller bandgap energy for the semiconductor.

It is possible to reduce the effective χ, and even make it negative, by depositing a thin film of Cs on the surface of a highly p-doped semiconductor such as GaAs. The Cs atoms easily donate their outer electron to the acceptor ions in the GaAs, creating a space-charge region near the surface, as indicated in Fig. 13-8. The electric field in this space-charge region bends the electron energy bands downward near the surface, since the electron's potential energy decreases as it moves opposite to the electric field direction. Electrons excited from the top of the valence band in the interior can then reach the vacuum level outside the material by absorbing a (virtual) photon of energy $h\nu_{th} < E_g$, and "tunnelling" through the thin surface layer. The effective electron affinity $\chi_{\text{eff}} = h\nu_{th} - E_g$ is then negative, leading to the term *negative electron affinity* (NEA) for these materials. This has become an important way of extending the response of photoemissive detectors to longer wavelengths.

Vacuum Photodiode

After electrons are ejected from a photoemissive material, they must somehow be collected to obtain a signal. A simple device that accomplishes this is the vacuum photodiode, il-

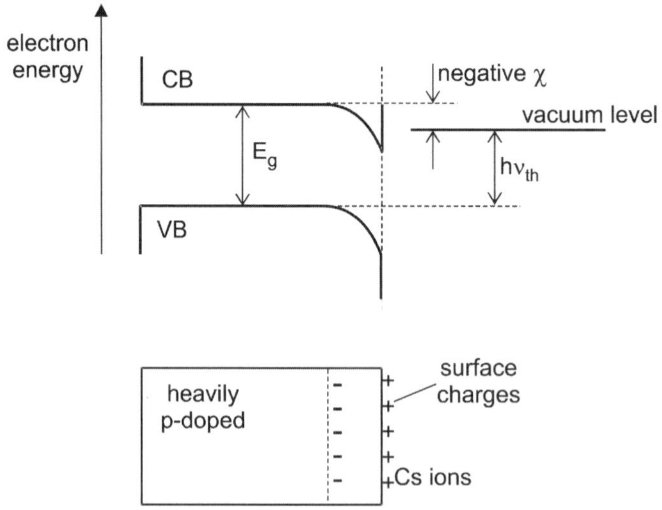

Figure 13-8 In a negative electron affinity (NEA) material, surface charges lower the energy bands near the surface, making the effective χ negative. This allows photoemission for $h\nu < E_g$.

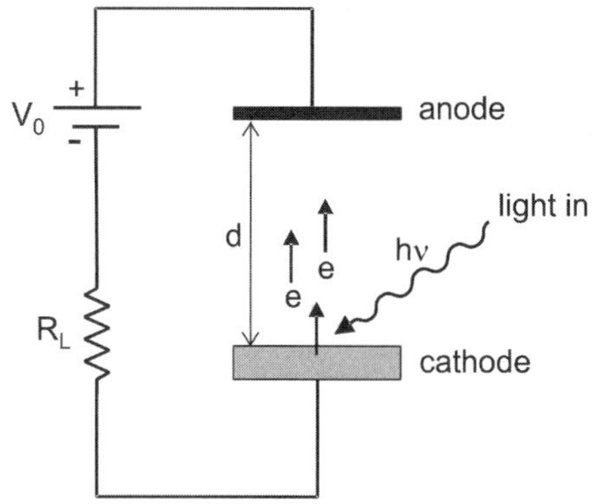

Figure 13-9 In a vacuum photodiode, electrons ejected by photoemission from a photocathode are collected by an anode held at high positive potential.

lustrated in Fig. 13-9. Two electrodes are placed in an evacuated tube, one of them the photoemissive material, and the other a collection electrode. A high voltage is applied between the electrodes, so that the photoemissive material is at lower potential (the *cathode*) and the collection electrode at higher potential (the *anode*). A cathode that responds to light by emitting electrons is termed a *photocathode.*

Electrons emitted by the photocathode will be accelerated by the electric field between the electrodes, causing a current in the external circuit. This current gives rise to a voltage across the series resistor R_L, which consitutes the measured signal. The current i generated depends on the incident light power P_{in} according to

$$\text{charge/time} = \left[\frac{\text{energy/time}}{\text{energy/photon}}\right]\left[\frac{\text{charge}}{\text{electron}}\right]\left[\frac{\text{electrons}}{\text{photon}}\right]$$

which can be written as

$$i = \frac{P_{\text{in}}}{h\nu} e\eta \tag{13-7}$$

In this equation, e is the magnitude of the electron charge and η is the efficiency with which incident photons are converted into ejected electrons. The *responsivity* of the detector is defined as

$$\mathcal{R} \equiv \frac{i}{P_{\text{in}}} = \frac{e\eta}{h\nu} \quad \text{(detector responsivity)} \tag{13-8}$$

which is a measure of the output (current) versus input (optical power) for the detector. The SI units for $\mathcal{R}$ are A/W.

EXAMPLE 13-1

Determine the responsivity of a photodetector of quantum efficiency η to light of wavelength 1300 nm. Evaluate this when $\eta = 0.2$.

Solution: The responsivity is

$$\mathcal{R} = \frac{e\eta}{h\nu} = \frac{e\lambda\eta}{hc} = \frac{(1.6 \times 10^{-19})(1.3 \times 10^{-6})\eta}{(6.63 \times 10^{-34})(3 \times 10^{8})} = 1.046\ \eta \text{ A/W}$$

Note that for wavelengths of interest for telecommunications, we have the simple approximation $\mathcal{R} \sim \eta$ [A/W]. Putting in $\eta = 0.2$ above gives $\mathcal{R} = 0.209$ A/W.

An interesting question about the current generated (termed the *photocurrent*) is this: what does the time dependence of the current pulse look like when a single electron is ejected from the photocathode and travels to the anode? It might be supposed that the current pulse is only observed when the electron arrives at the anode, where it is "collected." The correct answer at first seems a bit surprising: current flows constantly during the time that the electron is in transit between electrodes.

To see why this occurs, consider an electron moving toward the anode under the influence of an electric field, as shown in Fig. 13-10. The work done on the electron during a small time Δt while it moves a distance Δx is

$$\begin{aligned} \Delta W &= F\Delta x = eE\Delta x \\ &= eEv\Delta t \end{aligned}$$

where $v = \Delta x/\Delta t$ is the speed of the electron and $E = V_0/d$ is the magnitude of the electric field. Work is done at a rate $P = \Delta W/\Delta t = eEv$, and goes into the increased kinetic energy of the electron as it accelerates toward the anode. The source of this energy is the external circuit, which supplies an electrical power $V_0 i$ at a fixed voltage V_0. Setting the power supplied equal to the rate at which work is done, we have

$$P_{supplied} = V_0 i = eEv$$

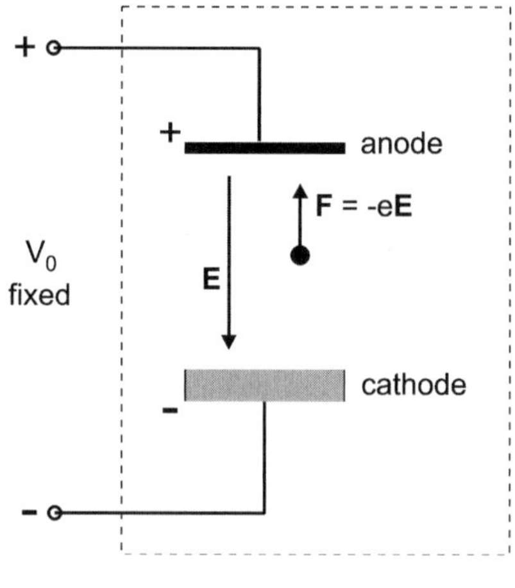

Figure 13-10 The current response *i*(*t*) for a single electron moving from photocathode to anode can be determined by defining a region of space that encloses the diode (dashed line), and applying the work-energy theorem to this volume.

or

$$i(t) = \frac{eE}{V_0} v(t) \tag{13-9}$$

This important result says that the actual time dependence of the current pulse follows the time dependence of the electron's velocity as it moves in an electric field E through a potential difference V_0. Because it is based on fundamental energy principles, this result applies quite generally to the various types of photodetectors that we will discuss in this chapter and the next.

An important application of Eq. (13-9) is in determining the time response of a photon detector. In the vacuum photodiode, electrons experience an acceleration $a = eE/m$, where m is the electron mass. We will assume for simplicity that the detector has plane parallel electrodes, in which case it is referred to as a *biplanar phototube.* In this case, E and a are both constant, so

$$v(t) = at = \frac{eE}{m} t \tag{13-10}$$

$$x(t) = \frac{1}{2} at^2 = \frac{1}{2} \frac{eE}{m} t^2 \tag{13-11}$$

For constant E, we also have $E = V_0/d$, where d is the spacing between the electrodes. Since $v(t)$ increases linearly with t, then according to Eq. (13-9), so too does the current $i(t)$. The current pulse, therefore, has the shape depicted in Fig. 13-11, coming to an end at $t = \tau$, when the electron reaches the anode. The response time τ can then be determined from Eq. (13-11) by writing

$$d = \frac{1}{2} \frac{eE}{m} \tau^2$$

which gives

$$\tau = \sqrt{\frac{2md}{eE}} = d\sqrt{\frac{2m}{eV_0}} \qquad \text{(vacuum photodiode response time)} \tag{13-12}$$

The time response is seen to be better (τ smaller) for large applied voltage V_0, and small electrode separation d. Making d too small, however, increases the capacitance be-

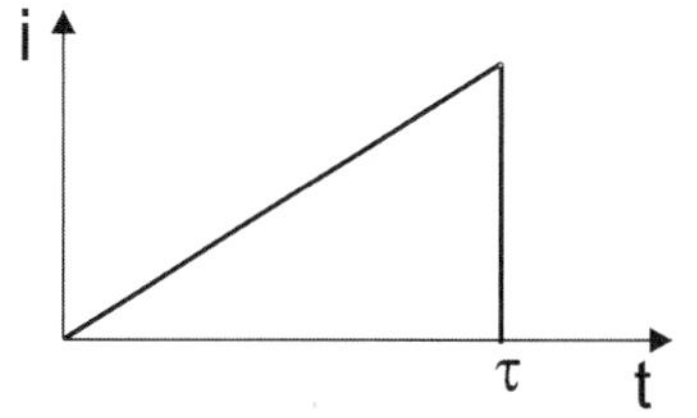

Figure 13-11 In a vacuum photodiode, the current increases linearly with time during the pulse.

tween the electrodes, degrading the time response. Therefore, high voltages (a few kV) are used for the best time response, which can be in the range 100–500 ps. The requirement of a high-voltage power supply limits the practical utility of vacuum photodiodes, although they are useful for specialized applications. They have a good response in the UV and good linearity over a wide range of incident light levels, making them useful for precise monitoring of fast high-power laser pulses.

Although vacuum photodiodes are not the most common detectors in use, their general operating principles are simple to understand and can be directly taken over into our discussion of other detector types. For example, one important parameter for any photodetector is the total charge sent around the external circuit in response to a single absorbed photon. The total charge Q moved during the time τ of the pulse is given by the area under the curve of Fig. 13-11. Using Eq. (13-9) and $E = V_0/d$, this is

$$\begin{aligned} Q &= \int_0^{\tau} i(t)\,dt = \frac{e}{d}\int_0^{\tau} v(t)\,dt \\ Q &= \frac{e}{d}\int_0^{d} dx = e \end{aligned} \qquad (13\text{-}13)$$

We therefore obtain the very satisfying result that the total charge sent around the circuit while one electron is making its transit between the electrodes is just e, the charge of one electron. In contrast to this, other detectors that we will discuss may have $Q < e$ or $Q > e$, making the vacuum photodiode a good point of comparison.

Photomultiplier

Although the vacuum photodiode has the advantage of simplicity and reliability for precise power measurements, it is not very sensitive to low light levels. One way to increase the sensitivity is to add an amplification section between the photocathode and anode, as illustrated in Fig. 13-12. This device is called a *photomultiplier* and works in the following way. Incident photons eject electrons from the photocathode just as in a vacuum photodiode. As the electrons move from the photocathode to the anode, they strike a series of secondary electrodes called *dynodes,* which are held at potentials intermediate between the cathode and anode. When an electron strikes a dynode, it ejects a number of addition-

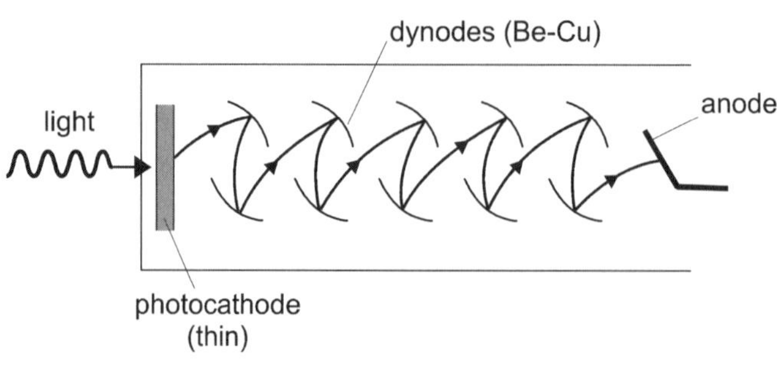

Figure 13-12 In a photomultiplier tube, electrons emitted from a photocathode strike a series of dynodes on their way to the anode. These collisions eject additional (secondary) electrons from each dynode, effectively amplifying the detector signal.

al electrons, each of which is then accelerated to collide with the next dynode. The result is an avalanche process, with the number of electrons increasing exponentially. If each dynode produces δ electrons when struck by a single electron, then for a cascade of N dynodes, the gain is

$$G = \frac{Q}{e} = \delta^N \tag{13-14}$$

For typical values $\delta \approx 5$ and $N \approx 10$ this results in $G \approx 10^7$, an enormous enhancement. This large gain makes the photomultiplier suitable for detecting very low levels of light, even down to the level of individual photons.

Figure 13-13 shows the typical electrical connections inside a photomultiplier tube. A negative high voltage (usually ~ 1 kV) is applied to the photocathode, and the proper potential for each dynode is maintained by a chain of equal-value resistors. The anode is approximately at ground potential, and attracts the large bunch of electrons emitted from the last dynode (which is at a potential $\approx -V_0/N$). While these electrons are in transit, they generate a current through the load resistor R_L, and the voltage across this resistor constitutes the detector signal. It is important to note that the two electrical connections for the signal output are both near ground potential and, therefore, safe to touch and connect to equipment. The dangerous high voltage remains safely inside the device.

The spectral response of the photomultiplier depends on the type of photocathode material used. The responsivity $\mathcal{R}$ for a few representative types are shown in Fig. 13-14. The $\mathcal{R}$ values given are for single electron emission from the photocathode, and must be multiplied by G to obtain the actual sensitivity of the photomultipler. The spectral response for certain metallic alloy photocathodes is denoted by an "S" number, according to the composition. For example, the composition Na_2KSb:Ce mentioned previously has an "S20" response, which is efficient and extends from the near UV region out to $\approx$ 800 nm in the IR region. This is perhaps the most popular photocathode type for measurements in the visible region.

Other metallic alloy compositions have a better response in either the UV or the IR regions. The first important commercial photocathode had the composition Ag-O-Cs, with a spectral response designated S1. Although the efficiency is relatively low, it has a response extending out past 1000 nm, which makes it still useful today. The S1 photocathode is typically cooled to reduce the noise associated with thermally emitted electrons.

The NEA GaAs photocathode has a very uniform response that extends throughout the visible region and out to $\approx$ 900 nm in the IR region. It is superior to the S1 photocathode

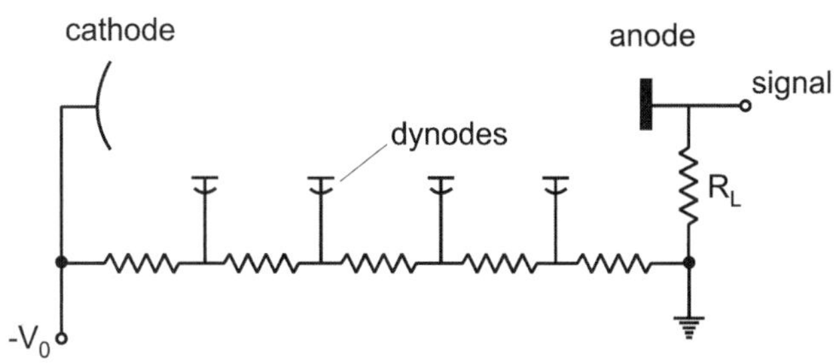

Figure 13-13 Typical electrical connections inside a photomultiplier. The dynode potentials are maintained by a resistor chain connected between anode and cathode.

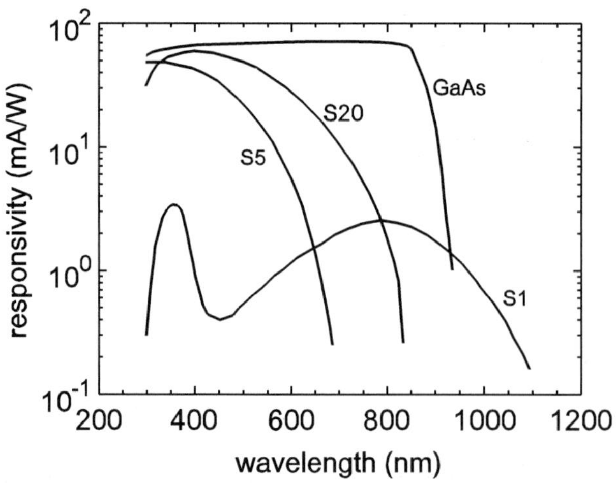

Figure 13-14 Typical spectral response $\mathcal{R}(\lambda)$ for some common photocathodes. In a photomultiplier, this must be multiplied by the gain $G = \delta^N$ to obtain the detector responsivity.

in that the efficiency is much higher, and it does not need cooling. However, it does not extend as far into the IR region as the S1. A similar photocathode, the NEA InGaAs, has a response that extends out past 1000 nm, and can therefore compete with the S1 there. It has the advantage over the S1 that there is less noise due to thermally emitted electrons, so cooling is often not required.

Although photomultipliers represent the ultimate in sensitivity for light detection, they have some disadvantages for everyday applications. Like the vacuum photodiode, they require a high voltage, which is inconvenient and entails safety concerns. The time response is slower than vacuum photodiodes, because of the longer transit time needed for electrons to go from the photocathode to the anode. There is also an inherent spreading of the electron bunches during the transit, because electrons ejected from the dynodes are emitted in different directions and take different paths down the tube. From a practical point of view, these detectors are fragile and easily damaged by excessive light intensity. They are also inherently "bulk" devices, not easily miniaturized for integrated optics applications. Nonetheless, they play an important role for applications such as fluorescence detection and light scattering, where sensitivity is of prime concern.

Photoconductive Detectors

The photon detectors discussed so far work by ejecting electrons from a photocathode material. It is also possible for an electron in the valence band (VB) to be promoted to the conduction band (CB) without being ejected from the material. This process is illustrated in Fig. 13-15, and might be thought of as an "internal photoelectric effect." Once the electron is in the CB, it becomes mobile, and contributes to the electrical conductivity. This increase in a material's conductivity upon absorption of light is termed *photoconductivity,* and is the operating principle of the photoconductive detector or *photocell.*

It is clear from Fig. 13-15 that for a photon to be absorbed, its energy $h\nu$ must be greater than the bandgap energy E_g of the material. Photons entering the material are absorbed with a probability α per unit length, where α is the attenuation coefficient. Figure

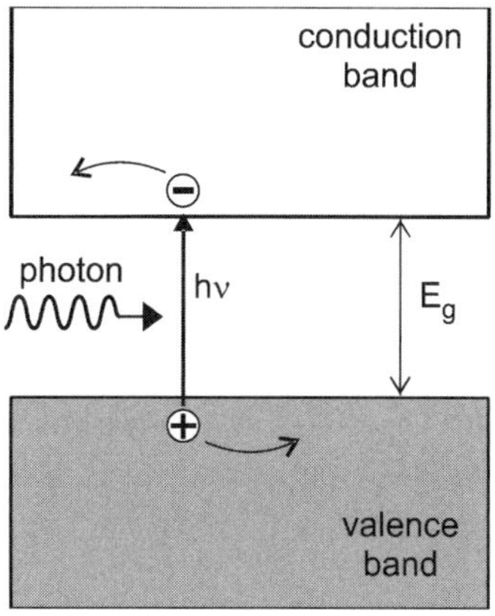

Figure 13-15 In the "internal photoelectric effect," electron–hole pairs are created by the absorption of a photon in a semiconductor.

13-16 shows the wavelength dependence of α for a few representative semiconductors. After light has propagated a distance x into the material, its intensity is reduced according to Beer's law, $I(x) = I(0)\exp(-\alpha x)$, where $I(0)$ is the intensity just inside the surface. If the material's total thickness in the direction of light propagation is d, then a fraction

$$\eta_{abs} = \frac{P_{abs}}{P_{in}} = (1 - R)(1 - e^{-\alpha d}) \tag{13-15}$$

of the incident light is absorbed. This expression takes into account the fraction R of the incident light that is reflected from the surface.

The behavior of a photocell can be understood by considering the simple model shown in Fig. 13-17. A uniform semiconductor of length L is irradiated from the side with light,

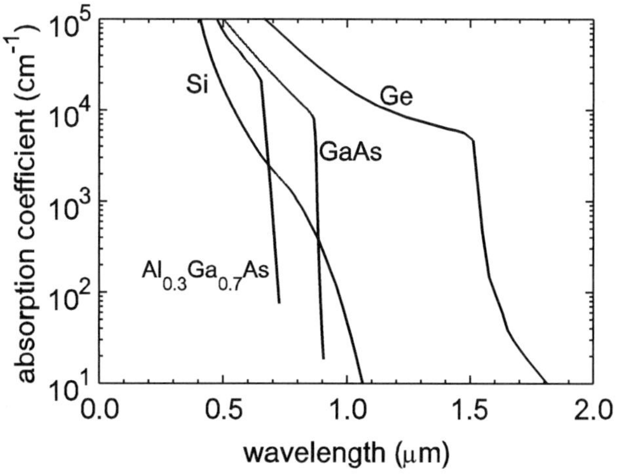

Figure 13-16 Spectral dependence of the absorption coefficient for some representative semiconductors.

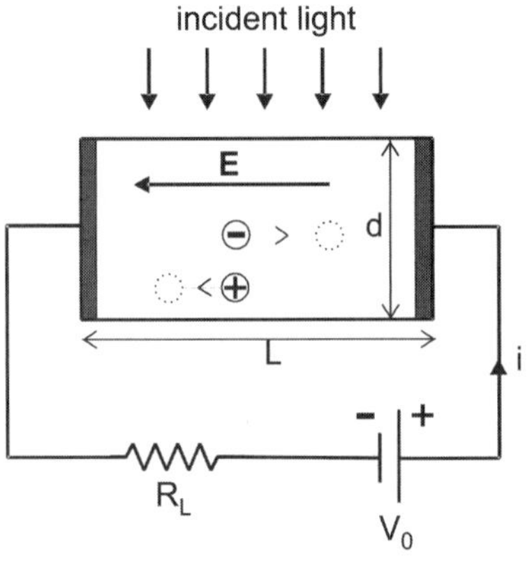

Figure 13-17 In a photocell, the charge carriers created by absorbed photons move under the influence of an applied electric field.

and photons that are absorbed create electron–hole pairs inside the semiconductor. A voltage V_0 is applied between the ends as shown, giving rise to an electric field $E = V_0/L$ in the material. Under the influence of this E field, the electrons and holes move in opposite directions, giving rise to a signal current i_s in the external circuit. This current induces a voltage drop i_sR_L in the series load resistor R_L, which constitutes the detector signal.

One important characteristic of the photocell is the presence of a background current i_0, even when there is no incident light. This occurs because the semiconductor intrinsically has some small but finite electrical conductivity, due to thermally generated electron–hole pairs. If the electrical resistance of the semiconductor is R_d, the background current is $i_0 = V_0/(R_L + R_d)$. The total current is then $i = i_0 + i_s$.

In practice, the background i_0 often dominates the signal current i_s, requiring special techniques for extracting the small signal from a large constant background. One method utilizes a rotating slotted blade to modulate the light intensity, creating a time-varying $i_s(t)$. The constant background i_0 can then be blocked by using a lock-in detector, which responds only to signals varying in time. Even when the background is suppressed in this way, however, it increases the noise of the detector, as we shall see in the next section.

To evaluate the induced signal current i_s for a single electron–hole pair, we can use Eq. (13-9), which was used previously for the vacuum photodiode. According to this equation, a signal current is produced whenever charge carriers move in the presence of an electric field. The difference here is that the electron velocity $v_e(t)$ is no longer linear with t, as it was for the vacuum photodiode. Collisions (with other electrons, phonons, and impurities) randomize the electron's motion, so that it moves with a constant average *drift velocity,* given by

$$v_e = \mu_e E \tag{13-16}$$

where μ_e is the electron *mobility.* A similar equation can be written for the motion of holes, $v_h = \mu_h E$, and both holes and electrons contribute to the signal current i_s. However, it is found in most materials that $\mu_h \ll \mu_e$, so the effect of hole motion is usually of minor importance. Also, holes commonly become trapped at impurity sites, where they become immobile and stop contributing to the photocurrent. Therefore, in the following treatment we will neglect the contribution of holes to i_s.

Using Eq. (13-16) in Eq. (13-9), along with $E = V_0/L$, we then have for the signal current

$$i_s(t) = \frac{e\mu_e E}{L} \tag{13-17}$$

This current is maintained as long as the electron remains in the conduction band, which is limited by the electron lifetime τ. The current pulse then has the time dependence shown in Fig. 13-18 (compare with Fig. 13-11 for the vacuum photodiode). Integrating this current over time gives the total charge in the current pulse,

$$Q = \int i_s(t)\, dt = \frac{e\mu_e E\tau}{L} = \frac{e\mu_e V_0 \tau}{L^2} \tag{13-18}$$

It is conventional to define the *photoconductive gain G* of the detector as

$$G \equiv \frac{Q}{e} = \frac{\mu_e V_0 \tau}{L^2} \qquad \text{(photoconductive gain)} \tag{13-19}$$

a definition similar to that of the photomultiplier gain.

This result shows that the detected charge (and hence the gain) increases as the applied voltage V_0 increases. An interesting feature of this equation is that G can be made larger than 1 for sufficiently large V_0. That is, the charge in the current pulse resulting from a single photon absorption can be greater than the charge of an electron! This is, at first glance, a rather puzzling conclusion. To see what is happening, consider the electron *transit time,* t_{tr}, defined as the time it takes for the electron to travel the entire length L of the semiconductor. Using Eq. (13-16) for v_e, this is

$$t_{tr} = \frac{L}{v_e} = \frac{L}{\mu_e E} = \frac{L^2}{\mu_e V_0} \qquad \text{(electron transit time)} \tag{13-20}$$

Combining this with Eq. (13-19) gives a simple expression for the photoconductive gain:

$$G = \frac{\tau}{t_{tr}} \tag{13-21}$$

From the above, we see that the photoconductive gain is greater than unity when the electron stays in the CB longer than the time it takes to traverse the semiconductor. It might be supposed that when the electron reaches the electrode at the semiconductor

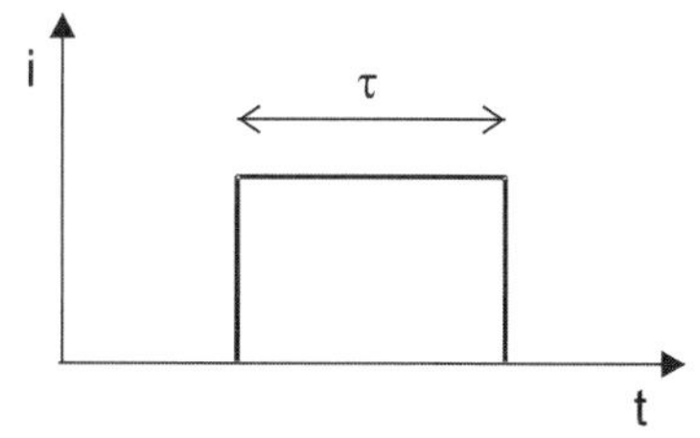

Figure 13-18 The current response for a single electron–hole pair created in the photocell semiconductor.

edge, the current pulse should end, since there is no further motion of charges in the electric field. The effective τ would then be limited by the transit time t_{tr}, and the gain limited to unity. This is indeed the case, if no additional electrons are released into the semiconductor to take the place of the one that has left.

If the contact between the electrodes and the semiconductor is *ohmic* (that is, obeying Ohm's law), then there is no barrier to the injection of new electrons into the semiconductor. In this case, electrons leaving the semiconductor are readily replenished by new ones, as depicted in Fig. 13-19. Electrons that are photoexcited from the VB to the CB can then effectively remain in the CB for a time longer than the electron transit time. This replenishment process will continue until an electron–hole recombination occurs, or until the hole that was initially created makes a complete transit across the semiconductor.

The photoconductive gain is linear with applied voltage V_0, according to Eq. (13-19). Higher voltage, therefore, gives higher gain and a larger signal. Figure 13-20 shows how this relation would be modified if the electron were not replenished at the electrodes. In that case, the gain would saturate at unity, when the applied voltage reaches $V_{sat} = L^2/(\mu\tau)$. Increasing the voltage beyond V_{sat} would improve the time response of the detector (which would then be limited by t_{tr}), but would not increase the signal. This feature is a characteristic of photodiode detectors, as we shall see in the next chapter. The p–n junctions in these devices present a barrier for injected charges, and no replenishment of the electron occurs.

The discussion so far has concerned the effective charge generated by a single absorbed photon. The number of photons striking the semiconductor per unit time will be $P_{in}/h\nu$, where P_{in} is the incident optical power and $h\nu$ is the photon energy. Therefore, the signal current (charge generated per unit time) will be

$$i_s = \frac{P_{in}}{h\nu}\eta_{abs}Ge \qquad \text{(photoconductive signal current)} \qquad (13\text{-}22)$$

where η_{abs} is given by Eq. (13-15). This should be compared with Eq. (13-7) for a vacuum photodiode.

Some common photoconductive materials, along with their range of spectral sensitivity, include CdS (400–700 nm), CdSe (500–900 nm), PbS (1–3.2 μm), and PbSe (1.5–5.2 μm). They are highly sensitive, and can operate at room temperature (although their noise properties are improved at low temperature; see next section). A key drawback for certain

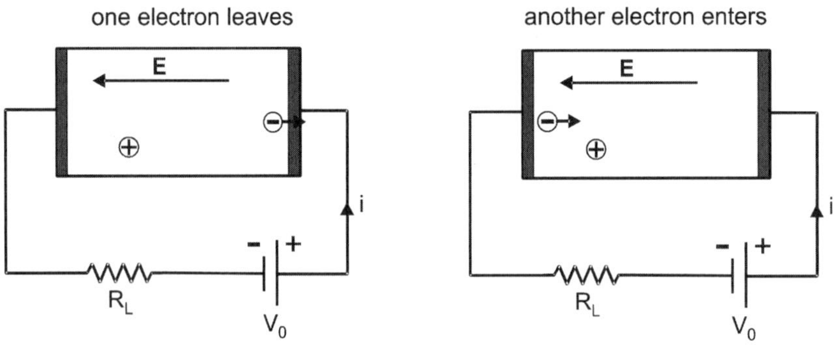

Figure 13-19 Gain is possible in a photocell because electrons leaving the semiconductor on one side are replenished by other electrons entering on the other side.

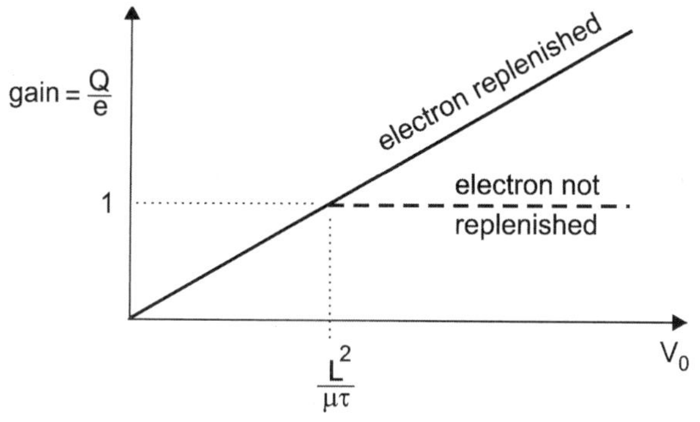

Figure 13-20 The photoconductive gain increases linearly with applied voltage, and can exceed unity if electrons are replenished after reaching the electrode.

applications is the slow response time, which is determined by the electron lifetime τ. This can range from 0.1 μs to 0.1 s, much too slow for high-speed data communications. Photocells therefore find their niche in photometry, thermometry, and other applications that require precise light-level measurements without the need for fast time response.

EXAMPLE 13-2

A CdS photocell has a separation between electrodes of 0.2 mm, and is biased with 1.2 V. Light with wavelength 550 nm and power 5 μW is incident on the photocell, which absorbs 75% of the incident light. (a) Determine the photoconductive gain of this detector, assuming that the electron lifetime in CdS is 1 ms and the electron mobility is 300 $cm^2/(Vs)$. (b) Determine the signal current generated in this photocell.

Solution: (a) The mobility in MKS units is 3×10^{-2} $m^2/(Vs)$, so

$$G = \frac{\tau\mu_e V_0}{L^2} = \frac{(10^{-3})(3 \times 10^{-2})(1.2)}{(2 \times 10^{-4})^2} = 900$$

(b) The photon energy of the incident light is

$$h\nu = \frac{hc}{\lambda} = \frac{(6.63 \times 10^{-34})(3 \times 10^8)}{550 \times 10^{-9}} = 3.62 \times 10^{-19}\ \text{J}$$

so the signal current is

$$i_s = \left(\frac{5 \times 10^{-6}}{3.62 \times 10^{-19}}\right)(0.75)(900)(1.6 \times 10^{-19}) = 1.49 \times 10^{-3}\ \text{A}$$

13-3. NOISE IN PHOTON DETECTORS

We saw in the previous section that the signal in a photocell can be made larger by the process of photoconductive gain. Although a larger signal is certainly desirable, the more important question for detector sensitivity is: how large is the signal compared to the noise? Noise is defined as random signal fluctuations that are not caused by or correlated with the physical quantity being measured. It is present in any practical detector system, and much of the work of detector design concerns reducing the noise. In this section, we discuss the origin of the two primary types of noise in detectors: shot noise and Johnson (or thermal) noise.

Shot Noise

Shot noise arises from the statistical process by which moving charges give rise to a current. Consider a current that is nominally constant in time, with the value $\bar{i}$. At a microscopic scale, the current consists of discrete charge carriers (electrons or holes) in motion, as illustrated in Fig. 13-21. For simplicity, in the following we will consider electrons to be the charge carrier, although the same arguments apply to holes. If a reference plane is drawn perpendicular to the current direction, the time at which the electrons cross this plane can be marked on a time axis, as in Fig. 13-21b. Because the motion of the electrons is a statistical process, they do not cross the plane at regular intervals, but rather with a distribution of arrival times.

If there is no correlation between the arrival times of different electrons, the arrival pattern is termed a *Poisson random process.* The probability $P(n)$ that n electrons will cross the plane in a time interval Δt is then given by the *Poisson distribution*

$$P(n) = \frac{(\bar{n})^n e^{-\bar{n}}}{n!} \qquad \text{(Poisson distribution)} \qquad (13\text{-}23)$$

where

$$\bar{n} = (\bar{i}/e)\,\Delta t \qquad (13\text{-}24)$$

is the average number of electrons crossing the plane during Δt, and $\bar{i}/e$ is the average number of electrons crossing the plane per unit time. For large $\bar{n}$, the function $P(n)$ can be closely approximated by a Gaussian, as depicted in Fig. 13-22. It has a well-defined maximum at $n = \bar{n}$, with an rms (root mean square) width δn given by

$$\delta n = \sqrt{\bar{n}} \qquad \text{(width of Poisson distribution)} \qquad (13\text{-}25)$$

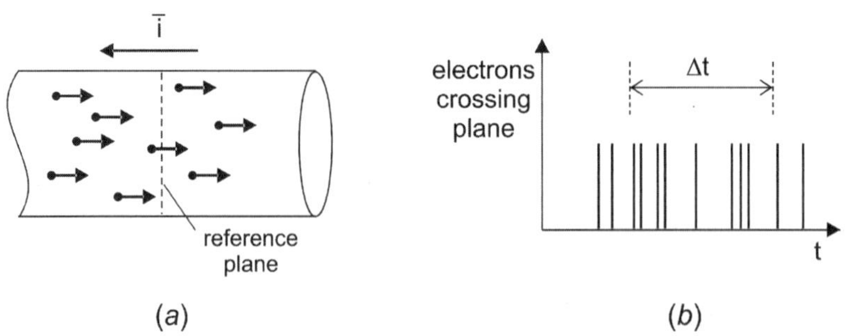

Figure 13-21 (a) The electrical current is the amount of charge passing a reference plane per unit time. (b) Electrons pass the reference plane at a statistical distribution of times, resulting in current noise.

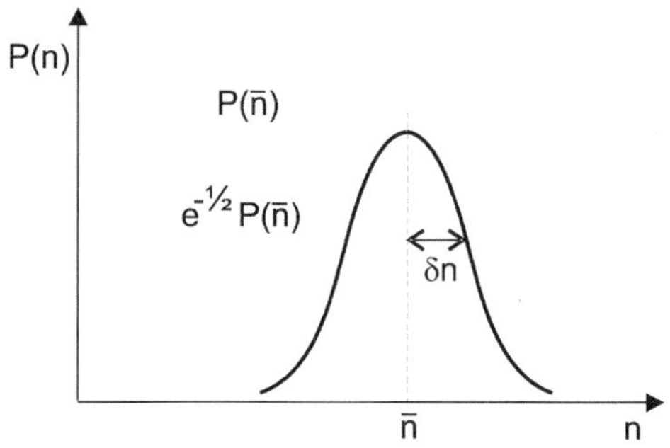

Figure 13-22 The number n of electrons crossing the reference plane in time Δt varies according to the Poisson distribution, with an average $\bar{n}$ and rms deviation δn.

Equation (13-25) is an important feature of the Poisson distribution. Stated loosely, it says that the width of the distribution (and the standard deviation of the equivalent Gaussian) equals the square root of the center, or most probable, value. The ratio of width to center value is

$$\frac{\delta n}{\bar{n}} = \frac{1}{\sqrt{\bar{n}}}$$

which means that the distribution gets sharper for larger $\bar{n}$.

If the number n of electrons crossing the reference plane in time Δt varies from one time interval Δt to the next, the instantaneous current

$$i = \frac{en}{\Delta t} \tag{13-26}$$

also varies in time. The current will fluctuate by an amount

$$\delta i = \frac{e}{\Delta t} \delta n$$

as illustrated in Fig. 13-23. Using Eqs. (13-24) and (13-25), this can be written as

$$\delta i = \frac{e}{\Delta t} \sqrt{\bar{n}} = \frac{e}{\Delta t} \sqrt{\left(\frac{\bar{i}}{e}\right) \Delta t} = \sqrt{\frac{e\bar{i}}{\Delta t}} \tag{13-27}$$

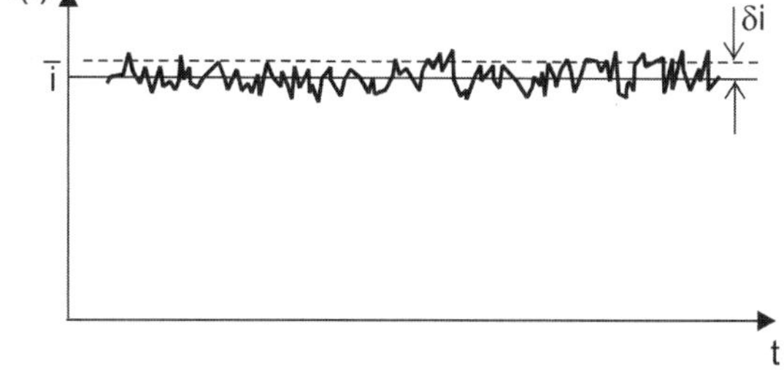

Figure 13-23 Because of shot noise, a current $\bar{i}$ exhibits fluctuations with an rms deviation δi.

An important implication of Eq. (13-27) is that the current noise increases as Δt is made smaller. We can interpret Δt as the measurement time, since the n electrons were determined to be crossing the reference plane during that time. Therefore, shorter measurement times give a higher noise, and vice versa. It is common to describe a measurement system in terms of its *bandwidth,* rather than its measurement time, the two being reciprocally related by the Fourier transform (see Appendix B). Denoting the bandwidth by B, we can write

$$B = \frac{\mathcal{K}}{\Delta t} \tag{13-28}$$

where $\mathcal{K} \sim 1$ is a constant of proportionality, the exact value of which depends on how the bandwidth and measurement time are defined. A rigorous analysis shows that the proper value here is $\mathcal{K} = 1/2$, so the rms current noise becomes

$$\delta i = \sqrt{2e\bar{i}B} \qquad \text{(shot noise)} \tag{13-29}$$

which is the widely used expression for current (shot) noise.

This result shows that the noise increases with the square root of the detector bandwidth. One way to reduce the noise in a detector, then, is to reduce the detection bandwidth with appropriate electrical filters. However, this has the side effect of restricting the range of modulation frequencies that can be detected, which limits the time response of the detector. This issue will be discussed further in Chapter 14.

According to Eq. (13-29), the shot noise also increases with increasing average current $\bar{i}$, but only as the square root of $\bar{i}$. Therefore, the current increases faster than the noise, and higher currents are expected to give a higher ratio of signal to noise. This is indeed the case, provided that the current $\bar{i}$ is truly a "signal" current.

In the case of the photocell, discussed in the previous section, we saw that most of the current through the device was not a signal at all, but rather a background or *dark current,* present even with no light incident on the detector. The shot noise from this dark current can easily overwhelm a weak signal current, and is one of the fundamental limitations of this type of detector.

One way to reduce the dark current in a photoconductor is to increase the series load resistance R_L. This is clear from Fig. 13-17, where the dark current is given by $i_0 = V_0/(R_L + R_d)$. However, increasing R_L introduces noise of another type, as we discuss in the next section.

Johnson Noise

Johnson noise refers to the voltage or current fluctuations in a resistor due to thermal agitation of the electrons in the material. It is also referred to as *thermal noise* or *Nyquist noise.* The effect was discovered experimentally in 1927 by Johnson, and soon thereafter explained theoretically by Nyquist using thermodynamic and statistical mechanical arguments. The original paper in *Physical Review* (Nyquist 1928) is well worth reading, even today, for an insightful derivation of the thermal noise formula. In the following, we give an alternative, simplified derivation of this formula.

Consider a resistor of resistance R connected through a switch to an LC oscillator circuit, as shown in Fig. 13-24. Assume that all parts of this circuit are at the same tempera-

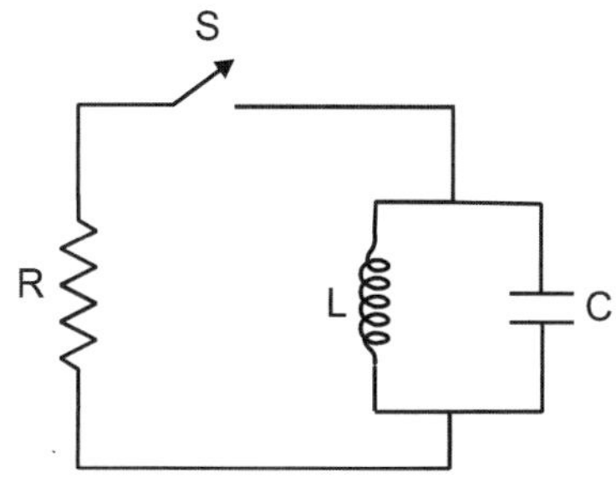

Figure 13-24 The Johnson noise generated by a resistor can be determined by considering the power fluctuations in this circuit, in which the switch is closed for a time interval Δt.

ture, and in thermal equilibrium with the surroundings. Initially, the switch is open, isolating the resistor from the *LC* oscillator. The *LC* circuit has a single mode of oscillation at frequency $\omega_0 = 1/\sqrt{LC}$, with the total energy of oscillation given by

$$E = \tfrac{1}{2} Li^2 + \tfrac{1}{2} CV^2$$

where V and i are the voltage across the capacitor and current through the inductor. In statistical mechanics, the principle of equipartition states that for every degree of freedom in a system (or, equivalently, for every term in the energy expression containing a normal coordinate squared) there is associated a thermal energy $k_BT/2$, where k_B is Boltzmann's constant and T is the absolute temperature. For the *LC* oscillator, there are two such terms, so the thermal energy is k_BT. This means that even without any external excitation, the voltage and current in the *LC* circuit will exhibit random fluctuations, with the energy fluctuating by $\delta E = k_BT$ on average.

The resistor is not a resonant system, and has no natural modes of oscillation. However, it is expected that the thermal agitation of electrons in the resistor will cause random current fluctuations and corresponding voltage fluctuations. To evaluate the magnitude of these fluctuations, assume that the switch is closed at time $t = 0$ and held closed for a time Δt. During this time, power can be transferred from the *LC* circuit to the resistor and from the resistor to the *LC* circuit. Since the system is in thermal equilibrium, the same power must flow either way on average. We can therefore determine the power that is generated in the resistor by calculating the power that is generated in the *LC* circuit and subsequently dissipated in the resistor.

The fluctuating energy δE in the *LC* circuit causes power to flow to the resistor at a rate

$$\delta P \sim \frac{\delta E}{\Delta t} \sim \frac{k_BT}{\Delta t} \tag{13-30}$$

which is equal to the rate at which power is dissipated in the resistor,

$$\delta P \sim \frac{V_N^2}{R} \tag{13-31}$$

Here V_N is the rms amplitude of voltage fluctuations across the resistor, which we call the noise voltage. Combining Eqs. (13-30) and (13-31), we find the noise voltage to be

$$V_N \sim \sqrt{\frac{k_B TR}{\Delta t}} \qquad (13\text{-}32)$$

This result says that the voltage fluctuations across the resistor become smaller as the measurement time Δt becomes longer. Physically, this comes about because the same average energy fluctuation is being spread out over a longer time interval, which reduces the power fluctuation and hence the noise voltage.

The noise voltage can be expressed in terms of the detection bandwidth rather than the measurement time, as was done for shot noise. Using Eq. (13-28) with a value $\mathcal{K} = 1/4$ yields

$$V_N = \sqrt{4k_B TRB} \qquad \text{(Johnson noise voltage)} \qquad (13\text{-}33)$$

which is the widely used expression for thermally induced noise in a resistor. As with shot noise, the Johnson noise increases with the square root of the detection bandwidth B, and can be reduced by appropriate electronic filters.

It is sometimes useful to express the Johnson noise in terms of current fluctuations rather than voltage fluctuations. Using $i_N^2 R = V_N^2/R$, this becomes

$$i_N = \sqrt{\frac{4k_B TB}{R}} \qquad \text{(Johnson noise current)} \qquad (13\text{-}34)$$

Comparing Eqs. (13-33) and (13-34), we see that increasing the resistance leads to increasing noise voltage, but decreasing noise current. These results will play an important role in characterizing the signal-to-noise properties of detectors, to be discussed in Chapter 14.

PROBLEMS

13.1 A thermoelectric-based power meter has a response time of 20 s and sensitivity (output voltage per unit incident power) of 90 mV/W. If the time response is reduced to 8 s by increasing the thermal conductance between the sensor element and heat sink, what will be the new sensitivity?

13.2 Repeat Problem 13.1 assuming that the reduction in response time comes from reducing the sensor element mass.

13.3 A laser beam is switched on at time $t = 0$, and is incident on a thermal power meter. If the detector response time is 15 s, at what time will the reading come to within 2% of the "true" value?

13.4 Sodium metal has a work function of 2.28 eV. (a) At what wavelength of incident light will electrons be ejected from the material? (b) If light of wavelength 450 nm is incident on the sodium, determine the maximum kinetic energy of the ejected electrons.

13.5 When Cs atoms are deposited on the surface of a GaAs photoelectric detector element, it is found that electrons are ejected for incident wavelengths shorter than 910 nm. Determine the electron affinity. Take $E_g = 1.425$ eV for GaAs.

13.6 A photodetector has a responsivity of 0.844 A/W at 1500 nm. (a) Determine the quantum efficiency. (b) Determine the responsivity for incident wavelengths 1300 and 800 nm.

13.7 In Problem 13.6, determine the detector photocurrent if the incident light has wavelength 1500 nm and power –33 dBm.

13.8 A vacuum photodiode operates at a voltage of 2.5 kV and has a response time of 270 ps. Determine the plate spacing, assuming that the time response is limited by the transit time.

13.9 Assume that the total time response of a vacuum photodiode is given by the sum of the transit time and the RC time constant. (a) Derive an expression for the plate separation d that minimizes the total response time. Put your answer in terms of the applied voltage V_0, the plate area A, and the load resistor R. (b) Calculate this optimum plate separation for parallel plates of diameter 2 cm, a voltage 2.5 kV, and a load resistance 50 Ω. Repeat for a load resistance 1 kΩ. (c) Determine the optimized total response time for the two values of R in part b.

13.10 A photomultipler with an S20 photocathode has a chain of eight dynodes that provide a gain of 2×10^6 at the operating voltage of 2 kV. Light of wavelength 700 nm and power 7 nW is incident on the photocathode. (a) For each electron incident on one of the dynodes, determine the average number of electrons leaving that dynode and moving on to the next one. (b) Determine the voltage difference between dynodes, and the kinetic energy that the electrons pick up in moving from one dynode to the next. (c) Determine the anode current and the voltage across the 1 kΩ load resistor. (d) Repeat part c if the wavelength is shifted to 600 nm with the same optical power.

13.11 A CdS photocell has a separation between electrodes of 300 μm, with electron lifetime and mobility of 3 ms and 300 $cm^2/(Vs)$, respectively. (a) What voltage must be applied between electrodes to generate a photoconductive gain of 500? (b) Determine the photocurrent that results when 2 μW of 500 nm light is absorbed in the photoconductor.

13.12 Consider the photocell of Problem 13.11, biased with battery and resistor as shown in Fig. 13-17. The photocell resistance with no light incident is 50 $k\Omega$, and the battery voltage is $V_0 = 9$ V. (a) Determine the load resistance R_L that will give a voltage of 1.2 V across the photocell. (b) Determine the steady-state "dark current" i_0. (c) Calculate the rms noise currents due to shot noise and Johnson noise for this circuit, assuming a detector bandwidth of 50 Hz. (d) Determine the minimum light power at 500 nm that can be detected, using the criterion that the signal current must be at least equal to the noise current. Assume that the light is fully absorbed.

Chapter 14

Photodiode Detectors

We saw in the previous chapter that photoconductive detectors suffer from two principle drawbacks: a poor time response, due to the long electron lifetime, and significant shot noise from the high level of dark current. Both of these problems are remedied in the *photodiode detector,* shown schematically in Fig. 14-1. In this device, light incident on the p–n junction of a semiconductor creates electron–hole pairs, which are swept out of the depletion region by the electric field there. Current flows in the external circuit only while charges are moving through this E field region, so the time response of the detector can be made quite fast. The p–n junction also provides a potential barrier for majority charge carriers, greatly reducing the amount of dark current and associated shot noise.

In this chapter, we discuss some important characteristics of photodiode detectors, including their behavior as electrical circuit elements, their time response, and their signal-to-noise properties.

14.1. BIASING THE PHOTODIODE

Although there are several types of circuits used to measure the photodiode signal current (see Section 14-5), the way that the photodiode is biased falls into one of two fundamental categories. In the *photovoltaic mode* (Fig. 14-2a), a load resistor R_L is directly connected across the photodiode, whereas in the *photoconductive mode* (Fig. 14-2b), the load resistor is connected through a series bias voltage V_B. In either case, the photocurrent generates a voltage V_R across the load resistor, which constitutes the detector output signal. The photoconductive mode we are discussing here should not be confused with the photoconductive-type detectors discussed in the previous chapter. The distinction is the presence or absence of a p–n junction in the device.

The current i in the circuit depends not only on the incident light intensity, but also on the values of R_L and V_B. To evaluate this current, we add up the potential changes around the circuit loop of Fig. 14-2b, and set the sum equal to zero (voltage loop law):

$$V_d + V_B + V_R = 0$$

We have adopted a sign convention for the diode voltage V_d and the resistor voltage V_R such that positive current flows into the positive side of each element. Of course, the actual values of the current or voltages may be either positive or negative. Writing $V_R = iR_L$ and solving for i, we have

$$i = \frac{-1}{R}(V_d + V_B) \qquad \text{(load line)} \tag{14-1}$$

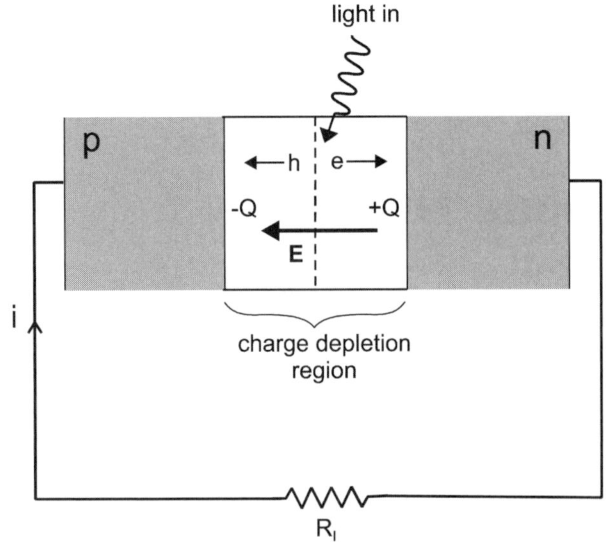

Figure 14-1 In a photodiode detector, the motion of electrons and holes across the charge depletion region causes a current in the external circuit. By convention, current *i* is defined as positive when it enters the p side of the diode. The photocurrent produced is therefore negative.

Equation (14-1) gives a relation between current i and diode voltage V_d that is imposed by the external circuit. Another relation between i and V_d comes from the internal constraints of the diode itself. The i–V relation for a semiconductor diode was given earlier in Eq. (10-21) and Fig. (10-14) for the case of no light absorption. When light is absorbed, the electron–hole pairs that are created cause an additional negative current, termed a *photocurrent.* The magnitude of the photocurrent is given by Eq. (13-7), which can be written here as

$$i_\lambda = \frac{P_{\text{in}}}{h\nu} e\eta_{\text{abs}} \qquad \text{(photocurrent)} \tag{14-2}$$

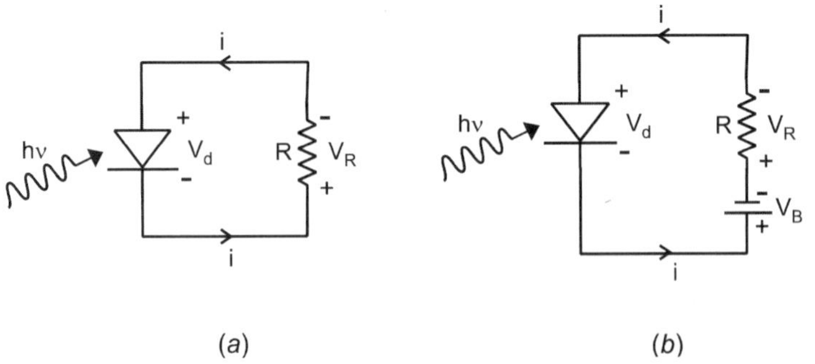

Figure 14-2 (a) In the photovoltaic mode, a load resistor is directly connected across the photodiode. (b) In the photoconductive mode, the load resistor in connected in series with a reverse-bias voltage.

where η_{abs} is the fraction of incident photons that are absorbed to create electron–hole pairs. This photocurrent adds to the diode current of Eq. (10-21) to give a total circuit current

$$i = i_0\left[\exp\left(\frac{eV_d}{\beta k_B T}\right) - 1\right] - i_\lambda \qquad (14\text{-}3)$$

According to Eqs. (14-2) and (14-3), the diode *i–V* curve is shifted downward (along the $-i$ axis) by an amount i_λ, which is proportional to the incident light intensity. A few representative *i–V* curves for a photodiode are shown in Fig. 14-3, for equally spaced values of light intensity. For self-consistency, the diode current i and voltage V_d must satisfy both Eq. (14-1) and the diode *i–V* relation simultaneously. The solution for i can easily be obtained graphically, by plotting Eq. (14-1) on the same graph as the diode *i–V* curves. This procedure, shown in Fig. 14-3, is known as a *load-line analysis,* and Eq. (14-1) is known as the *load line.* A similar analysis was discussed in Section 11-1 in connection with biasing an LED.

Since the photovoltaic mode is just a special case of the photoconductive mode, with $V_B = 0$, both circuits can be analyzed in the same fashion using Fig. 14-3. The intersection of the load line and the diode *i–V* curve corresponds to the *operating point* of the circuit, which gives the value of both i and V_d. For the photovoltaic mode, the load line passes through the origin, so the operating point is always in the fourth quadrant, with positive V_d and negative i. In the photoconductive mode, the intercept on the V_d axis is at $V_d = -V_B$, so the operating point can be either in the third or fourth quadrants. The current i is always negative, but V_d can be either positive or negative.

The photovoltaic and photoconductive modes each have advantages and disadvantages, depending on the application. For low-level light detection, the photovoltaic mode has higher ultimate sensitivity than the photoconductive mode. This is because under dark conditions (no incident light), the photovoltaic operating point is at $i = 0$, whereas the photoconductive mode is at $i = -i_0$, the reverse saturation current. This minimum current

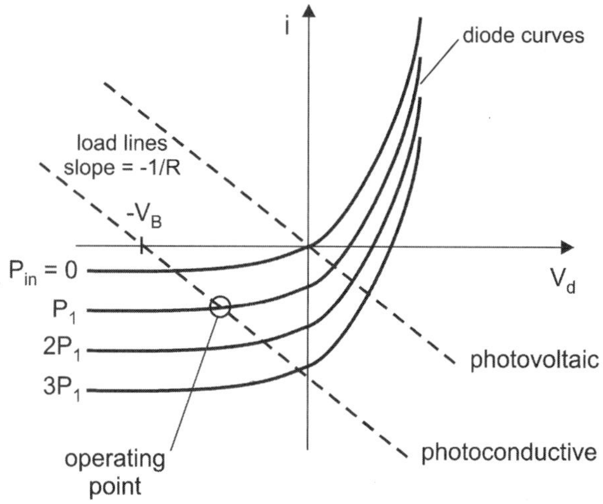

Figure 14-3 The operating point in a photodiode circuit is determined by the intersection between the load line and the diode *i–V* curve. Increasing optical power P_{in} shifts the *i–V* curve downward by an amount $i_\lambda \propto P_{in}$, moving the operating point down and to the right.

i_0 is termed the *dark current.* The shot noise from this dark current makes the photoconductive mode inherently more noisy. On the other hand, the photoconductive mode has a faster time response, and a linear response over a wider range of light intensities, as we shall see in the following sections.

One important application utilizing the photovoltaic mode is the *solar cell,* which converts optical power into electrical power. The electrical power supplied to the load resistor is $P_{\text{elec}} = i^2 R_L$, where i is determined by Eq. (14-3) with $V_d = -iR$. Under practical conditions of solar illumination, $i_\lambda \gg i_0$, and Eq. (14-3) can be approximated as

$$i \simeq i_0 \exp\left(\frac{-eiR}{\beta k_B T}\right) - i_\lambda \qquad \text{(solar cell current)} \qquad (14\text{-}4)$$

with i_λ given by Eq. (14-2). The efficiency

$$\eta_{sc} = \frac{P_{\text{elec}}}{P_{\text{in}}} = \frac{i^2 R}{P_{\text{in}}} \qquad \text{(solar cell efficiency)} \qquad (14\text{-}5)$$

of converting optical power into electrical power can then be calculated by solving Eq. (14-4) for i. Since this is an implicit equation for i, it must be solved numerically or graphically.

An important consideration for the solar cell is the choice of load resistance that maximizes the conversion efficiency. Since the load line in Fig. 14-3 has a slope $-1/R_L$, the operating point moves close to $i = 0$ for large R and $V_d = 0$ for small R. The power $|iV_d|$ delivered to the resistor will, therefore, have a maximum at some value of R. This optimum value of resistance can be determined graphically, or numerically as in the following example.

EXAMPLE 14-1

A silicon solar cell has an area of 4 cm^2, reverse saturation current density 1.5×10^{-8} A/cm^2, and diode ideality factor $\beta = 1$. Assume that light of intensity $I = 1000$ W/m^2 and average wavelength 500 nm is incident on the cell, and that 80% of the light is absorbed. Determine the optimum load resistance and power conversion efficiency. Repeat the calculation for $\beta = 2$.

Solution: The power striking the cell is (1000 W/m^2)(4×10^{-4} m^2) = 0.4 W. The photocurrent is then

$$i_\lambda = \frac{P_{\text{in}}\lambda}{hc} e\eta_{\text{abs}} = \frac{(0.4)(500 \times 10^{-9})(1.6 \times 10^{-19})(0.8)}{(6.63 \times 10^{-34})(3 \times 10^8)} = 0.129 \text{ A}$$

For room temperature (20°C = 293 K),

$$V_T \equiv \frac{k_B T}{e} = \frac{(1.38 \times 10^{-23})(293)}{1.6 \times 10^{-19}} = 0.0253 \text{ V}$$

is the "voltage equivalent of temperature." Putting this in Eq. (14-4) gives (for $\beta = 1$)

$$i = (6 \times 10^{-8}) \exp\left(\frac{-iR}{0.0253}\right) - 0.129$$

For a particular value of R, this equation is solved numerically for i, and the efficiency $\eta_{sc} = i^2R/P_{\text{in}}$ is calculated. By varying R, the graph shown in Fig. 14-4 is obtained. The optimum efficiency of 8.94% is obtained for $R = 2.5\Omega$.

If $\beta = 2$, Eq. (14-4) becomes

$$i = (6 \times 10^{-8}) \exp\left(\frac{-iR}{0.0506}\right) - 0.129$$

and the optimum efficiency increases to 17.9% at $R = 5\Omega$.

A more detailed model of solar cell efficiency would take into account the variation with wavelength of the optical power from the sun and the fraction of this light that is absorbed by the silicon. In practice, solar cells based on crystalline silicon can have efficiencies as high as 24% in the laboratory, with ~ 15% being typical in commercial devices. Thin films of amorphous silicon (atoms not ordered periodically) are inexpensive to manufacture but have lower efficiencies, typically 13% in the laboratory and 5–7% in commercial devices.

14-2. OUTPUT SATURATION

In the case of the solar cell just discussed, the primary goal is to convert as much optical power as possible into electrical power. When the photodiode is used as a light detector, however, it is generally more important that the detector output be linear with the incident light power. In this section, we examine the linearity of photodiode detector circuits using the two types of biasing modes.

Photovoltaic Mode

Consider first the photovoltaic bias mode shown in Fig. 14-2a. When R is very large (open-circuit condition), the load line is nearly horizontal, and the operating point is close to the V_d axis where $i \approx 0$. In that case, Eq. (14-3) becomes

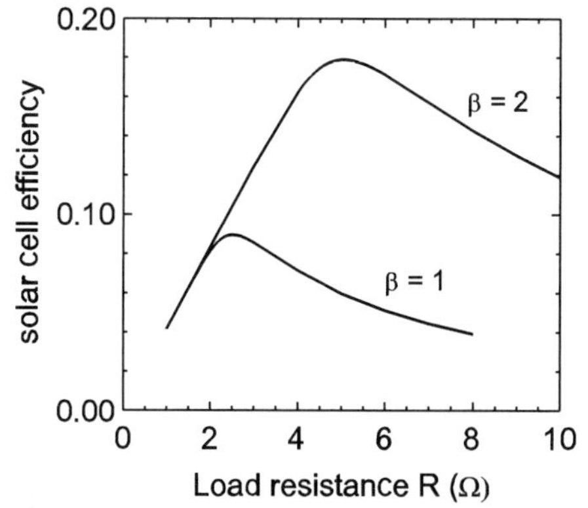

Figure 14-4 Variation of solar cell efficiency with load resistance for Example 14-1. Area of the cell is 4 cm^2. Optimum efficiency is higher and occurs at a higher load resistance when the diode ideality factor (β) is 2 rather than 1.

$$0 \simeq i_0\left[\exp\left(\frac{eV_d}{\beta k_B T}\right) - 1\right] - i_\lambda$$

with i_λ given by Eq. (14-2). Solving for the diode voltage gives

$$V_d = \frac{\beta k_B T}{e} \ln\left(1 + \frac{i_\lambda}{i_0}\right) \tag{14-6}$$

If the induced photocurrent is much greater than the dark current ($i_\lambda \gg i_0$), this becomes

$$V_d \simeq \beta V_T \ln\left(\frac{P_{\text{in}} e \eta_{\text{abs}}}{i_0 h\nu}\right) \quad \text{(open circuit, } i_\lambda \gg i_0\text{)} \tag{14-7}$$

where

$$V_T \equiv \frac{k_B T}{e} \quad \text{(voltage equivalent of temperature)} \tag{14-8}$$

The diode voltage, therefore, varies logarithmically with the incident power for $i_\lambda \gg i_0$. For $i_\lambda \ll i_0$, however,

$$V_d \simeq \beta V_T \frac{i_\lambda}{i_0} = \beta V_T \frac{P_{\text{in}} e \eta_{\text{abs}}}{i_0 h\nu} \tag{14-9}$$

Defining the quantity

$$R_{sh} = \frac{\beta V_T}{i_0} = \frac{\beta k_B T}{e i_0} \quad \text{(shunt resistance)} \tag{14-10}$$

the diode voltage can be written as

$$V_d \simeq R_{sh} i_\lambda = R_{sh}\left(\frac{P_{\text{in}} e \eta_{\text{abs}}}{h\nu}\right) \quad \text{(open circuit, } i_\lambda \ll i_0\text{)} \tag{14-11}$$

This result suggests that the photodiode can be modeled as an ideal current source connected in parallel with a resistor R_{sh}, as depicted in Fig. 14-5. Since $i = 0$ for an open cir-

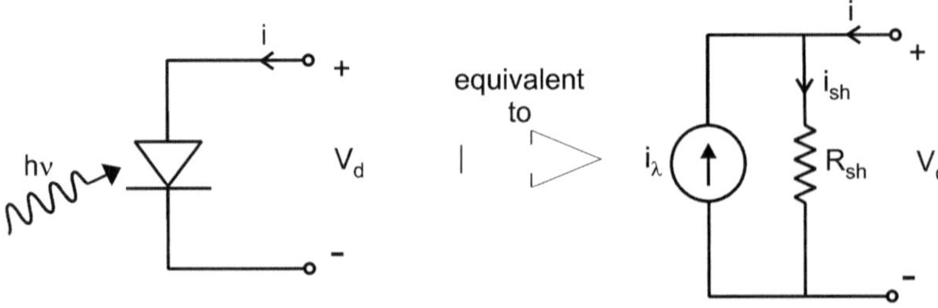

Figure 14-5 When $i_\lambda \ll i_0$, the photodiode can be modeled as an ideal current source in parallel with a shunt resistance R_{sh}.

cuit, the diode voltage becomes $V_d = i_\lambda R_{sh}$, in agreement with Eq. (14-11). Since R_{sh} appears in parallel with the current source, it is termed a *shunt resistance.* Higher values of R_{sh} are generally desirable, because the detector is then more sensitive to weak light signals (V_d large for small i_λ).

Values of shunt resistance vary widely, and are higher for wider bandgap materials, for which i_0 is smaller. R_{sh} is also higher at lower temperature (less thermal generation of electron–hole pairs) and for smaller junction area (since $i_0 = J_0 A$). For a typical room-temperature silicon photodiode with 1 cm^2 area, $R_{sh} \approx 10\ \mathrm{M}\Omega$.

The response of the open-circuit photodiode to varying optical powers can be summarized as follows. At low incident power levels the response is linear with power, whereas at high power levels the response becomes logarithmic. The transition between these two regimes corresponds to $i_\lambda \sim i_0$, which is equivalent to $i_\lambda R_{sh} \sim V_T$. This deviation from linearity at high optical powers is referred to as *saturation* of the output signal, and is generally to be avoided.

To increase the range of optical powers over which the photodiode response is linear, the load resistance R_L can be made small. This makes the load line in Fig. 14-3 nearly vertical, intersecting the diode curves close to the current axis ($V_d \approx 0$). Since the diode curves are approximately evenly spaced for $V_d \leq 0$, the operating point moves downward along the $-i$ axis in proportion to the optical power. The voltage across the resistor, $V_R = iR_L$, is therefore linear with the optical power, as desired.

This conclusion can also be arrived at analytically. If $V_d \ll V_T$, the approximation $e^x \approx 1 + x$ allows Eq. (14-3) to be written as

$$i \simeq i_0\left(\frac{V_d}{\beta V_T}\right) - i_\lambda$$

or

$$i \simeq \frac{V_d}{R_{sh}} - i_\lambda \tag{14-12}$$

Using $V_d = -iR_L$ for the photovoltaic mode (Fig. 14-2a) and solving for i gives

$$i \simeq \frac{-i_\lambda}{1 + R_L/R_{sh}} \tag{14-13}$$

Since $i_\lambda \propto P_{\text{in}}$, we conclude that $V_R = iR_L \propto P_{\text{in}}$, as desired.

The equations above can be understood in terms of the equivalent circuit shown in Fig. 14-6. As before, the photodiode is represented as an ideal current source shunted by the resistance R_{sh}. The load resistor R_L is now connected in parallel with both of these. Defining the current i_{sh} through the shunt resistance to be positive in the downward direction, we have by the junction rule

$$i + i_\lambda = i_{sh}$$

This is equivalent to Eq. (14-12), using $i_{sh} = V_d/R_{sh}$. It is left as an exercise to show that Eq. (14-13) can be derived from this equivalent circuit model.

When $R_L \ll R_{sh}$, Eq. (14-13) becomes $i \simeq -i_\lambda$, and the diode voltage is

$$V_d = -iR_L \simeq i_\lambda R_L$$

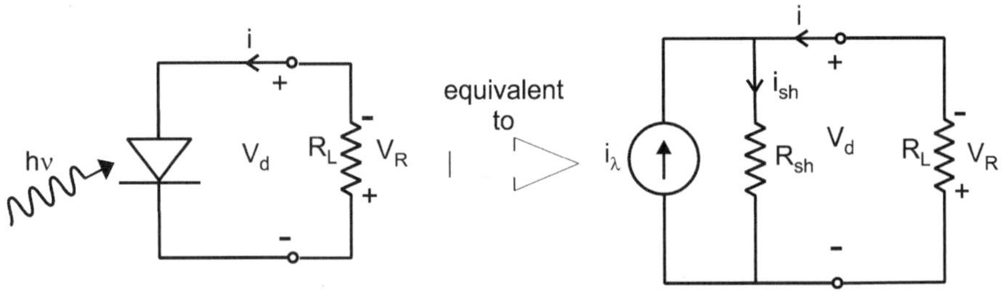

Figure 14-6 Equivalent circuit for a photodiode biased with load resistor R_L. This model is valid when $V_d \ll V_T$.

Defining the output of the detector as $V_{\text{out}} \equiv V_d$, we then have

$$V_{\text{out}} = V_d \simeq \frac{P_{\text{in}}}{h\nu} e\eta_{\text{abs}} R_L \qquad (V_d \ll V_T, R_L \ll R_{sh}) \tag{14-14}$$

where Eq. (14-2) has been used.

This result shows that under the two specified conditions the detector voltage V_d is linear not only with the incident optical power but also with the load resistance. In practice, it is easier to measure a larger voltage, so a larger R_L is desirable. However, as R_L is increased, one of these two conditions will eventually break down, and V_d will no longer increase with R_L. One possibility is that $R_L \ll R_{sh}$ breaks down, while the condition $V_d \ll V_T$ still holds. In the limit where $R_L \gg R_{sh}$, Eq. (14-13) gives

$$V_d = -iR_L \simeq i_\lambda R_{sh}$$

which becomes independent of R_L. This is the same result obtained in Eq. (14-11) for the open-circuit condition $R_L \to \infty$. Although the detector output V_{out} is no longer linear with R_L in this case, it is still linear with P_{in}.

The other possibility as R_L increases is that the condition $V_d \ll V_T$ breaks down first, in which case the exact expression in Eq. (14-3) must be used. Under these conditions, the output V_{out} is no longer linear with either R_L or P_{in}. According to Eq. (14-14), this saturation will occur at a certain value of the product $P_{\text{in}}R_L$. For higher R_L, saturation occurs at a lower P_{in}, and at higher P_{in}, saturation occurs at a lower R_L. There is, therefore, a trade-off between sensitivity (large output for small input) and dynamic range (range of inputs for which response is linear). The saturation with incident power for different values of load resistance is illustrated in Fig. 14-7.

Photoconductive Mode

Saturation behavior in the photoconductive mode can be understood by referring to the load-line analysis of Fig. 14-3. The load line has a slope $-1/R_L$, with an intercept on the voltage axis of $V_d = -V_B$. As the incident optical power increases, the operating point moves downward and to the right along the load line, decreasing the magnitude of reverse-bias voltage and increasing the magnitude of the current. Both the voltage and current change linearly with increasing optical power, until the operating point reaches the

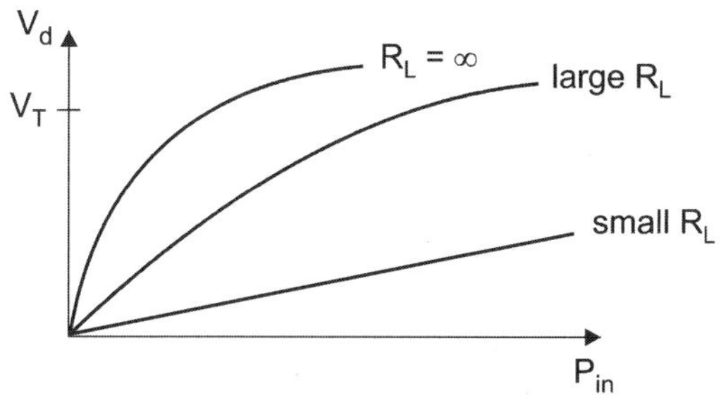

Figure 14-7 Diode voltage versus incident optical power for the photovoltaic mode. Smaller load resistance R_L gives a larger dynamic range but lower sensitivity. For $R_L \to \infty$, the effective resistance reaches the upper limit of R_{sh}.

current axis ($V_d = 0$). At that point, the detector circuit saturates, and the output is no longer linear with the incident optical power.

In the linear regime, we can obtain a simple analytical expression for the detector signal as follows. It is clear from Fig. 14-3 that $V_d < 0$ in the linear regime. Eq. (14-3) can then be written as

$$i = i_0\left[\exp\left(\frac{-|V_d|}{\beta V_T}\right) - 1\right] - i_\lambda$$

Unless the operating point is close to saturation, it is a good approximation that $|V_d| \gg V_T$. The exponential term above can then be neglected, giving

$$i \simeq -i_0 - i_\lambda \tag{14-15}$$

The detector output in the photoconductive mode is generally taken to be the voltage V_R across the load resistor. Since this will always be negative, we define the output as $V_{\text{out}} \equiv -V_R$ to give a positive number. Therefore,

$$V_{\text{out}} = -V_R = -iR_L \simeq (i_0 + i_\lambda)R_L \tag{14-16}$$

which can be written as

$$V_{\text{out}} \simeq i_0 R_L + \frac{P_{\text{in}}}{h\nu} e\eta_{\text{abs}} R_L \qquad (V_{\text{out}} < V_B) \tag{14-17}$$

using Eq. (14-2).

The detector output is seen to have two components, one proportional to the incident optical power, and the other independent of power. The component that varies with P_{in} is identical to the expression obtained in Eq. (14-14) for the photovoltaic mode. In both cases, the output voltage arises from the photocurrent i_λ flowing through load resistor R_L. The detector output can be expressed more compactly by defining the *responsivity* of the detector as

$$\mathcal{R} \equiv \frac{i_\lambda}{P_{\text{in}}} = \frac{e\eta_{\text{abs}}}{h\nu} \tag{14-18}$$

which is similar to the definition given in Eq. (13-8) for emissive-type photodetectors. The output in the photoconductive mode is then

$$V_{out} = (i_0 + \mathcal{R}P_{in})R_L \tag{14-19}$$

A similar expression applies to the photovoltaic mode, but without the i_0 term.

According to Eq. (14-17), the change in output voltage is linear with both R_L and P_{in}. However, this relation will only hold as long as $V_d < 0$, which requires that $V_{out} < V_B$. If P_{in} is increased above the point where $V_{out} \approx V_B$, the output saturates, and becomes approximately independent of P_{in}. This behavior is illustrated in Fig. 14-8 for two values of R_L. Larger R_L makes the detector more sensitive, since there is a larger output for a small value of P_{in}. However, this reduces the range of P_{in} over which the response is linear. The result is a sensitivity-dynamic range trade-off similar to that of the photovoltaic mode.

Although the photoconductive and photovoltaic modes have the similarities mentioned above, there are some significant differences. One difference is that saturation occurs at $V_{out} \approx V_B$ in the photoconductive mode, but at only $V_{out} \approx V_T$ in the photovoltaic mode. Since $V_T \approx 0.025$ V at room temperature, whereas V_B is typically several volts, the detector output in the photoconductive mode can be approximately two orders of magnitude larger than in the photovoltaic mode. This means that for the same detector sensitivity (same R_L), the dynamic range is approximately two orders of magnitude larger in the photoconductive mode than in the photovoltaic mode. This improved dynamic range is an important advantage of the photoconductive mode.

Another difference is that the photoconductive mode has a dark current, whereas the photovoltaic mode does not. The presence of dark current has two consequences. First, it contributes a constant background level that must be subtracted from the detector output in order to obtain the "true" signal (the signal arising from the incident light). Second, it contributes shot noise to the detector output, as discussed in Section 13-3. If the optical power is sufficiently large so that $i_\lambda \gg i_0$, then both of these effects become unimportant. In this large-signal regime, the photoconductive mode is the best choice for the detector circuit. However, if $i_\lambda \ll i_0$, then shot noise from the dark current can become a dominant source of detector noise. In this small-signal regime, the photovoltaic mode is a better choice, in order to obtain the best possible signal-to-noise ratio. The signal-to-noise properties of detector circuits are further discussed in Section 14-5.

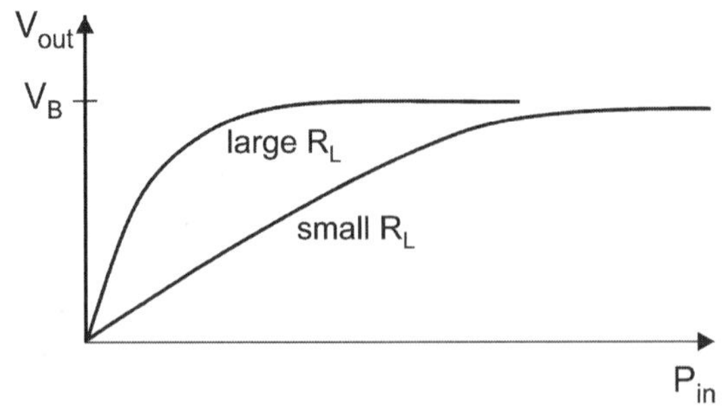

Figure 14-8 For a photodiode biased in the photoconductive mode, the detector response is linear for output voltages up to the reverse-bias voltage V_B. Larger load resistance R_L gives higher sensitivity but smaller dynamic range.

It should be emphasized that the dark current in a reverse-biased photodiode detector is much smaller and more well-defined than that in a photoconductive-type detector (one without a p–n junction). For example, a 1 cm^2 silicon photodiode has a typical dark current $i_0 \approx 1.5 \times 10^{-8}$ A, independent of reverse-bias voltage. In contrast, a CdS photocell has a background current that depends on the applied voltage, a typical value being $\sim 10^{-5}$ A for a similar cross-sectional area and applied bias of 10 V.

14-3. RESPONSE TIME

An important characteristic of any photodetector is its response time—the time it takes for the detector output to change in response to changes in the input light intensity. It was noted in Chapter 13 that the response time of photoconductive-type detectors is quite poor because of the electron replenishment process, which keeps the induced photocurrent flowing for the duration of the electron's lifetime. In photodiode detectors, this replenishment process is suppressed by the p–n junction, which presents a barrier to the movement of majority carriers. The response time is thereby significantly improved, since the time taken for charge carriers to move through the high-field region of the junction (the carrier *transit time*) can be much shorter than the carrier lifetime. This improved time response is countered in part, however, by the capacitance of the p–n junction and associated RC time constant. In this section, we consider the implications and relative importance of transit time and capacitance in determining the photodiode response time.

Junction Capacitance

The capacitance of a p–n junction can be evaluated by determining how the charge on either side of the junction changes in response to a changing diode voltage. We will assume for simplicity a highly doped p region and weakly doped n region, so $N_A \gg N_D$ as in Fig. 10-11. In this case, most of the charge depletion region is on the n side ($d \approx d_n \gg d_p$), and the junction width d is given by Eq. (10-20). When an external voltage V is applied to the diode, the internal potential V_0 is replaced by $V_0 - V$, giving

$$d = \sqrt{\frac{2\varepsilon(V_0 - V)}{eN_D}} \qquad \text{(p–n junction width)} \qquad (14\text{-}20)$$

where V is positive for forward bias and negative for reverse bias. A change ΔV in external voltage causes a change in junction width of

$$\Delta d = \frac{1}{2d}\left(\frac{-2\varepsilon}{eN_D}\right)\Delta V \qquad (14\text{-}21)$$

as can be verified by taking the differential of both sides of Eq. (14-20). Most of this change in width occurs on the n side of the junction, since $d_n \gg d_p$. Writing the charge of the uncovered ion cores in the *n* region as $Q = eN_D(Ad_n)$, the change in this charge is

$$\Delta Q = eN_D(A\Delta d) \qquad (14\text{-}22)$$

There is an equal but opposite change in the charge on the p side of the junction. Combining Eqs. (14-21) and (14-22), the junction capacitance is then

$$C \equiv \left| \frac{\Delta Q}{\Delta V} \right| = \frac{\varepsilon A}{d} \tag{14-23}$$

which is the familiar expression for the capacitance of a parallel-plate capacitor of plate area A and spacing d, separated by a medium with dielectric constant ε. Using Eq. (14-20) for d gives the result

$$C = A\sqrt{\frac{eN_D\varepsilon}{2(V_0 - V)}} \qquad \text{(p–n junction capacitance)} \tag{14-24}$$

For reverse-bias voltage $V = -V_B$, this becomes

$$C = A\sqrt{\frac{eN_D\varepsilon}{2(V_0 + V_B)}} \tag{14-25}$$

It is seen that the capacitance of the p–n junction is not constant, but rather decreases with increasing V_B. This is a consequence of the junction width d increasing with V_B.

The junction capacitance can be considered to be in parallel with the diodes in Fig. 14-2, which leads to a first-order RC circuit time response. If the incident power is suddenly switched from zero to some constant value at $t = 0$, the output voltage increases exponentially in time according to

$$V_{\text{out}}(t) = V_{\text{max}}(1 - e^{-t/RC}) \tag{14-26}$$

which is shown graphically in Fig. 14-9. The product RC is known as the *time constant* of the circuit, and has units of seconds with R in ohms and C in farads. The time constant measures how quickly the output responds to a changing input. For example, at time $t = RC$ the output has risen to 63.2% of the final steady-state value, and at time $t = 2RC$ it has risen to 86.5% of this value.

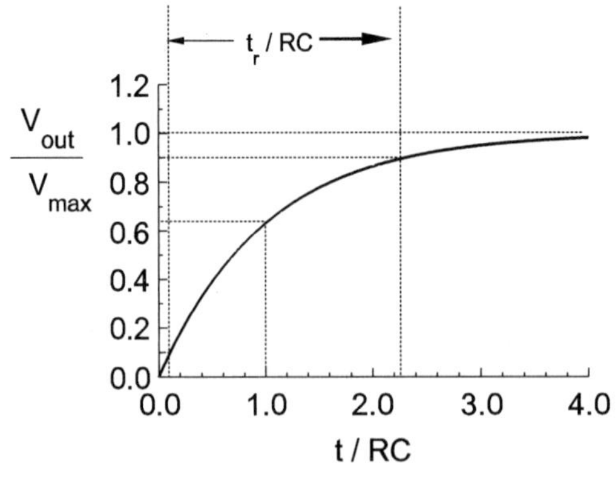

Figure 14-9 In an exponential rise, the time constant *RC* gives the time taken for the output to reach 63.2% of the final steady-state value. The rise time t_r is the time taken for the output to go from 10% to 90% of the final value.

An alternative measure of the time response is the *rise time,* defined as the time taken to rise from 10% to 90% of the final value. For the RC circuit, it is straightforward to show (see Problem 14.6) that

$$t_r = (\ln 9)RC = 2.2\ RC \qquad \text{(rise time for RC circuit)} \tag{14-27}$$

The rise time is a more general definition for time response than the time constant, because it applies equally well to an exponential or nonexponential time dependence. When the time response is nonexponential, Eq. (14-27) can be interpreted as defining an effective time constant in terms of the rise time.

The rise time or RC time constant characterizes the detector time response to a step-function intensity modulation, in which the incident light intensity is suddenly changed from one value to another. If instead the incident light intensity is sinusoidally modulated, then the output will be sinusoidally modulated as well. These two different types of modulation were discussed in connection with the LED time response (see Fig. 11-3). The output amplitude is approximately independent of frequency up to a limiting upper value, the bandwidth, above which the amplitude becomes smaller. Denoting the 3 dB electrical bandwidth as B, we have

$$B = \frac{1}{2\pi RC} \tag{14-28}$$

where Eq. (11-13) has been used with $\tau = RC$ and $B = f_e$. The bandwidth can be written in terms of the rise time as

$$B = \frac{2.2}{2\pi t_r} = \frac{0.35}{t_r} \tag{14-29}$$

using Eq. (14-27). This last expression can be taken as defining the 3 dB bandwidth in the case of nonexponential time response.

The above results show that a smaller capacitance leads to a faster time response and larger bandwidth, which is usually desirable for a photodetector. According to Eq. (14-25), there are several parameters that can be adjusted to reduce the capacitance. For example, the reverse-bias voltage V_B can be increased. This makes the photoconductive mode ($V_B > 0$) inherently faster than the photovoltaic mode ($V_B = 0$). Indeed, this is another characteristic advantage of the photoconductive mode, in addition to the increased dynamic range that was discussed earlier. There is a practical limit to V_B, however, due to electrical breakdown in the junction. Typical reverse-bias voltages are 5–10 V.

Another way to reduce the capacitance is to decrease the junction area A. Smaller detector areas, therefore, give a faster time response. The downside of this approach is that it may not be possible to direct all the available light onto the semiconductor material if A is too small. The ability to focus light onto a small detector area is governed by the brightness of the light source, as discussed in Appendix A (see also Chapter 15). High-brightness sources such as a laser can be focused to a very small area ($\sim \lambda^2$), which allows a small detector area to be used without loss of efficiency. Low-brightness sources such as an LED or incandenscent filament, however, can be imaged only onto a much larger area. If a photodiode with small A is used to detect light of low brightness, much of the light to be measured will not strike the detector's active area, making the detector less sensitive. This results in a sensitivity–speed trade-off, which must be optimized for best performance in a particular application.

The other parameter in Eq. (14-25) that can be adjusted to reduce the capacitance is N_D, the density of donors on the weakly doped n side of the junction. The capacitance is reduced when N_D is made smaller, because the junction width d then increases. This is one reason that in most photodiodes, one side of the junction is very weakly doped. There are other reasons for this also, as we will soon see.

Reducing the capacitance is one way to reduce the RC time constant, but it is not the only way. Reducing the load resistance R_L has the same effect, although this decreases the detector sensitivity, as seen in Eq. (14-19) and Fig. 14-8. Large R_L is best for high sensitivity, and small R_L is best for high speed, resulting in another sensitivity–speed trade-off. The dynamic range is also better for small R_L, as previously discussed. These various trade-offs are summarized in Fig. 14-10.

Carrier Transit Time

Single p–n Junction

When the RC time constant is made sufficiently small, the photodiode response time will be limited by the motion of charge carriers across the device. The fundamental principle needed for this analysis is given by Eq. (13-9), developed in connection with the vacuum photodiode. According to this result, which applies quite generally to any photon-type detector, the current pulse from a single photoexcited electron lasts as long as the electron is moving through a region with high electric field. A similar relation applies to photoexcited holes. Since the E field is high only in the depletion region (see Fig. 10-11), the current pulse will last a time

$$t_{tr} = \frac{d}{v} \tag{14-30}$$

known as the *transit time,* where d is the width of the depletion region and v is the velocity of the charge carrier. At low to moderate electric field strength, the electron velocity is $v_e = \mu_e E$, where μ_e is the electron *mobility,* and the hole velocity is $v_h = \mu_h E$. The time re-

	sensitivity	dynamic range	time response
large R	high	low	slow
small R	low	high	fast
large area	high	—	slow
small area	low	—	fast

Figure 14-10 Summary of performance trade-offs in choosing load resistor R_L and detector area A.

sponse will be limited by the charge carrier with the smallest mobility, which is usually a hole. In this case, the transit time becomes

$$t_{tr} = \frac{d}{\mu_h E} \quad \text{(transit time, low field)} \tag{14-31}$$

If the field is sufficiently high (above a value $E_{sat} \sim 2 \times 10^6$ V/m for holes in Si), then the carrier velocity is no longer proportional to E, saturating at the upper limit v_s (~ 10^5 m/s for holes in Si). Under high-field conditions, the transit time is

$$t_{tr} = \frac{d}{v_s} \quad \text{(transit time, high field)} \tag{14-32}$$

It would appear from Eq. (14-31) that a higher E field gives a shorter transit time. However, the width of the depletion region d increases with increasing E, and this tends to increase the transit time. To see how these two effects offset each other, we express both E and d in terms of the junction potential $V_0 + V_B$. Taking E as approximately constant across the depletion region,

$$E \simeq \frac{V_0 + V_B}{d}$$

the transit time then becomes

$$t_{tr} = \frac{d^2}{\mu_h(V_0+V_B)} = \frac{2\varepsilon}{\mu_h e N_D} \quad \text{(transit time, low field)} \tag{14-33}$$

where Eq. (14-20) has been used with $V = -V_B$.

This result shows that the transit time with low field is actually independent of the applied reverse bias. Apart from the choice of semiconductor material, which determines ε and μ_h, it depends only on the donor concentration N_D. Higher N_D would appear to be best for a fast time response (small t_{tr}). However, we found in Eq. (14-25) that the junction capacitance increases with higher N_D, resulting in a slower response. The time response will, therefore, be optimized when the contributions of capacitance and transit time to the response time are approximately equal. The value of N_D giving this optimum response depends on several other parameters, as shown in the following example.

EXAMPLE 14-2

A p^+–n silicon photodiode has an junction area of 1 mm^2, and a reverse bias of 10 V is applied through a 10 kΩ load resistor. Determine the doping level in the lightly doped n region that minimizes the response time, and determine the junction width for this doping level. Take the hole mobility in Si to be 5×10^{-2} m^2/Vs.

Solution: The optimum time response will occur when the capacitance rise time 2.2 $R_L C$ is approximately equal to transit time t_{tr}. Using Eqs. (14-25) and (14-33),

$$2.2 R_L A \sqrt{\frac{e\varepsilon_0\varepsilon_r N_D}{2V_B}} = \frac{2\varepsilon_0\varepsilon_r}{\mu_h e N_D}$$

where $\varepsilon_0 = 8.85 \times 10^{-12}$ F/m is the permitivity of free space, $\varepsilon_r = 11.9$ is the relative dielectric constant for silicon, and $V_0 + V_B \approx V_B = 10$ V. Solving this for N_D gives

$$N_D = \frac{1}{e}\left[\frac{1.65\,\varepsilon_0\varepsilon_r V_B}{R_L^2 A^2 \mu_h^2}\right]^{1/3} = \frac{1}{1.6 \times 10^{-19}}\left[\frac{1.65(8.85 \times 10^{-12})(11.9)(10)}{(10^4)^2(10^{-6})^2(5 \times 10^{-2})^2}\right]^{1/3}$$

$$N_D = 1.18 \times 10^{18}\ \text{m}^{-3} = 1.18 \times 10^{12}\ \text{cm}^{-3}$$

This is a very light doping level, only two orders of magnitude above the "intrinsic" carrier concentration in undoped silicon of ~ 1.4×10^{10} cm^{-3}. The junction width is then obtained from Eq. (14-20) as

$$d \simeq \sqrt{\frac{2(8.85 \times 10^{-12})(11.9)(10)}{(1.6 \times 10^{-19})(1.18 \times 10^{18})}} = 1.06 \times 10^{-4}\ \text{m} = 0.106\ \text{mm}$$

The electric field in the depletion region is $E \simeq 10$ V/10^{-4} m $= 10^5$ V/m. This is less than the saturating field value $E_{sat} \sim 2 \times 10^6$ V/m, which justifies using Eq. (14-33) for the transit time.

14-4. TYPES OF PHOTODIODES

PIN Photodiode

The analysis of the transit time response in the previous section assumed that the photoexcited electron–hole pairs are created in the depletion region, where there is a strong E field. This is not always the case, however, as illustrated in Fig. 14-11a. Some light that is incident on the highly doped p^+ side passes completely through the depletion region, and is absorbed in the n-type region on the other side. In this latter region, the E field is very small, because the high free-carrier concentration makes the electrical conductivity high (and E is small inside a good conductor). According to Eq. (13-9), there is little contribution to the photocurrent when E is small. Therefore, the photoexcited charge carriers created outside the depletion region do not contribute significantly to the photocurrent, as long as they remain outside the high-field depletion region.

The charge carriers created outside the depletion region do not remain motionless, however. Like any particles subject to random thermal motion, they spread out from their initial position in a process known as *diffusion.* Diffusion proceeds much more slowly than *drift,* which is the term given to the directed motion induced by an electric field. After spreading out by diffusion for a certain time, some holes initially generated in the n-type region will enter the depletion region, where they are quickly swept across by the high E field there. This results in an additional component to the photocurrent, delayed by the diffusion time. Holes generated at different distances from the edge of the depletion region have different diffusion times, which results in a diffusion "tail" in the photocurrent response to a square-wave light pulse. This type of signal distortion, depicted in Fig. 14-11b, is generally undesirable.

The solution to the problem of diffusing charge carriers is to simply eliminate the diffusion region. This can be accomplished by decreasing the donor concentration in the *n*

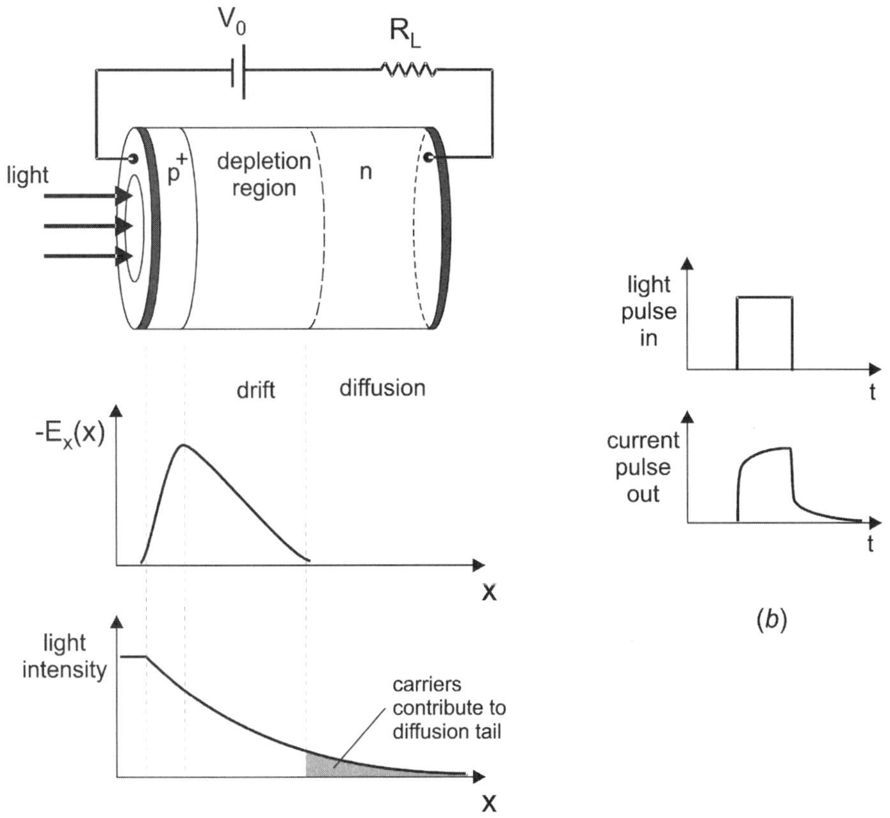

Figure 14-11 (a) In a simple p–n junction photodiode, charge carriers may be created in a high-field drift region or a low-field diffusion region. (b) Charge carriers created in the diffusion region give rise to a distortion in the photocurrent waveform.

region until the depletion region occupies nearly the entire space between the electrodes. As shown in Fig. 14-12, the E field then extends nearly all the way to the far electrode, so charge carriers generated anywhere in the material will be subject to drift rather than diffusion. At the far end is a thin, highly doped n region, needed to make good ohmic contact with the electrode. Since the middle region is very lightly doped (nearly intrinsic), it is labelled I, and the device is termed a *PIN photodiode.*

The PIN photodiode is the most commonly used photon detector today. It not only eliminates carrier diffusion, but also has the advantage that the depletion width d is fixed by the geometry of the device. The ability to adjust d by design, rather than applied voltage, allows the photodiode's performance to be optimized for specific applications. For example, making d larger increases the path length for absorption of light, which increases the efficiency η_{abs} with which light is absorbed. This is especially important for wavelengths near the semiconductor's bandgap, where $\alpha d \ll 1$. In this regime, Eq. (13-15) gives $\eta_{abs} \simeq (1 - R)\alpha d$.

On the other hand, a larger d degrades the time response by increasing the transit time. In Eqs. (14-31) and (14-32), d should now be considered a constant, independent of applied voltage. The transit time is therefore minimized by increasing E to the saturating value E_s.

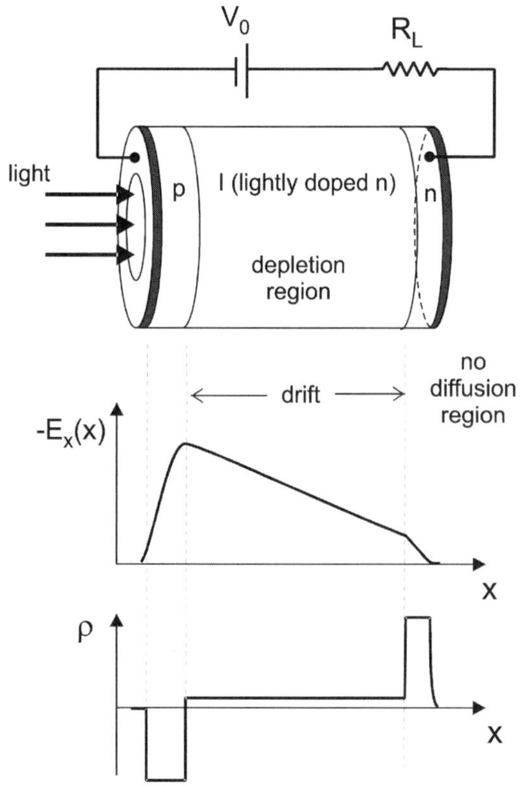

Figure 14-12 (a) In a PIN photodiode, charge carriers are mostly created in the high-field drift region, which extends almost to the far electrode. The lightly n-doped "intrinsic" region has a nearly constant *E* field, which sweeps charge carriers through the device without diffusion.

EXAMPLE 14-3

A silicon PIN photodiode has an intrinsic region of thickness 0.1 mm. Determine the minimum rise time for the detector, its corresponding bandwidth, and the required reverse-bias voltage. Repeat for an intrinsic region of thickness of 10 μm.

Solution: The minimum transit time for holes in silicon is

$$t_{tr} \simeq \frac{d}{v_s} = \frac{10^{-4}\text{ m}}{10^5\text{ m/s}} = 10^{-9}\text{ s} = 1\text{ ns}$$

Taking this as the rise time, the corresponding bandwidth is

$$B = \frac{0.35}{t_r} = \frac{0.35}{10^{-9}\text{ s}} = 350\text{ MHz}$$

The required *E* field is $E = E_s \simeq 2 \times 10^6$ V/m, so

$$\Delta V = Ed = (2 \times 10^6\text{ V/m})(10^{-4}\text{ m}) = 200\text{ V}$$

which is an inconveniently high voltage. Repeating the calculation for $d = 10\ \mu$m gives $t_{tr} = 0.1$ ns, $B = 3.5$ GHz, and $\Delta V = 20$ V. This is a much improved time response, and occurs with a more convenient bias voltage.

The above example shows that in terms of transit time, a thinner intrinsic region is preferable. However, if d is made too small, capacitive effects become important. There is, therefore, an optimum value of d that minimizes the overall response time (see Problem 14.8). Another problem with small d is illustrated by the following example.

EXAMPLE 14-4

For the silicon PIN photodiodes of Example 14-3, determine the absorption efficiency for 860 nm light. At this wavelength, the absorption coefficient in silicon is 335 cm^{-1} and the reflectivity (from air) is 32%.

Solution: For $d = 0.1$ mm,

$$\alpha d = (335\ \text{cm}^{-1})(10^{-2}\ \text{cm}) = 3.35$$

Eq. (13-15) then gives

$$\eta_{\text{abs}} = (1 - 0.32)(1 - e^{-3.35}) = 0.656$$

Note that this neglects light reflected back from the far end of the Si material, so it somewhat underestimates η_{abs}. Repeating the calculation for $d = 10\ \mu$m gives $\alpha d = 0.335$, so

$$\eta_{\text{abs}} = (1 - 0.32)(1 - e^{-0.335}) = 0.194$$

The thinner intrinsic region is seen to be less efficient at absorbing the incident light. There is, therefore, a trade-off between detector speed and sensitivity. A PIN photodiode can be optimized for either of these, depending on the application.

Avalanche Photodiode

If a small load resistance R_L is used to increase the frequency bandwidth of a PIN photodiode, the signal voltage may be quite small, requiring amplification. This can be accomplished with electronic amplifiers, but these introduce their own sources of noise, and it is sometimes desirable to increase the signal generated by the detector, before amplification. One way to do this is through the avalanche multiplication process, depicted in Fig. 14-13a. This is the solid-state analog of the electron multiplication that takes place in a photomultiplier tube. An electron is accelerated by the E field in the depletion region of a reverse-biased p–n junction, and gains kinetic energy in proportion to the distance traveled. When the electron's kinetic energy is high enough, it can collide with an atom in the semiconductor and create an additional electron–hole pair, a process termed *impact ion-*

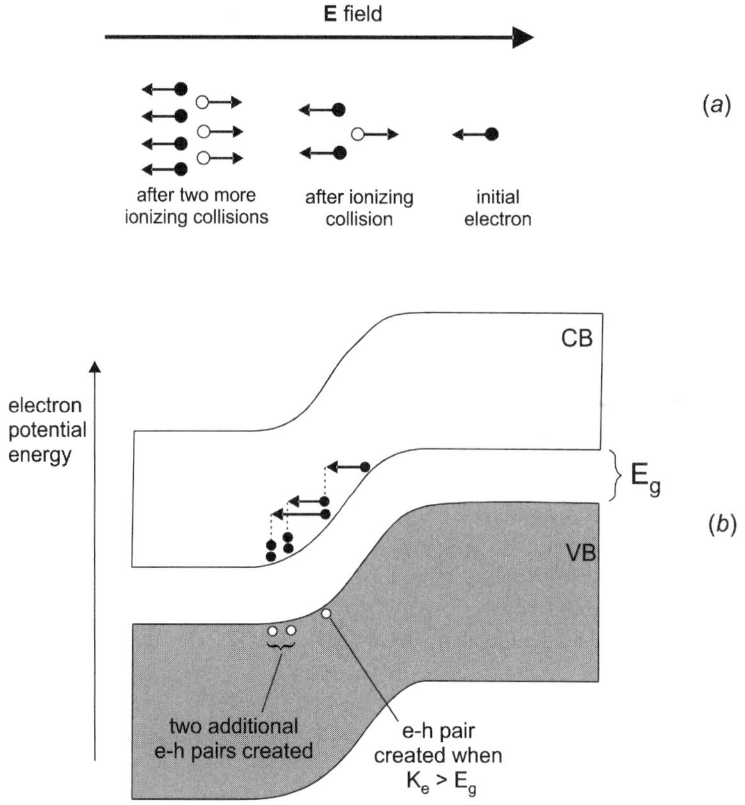

Figure 14-13 (a) An electron accelerated in a strong E field creates an additional electron–hole pair by impact ionization. Both the new electron and the original one then create additional electron–hole pairs, resulting in avalanche multiplication. (b) Impact ionization can occur when an electron in the CB picks up kinetic energy greater than the band-gap energy.

ization. There are now two electrons, and as each one of these accelerates in the E field, it can create yet another electron–hole pair by the same mechanism. There are now a total of four electrons, each of which can create another one to give eight, and so on. The result is *avalanche multiplication,* in which the number of charge carriers increases exponentially with distance traveled.

The creation of electron–hole pairs by impact ionization can be understood in terms of the energy band picture of Fig. 14-13b. After moving a distance Δx in a direction opposite to the E field, the electron loses an amount of potential energy of magnitude $eE\Delta x$. If no energy is lost to other processes, the electron then gains this same amount of kinetic energy K. When $K > E_g$, it becomes energetically possible for a collisional-energy transfer process to take place, in which the electron kinetic energy decreases by $\Delta K = -E_g$, while at the same time the potential energy of a valence electron is increased by this same amount. Increasing the potential energy of a valence electron by an amount E_g corresponds, in the band picture, to taking an electron out of the valence band and placing it in the conduction band, that is, creation of an electron–hole pair.

The energy needed to create an electron–hole pair is usually much less than the potential energy change of an electron as it moves across a reverse-biased p–n junction. For example, a typical band-gap energy is $E_g \sim 1$–2 eV, whereas a 10 V reverse bias corresponds to a potential energy change of 10 eV. Since the electron acquires sufficient energy to create several electron–hole pairs as it moves through this potential difference, it might seem that avalanche multiplication should occur readily in most reverse-biased photodiodes. However, the avalanche phenomenon actually plays a minor role in conventional PIN photodiodes. The explanation for this is that the electron undergoes nonionizing collisions as well as ionizing collisions as it moves through the electric field. For example, the electron can scatter off the thermally induced vibrations in the material (lattice phonons), giving up kinetic energy to heat. At moderate electric field values, these nonionizing collisions prevent the electron's kinetic energy from reaching the threshold value $K = E_g$. When the electric field is sufficiently high, however, the electron can pick up kinetic energy $K > E_g$ before a nonionizing collision occurs, and the avalanche mechanism becomes more efficient.

The electric field at which the avalanche mechanism becomes important depends on the material, being higher in wider-band-gap materials that have a higher threshold kinetic energy. For silicon, an electric field $E \sim 5 \times 10^7$ V/m over a path length of ~ 2 μm is needed for efficient avalanche multiplication. This corresponds to a potential difference $\Delta V \sim 100$ V, much higher than the reverse bias of a typical PIN photodiode.

A photodiode utilizing avalanche multiplication to achieve gain is termed an *avalanche photodiode,* or APD. The structure of an APD, depicted in Fig. 14-14, differs from that of the PIN photodiode (Fig. 14-12) in two ways. First, light enters through a

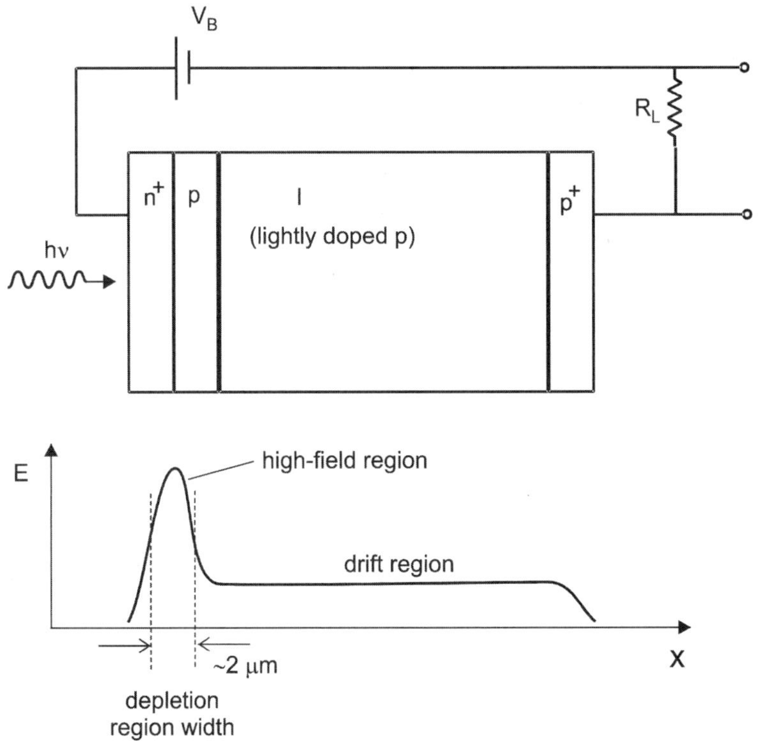

Figure 14-14 In an avalanche photodiode (APD), electrons photoexcited in a nearly intrinsic region are swept out by a small electric field there, and injected into a high-field region between highly doped n and p layers. Avalanche multiplication occurs primarily in the high-field region.

highly doped n-type layer rather than a p-type layer. Second, an additional p-type layer has been added between the highly doped n layer and the nearly intrinsic layer. The left-most n and p layers are highly (and nearly equally) doped, so the junction width between them is small ($d \sim 2$ μm), and the high electric field is mostly confined to this region. In the adjacent intrinsic region (actually lightly doped p-type), there is a much smaller and nearly uniform E field, which extends out to the highly doped p^+ region on the right.

The APD operates in the following way. Light passes through the thin n^+ and p layers and is absorbed in the much thicker intrinsic region. Electrons and holes created by photoabsorption then drift in opposite directions under the influence of the E field, electrons to the left and holes to the right. The electrons eventually make it to the high-field region, where they undergo avalanche multiplication. Holes do not initiate the avalanche in this scheme, but those created by impact ionization can contribute to its development. In silicon, however, holes are much less efficient at causing ionization than are electrons, and therefore make only a minor contribution to the amplification.

In principle, an APD could be constructed with a p^+–n–i–n^+ structure, instead of the n^+–p–i–p^+ shown in Fig. 14-14. However, in this case it would be the holes that are injected into the high-field region, and in silicon this would result in weak amplification. For this reason the n^+–p–i–p^+ structure is always used for silicon APD's. This asymmetry between electron and hole ionization probabilities also has implications for the signal-to-noise properties of the APD. Since avalanche multiplication is a statistical process, there is less statistical variation in output current (i.e., less noise) when only one type of charge carrier contributes to the avalanche. For this reason, germanium APDs, in which the electrons and holes have nearly equal ionization probabilities, are inherently more noisy than silicon APDs.

The effect of avalanche multiplication can be characterized by the multiplication factor M, defined as the ratio of photocurrent with amplification to photocurrent without amplification. The detector output voltage is then still given by Eq. (14-19), with Eq. (14-18) replaced by

$$\mathcal{R} \equiv \frac{i_\lambda}{P_{\text{in}}} = \frac{Me\eta_{\text{abs}}}{h\nu} \qquad \text{(APD responsivity)} \qquad (14\text{-}34)$$

M can be as high as 100 in a silicon APD, but is more typically ~ 10 in a germanium APD.

The operation of the APD depends on the proper electric field profile within the device, and this in turn requires the proper bias voltage. The relation between $E(x)$ and bias voltage (magnitudes only) is

$$\int E\,dx = V_B$$

where we have neglected the built-in potential V_0 compared with V_B. As the bias voltage increases, the area under the $E(x)$ curve increases proportionately, as illustrated in Fig. 14-15a. At some critical voltage, the depletion region "reaches through" to the highly doped p^+ region, with the electric field extending uniformly throughout the intrinsic region. Electrons generated anywhere in the intrinsic region are then efficiently swept out and injected into the high-field region for amplification. A device biased in this way is termed a *reach-through APD*.

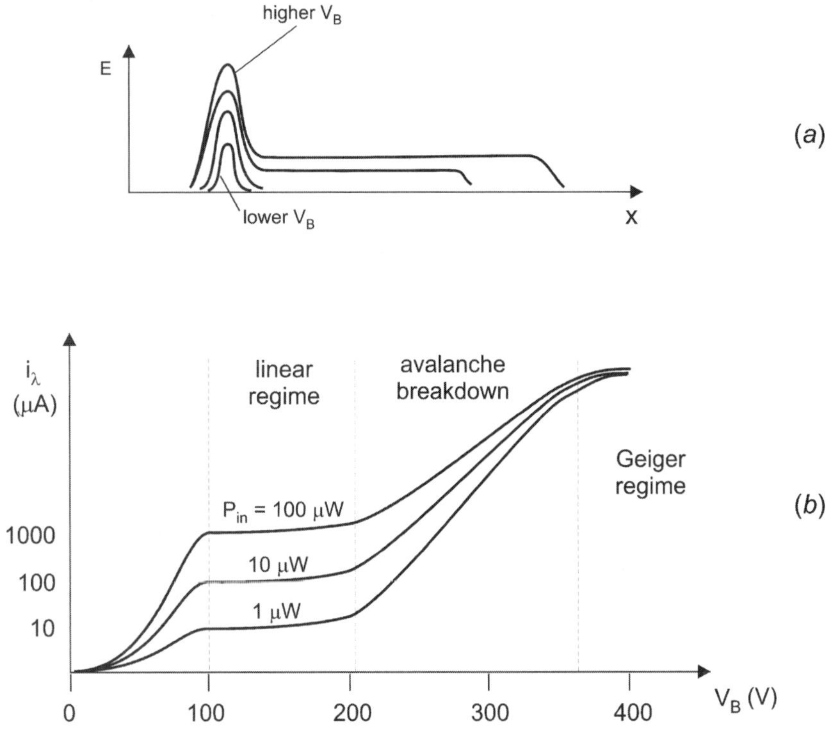

Figure 14-15 (a) The area under $E(x)$ increases with V_B until the field "reaches through" the intrinsic region. (b) Photocurrent i_λ versus reverse-bias voltage for three values of incident light power P_{in}, assuming $\mathcal{R}$ = 10 mA/mW. In the linear regime (after reach-through), $i_\lambda = \mathcal{R}P_{in}$, but in the Geiger regime i_λ becomes independent of P_{in}.

Fig. 14-15b shows a typical variation of responsivity with applied bias voltage for a reach-through APD. Below the threshold value (usually ~ 100–200 V), $\mathcal{R}$ increases with V_B as the E field starts to extend into the intrinsic region. There is a linear operating region above this, in which all photogenerated electrons are collected for amplification. In this region, the avalanche is well behaved, and the detector output is proportional to the incident light power. At still higher bias voltage, the avalanche process becomes uncontrolled, and *avalanche breakdown* ensues. In this situation, the detector output rises to a saturation value which is independent of the number of photons absorbed. This is the *Geiger mode* regime, analogous in operation to the Geiger counter used to detect nuclear radiation. The APD Geiger mode is useful when the purpose is to determine whether any photons are present, rather than to determine their number. It has applications in *photon counting,* in which the arrival time of individual photons is measured.

The responsivity of an APD generally decreases with increasing temperature, due to the increasing probability of electrons making nonionizing collisions with phonons. These collisions take away some of the kinetic energy gained by an electron that might otherwise be available for creating additional electron–hole pairs. In order to stabilize the gain of APD detectors in a changing ambient temperature, then, temperature-control circuitry is needed. Despite these complications, and the need for high-voltage bias, the improved responsivity of the APD makes it an attractive choice for applications limited by a weak light signal, such as in fiber optic communications.

Schottky Photodiode

The PIN and avalanche photodiodes discussed in the previous two sections are both p–n-junction-based devices. In contrast, the *Schottky photodiode* utilizes a metal–semiconductor junction to separate and collect the photogenerated charge carriers. Fig. 14-16 illustrates the operation of a Schottky photodiode for the metal–n–n^+ configuration, the most common type. Photons pass through a partially transparent metallic layer (often gold), and are absorbed in the n-type semiconductor. Charge carriers generated within the depletion region are efficiently swept out by the built-in electric field (see Fig. 10-17), giving rise to a photocurrent i_λ. Just as for a PIN photodiode, the diffusion tail in the time response can be minimized by adjusting the donor concentration N_D in the n-type region so that the depletion region extends all the way through to the n^+ layer. Apart from the lack of a p-type layer, the structure and operation of a Schottky photodiode is similar to that of a PIN photodiode.

Schottky photodiodes have some advantages over PIN photodiodes for certain applications. One advantage is a practical issue in manufacturing the devices. In connecting the Schottky photodiode with wires in the external circuit, only one metal–semiconductor connection needs to be made (metal–n^+), and ohmic contacts are readily formed for such a junction. Another advantage of the Schottky photodiode is an improved time response. Since it lacks the p-type layer of a PIN photodiode, there is no remnant diffusion tail arising from charge carriers generated in the p-type layer. This becomes especially important at short wavelengths, at which the large absorption coefficient would result in a significant fraction of the light being absorbed in the thin p-type layer of a PIN photodiode. Schottky photodiodes with bandwidths in the range 25–60 GHz are commercially available.

A further advantage of the Schottky photodiode is that metal junctions can be made with a wide variety of semiconductors, including those with wide band gap E_g, such as

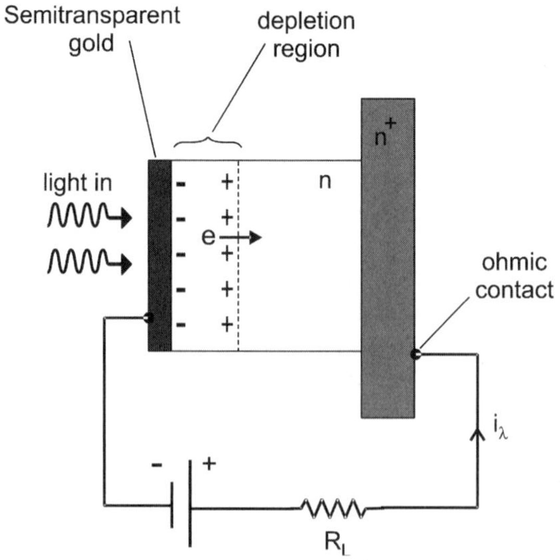

Figure 14-16 In a Schottky photodiode, light is absorbed in the depletion region of an n-type semiconductor after passing through a semitransparent metallic film.

SiC, GaN, and AlGaN. These wide-band-gap detectors have found use as "solar blind" sensors, which respond only to wavelengths in the UV region. They are inherently insensitive to sunlight in the visible and infrared regions where $h\nu < E_g$, and have applications such as flame sensors.

Schottky photodiodes do have some disadvantages, however. They tend to be less efficient than PIN photodiodes at longer wavelengths, due to reflection and absorption of light in the metal layer. To reduce light reflection, an antireflection coating is often applied, but this complicates the manufacture of such devices. Schottky photodiodes are primarily used for detecting blue or UV wavelengths, or in high-speed applications, where some loss in efficiency can be tolerated.

14-5. SIGNAL-TO-NOISE RATIO

The detectability of a small signal depends on how large it is compared to the noise. This is usually expressed by the *signal-to-noise ratio* (SNR), defined as the ratio of electrical signal power to electrical noise power. Taking the electrical signal power in the circuits of Fig. 14-2 as that due to the photocurrent i_λ, we have

$$P_{\text{sig}} = i_\lambda^2 R_L = \left(\frac{P_{\text{in}}\eta_{\text{abs}}e}{h\nu}\right)^2 R_L \qquad \text{(electrical signal power)} \tag{14-35}$$

where Eq. (14-2) has been used. Note that this expression applies only when the detector circuit is well below saturation. Also, in the photoconductive mode the "signal" current is defined as the measured current minus the dark current. The two contributions to the noise were discussed in Section 13-3. For shot noise, Eq. (13-29) gives the electrical noise power as

$$P_{\text{shot}} = (i_N)^2 R_L = 2e(i_\lambda + i_0)BR_L \qquad \text{(shot noise electrical power)} \tag{14-36}$$

where the total current $\bar{i}$ consists of both signal current i_λ and dark current i_0. For thermal noise, Eq. (13-33) gives the electrical noise power as

$$P_{\text{therm}} = \frac{V_N^2}{R_L} = 4k_BTB \qquad \text{(thermal noise electrical power)} \tag{14-37}$$

Note that the thermal noise power is independent of R_L, whereas the shot noise power increases linearly with R_L. Therefore, the dominant source of noise tends to be shot noise for large R_L, and thermal noise for small R_L.

Using the above equations, the signal-to-noise ratio can be written as

$$\text{SNR} = \frac{P_{\text{sig}}}{P_{\text{shot}} + P_{\text{therm}}} = \frac{i_\lambda^2 R_L}{2e(i_\lambda + i_0)BR_L + 4k_BTB} \tag{14-38}$$

where the signal current i_λ is related to the incident optical power P_{in} by $i_\lambda = (P_{\text{in}}/h\nu)\,\eta_{\text{abs}}e$. Since the SNR is a ratio of signal and noise powers, and the power is proportional to the square of voltage or current, we can write

$$\text{SNR} = \left(\frac{i_\lambda}{i_N}\right)^2 = \left(\frac{V_{\text{sig}}}{V_N}\right)^2 \tag{14-39}$$

where $V_{sig} = i_\lambda R_L$. It is therefore the square root of the SNR that gives the ratio of signal amplitude to rms noise amplitude.

It is useful to consider the following limiting cases, in which one source of noise dominates.

1. Large signal. When $i_\lambda \gg i_0$, and $i_\lambda R_L \gg V_T$ (recall $V_T \equiv k_B T/e$ is the voltage equivalent of temperature), we have

$$\text{SNR} \approx \frac{i_\lambda^2 R_L}{2ei_\lambda BR_L} = \frac{i_\lambda}{2eB}$$

The noise here is dominated by the shot noise from the signal current, and the SNR is independent of load resistance. Since $1/B$ corresponds to the measurement time, this result says that the SNR is roughly the number of charge carriers produced during this measurement time. Since $i_\lambda \propto P_{in}$, the SNR increases linearly with the incident optical power.

2. Small signal, large R_L. When $i_\lambda \ll i_0$ and $i_0 R_L \gg V_T$,

$$\text{SNR} \approx \frac{i_\lambda^2 R_L}{2ei_0 BR_L} = \frac{i_\lambda^2}{2ei_0 B}$$

In this regime, the SNR is limited by shot noise from the dark current i_0, and is again independent of load resistance. Note that SNR here increases with the square of the incident optical power.

3. Small signal, small R_L. When $i_\lambda \ll i_0$ and $i_0 R_L \ll V_T$,

$$\text{SNR} \approx \frac{i_\lambda^2 R_L}{4k_B TB}$$

In this regime, the SNR is limited by thermal noise from the load resistor R_L. This situation is often encountered in practice, and leads to a trade-off of SNR with detector response time. Increasing R_L improves the SNR, but degrades the response time due to RC time constant effects. Decreasing R_L improves the response time, but at a sacrifice in SNR ratio.

When the signal becomes equal to the noise (SNR = 1), it is barely discernible, and this can be considered to be the criterion for signal detectability. The optical power that gives SNR = 1 is known as the *noise equivalent power,* or NEP, and is a measure of the detector's sensitivity. In the limiting case #2 above, where shot noise from the dark current dominates, the NEP is found by setting SNR = 1 and using Eq. (14-2) with P_{in} = NEP. This gives

$$1 = \frac{(\text{NEP}\,\eta_{abs} e/h\nu)^2}{2ei_0 B}$$

or

$$\text{NEP} = \frac{h\nu}{\eta_{abs} e}\sqrt{2ei_0 B} \tag{14-40}$$

The MKS unit for NEP is watts, since it is an optical power. An alternative unit commonly used for optical power is the dBm, defined as the power in dB relative to 1 mW. Thus, for an optical power P measured in mW,

$$\text{optical power in dBm} = 10 \log_{10}\left(\frac{P}{1 \text{ mW}}\right) \tag{14-41}$$

For example, an optical power of –20 dBm is 0.01 mW, whereas an optical power of +20 dBm is 100 mW.

It is useful to separate the NEP for a detector into those factors that are fundamental and those that can be adjusted by the device geometry or detector circuit. The dark current i_0, for example, is not fundamental, since it is proportional to the area of the p–n junction. The fundamental parameter is the dark current density J_0, which depends on the material used and the temperature, but not on the device geometry. Writing $i_0 = J_0 A$, we obtain

$$\text{NEP} = \frac{h\nu}{\eta_{\text{abs}}}\sqrt{\frac{2J_0 AB}{e}} \tag{14-42}$$

This shows that the minimum power that can be detected is proportional to $\sqrt{AB}$. The detector can be made more sensitive by decreasing the junction area A, or by decreasing the detection circuit bandwidth. To obtain a parameter that is independent of B and A, the NEP can be divided by $\sqrt{AB}$. It is conventional to define the reciprical of this as a figure of merit, since it is then a larger number for a better (more sensitive) detector. Designating this figure of merit as D^* (pronounced "dee star"), we have

$$D^* \equiv \frac{\sqrt{AB}}{\text{NEP}} \tag{14-43}$$

which becomes

$$\begin{aligned} D^* &= \frac{\eta_{\text{abs}}}{h\nu}\sqrt{\frac{e}{2J_0}} \\ &= \frac{\mathcal{R}}{\sqrt{2eJ_0}} \end{aligned} \tag{14-44}$$

where Eq. (14-18) was used in the last step.

The D^* parameter provides a good way of comparing the ultimate sensitivity limits for different types of detectors. According to Eq. (14-44), D^* varies inversely as the square root of dark current density, so that narrow-band-gap materials (with high J_0) have a smaller D^* than wider-band-gap materials. Since a narrower band gap is required to detect light of longer-wavelength, D^* is inherently smaller for longer wavelength detectors, all other things being equal. D^* also depends on wavelength through the responsivity $\mathcal{R}(\lambda)$. Fig. 14-17 shows the wavelength dependence of D^* for some common detector materials. Note that since J_0 decreases with decreasing temperature, the D^* for a given photodetector can generally be improved by lowering the temperature of the semiconductor element.

14-6. DETECTOR CIRCUITS

So far, we have considered only the simple detector circuits shown in Fig. 14-2, in which the output is taken as the voltage across the series load resistor R_L. Here, we consider two types of detector circuits that provide not only the proper bias for the diode, but also a degree of amplification.

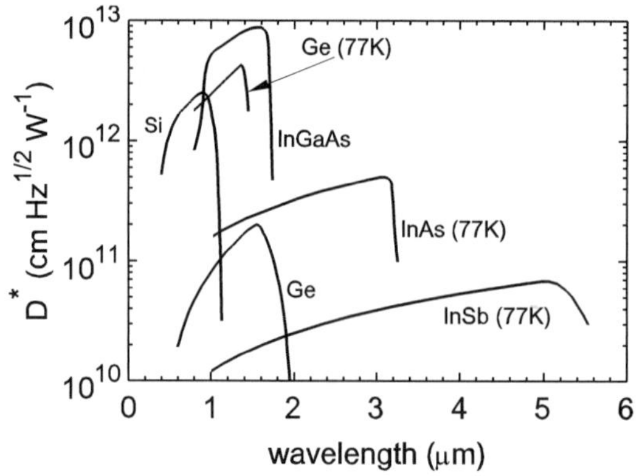

Figure 14-17 The specific detectivity parameter D^* for some respresentative photodetector materials. The maximum possible D^* decreases at longer wavelength because the dark current is higher for narrower-band-gap semiconductors.

High-Impedance Amplifier

Figure 14-18 shows one scheme for amplifying the signal in a photodiode circuit. This is basically the same as the circuits of Fig. 14-2, except that an additional amplification stage has been added with an FET (field effect transistor). The voltage generated across the load resistor is applied between the gate (G) and source (S) of the FET, and this results in an amplified output voltage between the source and drain (D). For the best possible SNR, the photodiode can be operated in the photovoltaic mode, where $V_B = 0$ and there is no dark current. The signal output will be proportional to R_L (below saturation), so higher R_L is best for detecting very small signals. This circuit is a good choice when the best possible SNR is desired.

If time response is important, however, this is not the best circuit to use. The large load resistance, in combination with the diode capacitance C_{diode}, gives a 3 dB electrical bandwidth:

$$B = \frac{1}{2\pi R_L C_{\text{diode}}} \quad \text{(high-impedance amplifier bandwidth)} \qquad (14\text{-}45)$$

This circuit, then, suffers from the same sensitivity/time response trade-off that we discussed earlier.

Transimpedance Amplifier

When response time is important, a better choice for detector circuit is the one shown in Fig. 14-19. This circuit uses an operational amplifier (op-amp) to convert the photocurrent i_λ directly into an output voltage, hence the term *transimpedance amplifier.* The op-amp has the property that the two input terminals are held at nearly the same potential (virtual ground), while at the same time very little current is allowed to flow into or out of either terminal. For the purpose of biasing the photodiode, then, the op-amp input acts like a short circuit ($R_L = 0$), which keeps the diode below saturation for any level of light input. Any photocurrent must flow not through the input terminals of the op-amp, but

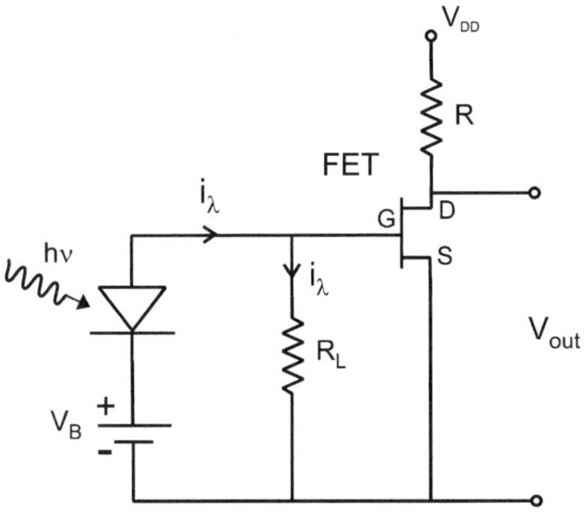

Figure 14-18 High-impedance FET amplifier circuit for the photodiode.

rather through the feedback resistor R_F, which is connected between the input and output of the op-amp. The output voltage is then equal to the voltage across this feedback resistor, $V_{out} = i_\lambda R_F$, since both input terminals of the op-amp are at ground potential. The circuit acts as if $R_L = R_F$ in terms of the output voltage, but it acts as if $R_L = 0$ in terms of biasing the photodiode. This gives the best possible linearity and dynamic range for the photodiode, and the output voltage is in fact limited only by the maximum output voltage of the op-amp (typically ~ 10 V).

A further advantage of the transimpedance amplifier is seen in the time response. Since the diode voltage is held essentially constant, the capacitance of the diode's p–n junction no longer limits the time response. Instead, it is the feedback capacitance C_F characteristic of the op-amp that matters, and the bandwidth becomes

$$B = \frac{1}{2\pi R_F C_F} \qquad \text{(transimpedance amplifier bandwidth)} \qquad (14\text{-}46)$$

Since C_F can be much smaller than C_{diode}, the transimpedance amplifier can have a much higher bandwidth for the same sensitivity ($R_F = R_L$). This type of detector circuit is the preferred one in many situations, because of the advantages of high speed and large dynamic range. The only drawback is in obtaining the best possible SNR for weak signals, in which case a photodiode in the photovoltaic mode followed by a high-impedance amplifier is the best choice.

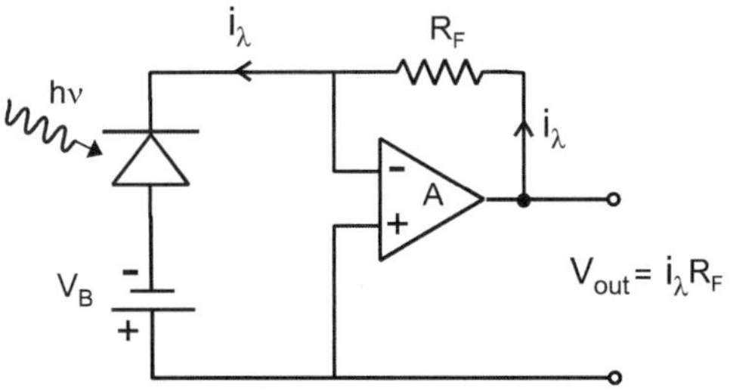

Figure 14-19 Transimpedance amplifier circuit for the photodiode.

PROBLEMS

14.1 A photodiode with responsivity 0.3 A/W and dark current 2 nA is biased in the photoconductive mode, with a 9 V battery and 500 kΩ resistor. Make a sketch like that of Fig. 14-3, showing the load line and the diode curves for incident powers from zero to 100 μW in steps of 20 μW. Circle the operating point for an incident power of 40 μW, and determine the approximate diode voltage from the graph.

14.2 For Problem 14.1, make a sketch of the output voltage (across the resistor) versus the incident optical power, for the range 0 to 100 μW. At what optical power does the detector response saturate?

14.3 The photodiode of Problem 14.1 is removed from the circuit and operated in the photovoltaic mode. (a) Determine the shunt resistance assuming $\beta = 2$. (b) Under open circuit conditions (no load resistor), what incident optical power will result in saturation of the output voltage? (c) A load resistor is now added to increase the dynamic range. What value of load resistance is needed so that optical powers up to 20 μW can be detected without saturation?

14.4 Show that Eq. (14-13) can be obtained from the equivalent circuit model shown in Fig. 14-6.

14.5 A silicon solar cell has area 50 cm^2, reverse-saturation current 0.75 μA, $\beta = 2$. The electrical power generated in the 0.4 Ω load resistor is 894 mW. Determine (a) The circuit current i, (b) the photocurrent i_λ, (c) the optical power incident on the cell, assuming that 80% is absorbed, and (d) the optical-to-electrical conversion efficiency of the cell. Assume the temperature remains near 300 K. Assume $\lambda = 500$ nm for the incident light.

14.6 Using Eq. (14-26) for an exponential voltage rise, show that the rise time (10% to 90% points) is given by $t_r \simeq 2.2RC$.

14.7 A silicon p–n junction photodiode has junction area 1 cm^2, and doping levels 10^{14} and 10^{16} cm^{-3} on the n and p sides, respectively. It is reverse biased with 15 V and a 10 kΩ load resistor is used. (a) Determine the 3 dB electrical bandwidth due to the RC time constant. (b) Determine the bandwidth due to the hole transit time. (c) What is the limiting bandwidth in this case?

14.8 Assume the total response time of a silicon photodiode can be taken as the sum of the transit time (limited by saturation velocity 10^5 m/s) and the RC rise time t_r. Derive an expression for the optimum intrinsic region thickness d. If the load resistance is 50 Ω and the detector area is 0.01 mm^2, calculate d and the resulting detector bandwidth.

14.9 A high-speed germanium PIN photodiode has a depletion width of 10 μm and a reverse-bias voltage of 10 V. The hole mobility in Ge is $\simeq 0.2$ $m^2/(Vs)$, the saturation velocity is $\simeq 7 \times 10^4$ m/s, and the refractive index at 1300 nm is $\simeq 4.3$. (a) Determine the transit time limit to the response time, and calculate the corresponding 3 dB electrical bandwidth. (b) If light of wavelength 1300 nm is detected, determine the fraction of incident light that is absorbed in the depletion region (include the reflection loss from the air–Ge interface). (c) Repeat part b if the detected wavelength is 1600 nm. See Fig. 13-16 for Ge absorption coefficient.

14.10 A silicon APD has a responsivity of 20 A/W at the detection wavelength of 850 nm, and the absorption efficiency is 0.7. Determine the avalanche gain.

14.11 A silicon photodiode is configured as shown in Fig. 14-18 with a 90 V bias voltage. The light to be detected has intensity 20 $\mu W/cm^2$ and wavelength 920 nm. Relevant material properties for the detector are: absorption efficiency = 0.18, dark current density at room temperature = 15 nA/cm^2, charge carrier mobility = 0.048 m^2/Vs, and carrier saturation velocity = 10^5 m/s. At the applied bias voltage, it is known that the width of the depletion region is 0.2 mm. (a) If the photocurrent is 150 nA, what is the area of the detector? (b) If the load resistor is 100 kΩ, determine the RC time constant of the circuit, and the corresponding 3 dB bandwidth. Video requires a bandwidth of about 2 MHz. Will the circuit be suitable for video applications? (c) Determine the transit-time response for the circuit, and compare it with the RC time constant. Which is the primary limit to the bandwidth in this circuit? (d) Repeat part b assuming a load resistance of 10 kΩ. Is the circuit now suitable for video applications?

14.12 In Problem 14.11, check to see that there is sufficient signal-to-noise (S/N) ratio. Using the results for the 10 kΩ load resistor, determine the power S/N ratio, and also find the ratio of the rms deviation in signal voltage to the average signal voltage (express as a percentage).

14.13 The photodetector of Problem 14.11 is now used for low-level dc light-level measurements. Assume that in this application the effective bandwidth is 1 Hz. (a) Determine how large the load resistor must be in order for the noise to be dominated by dark current shot noise rather than by thermal noise. Take as the criterion that the shot noise power is five times the thermal noise power. (b) In the limiting case described in part a, determine the noise equivalent power (NEP) for the detector (in units of watts). (c) For the conditions described in part a, determine the minimum light intensity that can be detected with this detector, taking as the criterion that the signal voltage must be 10 times the rms noise voltage.

14.14 Consider the transimpedance amplifier optical receiver shown in Fig. 14-19. The feedback resistance is 10 kΩ and the feedback capacitance is 0.2 pF. The diode's capacitance is 5 pF, and its responsivity is 0.5 A/W. The incident optical power is 0.5 mW. (a) Compute the signal current. (b) Compute the receiver's output voltage. (c) Compute the receiver's 3 dB electrical bandwidth. (d) Compute the rms thermal-noise current generated in the feedback resistor, assuming a temperature of 300K. (e) Assuming no dark current, and an ideal (noiseless) amplifier, compute the output SNR, expressed in dB. The actual SNR will be somewhat lower due to noise introduced by the amplifier.

14.15 Use the data in Fig. 14-17 to determine the following: (a) The minimum optical power at 900 nm that can be detected (SNR = 1) by a Si photodiode of area 0.02 cm^2 in a 1 Hz bandwidth, (b) the dark-current density of an InAs detector at 77 K, assuming that the absorption efficiency is near unity for $\lambda = 2.8$ μm, and (c) the dark current and minimum detectable power for an InGaAs detector of area 0.02 cm^2, operating at 1550 nm in a 1 Hz bandwidth. Assume an absorption efficiency near unity at 1550 nm.

Chapter 15

Lasers and Coherent Light

In the 1960s when lasers were first being developed, it was often said jokingly that the laser was a solution in search of a problem. It was a novel device with interesting properties, but it was not clear how it would be used in practice. Today, of course, the situation is quite different, and the laser has become an enabling technology with applications as diverse as point-of-sale bar-code scanners, reading and writing data on CDs and DVDs, creating masks for photolithography on integrated circuit chips, optical communications, precision cutting of materials for manufacturing, and laser surgery. It is no exageration to say that without the laser, our modern technological world would be nothing like it is today.

What then makes laser light so special? The short answer is that laser light is coherent. To understand what this means, we will start this chapter with a brief look at the operating principles of a laser. This will be followed by a more detailed look at the nature of coherent light and the importance of coherence for laser applications.

15-1. OVERVIEW OF LASER OPERATION

Figure 15-1 illustrates the three basic elements required for laser action. A *gain medium* amplifies the light, mirrors (or other reflective devices) provide *optical feedback,* and there must be some *pumping mechanism* to supply energy to the laser. The mirrors are arranged to circulate the light back and forth through the gain medium, forming an optical *cavity* or *optical resonator.* This constitutes optical feedback in the sense that some of the amplified output is "fed back" to become input for additional amplification. The combination of gain and feedback is familiar in electrical circuits, and gives rise to electrical oscillations. In fact, the laser is quite similar conceptually to an electrical oscillator with a very high frequency (~ 10^{14} Hz). Just as an electrical oscillator needs to be "plugged in," or supplied with energy, so does a laser need to be supplied with energy via the pump.

Although electrical and optical oscillators are similar in overall concept, they differ considerably in the details of the three basic elements. For example, electrical feedback can be implemented by simply wiring a resistor between the output and inputs of an amplifier. For optical feedback, on the other hand, careful consideration has to be given to the design of appropriate optical resonators, and this will be discussed in Chapter 16.

The mechanism of amplification is perhaps the most fundamental difference between electrical and optical oscillators. In a laser, amplification occurs by *stimulated emission,* a process first proposed by Albert Einstein in 1917. The basic idea of stimulated emission can be understood by considering the three ways that light interacts with an atom, as illustrated in Fig. 15-2. In *absorption,* an atom initially in the ground state (lowest energy level) is raised to a higher energy level (excited state), thereby destroying (absorbing) the in-

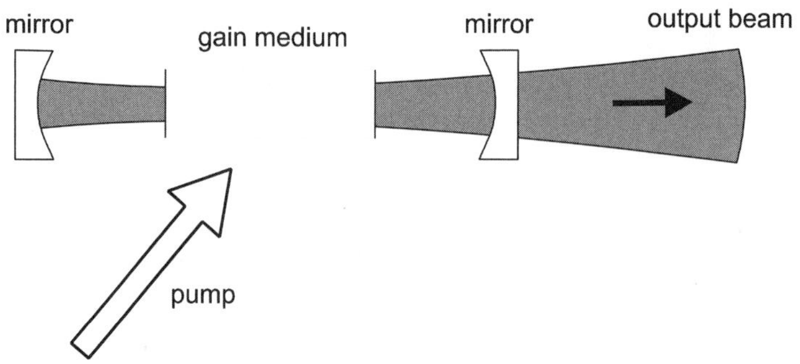

Figure 15-1 Basic elements of a laser.

cident photon. The inverse process is *spontaneous emission,* in which an atom initially in the excited state falls back to the ground state, creating (emitting) a photon. In the third process (stimulated emission), there is already an incident photon and the atom is also already in the excited state. The atom can then be "stimulated" to emit another photon, virtually identical to the photon that was originally incident on the atom—an "optical clone," if you like. This duplication of photons constitutes amplification, since the greater number of photons corresponds to more energy in the light wave.

In the stimulated emission process, total energy must be conserved, as in any physical process in which the particle masses do not change. The increasing optical energy comes from the energy stored in the atoms, and in order to continually amplify the light wave, energy must be continually given to the atoms. This transfer of energy to the atoms is the pumping process, which can take many different forms. In Chapter 23, we will survey the different types of lasers, and see how electrical, optical, or other types of energy sources can be used to pump a laser.

15-2. OPTICAL COHERENCE

The stimulated emission process that gives rise to optical amplification has another important consequence. Because the newly created photon is identical to the original photon, the E fields of the photons reinforce each other and the resulting light is *coherent.* The idea of coherence can be understood by considering the analogy of a marchers in a parade, as illustrated in Fig. 15-3. In coherent marching, each person within a given row

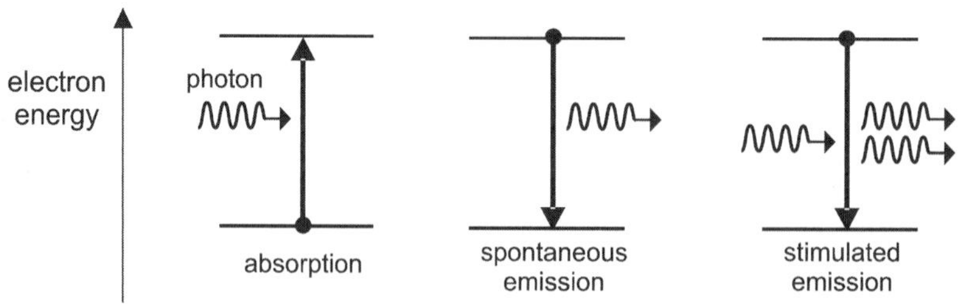

Figure 15-2 Stimulated emission, the basis of optical amplification, is one of the three ways that light can interact with atoms.

Figure 15-3 Coherent light is analogous to a group of marchers who are raising their left legs at the same time (correlated motion). (Photo courtesy of The University of Michigan Marching Band.)

or column is raising his or her left leg at the same time. Coherent marching leads to a kind of predictive power: I know that if I am raising my left leg, then everyone else in my row is also raising their left leg. Incoherent marching lacks this predictive power, and I cannot say, based on the state of my own leg, what other marchers' legs are doing. The marching is called partially coherent if there is a limited predictive power, allowing me to predict whether other marchers are raising their left legs only within a certain vicinity.

The idea of optical coherence is similar to that of coherent marching. For coherent light, if I know the value of E at one point in space, I can predict the value of E at other points in space. We say that there is a *correlation* between values of E at different points in space. The simple sinusoidal plane wave of Eq. (2-2), for example, is perfectly coherent, since the values of E are highly correlated in all directions. A wave that is only partially coherent can be characterized by its degree of coherence in two distinct directions: perpendicular to the wave front (i.e., in the direction of wave propagation), and parallel to the wave front. These two types of coherence are considered next in some detail.

Temporal Coherence

The degree of coherence in the direction of wave propagation is referred to as *longitudinal* or *temporal* coherence. In the marching analogy, this corresponds to whether everyone in a given column is raising their left leg at the same time. Perfect longitudinal coherence for an optical wave implies that the planes of constant phase are uniformly spaced without interuption. These planes of constant phase move with the speed of the wave, and if observed from a fixed point in space, the E field will be seen to oscillate uniformly in time without any phase interuptions. The fixed observer could then predict the value of E at any future time, which is why this is called temporal coherence.

In practice, light sources never have perfect temporal coherence. The degree of coherence will be limited either by the finite duration of the light, or by interuptions in the phase of the wave. The average time between phase interuptions is termed the *coherence time* T_c, as illustrated in Fig. 15-4. In the marching analogy, this corresponds to one of the marchers misstepping, raising a right leg instead of a left, with marchers in back also raising a right leg, while marchers in front continue to raise a left leg. This abrupt shift from left to right leg is analogous to a phase interuption in an optical wave. The average distance from one phase interruption to the next in a wave is called the *longitudinal coherence length,* L_c, and can be thought of as the average distance (in the direction of propagation) over which the wave is coherent. Since the wave (in vacuum) propagates with speed c, the coherence length is related to the coherence time by

$$L_c = cT_c \qquad \text{(longitudinal coherence length)} \qquad (15\text{-}1)$$

The high degree of coherence of laser light is important for applications such as holography that involve interference of two beams. To create a hologram, a single laser beam is split into two parts with a beam splitter, as shown in Fig. 15-5. One part of the beam travels a distance $L_1 + L_2$ to reach a photographic plate, while the other part travels a distance L_3 to the object being recorded. Light scattered from the object travels a distance L_4 to the plate, where it interferes with light from the first part of the beam, creating a hologram. In order for light from the two paths to interfere constructively and destructively at the photographic plate, it is necessary that the path difference be less than the coherence length, that is, $|(L_3 + L_4) - (L_1 + L_2)| < L_c$. For centimeter-scale objects, it is thus necessary for the coherence length to be at least a few centimeters. He–Ne lasers typically have $L_c \sim 10$–20 cm, which is adequate for holography. Semiconductor diode lasers such as GaAs, however, have $L_c \sim 1$ mm, and are generally unsuitable for holography. The coherence length of a laser can be increased by reducing the spectral width $\Delta\nu$, and we discuss some techniques for doing this in Chpater 21.

The high degree of coherence has an important consequence for the spread of wavelengths in laser light, known as its *spectral width* or *linewidth.* According to the Fourier transform principle (see Appendix B), the time-dependent waveform in Fig. 15-4 can be obtained by adding together an infinite number of pure sinusoidal components, having the distribution of frequencies ν shown in Fig. 15-6. This curve is a distribution function, and shows the relative number of sine wave components needed in a small frequency interval

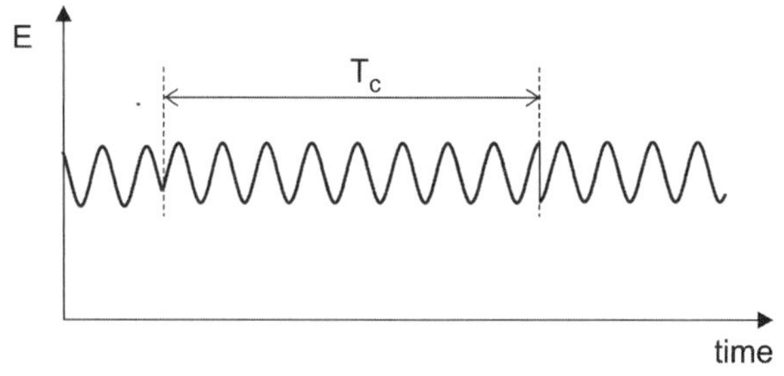

Figure 15-4 Coherence time T_c is the time between phase interuptions.

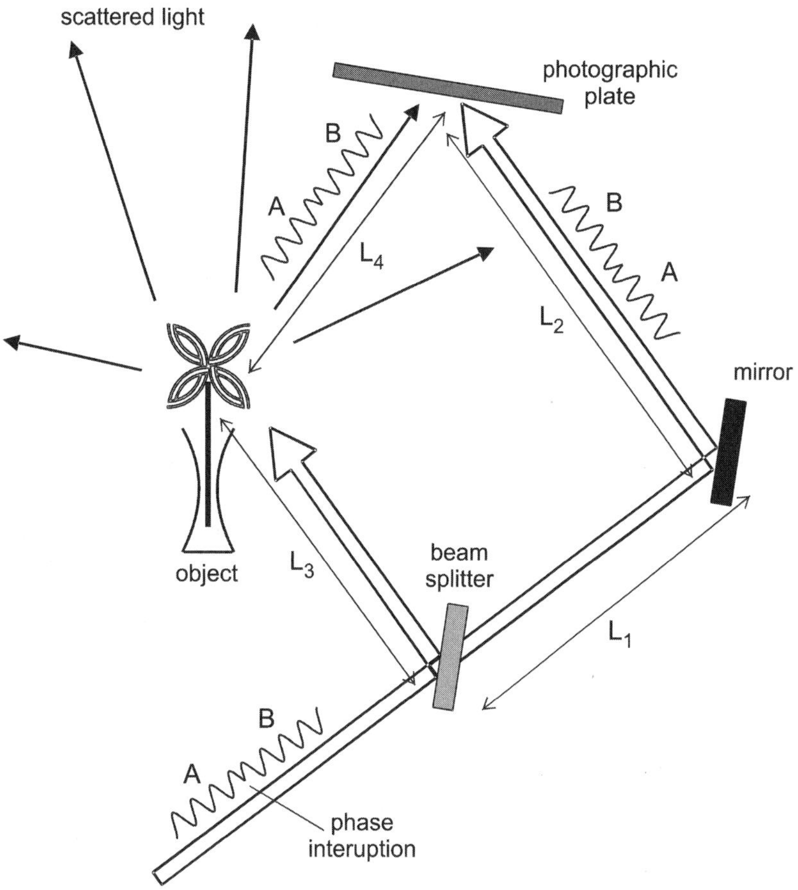

Figure 15-5 In holography, a wave of finite coherence length (section A uncorrelated in phase with section B) is split by a beam splitter and sent along two paths to the photographic plate. If the length of one path is sufficiently different from the length of the other path, the two uncorrelated wave sections A and B will arrive at the plate together. No interference pattern is recorded in this case. If the path lengths are carefully adjusted to be the same, the two A sections arrive together and an intefer-ence pattern is produced.

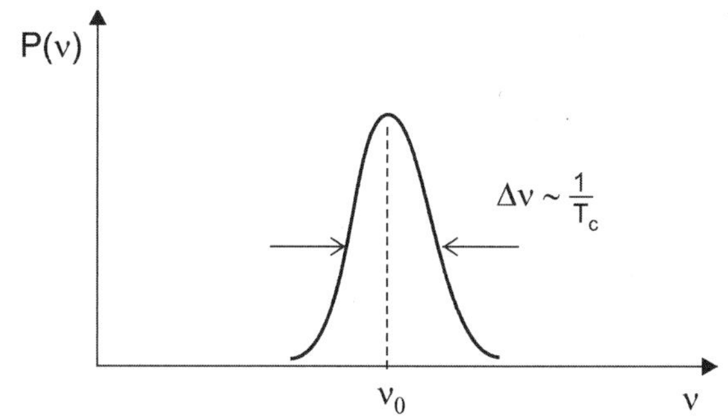

Figure 15-6 Power spectral distribution for light with coherence time T_c.

$d\nu$ about ν. It is peaked at the frequency $\nu_0 = 1/T$ of the uninterupted portion of $E(t)$ in Fig. 15-4, and has a spectral width $\Delta\nu$ given by

$$\Delta\nu \simeq \frac{1}{T_c} \qquad \text{(spectral width)} \tag{15-2}$$

The exact factor required to make Eq. (15-2) an equality depends on the shape of the distribution function as well as the definition of width, but is close to unity and does not concern us here. The important point is that the spectral width is narrow when the coherence time is long. The closer the wave is to a pure sine wave (fewer phase interuptions), the closer it is also to being a single frequency (narrower spectral width). Light that is nearly single frequency is termed *monochromatic,* and in the visible region would appear as a single pure color. Laser light, then, can be said to be both highly coherent and highly monochromatic.

The monochromatic nature of laser light is important for a number of applications. We have already seen (Chapter 6) that intramodal dispersion in an optical fiber is proportional to the spectral width, and limits the maximum bit rate in optical communications. This makes the laser, with its narrow spectral width, an ideal light source for high-speed communications. Narrow laser linewidths also allow signals at different wavelengths to be combined for transmission and later separated, a technique known as *wavelength division multiplexing* or WDM (see Chapter 24). Optical spectroscopy, the detailed study of the absorption and emission spectra of materials, benefits greatly from laser excitation as well.

EXAMPLE 15-1

A semiconductor laser operates at a free-space wavelength of 790 nm, and has a longitudinal coherence length of 1 mm. Determine the linewidth in terms of both frequency and wavelength.

Solution: The frequency linewidth is

$$\Delta\nu \simeq \frac{1}{T_c} = \frac{c}{L_c} = \frac{3 \times 10^8}{10^{-3}} = 3 \times 10^{11} \text{ Hz}$$

The wavelength linewidth is obtained by taking the differential of $\lambda = c/\nu$,

$$\Delta\lambda = \frac{c}{\nu^2}\Delta\nu = \frac{\lambda^2}{c}\Delta\nu = \frac{(790 \times 10^{-9})^2}{3 \times 10^8}(3 \times 10^{11}) = 6.24 \times 10^{-10} \text{ m}$$

The linewidth is then 0.62 nm.

Spatial Coherence

The degree of coherence along a wave front (perpendicular to the direction of wave propagation) is referred to as *transverse* or *spatial coherence.* In the marching analogy, this corresponds to whether everyone in a given row is raising their left leg at the same time. Perfect spatial coherence for an optical wave means that the wavefronts are continuous, with no interuptions in phase. The E field at two points along a wavefront, such as points

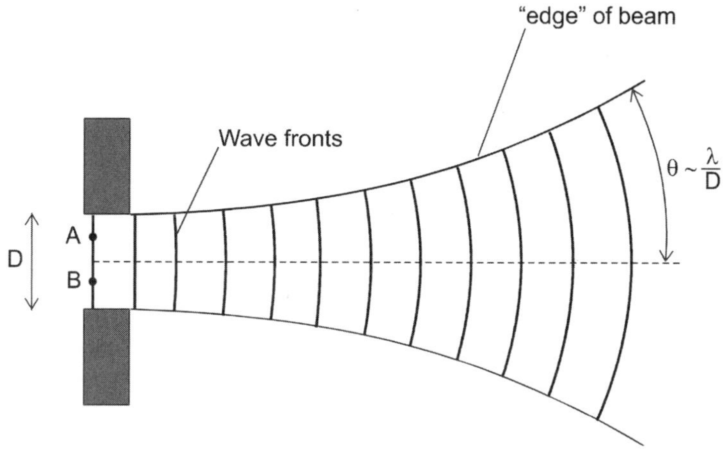

Figure 15-7 Spatially coherent light has smooth and continuous wave fronts, and the *E* fields at points A and B are correlated.

A and B in Fig. 15-7, are then correlated; knowing the field at one point allows the field at the other point to be predicted.

Spatial coherence has important consequences for the directionality of laser beams. If perfectly coherent light passes through an aperture of diameter D, it diverges with a half-angle $\theta \sim \lambda/D$ [Eq. (2-25)] for distances $z \gg D$ from the aperture. Larger beam diameters correspond to smaller divergence angles, that is, more directional beams. Light that is only partially coherent can be characterized by a transverse coherence length D_c, shown in Fig. 15-8, which is the maximum separation between two points along the wavefront, for which the fields at the two points are correlated. The diffraction pattern for partially coherent light is similar to that of coherent light sent through an aperture of diameter D_c, with a resulting divergence angle

$$\theta \sim \frac{\lambda}{D_c} \qquad \text{(partial coherence)} \tag{15-3}$$

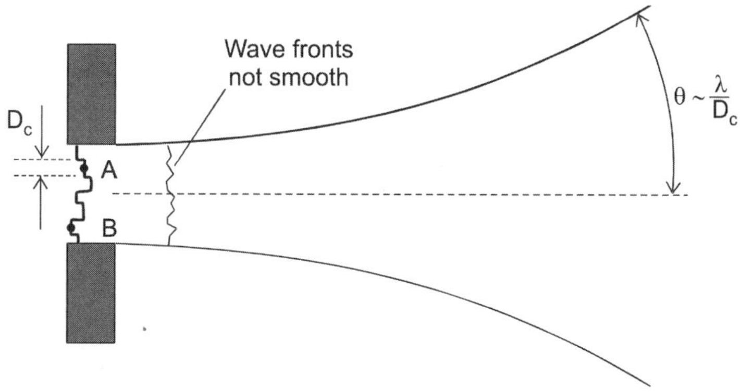

Figure 15-8 Light with partial spatial coherence has interuptions in phase along the wave fronts, and the *E* fields at points A and B are uncorrelated for separation greater than D_c. The divergence angle is similar to that of a coherent beam passing through an aperture of width D_c.

Incoherent light has a very small value of D_c, resulting in light that spreads out very quickly. Partially coherent light becomes more directional as the coherence length D_c is increased, until $D_c = D$, at which point the beam of light is said to be *diffraction limited.* It is important to realize that Eq. (2-25) is only valid for a diffraction-limited beam.

A figure of merit that is often used to describe a partially coherent beam is the M^2 parameter, defined as the ratio of its divergence to that of a perfectly coherent beam. It can be related to the spatial coherence length by

$$M^2 \equiv \frac{\theta}{\lambda/D} = \frac{\lambda/D_c}{\lambda/D} = \frac{D}{D_c} \qquad (15\text{-}4)$$

Since $D_c \leq D$, then $M^2 \geq 1$, with $M^2 = 1$ corresponding to a perfectly coherent beam.

Brightness

The highly directional nature of laser light has important implications for the *brightness* of the light. As discussed in Chapter 12 and Appendix A, the brightness of a light source is the power emitted per unit solid angle, per unit emitting area. In the case of a perfectly coherent laser beam of diameter D, the emitting area is $A_s \sim D^2$ and the solid angle of emission is $\Delta\Omega \simeq \pi\theta^2$ (see Appendix A). Using $\theta \sim \lambda/D$ for coherent light, the brightness can be written as

$$B = \frac{P}{A_s\Delta\Omega} = \frac{P}{D^2\pi\left(\frac{\lambda}{D}\right)^2} = \frac{P}{\pi\lambda^2} \qquad (15\text{-}5)$$

For light that is only partially coherent, $\theta \sim \lambda/D_c$, and the brightness is reduced by the factor $(D_c/D)^2$. This derivation only gives the order of magnitude of the laser's brightness, and the factor of $1/\pi$ should not be taken too seriously. More exact expressions for the spatial variation of intensity in a laser beam will be presented in Chapter 17. The important point here is that the brightness of a laser is independent of the beam diameter, depending only on the power and the wavelength. Shorter-wavelength lasers have a greater brightness, for the same optical power.

Brightness is an important parameter because it characterizes the degree to which light can be focused to a point. Say that light from a source with emitting area A_s and brightness B is imaged with a lens into a spot of area A_s', as shown in Fig. 15-9. According to the *brightness theorem* (see Appendix A), the brightness of an optical beam is not changed by passing through any combination of lenses, mirrors, or other passive optical elements. Therefore,

$$B = \frac{P}{A_s\Omega} = \frac{P}{A_s'\Omega'} \qquad \text{(brightness theorem)} \qquad (15\text{-}6)$$

where Ω and Ω' are the solid angles corresponding to the linear angles θ and θ', respectively. In terms of the source brightness B, the intensity of the beam at the focus is

$$I' = B\Omega' \qquad \text{(intensity of focused beam)} \qquad (15\text{-}7)$$

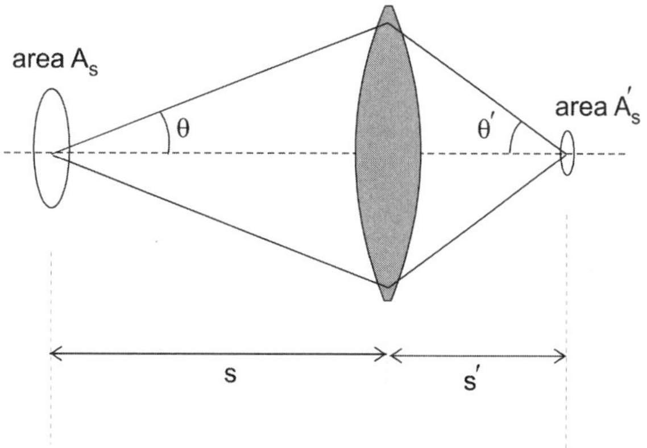

Figure 15-9 Light is emitted from area A_s into a cone of half-angle θ, and focused with a lens onto area A_s' in a cone of half-angle θ'.

To achieve the highest intensity, Ω' (and hence θ') should be made as large as possible. Values of θ' much higher than 45° give rise to significant aberations (distortions in the image), so a practical maximum value for the solid angle is $\Omega' = 2\pi(1 - \cos 45°) \sim 2$ sr. From Eq. (15-7), the maximum intensity at the focus is then $I'_{max} \sim 2B$. The brightness of an optical beam is, therefore, seen to be roughly equivalent to the maximum intensity that can be obtained by focusing the beam.

The ability to be focused to a small point is one of laser lights' greatest advantages over everyday incoherent light. It is this property that makes laser light ideal for coupling into the small cores of single-mode fibers. For high-power lasers, the high intensities at the focus can produce enough heat in a small volume to melt materials, enabling applications such as laser machining and laser surgery. These high intensities can also give rise to varioius nonlinear effects, which are discussed in Chapter 9.

EXAMPLE 15-2

Calculate the brightness of:

(a) An LED emitting 0.1 mW from a square area 0.2 mm on a side, into a cone of half-angle 60° (assume a uniform distribution within this cone).

(b) A He–Ne laser emitting 1 mW of light at 633 nm.

(c) The sun, which emits $\sim 4 \times 10^{26}$ W from its entire surface of radius 7×10^8 m.

Solution:
(a) The solid angle for a cone of half-angle 60° is

$$\Omega = 2\pi(1 - \cos 60°) = \pi$$

giving a brightness for the LED of

$$B \simeq \frac{10^{-4}}{(4 \times 10^{-8})\pi} \simeq 800 \frac{\text{W}}{\text{m}^2\ \text{sr}}$$

(b) For the laser, Eq. (15-5) gives

$$B \sim \frac{10^{-3}}{\pi(633 \times 10^{-9})^2} \simeq 8 \times 10^8 \frac{\text{W}}{\text{m}^2\ \text{sr}}$$

Note that the laser is some six orders of magnitude brighter than the LED.

(c) The sun emits its power uniformly into 4π sr, giving

$$B \simeq \frac{4 \times 10^{26}}{4\pi[4\pi(7 \times 10^8)^2]} \simeq 5.1 \times 10^6 \frac{\text{W}}{\text{m}^2\ \text{sr}}$$

The rather counterintuitive result is that the brightness of the sun is about two orders of magnitude smaller than that of a typical low-power HeNe laser! This illustrates one of the remarkable properties of laser light, its high brightness, which is directly related to its spatial coherence and directionality.

In this chapter, several aspects of the coherence properties of laser light have been presented. The interrelation of these different types of coherence are summarized in Fig. 15-10.

PROBLEMS

15.1 The light emission from an individual excited atom in the gas phase often lasts for a few nanoseconds (10^{-9} s), referred to as the lifetime of the emission. Taking 2 ns as the coherence time, compute the coherence length for light emitted by atoms in a gas-discharge tube.

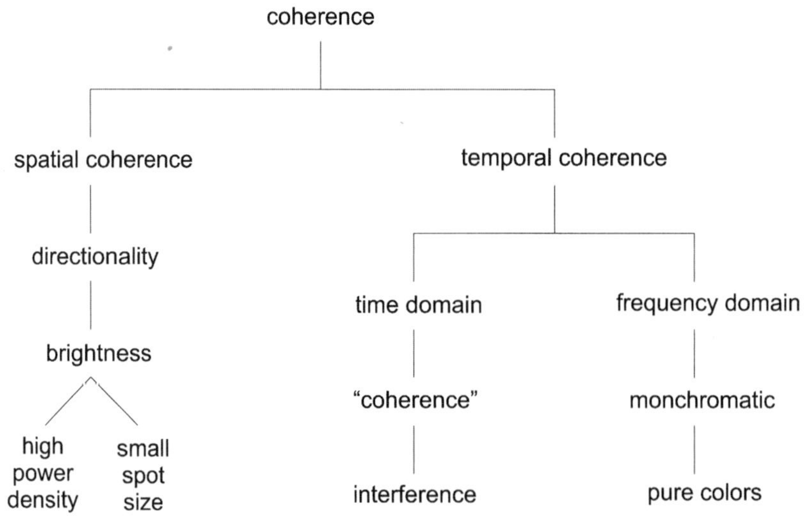

Figure 15-10 Summary of different types of coherence for laser light.

15.2 An optical pulse at 800 nm originally has a spectral width of 2 nm. After passing through a narrow band pass optical filter, the spectral width is reduced to 5 pm. Determine the longitidinal coherence length of the light before and after the filter.

15.3 In a holography setup such as that of Fig. 15-5, the distance $L_1 + L_2 = 33.6$ cm, and the distance $L_3 + L_4 = 32.9$ cm. What must be the frequency width of the laser used? If the laser is a He–Ne with $\lambda = 632.8$ nm, what is the corresponding width in wavelength?

15.4 A laser of wavelength 1030 nm has an initial beam diameter of 1.5 mm, and the divergence half-angle of the conical beam is 1.6°. Determine the spatial coherence length of the beam

15.5 A Nd:YAG laser beam is perfectly coherent, with initial beam diameter of 0.8 mm. If the laser is on the ground and directed up at a plane flying over at an altitude of 29,000 feet, what is the size of the beam when it hits the plane?

15.6 An argon ion laser with power 1.5 W at 488 nm is focused with a lens to the smallest practical spot size. Estimate the beam intensity and the electric field amplitude at the focus point.

15.7 A beam of light with wavelength 1.9 μm and $M^2 = 10$ passes through an aperture of diameter 12 mm. Determine the angular divergence of the beam in degrees (give the cone half-angle), and calculate the diameter of the beam at a distance of 5 m from the aperture.

15.8 According to Stefan's law, the total power radiated per unit surface area from a blackbody (a perfect absorber at all wavelengths) is given by σT^4, where $\sigma = 5.67 \times 10^{-8}$ W · m^{-2} · K^{-4} and T is the absolute temperature. (a) Use this to derive an expression for the brightness of a blackbody emitter, at a distance much greater than the dimensions of the emitting object. (b) Determine the brightness of a lightbulb, treating the filament as a blackbody at temperature 2700 K. Compare this with the brightness calculated in Example 15-2 for the LED, the He–Ne laser, and the sun.

15.9 Use the results of Problem 15.8 to determine the total power radiated from a tungsten filament in a lightbulb, if the filament is wound in the shape of a cylinder of diameter 1 mm and length 3 mm. Take the filament temperature to be 2700 K, and assume that it can be treated as a perfect blackbody. (In practice, the emission efficiency will be somewhat lower than this because tungsten is not really a perfect blackbody.)

Chapter 16

Optical Resonators

One of the three major components of a laser is the optical feedback mechanism, consisting of mirrors or other reflective elements. It is this optical feedback, in combination with optical amplification from stimulated emission, that gives rise to coherent laser oscillations. The simplest arrangement for optical feedback is a pair of mirrors on either side of the gain medium, forming an optical cavity, or optical resonator. The importance of the optical resonator goes beyond simply providing feedback, however. In this chapter, we explore in some detail the effect that the optical resonator has in shaping the frequency spectrum of the emitted laser light.

16-1. MODE FREQUENCIES

Consider the simplified view of an optical resonator shown in Fig. 16-1, with two mirrors separated by a distance L. In general, light can propagate in any direction in between the mirrors, but light that does not propagate close to the resonator axis (i.e., perpendicular to the mirror surfaces) is soon lost from the resonator and is not effective in providing optical feedback. To a first approximation, then, the optical resonator can be analyzed by considering waves only in one dimension.

1-D Treatment

Taking the resonator axis to be in the x direction, we consider electromagnetic plane waves that propagate between the mirrors in the form of Eq. (2-3), $E(x, t) = E_0 \cos(kx - \omega t)$. If the mirrors are highly reflecting, a wave starting at position A will be reflected back and forth between the mirrors many times, and the total E field at the point A will be determined by the superposition, or interference, of the E fields from the many different reflected waves. In general, the phase of the various reflected waves will be different when they reach point A, and the superposition gives rise to *destructive interference.* This levels off the peaks and valleys of the E field distribution, leading to a uniform light intensity within the cavity. However, if the phase of the E field is the same after propagating the round-trip distance $2L$, that is, if

$$E(x + 2L, t) = E(x, t) \tag{16-1}$$

then the reflected waves will reinforce one another, resulting in *constructive interference.* Since the cos function has a periodicity of 2π, this condition is equivalent to

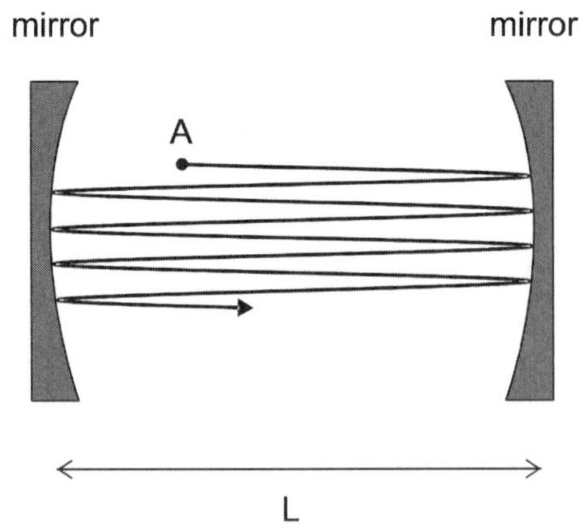

Figure 16-1 Optical cavity of length L with nearly flat mirrors.

$$
\begin{aligned}
k\,2L &= m2\pi \\
\frac{2\pi}{\lambda}\,2L &= m2\pi \qquad (16\text{-}2)\\
2L &= m\lambda
\end{aligned}
$$

where m is an integer and λ is the wavelength of light in the medium. This equation says that for the waves to add constructively, an integer number of wavelengths must fit into the round-trip distance $2L$. This makes sense physically, since the wavelength is the repeat distance for the traveling wave. The optical frequencies that give constructive interference are then

$$\nu_m = \frac{c/n}{\lambda} = m\frac{c}{2nL} \qquad \text{(mode frequencies)} \qquad (16\text{-}3)$$

where n is the refractive index of the medium inside the cavity.

At the frequencies given by Eq. (16-3), the reinforcement of the many reflected waves gives rise to a large E field amplitude inside the cavity. This increased amplitude due to multiple reflections is termed resonant enhancement, and the frequencies at which it occurs are called the *resonant frequencies* or *mode frequencies* of the cavity. The physical significance of the mode frequencies is that optical power can be stored in the laser cavity only at these particular frequencies. According to Eq. (16-3), the mode frequencies are all multiples, or harmonics, of a base frequency $c/(2nL)$. The frequency distribution of optical power stored in the resonator cavity is then a "comb spectrum," as illustrated in Fig. 16-2, with the mode frequencies evenly spaced by $c/(2nL)$.

At the resonant frequencies, there are traveling waves moving both left and right in the cavity, which combine to give the total E field at each point. The waves moving in the $+x$ and $-x$ directions can be written as

$$
\begin{aligned}
E_+(x, t) &= E_0 \cos(kx - \omega t + \phi) \\
E_-(x, t) &= E_0 \cos(kx + \omega t + \phi)
\end{aligned}
\qquad (16\text{-}4)
$$

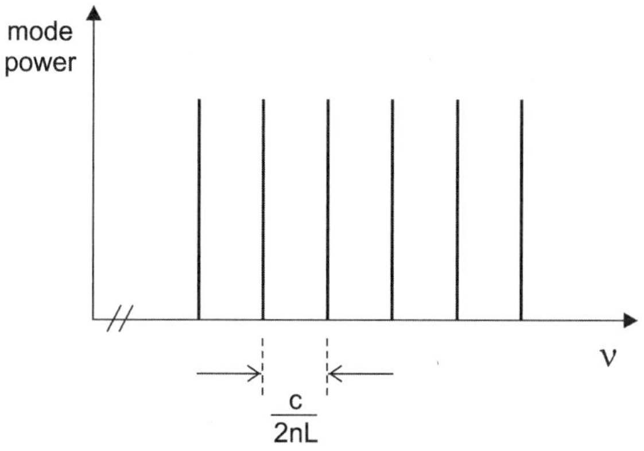

Figure 16-2 Power spectrum for light in a resonant cavity of length L.

where ϕ is a phase constant chosen to match the boundary conditions at the mirrors (for example, $E = 0$ at a metallic mirror). Using the trigonometric identity $\cos(A + B) = \cos A \cos B - \sin A \sin B$ it can be shown that (see Problem 16.1)

$$\begin{aligned} E(x, t) &= E_+(x, t) + E_-(x, t) \\ &= 2\, E_0 \cos(kx + \phi) \cos(\omega t) \end{aligned} \tag{16-5}$$

The spatial and temporal dependence of E given in Eq. (16-5) is that of a *standing wave,* as illustrated in Fig. 16-3.

At a particular value of x, the motion varies in time as $\cos \omega t$, with an amplitude given by $2\, E_0 \cos(kx + \phi)$. The amplitude becomes zero at certain locations, known as *nodes.* At the nodes, the E field and associated electromagnetic energy density ρ (Eq. 2-9) are both zero at all times. For a mode number m, there are $m - 1$ nodes between the cavity mirrors.

EXAMPLE 16-1

Estimate the mode number and mode spacing for an Ar ion laser oscillating at 514 nm in a cavity of length 1 m. Assume $n = 1$.

Solution: The frequency of the laser light is

$$\nu = \frac{3 \times 10^8}{514 \times 10^{-9}} = 5.84 \times 10^{14} \text{ Hz}$$

and the mode spacing is

$$\frac{c}{2L} = \frac{3 \times 10^8}{(2)(1)} = 1.5 \times 10^8 \text{ Hz}$$

The mode number is then

$$m = \frac{5.84 \times 10^{14}}{1.5 \times 10^8} \simeq 3.89 \times 10^6$$

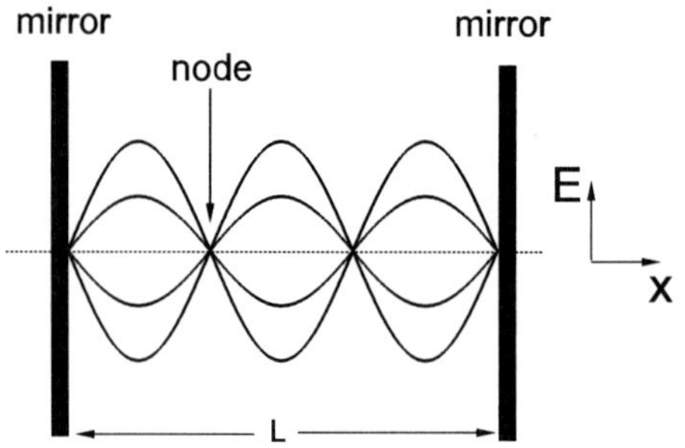

Figure 16-3 Standing wave pattern in laser cavity for $m = 3$, showing $E(x)$ at four values of t.

3-D Treatment

The mode spacing for a three-dimensional cavity can be obtained by extending the 1-D analysis of the previous section. We save for the next chapter a detailed discussion of the stable resonator modes in a laser cavity, and focus here on the general problem of finding the frequencies of modes in an enclosed cavity. Taking the cavity to be a cube of side L, the plane waves that can propagate in the cavity have a wave vector $\mathbf{k}$ with three components k_x, k_y, and k_z. The condition of Eq. (16-2) now applies to each of these components separately:

$$k_x = m_x \frac{\pi}{L}$$
$$k_y = m_y \frac{\pi}{L} \qquad (16\text{-}6)$$
$$k_z = m_z \frac{\pi}{L}$$

where m_x, m_y, and m_z are positive integers. Negative integers represent the same mode as the corresponding positive integer, because a given mode consists of the combination of traveling waves moving in opposite directions. The different modes can be represented as points in a three-dimensional "k space," as shown in Fig. 16-4, with a spacing between points of π/L. The density of modes is then $(L/\pi)^3$ modes per unit volume of k space. It will prove useful to obtain an expression for the mode density in frequency space for three diminsions. To do this, we note that the frequency ν is related to the magnitude of the wave vector k by $k = 2\pi\nu/c$. For notational convenience, we will let $n = 1$ in the following discussion. In a medium with index of refraction n, the formulae can be generalized by making the substitution $c \rightarrow c/n$. The procedure will be to count the number of modes having frequency less than some value ν, and from this to determine the number of modes in a small range of frequencies $d\nu$ around ν.

The number of modes with frequencies up to some value ν is the same as the number of modes with wave vector magnitudes up to a value $k = 2\pi\nu/c$. The surface in k space

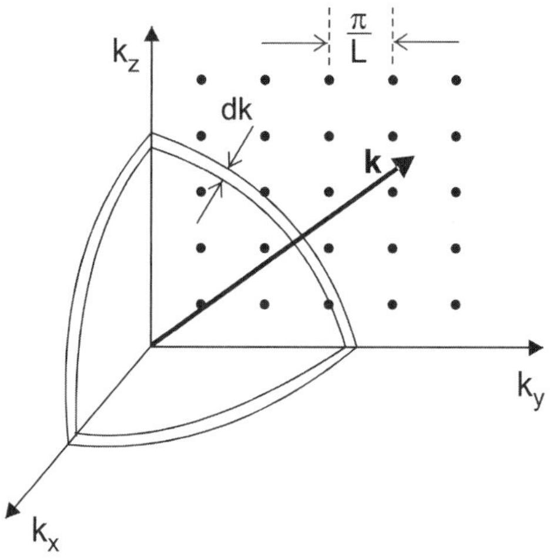

Figure 16-4 Modes in k space.

corresponding to the maximum k value is a sphere of radius k, centered at the origin. The number of distinct modes inside this sphere is then

$$N = \left(\frac{L}{\pi}\right)^3 \times \frac{4}{3}\pi k^3 \times \frac{1}{8} \times 2 \tag{16-7}$$

where the first factor is the number of modes per volume of k space, the second factor is the volume of a sphere of radius k in k space, the factor of 1/8 comes from only considering points with positive k_x, k_y, and k_z, and the factor of 2 comes from the two possible polarizations for each spatial mode. The number of modes having frequency between 0 and ν is then

$$N = \frac{8\pi\nu^3}{3c^3} L^3 \tag{16-8}$$

where $L^3 = V$ is the physical volume of the cavity.

The many modes counted in Eq. (16-8) span a very large frequency range, and most do not interact with the atoms in a laser cavity. The most relevant quantity is the number of modes contained within a small frequency interval $d\nu$ about the center frequency ν. The *spectral mode density* $\beta_\nu(\nu)$ is defined as the number of modes per unit frequency interval, per unit volume V, which from Eq. (16-8) is

$$\begin{aligned} \beta_\nu(\nu) &\equiv \frac{1}{V}\frac{dN}{d\nu} \\ &= \frac{8\pi\nu^2}{c^3} \end{aligned} \tag{16-9}$$

In a small frequency interval $\Delta\nu$, the number of cavity modes is then $\Delta N \approx \beta_\nu V \Delta\nu$. This result will prove to be useful in Chapter 18 when we consider the interaction of atoms with the modes in a laser cavity.

16-2. MODE WIDTH

In the preceding analysis of 1-D resonator modes, we assumed that the modes were perfectly sharp, with well-defined frequencies given by Eq. (16-3). In practice, there is always some spectral broadening of the modes, due to the finite reflectivity of the mirrors. In optics textbooks, the spectral shape of the modes between two parallel mirrors is usually derived by considering the interference of the many reflected beams. For understanding the properties of laser cavities, however, more physical insight can be obtained by considering the time dependence of light intensity in the cavity, and then relating this to the frequency spectrum.

Photon Lifetime

We consider here a laser cavity with no optical gain, which is termed a *passive optical resonator.* Light that happens to be inside the resonator will bounce back and forth between the mirrors, losing energy at each bounce. The rate at which light intensity decays can be determined by considering the loss of intensity in one round-trip through the resonator. Assume that the light has initial intensity I at point A in the cavity, as shown in Fig. 16-5. After reflecting from the right mirror with reflection coefficient R_2, the intensity is R_2I, and after a further reflection from the left mirror the intensity is R_1R_2I. The change in intensity in one round-trip distance $2L$ is then

$$\begin{aligned} \Delta I &= I(t+\Delta t) - I(t) \\ &= I(t)[R_1R_2 - 1] \end{aligned} \tag{16-10}$$

where $\Delta t = 2L/c$ is the round-trip time. In this section we will take $n = 1$ for simplicity, but the results can be generalized by replacing $c \to c/n$ in each formula. The time rate of change in intensity is then

$$\frac{\Delta I(t)}{\Delta t} = -\frac{1 - R_1R_2}{2L/c} I(t) \tag{16-11}$$

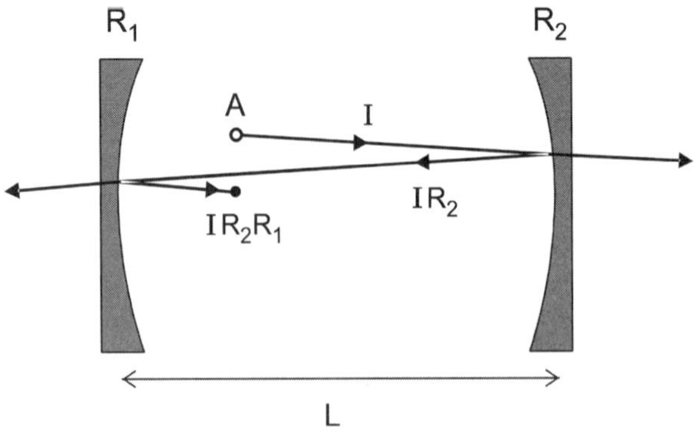

Figure 16-5 Light decreases in intensity during one round-trip through resonator due to mirror reflectivities R_1 and R_2 less than unity.

In laser cavities, the mirror reflectivities are usually high, so the fractional loss per round-trip is $\ll 1$. In this case, $I(t)$ can be approximated as a continuous function, and Eq. (16-11) becomes

$$\frac{dI}{dt} = -\frac{1}{\tau_c} I(t) \tag{16-12}$$

where the *photon lifetime* or *cavity lifetime* τ_c is defined as the time for the light intensity to decay to $1/e$ of its initial value. For small loss per round-trip, we have

$$\tau_c \simeq \frac{2L}{c(1 - R_1 R_2)} \qquad \text{(photon lifetime)} \tag{16-13}$$

The solution of Eq. (16-12) is

$$I(t) = I_0 e^{-t/\tau_c} \tag{16-14}$$

which can be easily verified by substitution. The light intensity in the cavity decays exponentially in time, with a decay time equal to the photon lifetime τ_c. The measurement of this decay time is one method of making accurate determinations of mirror reflectivities close to 1. In the *ring-down technique,* a short pulse is sent into the cavity, and the light exiting the cavity is monitored versus time. Mirror reflectivities are determined from the measured cavity lifetime using Eq. (16-13).

The frequency spectrum of the modes is determined from the time decay using the time–frequency uncertainty relation. The time dependence of E is that of a damped sinusoid,

$$E(t) = E_0 e^{-(t/2\tau_c)} \cos \omega t \tag{16-15}$$

as illustrated in Fig. 16-6. The time constant for the decay of $E(t)$ is $2\tau_c$ because $E \propto \sqrt{I(t)}$ (Eq. 2-9). In Appendix B it is shown that this type of time decay is characterized by the uncertainty relation

$$\Delta\omega_{1/2}\, \tau_c \simeq 1 \qquad \text{(uncertainty relation)} \tag{16-16}$$

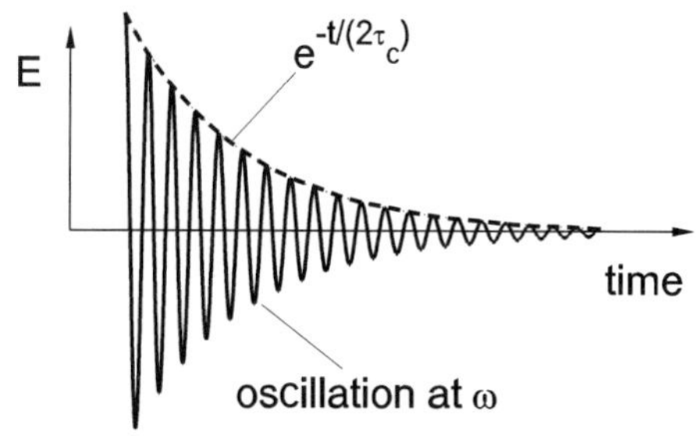

Figure 16-6 The E field oscillations in the cavity at angular frequency ω decay exponentially with time constant $2\tau_c$. In this plot, $\omega\tau_c = 5\pi$.

where $\Delta\omega_{1/2}$ is the angular frequency *full width at half maximum* (FWHM). Using $\Delta\nu_{1/2} = \Delta\omega_{1/2}/(2\pi)$, the frequency width of the modes can be written as

$$\Delta\nu_{1/2} \simeq \frac{1}{2\pi}(1 - R_1R_2)\frac{c}{2L} \qquad \text{(frequency width of mode)} \qquad (16\text{-}17)$$

This expression assumes high mirror reflectivities, and is accurate within ~ 10% for $R_1R_2 \gtrsim 0.80$. An expression valid for lower R is derived in Problem 16.6.

The frequency distribution of light intensity in the laser cavity is illustrated in Fig. 16-7, with the modes of width $\Delta\nu_{1/2}$ separated by $c/(2L)$. Although we have introduced them from the classical physics point of view, these cavity modes can be thought of as quantum states of the electromagnetic field. The photon, which is the quantum of the electromagnetic field, can be though of as "occupying" these cavity mode states, just as an electron occupies various quantum states in an atom or solid. From this viewpoint, the uncertainty relation in Eq. (16-16) becomes the Heisenberg uncertainty principle relating energy and time, $\Delta(\hbar\omega)\ \Delta t \simeq \hbar$. The energy of the photon $\hbar\omega$ is uncertain because it is uncertain when during the time τ_c the photon leaves the cavity.

Quality Factor *Q*

The time decay of the E field in an optical resonator is similar to that of a damped harmonic oscillator, and the terminology that is used to describe the sharpness of a resonance in the damped harmonic oscillator can also be applied to the optical resonator. The *quality factor Q* of a resonance is defined as the center frequency divided by the width, or

$$Q \equiv \frac{\nu}{\Delta\nu_{1/2}} \qquad \text{(quality factor of resonance)} \qquad (16\text{-}18)$$

which can be written here as

$$Q \simeq \frac{\nu(2L)(2\pi)}{(1 - R_1R_2)c} = \frac{4\pi L}{\lambda(1 - R_1R_2)} \qquad (16\text{-}19)$$

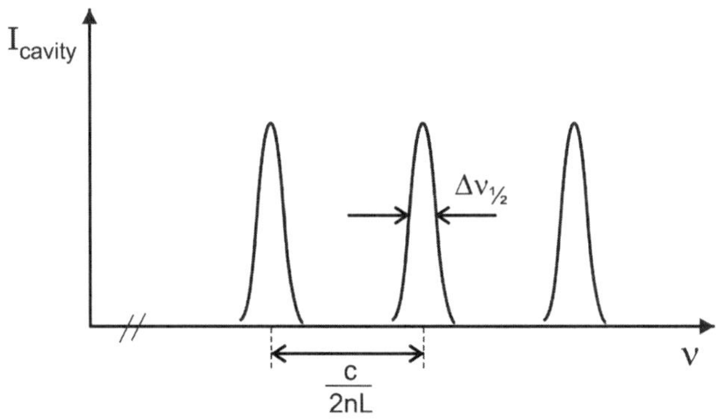

Figure 16-7 Cavity modes have width $\Delta\nu_{1/2}$ and spacing $c/(2L)$.

It is seen from Eq. (16-19) that the relative sharpness of the modes is greatest for very high reflectivity mirrors and long cavity lengths. To achieve very narrow laser linewidths, with correspondingly long coherence times T_c (see Chapter 15), the mirror reflectivities should be high. Most lasers in the visible and near IR regions use special mirrors made with multilayer dielectric thin films, rather than metallic mirrors, because they can be made to have a very high reflectivity over some range of wavelengths. Ordinary aluminum mirrors typically have $R \leq 0.90$, and are not often used in laser resonators.

Cavity Finesse

As the laser cavity length L increases, the modes become narrower, but the spacing between modes also decreases. A useful parameter that gives the mode width compared with the mode spacing is the *finesse,* defined by

$$\mathcal{F} \equiv \frac{\text{mode spacing}}{\text{mode width}} = \frac{c/(2L)}{\Delta\nu_{1/2}} \qquad \text{(finesse of cavity)} \tag{16-20}$$

Using Eq. (16-17), the finesse can be written as

$$\mathcal{F} \simeq \frac{2\pi}{1 - R_1R_2} \tag{16-21}$$

which is valid for high-reflectivity mirrors. Note that the finesse is independent of the cavity length, depending only on the mirror reflectivities. The finesse and cavity Q can be related using Eqs. (16-3), (16-19), and (16-21), giving

$$Q = \mathcal{F}\frac{\nu}{c/(2L)} = m\mathcal{F} \tag{16-22}$$

where m is the mode number. The three quantities Q, $\mathcal{F}$, and $\Delta\nu_{1/2}$ are thus equivalent ways of describing the spectral width of the cavity modes.

EXAMPLE 16-2

A He–Ne laser cavity is 1 m long with mirror reflectivities of 0.99, and operates at 632.8 nm. Determine the cavity Q, the finesse, the mode number, and the frequency width of a cavity mode in this laser. Assume the index of refraction is $n = 1$. Also, if the laser light were confined to a single cavity mode, what would be the coherence length of the light?

Solution: The frequency of the light is

$$\nu = \frac{3 \times 10^8}{632.8 \times 10^{-9}} = 4.74 \times 10^{14}\ \text{Hz}$$

and the mode number is

$$m = \frac{\nu}{c/(2L)} = \frac{2L}{\lambda} = \frac{2(1)}{632.8 \times 10^{-9}} = 3.16 \times 10^6$$

The finesse is

$$\mathcal{F} = \frac{2\pi}{1 - (0.99)^2} \simeq 316$$

and the quality factor is

$$Q = m\mathcal{F} \simeq (3.16 \times 10^6)(316) = 9.98 \times 10^8$$

The mode width can be found by either

$$\Delta\nu_{1/2} = \frac{\nu}{Q} = \frac{4.74 \times 10^{14}}{9.98 \times 10^8} = 4.75 \times 10^5 \text{ Hz}$$

or

$$\Delta\nu_{1/2} \simeq \frac{1}{2\pi}(1 - [0.99]^2)\,\frac{3 \times 10^8}{2(1)} = 4.75 \times 10^5 \text{ Hz}$$

The coherence length is

$$L_c = cT_c = \frac{c}{\Delta\nu_{1/2}} = \frac{3 \times 10^8}{4.75 \times 10^5} \simeq 630 \text{ m}$$

Ordinary He–Ne lasers oscillate on more than one mode, and the coherence length is much less than this.

16-3. FABRY–PEROT INTERFEROMETER

In the previous sections, we considered a pair of mirrors as a way of providing feedback for a laser, confining light inside the optical cavity. Another application for such a resonator is to act as an optical frequency filter, in which case it is called a *Fabry–Perot interferometer*. In this application, light of intensity I_{in} is incident externally on one side of the resonator, as in Fig. 16-8, and after multiple reflections within the cavity, light of intensity I_{out} exits through the other side. The transmission efficiency is defined as $T = I_{out}/I_{in}$, and varies with frequency as shown in Fig. 16-9. At the mode frequencies for the cavity, nearly all of the incident light is transmitted ($T \simeq 1$), whereas for frequencies off resonance very little light is transmitted. The full width at half maximum (FWHM) of each transmission peak is $\Delta\nu_{1/2}$ [Eq. (16-17)], which becomes very small for high-reflectivity mirrors. In effect, the Fabry–Perot is a narrow band pass optical filter, with a regular array of transmission peaks spaced by $c/(2L)$, giving a comb-shaped frequency spectrum.

It may seem puzzling at first that the transmission of the Fabry–Perot interfermometer can be 100% when the mirror reflectivites are very high, since these high mirror reflectivites should prevent most light from passing through. The resolution to this apparent paradox is to realize that at resonance, the light intensity inside the cavity builds up to a value much higher than that of the incident light. For example, if $R_1 = R_2 = 0.99$, then only 1% of the light I_{cav} circulating inside the cavity is transmitted through the output mirror R_2

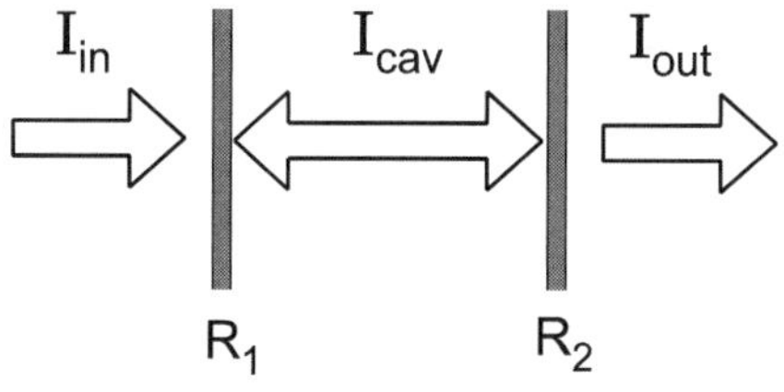

Figure 16-8 Transmission I_{out}/I_{in} through a Fabry–Perot interferometer is high at resonance, where the optical intensity I_{cav} inside the cavity builds up due to constructive interference of the multiple reflections.

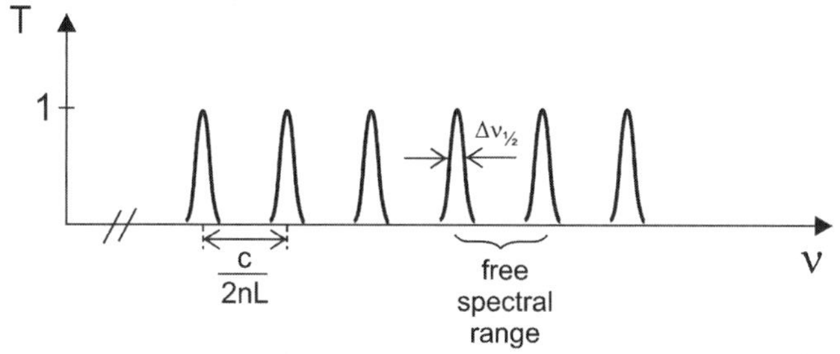

Figure 16-9 Transmission spectrum of the Fabry–Perot interferometer.

in each bounce. But if the light intensity inside the cavity at resonance builds up to a value $I_{cav} \simeq 100\, I_{in}$, then $I_{out} = (0.01)\, I_{cav} \simeq I_{in}$. Off resonance, the intensity inside the cavity remains low, and $T \ll 1$.

One application of the Fabry–Perot interferometer is in high-resolution optical spectroscopy. Fig. 16-10 shows a representative *fluorescence spectrum* (frequency distribution of emitted light) for an atomic transition, along with the Fabry–Perot transmission peaks in the vicinity of the fluorescence. If the emitted light is made to pass through the Fabry–Perot interferometer before detection, the detected signal will be the product of the fluorescence intensity I_f and the Fabry–Perot transmission T. The detected signal then corresponds to the part of the fluorescence spectrum that lines up with one or more of the cavity modes of the Fabry–Perot interferometer. The mode frequencies can be tuned continuously by varying L with a piezoelectric transducer, and the fluorescence spectrum can then be mapped out by scanning a single mode across the spectrum.

Light from different parts of the spectrum may be detected simultaneously in different *orders,* or mode numbers m, depending on the width of the fluorescence spectrum compared with the mode spacing $c/(2L)$. Since the mode spacing is the frequency range over which there are no interfering orders in the measured spectrum, it is also referred to as the *free spectral range.* If the medium between the mirrors has a refractive index n, the free spectral range is $c/(2nL)$. The mode frequencies can be swept by changing n as well as L.

EXAMPLE 16-3

A Fabry–Perot interferometer uses mirrors with reflectivity 0.99 spaced by 1 mm, with an air gap between them. (a) Determine the frequency resolution when measuring the sodium "D" spectral line at 589 nm. (b) Over what wavelength range is the measured spectrum free from overlapping orders?

Solution:

(a) From Eq. (16-17) the mode width is

$$\Delta\nu_{1/2} \simeq \frac{1}{2\pi}(1 - [0.99]^2)\frac{3 \times 10^8}{2(1 \times 10^{-3})} = 475 \text{ MHz}$$

(b) The free spectral range is

$$\Delta\nu_{FSR} = \frac{3 \times 10^8}{2(1 \times 10^{-3})} = 1.5 \times 10^{11} \text{ Hz} = 150 \text{ GHz}$$

In terms of wavelength this is

$$\Delta\lambda_{FSR} = \frac{\lambda^2}{c}\Delta\nu_{FSR} = \frac{(589 \times 10^{-9})^2}{3 \times 10^8}(1.50 \times 10^{11}) = 1.73 \times 10^{-10} \text{ m}$$

This Fabry–Perot interferometer can, therefore, only be scanned over 0.173 nm before overlapping orders appear in the spectrum. This example illustrates both the advantages and disadvantages of the Fabry–Perot interferometer. Very high resolution can be obtained, but at the expense of a limited scanning range.

PROBLEMS

16.1 Show that Eq. (16-5) follows from Eq. (16-4) using the trigonometric identity $\cos(A + B) = \cos A \cos B - \sin A \sin B$.

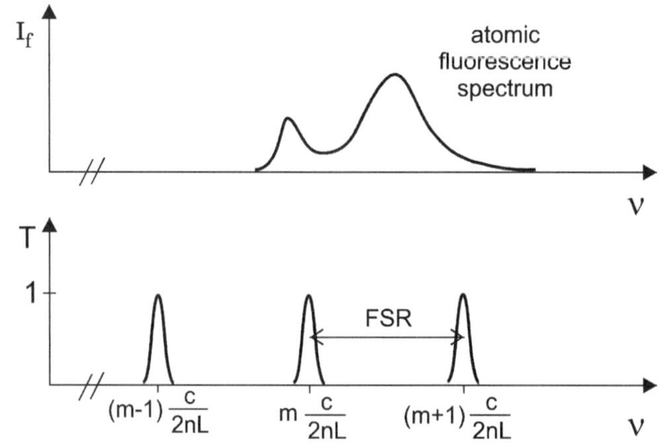

Figure 16-10 The Fabry–Perot interferometer can be used to analyze an optical frequency spectrum when the mode frequencies are continuously varied.

16.2 A GaAs diode laser has a cavity formed by the Fresnel reflections from the end facets of the GaAs chip, which is 0.8 mm long. If the laser wavelength (in air) is ≈ 850 nm, determine the approximate mode number and the mode spacing (in nm) for this laser.

16.3 A He–Ne laser cavity has a spacing of 15 cm between the mirrors, and the optical mode in the cavity has a diameter of ≈ 3 mm. (a) Determine the frequency difference between adjacent laser modes. (b) Determine the frequency difference between all possible cavity modes contained within the laser cavity volume. (c) The He–Ne gas mixture provides optical gain over a frequency width of ≈ 1.5 GHz. Compare the number of laser modes that are within this width to the total number of cavity modes within this width.

16.4 A ring-down measurement is made on an optical cavity with two identical high-reflectivity mirrors spaced by 45 cm in air. When a short pulse is sent into the cavity, the pulse intensity is observed to decay to 20% of its initial value in a time of 806 ns. Determine the mirror reflectivity to three significant figures.

16.5 In deriving Eq. (16-13) for the photon lifetime, it was assumed that the fractional loss per round trip is small, that is, $1 - R_1R_2 \ll 1$. If this condition does not hold, an alternative expression can be obtained that is valid for smaller R. (a) Show that $(R_1R_2)^p$ is the fraction of light remaining in the cavity after p complete round-trips, and set this equal to e^{-1} to show that the cavity lifetime (time required for the light to decay to e^{-1} of its initial value) is

$$\tau_c = \frac{2nL/c}{\ln(1/R_1R_2)}$$

(b) Show that this reduces to Eq. (16-13) for $R_1R_2 \approx 1$. (c) What is the percentage difference between the two expressions for τ_c when $R_1R_2 = 0.8$?

16.6 Using the expression for cavity lifetime from Problem 16.5, derive expressions for mode width, cavity Q, and cavity finesse that are valid for small R.

16.7 A semiconductor cavity is formed by cleaving the ends of a semiconductor chip so they are nearly parallel. Instead of external mirrors, the cavity relies on Fresnel reflection from the semiconductor–air interface. Assume an index of refraction 3.5, cavity length 0.8 mm, and laser wavelength 830 nm. (a) Use the results of Problem 16.5 to calculate the spacing and width of the longitudinal cavity modes (both in frequency and in wavelength). (b) Use the results of Problem 16.6 to calculate the Q and finesse of the cavity.

16.8 The semiconductor laser of Problem 16.7 is now modified to use external mirrors for the optical cavity. One mirror has $R = 0.98$, the other mirror has $R = 0.95$, and they are deposited directly on the ends of the semiconductor chip. Determine (a) the spacing between cavity modes, (b) the spectral width of the cavity modes, and (c) the cavity finesse.

16.9 Consider a variation of the laser of Problem 16.8, in which the two mirrors are freestanding and separated by 5 cm in air. The semiconductor (still of length 0.8 mm) between the mirrors is slightly tilted so that any Fresnel reflection from the semiconductor–air interface is lost from the cavity. Determine (a) the spacing between cavity modes, (b) the spectral width of the cavity modes, and (c) the cavity finesse.

16.10 An air-spaced Fabry–Perot interferometer has mirror spacing 0.15 mm and mirror reflectivities $R = 0.99$. It is used to measure the spectrum of the sodium doublet, which consists of two closely spaced emission lines at 588.995 and 589.592 nm. (a) Determine the mode number of the FP resonance. (b) By how much must the plate spacing be changed in order to scan a single mode from one of the lines to the other? (c) By how much must the plate spacing be changed so that the same emission line is seen again in a different order? If the plate spacing was increased, is the new mode number higher or lower? (d) What is the wavelength resolution of the resulting spectrum?

Chapter 17

Gaussian Beam Optics

In the last chapter, we concentrated on the confinement of laser light in the axial direction, along a line between the cavity mirrors. It was found that this confinement results in cavity modes that are standing waves in the longitudinal (axial) direction, with frequencies that depend on the mirror–mirror separation. In this chapter, we consider the distribution of light in the transverse direction, perpendicular to the cavity axis. It might be thought that to confine light in the transverse direction, mirrors would be needed along the sides of a laser cavity. We will see, however, that there is a solution to Maxwell's equations, the Gaussian beam, which provides a natural transverse confinement without the need for side mirrors. We will also explore the manipulation of these beams with lenses to focus or collimate the laser light.

17-1. GAUSSIAN BEAMS IN FREE SPACE

We begin by considering how a beam of light may propagate in free space, so that it might be naturally confined between the mirrors of a laser cavity. There are many possible solutions to Maxwell's equations in free space, the plane wave (Eq. 2-4) being the simplest. The plane wave has an infinite extent in the transverse direction, however, and is not a good candidate for the true 3-D modes in a laser cavity. We would like instead a solution that is confined laterally to some extent. In general, when a beam of light is confined to a diameter D in the transverse direction, it spreads out with a divergence angle $\theta \sim \lambda/D$ due to diffraction, as illustrated in Fig. 17-1.

The exact angular distribution of the diffracted light depends on the manner in which the beam's intensity goes to zero in the transverse direction. If the beam's intensity cuts off sharply, as it would, for example, when a plane wave passes through a circular aperture, the light intensity far from the aperture undergoes oscillations in the transverse direction, with the angle to the first minimum given by Eq. (2-26). These oscillations can be reduced by *apodization* of the aperture, that is, making the intensity transmitted through the aperture cut off more gradually. It turns out that if the electric field falls off as the Gaussian function $\exp(-r^2/w^2)$, with r the radial distance perpendicular to the z axis, the oscillations are completely eliminated. The parameter w is the value of r where the Gaussian function is a factor of $1/e$ times its maximum, as shown in Fig. 17-2. A beam having a Gaussian profile at one location will have a Gaussian profile for all positions z along the direction of propagation, and is called a *Gaussian beam.* The Gaussian beam has the smallest possible angular spread for a beam of a given initial diameter, and is the most fundamental light distribution produced by a laser. Because the

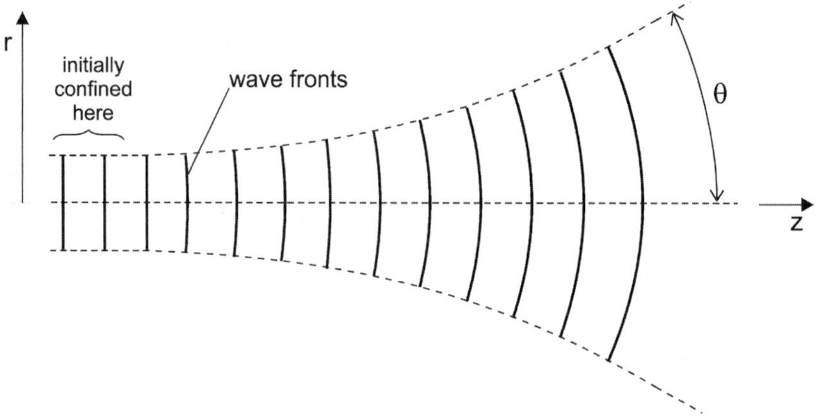

Figure 17-1 Diffraction of a beam that is initially confined laterally.

light is coherent and spreads out only due to diffraction, the beam is referred to as *diffraction limited.*

Intensity Distribution

The spatial distribution of the electric field magnitude for the Gaussian beam is given by

$$E(r, z) = E_0 \frac{w_0}{w(z)} e^{-r^2/w^2(z)} \qquad \text{(Gaussian beam)} \qquad (17\text{-}1)$$

where $r = \sqrt{x^2 + y^2}$ is the radial distance from the z axis, and z is the distance along the direction of propagation. The parameter $w(z)$ is the *spot size,* which is (loosely speaking) the "radius" of the beam at the position z. As the beam spreads out, the curvature of the wave fronts also changes, as shown in Fig. 17-1. A given wave front can be considered to be approximately spherical for small distances from the z axis (the *paraxial approximation*),

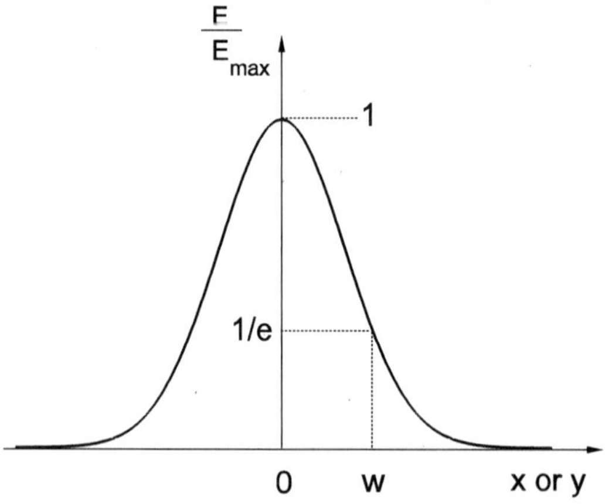

Figure 17-2 Variation of electric field in the radial direction for a Gaussian beam.

with a radius of curvature $R(z)$. The spot size and wave front radius can be shown to vary with propagation distance z as (Siegman 1986)

$$w^2(z) = w_0^2\left[1 + \left(\frac{z}{z_0}\right)^2\right] \tag{17-2}$$

$$R(z) = z\left[1 + \left(\frac{z_0}{z}\right)^2\right] \tag{17-3}$$

where the *beam waist* w_0 is the minimum value of the spot size $w(z)$, located at $z = 0$. The parameter z_0 is termed the *Rayleigh range,* defined as

$$z_0 = \frac{\pi w_0^2}{\lambda} \qquad \text{(Rayleigh range)} \tag{17-4}$$

Only two parameters are needed to completely specify a Gaussian beam of a given wavelength: the beam waist size w_0 and the location of the beam waist along the z axis. The variation of spot size with z is illustrated in Fig. 17-3, showing the key parameters. At $z = 0$ the spot size has its minimum value w_0, and it increases to $w = w_0\sqrt{2}$ at $z = z_0$. Note that since the area of the beam is $\propto w^2$, the area increases to twice its minimum value at the Rayleigh range distance.

For $z \gg z_0$, Eq. (17-2) becomes $w(z) \simeq w_0\, z/z_0$, which leads to a linear divergence angle $\theta = w(z)/z \simeq w_0/z_0$. This can be written using Eq. (17-4) as

$$\theta \simeq \frac{\lambda}{\pi w_0} \qquad \text{(Gaussian beam divergence)} \tag{17-5}$$

which is the half-cone divergence angle defined in Fig. 17-3. Eq. (17-5) should be compared to the similar expressions in Eqs. (2-25) and (2-26) for the divergence of a diffraction-limited beam of diameter D. The advantage of the formula in Eq. (17-5) is that the beam edges have been defined precisely by the spot size w in Eq. (17-1), which allows a quantitative treatment of beam spreading.

Peak Intensity

It is useful to relate the peak intensity on axis to the total power propagating in the Gaussian beam. Since intensity is power per unit area, the total power is found by integrating the in-

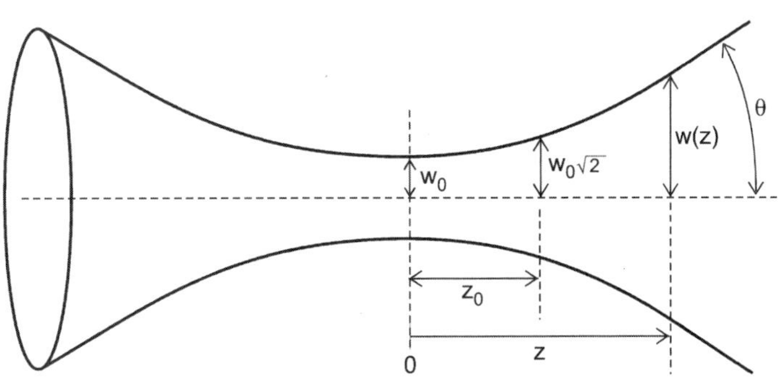

Figure 17-3 Variation of beam width with z for Gaussian beam.

tensity over the area of the beam for a particular value of z. For a circularly symmetric beam, the integration can be performed by considering concentric rings of radius r and thickness dr, as shown in Fig. 17-4. Using Eqs. (2-9) and (17-1), the power can be written as

$$\begin{aligned} P &= \int I(r,z)\, dA \\ &= \int_0^\infty I(r,z)(2\pi r)\, dr \\ &= 2\pi \frac{1}{2} cn\varepsilon_0 \int_0^\infty E^2(r,z)\, r\, dr \\ &= \pi cn\varepsilon_0 E_0^2 \frac{w_0^2}{w^2(z)} \int_0^\infty e^{-2r^2/w^2(z)}\, r\, dr \end{aligned} \tag{17-6}$$

where n is the refractive index of the medium. The integral in Eq. (17-6) is easily evaluated using the substitution $u \equiv 2r^2/w^2(z)$, with the result

$$P = \frac{\pi}{2} w_0^2 \frac{1}{2} cn\varepsilon_0 E_0^2 \tag{17-7}$$

Note that the total beam power is independent of z, since only the constants E_0 and w_0 appear in Eq. (17-7). This is to be expected, since the energy in the beam is not disappearing, but simply spreading out as it propagates. Since E_0 is the field amplitude at the center of the beam waist ($r = 0$, $z = 0$), the intensity which occurs there has the value given by Eq. (2-9) as

$$I_{max} = \tfrac{1}{2} cn\varepsilon_0 E_0^2 \qquad \text{(center of beam waist)} \tag{17-8}$$

Combining this with Eq. (17-7) gives I_{max} in terms of the beam power,

$$I_{max} = \frac{P}{\frac{1}{2}\pi w_0^2} \tag{17-9}$$

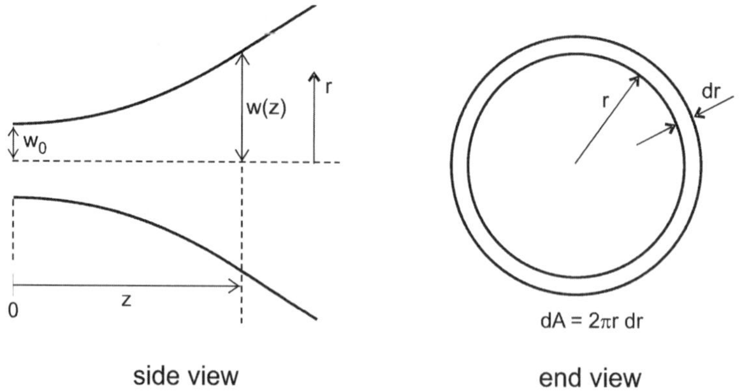

Figure 17-4 Differential area for integrating beam intensity is a ring of radius *r* and thickness *dr*, with area $2\pi r\, dr$.

The beam intensity on axis can be determined for arbitrary z by setting $r = 0$ in Eq. (17-1):

$$E(0, z)w(z) = E_0 w_0 \tag{17-10}$$

where $E(0, z)$ is the field amplitude on axis. Combining this with Eqs. (17-7) and (2-9) gives for the on-axis intensity

$$I(0, z) = \frac{P}{\frac{1}{2}\pi w^2(z)} \quad \text{(on-beam axis)} \tag{17-11}$$

For $z = 0$ this reduces to Eq. (17-9), where $I(0, 0) = I_{max}$.

Since intensity is power per unit area, the denominator of Eq. (17-11) can be interpreted as an effective area for the beam at position z. The spot size w is often taken loosely as the "beam radius," in which case the "area" of the beam would be πw^2, so the effective area is half this value. Alternatively, the "radius" of the beam might be considered to be $a = w/\sqrt{2}$, since the intensity falls off with r as $\exp(-2r^2/w^2) = \exp(-r^2/a^2)$. In that case, the effective area of the beam is the same as the "beam area" πa^2, an intuitively satisfying result.

17-2. GAUSSIAN BEAMS IN A LASER CAVITY

We have seen that the Gaussian beam is a good candidate for the distribution of light in a laser cavity, being self-confined in the transverse direction. The question that we ask now is: what particular Gaussian beam will be produced by a given laser cavity? The answer, very crudely, is "the one that fits in the cavity." By "fit," we mean that the wave fronts line up with the mirror surfaces at the cavity ends, as shown in Fig. 17-5. A beam that is diverging when it strikes one of the mirrors then becomes a converging beam after reflection, exactly retracing the incident beam profile and creating a standing wave. Taking r_1 and r_2 to be the radii of curvatures of the left and right mirrors, this condition amounts to requiring that the Gaussian beam radius of curvature R [Eq. (17-3)] be equal to r_1 at the left mirror, and r_2 at the right mirror. To simplify the treatment, we will consider first the symmetric resonator in which $r_1 = r_2$.

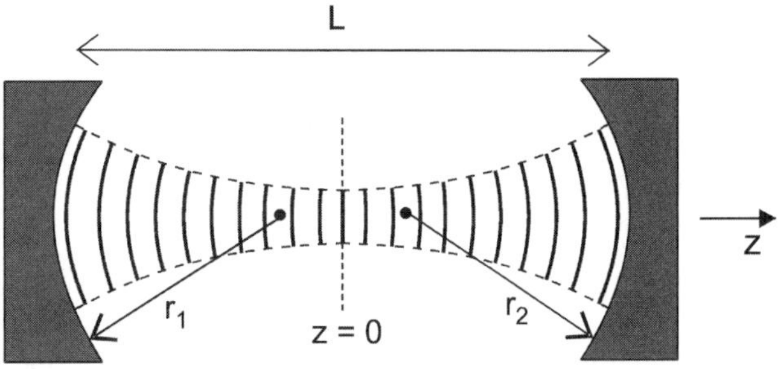

Figure 17-5 A Gaussian beam "fits" inside a laser cavity when the wave fronts match up with the mirror surfaces.

Stability Criterion in a Symmetric Resonator

For the symmetric resonator, $r_1 = r_2 \equiv r$, where r is positive for concave mirrors (as shown in Fig. 17-5). The beam waist is located in the center, a distance $L/2$ from each mirror. Setting $z = L/2$ in Eq. (17-3), the wave front matching condition becomes

$$R(L/2) = \frac{L}{2}\left[1 + \left(\frac{z_0}{L/2}\right)^2\right] = r \tag{17-12}$$

which can be solved for the Rayleigh range of the beam,

$$z_0 = \frac{L}{2}\sqrt{\frac{2r}{L} - 1} \tag{17-13}$$

The beam waist w_0 can be determined from Eq. (17-4):

$$z_0 = \frac{\pi w_0^2}{\lambda} = \frac{L}{2}\sqrt{\frac{2r}{L} - 1}$$

which gives

$$w_0^2 = \frac{\lambda L}{2\pi}\sqrt{\frac{2r}{L} - 1} \qquad \text{(spot size at center)} \tag{17-14}$$

Note that Eqs. (17-13) and (17-14) only give a real result if $2r > L$. A cavity with $2r < L$ does not support a Gaussian beam mode, and is termed an *unstable resonator.* Such resonators are not generally useful for continuous wave (CW) lasers, although they have some applications for pulsed lasers.

The stability of cavity modes can be further evaluated by calculating the beam spot size w at the position of the mirrors. At $z = L/2$, Eq. (17-2) gives

$$w^2(L/2) = w_0^2\left[1 + \left(\frac{L}{2z_0}\right)^2\right]$$

$$= w_0^2\left[1 + \frac{1}{\frac{2r}{L} - 1}\right]$$

which can be manipulated to give

$$w^2(L/2) = \frac{\lambda L}{2\pi}\frac{2r/L}{\sqrt{\frac{2r}{L} - 1}} \qquad \text{(spot size on mirrors)} \tag{17-15}$$

Again, a real value of w is obtained only for $r > L/2$. Fig. 17-6 shows how the spot sizes from Eqs. (17-14) and (17-15) vary with the ratio $r/(L/2)$. As $r/(L/2) \to 1$, the spot size on the mirrors gets very large, while the waist size gets very small. Such a cavity is termed *concentric,* because the centers of curvature for the two mirrors coincide at the middle of the cavity. When the spot size on the mirrors becomes much larger than the physical size of the mirrors, the cavity mode will have high loss, and is only marginally stable.

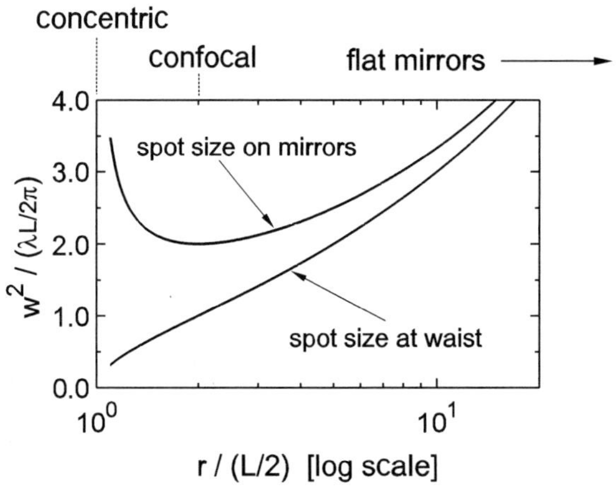

Figure 17-6 Square of Gaussian beam spot size on the mirrors of a symmetric cavity and at the beam waist, versus the mirror radius of curvature. The square of the spot size is normalized to $\lambda L/(2\pi)$, and the radius is normalized to $L/2$.

The spot size on the mirrors also becomes very large as $r \to \infty$, which corresponds to flat mirrors. In fact, both $w^2(L/2)$ and w_0^2 tend to the same limiting value of $(\lambda/\pi)\sqrt{rL/2}$ as $r \to \infty$. A cavity with perfectly flat mirrors will be very lossy because the mode width is much larger than the mirror diameters.

In between the two extremes of concentric and flat-mirror cavities, there are many choices for $r/(2L)$ that result in stable, low-loss laser cavities. A cavity with $r = L$ is the most robust, having the smallest possible spot size on the mirrors for a given cavity length. This is termed the *confocal cavity,* since the focal points for the two mirrors coincide (the focal length of a curved mirror being $r/2$). For this condition, the cavity length is twice the Rayleigh range, $2z_0 = L$, and the parameter $b = 2z_0$ is, therefore, sometimes referred to as the *confocal parameter.*

In the confocal cavity, the beam waist is $w_0 = \sqrt{\lambda L/(2\pi)}$, and the spot size on the mirrors is $w(L/2) = \sqrt{\lambda L/\pi}$. If the mirror diameter $2a$ (radius a) is sufficiently large, most of the light in the Gaussian beam will be reflected by the mirrors, resulting in low loss. A measure of the relative mirror diameter is given by the *Fresnel number* $N_F \equiv a^2/(\lambda L)$. Cavities with $N_F > 1$ have little loss due to diffraction around the mirror edges.

Stability Criterion in an Asymmetric Resonator

If the laser cavity has mirrors with different radii of curvature, a similar method can be used to find the Gaussian beam that "fits" into the cavity. In this case, the beam waist will no longer be in the middle, and there are now two parameters to solve for: the position and size of the beam waist. Two equations analogous to Eq. (17-12) can be written down to solve for these two unknowns, one for each mirror. The analysis is rather messy and does not provide much insight, so we simply quote the results here (Saleh and Teich 1991). Defining the g parameter for each mirror as

$$\begin{aligned} g_1 &\equiv 1 - L/r_1 \\ g_2 &\equiv 1 - L/r_2 \end{aligned} \tag{17-16}$$

the condition for stability becomes

$$0 \le g_1 g_2 \le 1 \qquad \text{(stability condition)} \tag{17-17}$$

where r_1 and r_2 are taken as positive for concave mirrors. In the case of a symmetrical resonator, this condition reduces to $g^2 < 1$, which is equivalent to $2r > L$ (see Problem 17.8). The concentric cavity corresponds to $g = -1$, the confocal cavity to $g = 0$, and the flat mirror cavity to $g = 1$.

Another special case is a cavity with one flat mirror and one concave mirror, shown in Fig. 17-7. In this case, $g_1 = 1$, so the stability condition in Eq. (17-17) reduces to $0 \le g_2 \le 1$, which is equivalent to $r_2 > L$ (see Problem 17.8). Some insight can be gained by considering this to be one-half of an equivalent symmetric resonator of length $L' = 2L$, as shown in the lower part of Fig. 17-7. The Gaussian beam solutions for the two situations are expected to be the same, because the boundary conditions (conditions on the wave front curvature) are the same. The Gaussian beam parameters for the resonator in the top part of Fig. 17-7 can thus be obtained from Eqs. (17-13), (17-14), and (17-15), by making the substitution $L \to L' = 2L$. Real solutions are then obtained only for $r_2 > L$, in agreement with the stability condition derived from Eq. (17-17).

Higher-Order Modes

So far, we have considered only a single solution of Maxwell's equations, the Gaussian beam, which has the smoothest possible variation in intensity perpendicular to the beam

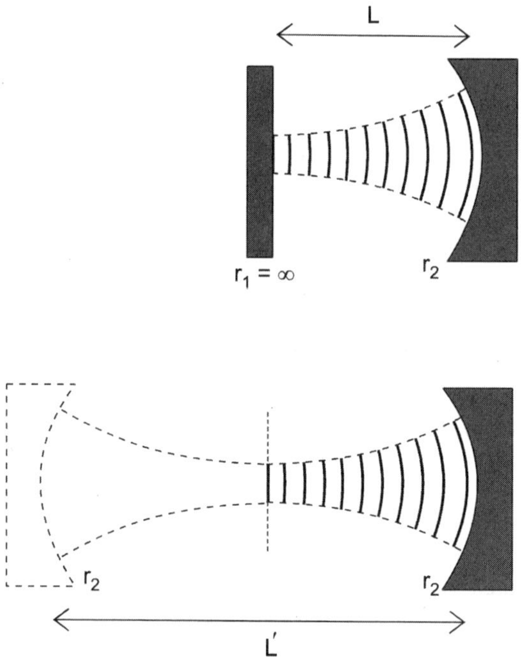

Figure 17-7 An asymmetric resonator of length L with plane and curved mirrors is equivalent to a symmetric resonator of length $L' = 2L$.

axis. The Gaussian beam is actually just one of a class of *transverse electromagnetic* (TEM) waves, known as *Hermite–Gaussian modes,* which are also solutions of Maxwell's equations in a laser cavity. Since these modes are to be confined in three dimensions inside the cavity, there should be a unique set of three integer labels for each mode [see Eqs. (16-6)]. One label is the longitudinal mode number, designated here as the integer number q, which characterizes the confinement along the resonator axis. This corresponds to the integer m used in our previous 1-D treatment [see Eq. (16-3)], and gives the 1-D mode frequencies $\nu = qc/(2L)$ (as usual, replacing $c \to c/n$ for refractive index n). The other two integers l and m characterize the distribution of light perpendicular to the cavity axis, with the transverse modes labeled TEM_{lm}

The Hermite–Gaussian modes are similar to Gaussian beams, except that the electric field amplitude is modulated in the transverse direction according to

$$E(x, y) = AH_l\left(\frac{\sqrt{2}}{w}\,x\right)H_m\left(\frac{\sqrt{2}}{w}\,y\right)e^{-(x^2+y^2)/w^2} \tag{17-18}$$

where w is the spot size at position z, and the $H_m(u)$ are *Hermite polynomials.* For our purpose, it is sufficient to know that these are well-known polynomials of order m, which are the solution to a particular differential equation. For example, these functions turn up in the solution of the harmonic oscillator problem in quantum mechanics. The first few Hermite polynomials are

$$\begin{aligned} H_0(u) &= 1 \\ H_1(u) &= 2u \\ H_2(u) &= 4u^2 - 2 \\ H_3(u) &= 8u^3 - 12u \end{aligned} \tag{17-19}$$

The lowest-order mode, TEM_{00}, is just the same Gaussian mode we have considered previously, since $H_l(u) = H_m(u) = H_0(u) = 1$. The next-highest mode, TEM_{10}, has its Gaussian envelope multiplied by $H_1(x\sqrt{2}/w) \propto x$, giving a double-peaked structure as shown in Fig. 17-8. This transverse profile has one zero-intensity point ($x = 0$) and two intensity maxima. In general, for a mode of order l in the x direction, there are l points with

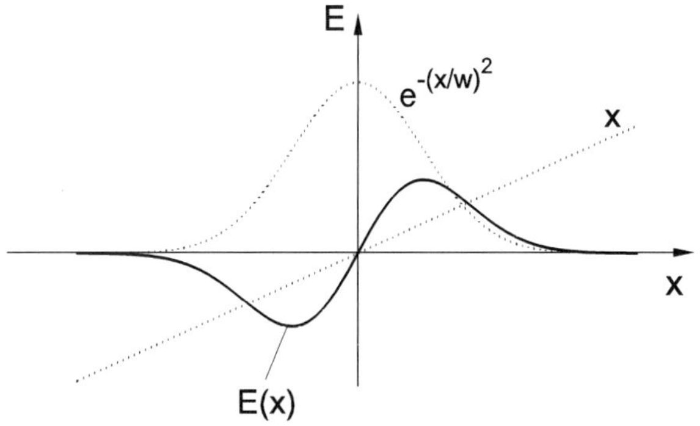

Figure 17-8 Multiplying a Gaussian function by $H_1(x\sqrt{2}/w) \propto x$ creates a double-peaked intensity distribution. This is an example of a Hermite–Gaussian mode.

intensity zeros and $l + 1$ points with intensity maxima. Similarly there are m zeros and $m + 1$ maxima in the y direction. Representative sketches for a few low-order Hermite–Gaussian profiles are given in Fig. 17-9.

The frequencies of the allowed Hermite–Gaussian modes depend on q, l, and m according to

$$\nu_{qlm} = \frac{c}{2L}\left[q + \frac{1+l+m}{\pi}\cos^{-1}(g_1 g_2)^{1/2}\right] \tag{17-20}$$

(see, for example, Hawkes and Latimer 1995), with g_1 and g_2 defined in Eq. (17-16). The TEM_{00} mode still has longitudinal modes spaced by $c/(2L)$, as in the 1-D treatment, although there is a small frequency shift that is the same for each mode. Since the mode number q for laser light is usually very large, this shift generally has no practical effect.

Higher-order (transverse) modes with $l \neq 0$ and/or $m \neq 0$ give rise to additional mode frequencies that depend on the curvature of the mirrors. In the limiting case of flat mirrors, where $g_1 = g_2 = 1$, the allowed frequencies become $\nu_{qlm} = qc/(2L)$, identical to the 1-D result and independent of l and m. The higher-order modes may still be present here, but they all have the same frequency as the lowest-order (Gaussian) mode. Modes such as this, which have different spatial distributions but the same frequency, are termed *degenerate* modes.

As the mirrors are made slightly curved, the degeneracy of the transverse modes is lifted, and modes with $l, m \neq 0$ appear at slightly higher frequency than the corresponding Gaussian modes. The mode spectrum is illustrated in Fig. 17-10, with the Gaussian (longitudinal) modes separated by $c/(2L)$. Continuing to increase the mirror curvature causes the spacing between transverse modes to increase, eventually reaching a spacing of $c/(4L)$ for the confocal resonator condition $g_1 = g_2 = 0$. In this case, modes with different l and m overlap (are degenerate with) modes with different q, and it becomes difficult to identify the modes by their frequency spectrum.

In some ways, the spectrum for the confocal cavity is quite simple, since all adjacent resonator frequencies are separated by $c/(4L)$, half the mode separation obtained in the 1-D treatment. However, this simplicity belies the complication that there are different combinations of Hermite–Gaussian modes that can lead to the same spectrum. Also, it is still true that the spacing between adjacent longitudinal modes (where q differs by 1) is $c/(2L)$.

The higher-order modes are wider than the fundamental Gaussian beam, even though the spot size w is the same. This is because the Hermite polynomials act as weighting factors that distort the Gaussian profile so as to enhance the part of the beam away from the axis. The effective beam radius w_{eff} will be a factor of M larger than the spot size w, with

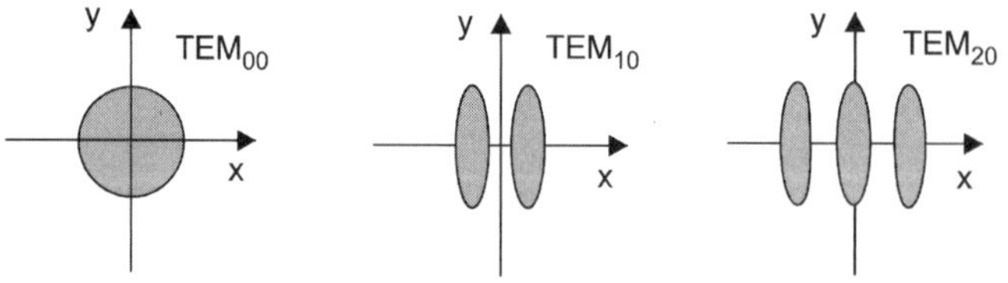

Figure 17-9 Representative intensity distributions for TEM_{lm} modes.

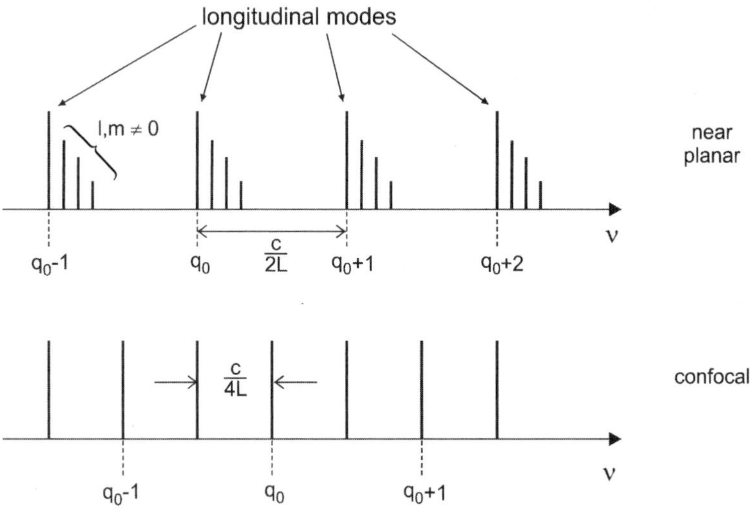

Figure 17-10 Frequency spectrum for near-planar and confocal laser cavities. For the confocal cavity, a given frequency may have contributions from more than one *qlm* mode.

M getting larger as the transverse order of the mode increases. In the far field ($z \gg z_0$), after the beam has left the laser resonator, the spot size is

$$w \simeq \frac{\lambda}{\pi w_0} z \tag{17-21}$$

where w_0 is the waist size of the beam. Since the effective spot size at the waist is also enhanced by a factor of M, we can write

$$w_{\text{eff}} = Mw(z)$$
$$w_{0,\text{eff}} = Mw_0$$

Combining the above with Eq. (17-21) then yields

$$w_{\text{eff}} \simeq M^2 \frac{\lambda}{\pi w_{0,\text{eff}}} z$$

Comparing this with Eq. (17-21), it can be seen that the multimode beam diverges a factor M^2 more strongly than would be expected for a Gaussian beam with the same effective waist size. This can be expressed in terms of an effective divergence angle for the multimode beam, $\theta_{\text{eff}} = w_{\text{eff}}/z$, which is

$$\theta_{\text{eff}} \simeq M^2 \frac{\lambda}{\pi\, w_{0,\text{eff}}} \quad \text{(divergence of multimode beam)} \tag{17-22}$$

The M^2 parameter was introduced in Chapter 15 as a way of describing the divergence of partially coherent light. We see here a related application for this parameter, as a way to characterize the divergence of a multimode laser beam in terms of its effective beam waist. This parameter is often quoted in the specifications for commercial lasers, to show how close the laser beam comes to being "diffraction limited."

17-3. GAUSSIAN BEAMS PASSING THROUGH A LENS

Once light comes out of a laser cavity, it is important to be able to manipulate it for a particular application. Generally, applications fall into one of two broad categories: (1) those in which the light is to be concentrated at a point (a focused beam), and (2) those in which the light is to travel in a straight line for a long distance (a collimated beam). To achieve either of these results, the light may be passed through a lens or reflected off a curved mirror. Since a mirror with radius of curvature r has the same effect as a lens of focal length $f = r/2$, our discussion can be confined to the effect of lenses, without loss of generality.

A Gaussian beam incident on a lens from the left, as shown in Fig. 17-11, will in general be transformed into a different Gaussian beam to the right of the lens. To completely specify the new Gaussian beam, two things must be known: the new beam waist size w_{02}, and the distance d_2 between the lens and the beam waist. These two parameters can be determined if both the beam spot size w_2 and the wave front radius of curvature R_2 are both known just after the lens. It is clear that the spot sizes just before and after the lens are equal, $w_2 = w_1$, because otherwise the flow of optical energy would be discontinuous, with energy suddenly appearing or disappearing. It only remains, then, to see how the wave front radius of curvature changes as it passes through the lens.

To see how a lens modifies an incident wave front, consider a point source of light at a distance s_1 to the left of a thin lens, as shown in Fig. 17-12. Spherical waves radiate outward from this point source, and have a radius of curvature R_1 upon reaching the lens. After passing through the lens, the radius of curvature of the wavefront becomes R_2, and the wave converges to a point at a distance s_2 to the right of the lens. We adopt a sign convention in which R is positive for a diverging wavefront, and negative for a converging wavefront, consistent with the sign of R in Eq. (17-3) when the wave is moving in the $+z$ direction. With this sign convention, $R_1 = s_1$ and $R_2 = -s_2$. The object distance s_1 and image distance s_2 for a lens of focal length f are related by the *lens equation,* given earlier in Eq. (2-32) as

$$\frac{1}{s_1} + \frac{1}{s_2} = \frac{1}{f} \qquad \text{(lens equation for imaging)} \qquad (17\text{-}23)$$

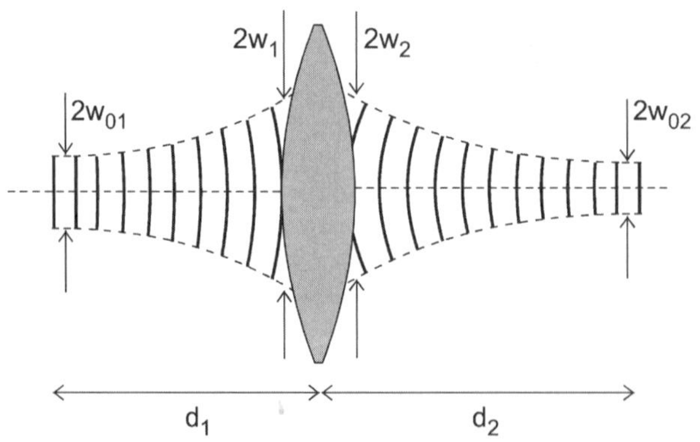

Figure 17-11 A lens transforms one Gaussian beam into another Gaussian beam.

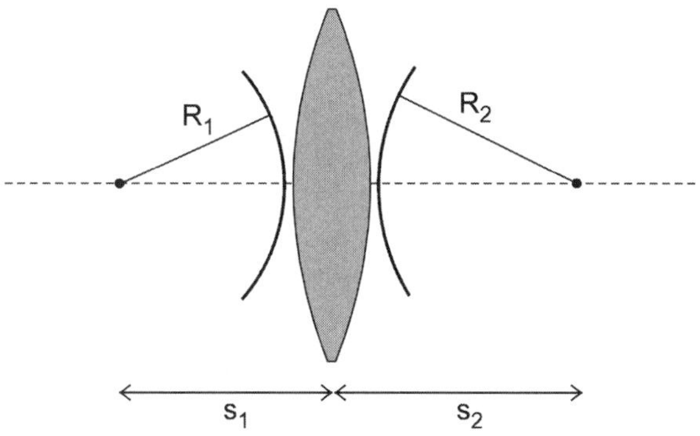

Figure 17-12 A point source a distance s_1 from a thin lens is imaged onto a point a distance s_2 to the right of the lens.

which can be written in terms of wave front curvatures as

$$\frac{1}{R_1} - \frac{1}{R_2} = \frac{1}{f} \quad \text{(lens equation for wavefronts)} \tag{17-24}$$

The effect of a thin lens on a wave front is, therefore, to change the radius of curvature according to Eq. (17-24). A diverging beam with positive R_1 becomes less diverging after passing through a weak positive lens ($f > 0$), and will become converging for a sufficiently strong positive lens (f sufficiently small). Negative lenses ($f < 0$) cause the divergence to increase, or if sufficiently strong can cause a converging beam to become diverging. In each case, Eq. (17-24) can be applied to find the wave front curvature after the lens, provided that the sign convention for R is followed.

The general procedure for finding the new Gaussian beam parameters in terms of those of the incident beam consists of three parts. First, Eqs. (17-2) and (17-3) are used to determine the spot size w_1 and wave front curvature R_1 just to the left of the lens. Next, Eq. (17-24) and the condition $w_2 = w_1$ are used to determine w_2 and R_2 just to the right of the lens. Finally, Eqs. (17-2) and (17-3) are used again, to determine the beam waist size w_{02} and its distance d_2 from the lens. For the general case, the algebra becomes very messy with this approach, and it is most suitable for computer calculations. To obtain some insight and develop useful simplified equations, we consider next an approximate treatment of beam focusing and beam collimation.

Gaussian Beam Focusing

Consider a collimated beam with plane parallel wave fronts incident on a lens of radius a, as shown in Fig. 17-13. If the incident beam does not fill the area of the lens, then a will be taken as the incident beam waist w_{01} just before the lens. The light is brought to a focus at a distance d_2 to the right of the lens, where a new beam waist w_{02} is created. We make the approximation that $w_{02} \ll a$, a reasonable assumption since the whole point of the focusing is to make the waist size w_{02} as small as possible. In that case, Eq. (17-2) evaluated at $z = -d_2$ with respect to the second beam waist becomes

$$a^2 = w_{02}^2\left[1 + \left(\frac{-d_2}{z_{02}}\right)^2\right]$$

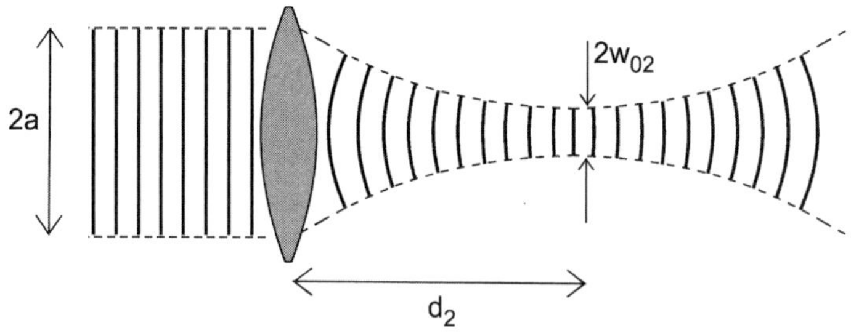

Figure 17-13 Collimated light is brought to a focus at a distance d_2 from a lens of focal length f.

where $z_{02} = \pi w_{02}^2/\lambda$ as in Eq. (17-4). Since $w_{02} \ll a$, the 1 in the above equation can be neglected compared with $(d_2/z_{02})^2$, resulting in

$$\begin{aligned} a &= w_{02}\left(\frac{d_2\lambda}{\pi w_{02}^2}\right) \\ &= d_2\frac{\lambda}{\pi w_{02}} \end{aligned} \tag{17-25}$$

We expect that d_2 is approximately the focal length f of the lens, since the focal length is defined in geometric optics as the distance from the lens to the point where parallel rays would be focused. To get this result from the Gaussian beam equations, Eq. (17-24) is first used to find the wave front curvature R_2 just after the lens. Since $R_1 = \infty$ (planar wave fronts), this gives $R_2 = -f$. Eq. (17-3) is then evaluated at $z = -d_2$ to give

$$\begin{aligned} -f &= -d_2\left[1 + \left(\frac{z_{02}}{d_2}\right)^2\right] \\ f &\simeq d_2 \end{aligned} \tag{17-26}$$

where the approximation $d_2 \gg z_{02}$ has again been used.

An expression for the size of the new beam waist can now be obtained by combining Eqs. (17-25) and (17-26),

$$w_{02} \simeq \frac{\lambda f}{\pi a} \quad \text{(waist size at focus)} \tag{17-27}$$

This simple but very useful result gives the spot size produced when a beam of wavelength λ and radius a is focused by a lens of focal length f. It is valid provided that $w_{02} \ll a$, or $\lambda f \ll \pi a^2$, generally an excellent approximation.

It is often the goal to obtain the smallest focus spot size possible. According to Eq. (17-27), this is achieved by using a short wavelength of light, a short-focal-length lens, and a large beam radius and lens diameter (note that a is the *smaller* of these two). Assuming that the incident beam can be made as large as desired by collimation (see next section), the limiting value of a will be $D/2$, where D is the lens diameter. The minimum spot size at the focus is then

$$w_{02} \simeq \frac{2\lambda f}{\pi D} = \frac{2}{\pi}\lambda F\# \tag{17-28}$$

where F# ≡ f/D is known as the *F number*. In a camera in which the limiting aperture has diameter D, this is also referred to as the *f stop,* designated as f/5 for F# = 5, for example. Eq. (17-28) says that, fundamentally, the minimum spot size is determined by the wavelength of light, and the F# of the lens. Small F numbers (below 2) are difficult to make without significant abberations, and the paraxial (small angle) approximation starts to break down. Since the paraxial approximation is always an underlying assumption in Gaussian beam optics, Eq. (17-28) must be used with some caution for F# < 2.

EXAMPLE 17-1

A beam from an argon laser has a diameter of 1.0 mm, and is focused by a 10 cm focal length lens with diameter 2.5 cm. Determine the spot size at the focus of the lens. Repeat the calculation if the beam is first expanded to fill the entire lens area. The wavelength of the light is 514.5 nm.

Solution: Focusing the original beam gives

$$w_{02} = \frac{(514.5 \times 10^{-9})(0.10)}{\pi(0.5 \times 10^{-3})} = 3.3 \times 10^{-5} \text{ m} = 33 \ \mu\text{m}$$

If the beam is first expanded to diameter 25 mm,

$$w_{02} = \frac{(514.5 \times 10^{-9})(0.10)}{\pi(12.5 \times 10^{-3})} = 1.3 \times 10^{-6} \text{ m} = 1.3 \ \mu\text{m}$$

This illustrates the importance of utilizing the entire lens area for achieving the smallest spot size at the focus.

Certain applications require that the light be focused to a particular spot size, which is not necessarily the minimum possible. For example, in coupling laser light into the core of an optical fiber, the spot size of the focused Gaussian beam should match the spot size of the (approximately) Gaussian beam profile of the fiber. This is referred to as *mode matching,* and results in a high coupling efficiency.

EXAMPLE 17-2

Laser light at 1500 nm is to be coupled into a fiber using a lens of focal length 6 cm. The fiber is step index with core radius 2.5 μm and numerical aperture 0.22. Determine the beam diameter incident on the lens that will give the best coupling efficiency.

Solution: The V parameter for this fiber at 1500 nm is

$$V = \frac{2\pi(2.5)}{1.5}(0.22) = 2.304$$

and the spot size of the approximately Gaussian mode profile is [using Eq. (4-18)]

$$w = a\left(0.65 + \frac{1.619}{V^{1.5}} + \frac{2.879}{V^6}\right) = 1.132a = 2.83\ \mu\text{m}$$

Using Eq. (17-28), the optimum beam diameter is then

$$D = \frac{2(1.5)(6)}{\pi(2.83)} = 2.02\ \text{cm}$$

Gaussian Beam Collimation

The opposite of beam focusing is beam *collimation,* in which the curved wavefronts of an initially diverging beam are transformed into the plane wavefronts of a collimated beam. This is illustrated in Fig. 17-14, where a Gaussian beam with waist w_{01} is transformed by the lens into a Gaussian beam with a larger waist w_{02}. Since the far-field divergence angle is $\theta \simeq \lambda/(\pi w_{02})$ according to Eq. (17-5), the larger waist of the new beam will give it a smaller divergence angle, which makes it more collimated. The Rayleigh range $z_{02} = \pi w^2_{02}/\lambda$ is also larger, so the beam diameter will remain approximately constant for a greater distance. It should be kept in mind that when we say a beam is "collimated," it is really a matter of degree. No beam is perfectly collimated, unless it is infinitely wide. Beam collimation has many practical applications. One application that we will consider in Chapter 24 is free-space optical communications, in which optical data is sent over kilometer-scale distances using a collimated beam in free space.

A practical question is where to put a lens of focal length f so as to collimate the beam. More specifically, what should be the distance d_1 between the lens and the beam waist of the original beam? Since the wavefront curvature for the collimated beam just after the lens is $R_2 = \infty$, Eq. (17-24) gives $R_1 = f$ for the wavefront curvature just before the lens. Using Eq. (17-3) to evaluate R for the original beam at $z = d_1$, we have

$$f = d_1\left[1 + \left(\frac{z_{01}}{d_1}\right)^2\right] \tag{17-29}$$

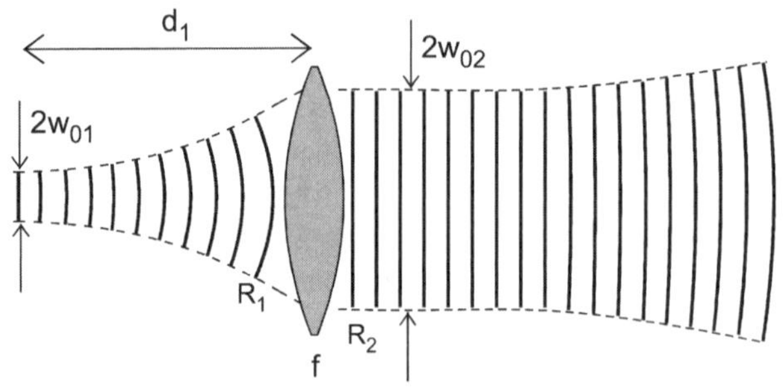

Figure 17-14 Light from a diverging beam is collimated by placing a lens of focal length f a distance d_1 from the beam waist.

where z_{01} is the Rayleigh range of the incident beam. Solving this equation for d_1 gives

$$d_1 = f\left[\frac{1}{2} \pm \frac{1}{2}\sqrt{1-\left(\frac{2z_{01}}{f}\right)^2}\right] \tag{17-30}$$

where the positive sign corresponds to the physically relevant solution. If $f \gg z_{01}$, Eq. (17-30) reduces to the simple result $d_1 \simeq f$. This agrees with the expectation from geometric optics that a point source (the beam waist) located at the focal point to the left of the lens will give rise to rays parallel to the optical axis to the right of the lens. As the focal length becomes smaller in comparison with z_{01}, however, geometric optics becomes increasingly inadequate, and the required value of d_1 becomes smaller than f. For values of $f < 2z_{01}$, there is no lens position that will give a collimated beam.

A *beam expander* is a device in which one collimated beam is converted into another collimated beam with an increased diameter. One way to accomplish this is shown in Fig. 17-15, in which the initial beam is first focused with a lens of focal length f_a, and the diverging beam that results is then collimated with a second lens of focal length f_b. The divergence of the beam between the two lenses is

$$\theta \simeq \frac{\lambda}{\pi w_{02}} = \frac{w_{03}}{f_b} = \frac{w_{01}}{f_a} \tag{17-31}$$

where it has been assumed that $f_a, f_b \gg z_{02}$. The final beam size w_{03} is then

$$w_{03} \simeq \frac{f_b}{f_a} w_{01} \qquad \text{(beam expander)} \tag{17-32}$$

which is larger than the initial beam size w_{01} by a factor (f_b/f_a). A beam expander can also be constructed with the first lens a diverging lens, $f_a < 0$. It is left as an exercise to show that in this case the beam is expanded by the same factor $f_b/|f_a|$, but the lenses should be separated by a distance $L = f_b - |f_a|$.

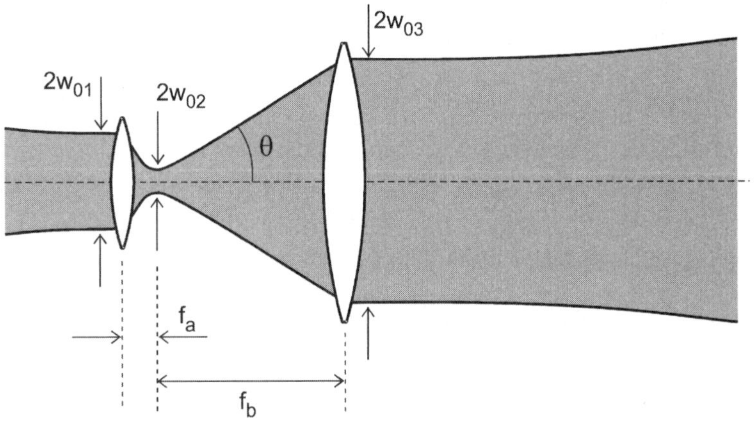

Figure 17-15 A beam expander is formed by separating two lenses with focal lengths f_a and f_b by a distance $L = f_a + f_b$.

PROBLEMS

17.1 A Gaussian beam of wavelength 720 nm in air has a 2 mm beam waist located at $z = 0$. (a) Determine the z position at which the spot size is 4 mm. (b) What is the radius of curvature of the beam's wavefront at that position.

17.2 What fraction of the energy in a Gaussian beam is at a distance from the beam axis greater than the spot size w?

17.3 Determine the on-axis brightness of a Gaussian beam of power P and wavelength λ, for $z \gg z_0$. Take the effective source area as πw_0^2. How does this expression compare with that for the brightness of a laser given in Eq. (15-5)?

17.4 A Gaussian beam has a beam waist of 176 μm, and the spot size is 293 μm a distance of 20 cm from the beam waist. Determine the wavelength of the beam.

17.5 An argon ion laser emits a Gaussian beam with power 3 W, wavelength 514.5 nm, and beam waist 0.5 mm. (a) Determine the on-axis beam intensity at a distance of 50 m from the laser. (b) If this light is collected with a lens of focal length 5 cm and diameter 1 cm, calculate the intensity of light at the focus. (c) Determine the brightness of the light at the focus. (d) Compare this with the brightness of the original laser beam, and comment on any difference.

17.6 A laser with wavelength 900 nm has a symmetric confocal cavity, and the spot size on the mirrors is 0.3 mm. (a) Determine the cavity length. (b) Determine the beam waist size. (c) Determine the mode spacing (consider transverse as well as longitudinal modes).

17.7 A symmetric laser cavity has mirrors of 10 m radius of curvature separated by 20 cm. The laser operates at 800 nm. (a) What is the beam waist size? (b) What mirror diameter is needed so there is little loss due to diffraction around the mirror edges? (c) Determine the longitudinal mode spacing. (d) What is the spacing between the lowest-order transverse modes, as a fraction of the longitudinal mode spacing?

17.8 Show that for the symmetrical resonator, Eq. (17-17) yields the condition $2r > L$, where L is the cavity length and r is the radius of curvature of either mirror. Also show that for a cavity with one flat mirror and one concave mirror (Fig. 17-7), the condition for stability becomes $r_2 > L$, where r_2 is the radius of curvature of the concave mirror.

17.9 The radius of the beam from a multimode laser is 3 mm just after it comes out of the laser. The wavelength is 1.5 μm, and the beam divergence (half-angle) is 1.4 mrad. Determine the M^2 parameter for this laser.

17.10 Sketch the electric field distribution in the x direction for the third-order Hermite–Gaussian mode, with $l = 3$. Include the spot size w on the x axis as a reference point. How does the "full width at half maximum" for this distribution compare with that of the lowest-order Gaussian mode?

17.11 A Nd:YAG laser with optical power 150 W at 1064 nm is used for laser machining. The beam out of the laser has diameter 3 mm (twice the spot size), and it is focused onto the work piece with a lens of focal length 25 mm. (a) Determine the spot size at the focus of the lens. (b) Determine the intensity at the focus. (c) If it is required that this intensity be maintained within a range of ±20%, determine how much the distance between the lens and work piece can be allowed to vary during the machining process.

17.12 Consider a beam expander similar to that of Fig. 17-15, except that the first lens is a diverging lens with focal length f_a a negative number. Show that the beam is expanded by the factor $|f_b/f_a|$ when the lenses are separated by a distance $L = f_b - |f_a|$. This variation is called a *Galilean telescope,* and has the advantage that it is more compact.

17.13 A Gaussian beam of waist w_{01} passes through a lens of focal length f, and converges to a focus at a second beam waist w_{02}, as shown in Fig. 17-11. The distances from the beam waists to the lens are d_1 and d_2. (a) Assuming that $d_1 \gg z_{01}$ and $d_2 \gg z_{02}$, determine the ratio of beam waists w_{02}/w_{01}. (b) Compare this result with the image/object height relation from geometric optics given in Eq. (2-30). (c) Show what effect, if any, the lens diameter has on your result in part a.

17.14 A laser cavity consists of two plane mirrors separated by a distance L, with a positive lens of focal length f placed halfway between them. Derive an expression for the spot size of the beam at the position of the lens, and at the mirrors. Discuss the stability conditions for this cavity.

Chapter 18

Stimulated Emission and Optical Gain

In the previous three chapters, we considered the special properties of laser light, and the way a laser beam propagates both inside and outside of a laser cavity. The laser cavity provides optical feedback, and is one of the three major components of a laser. We now turn to a discussion of the second essential component, the *gain medium.* This is the material between the cavity mirrors that amplifies the light as it propagates back and forth between the mirrors. The amplification occurs by stimulated emission, which we briefly discussed in Chapter 15. In this chapter, we consider stimulated emission in more detail, and show how it is related to the gain coefficient of an optical amplifier or laser.

18-1. TRANSITION RATES

Stimulated emission is one of the three fundamental processes by which an atom can make a radiative transition between two energy levels. As illustrated in Fig. 15-2, it occurs when a photon is incident on an atom that is initially in the upper of two energy levels. For stimulated emission to occur, the *resonance* condition $h\nu = E_{21}$ must be satisfied, where $h\nu$ is the photon energy and $E_{21} = E_2 - E_1$ is the energy difference between the two levels. If the photon energy were perfectly well defined (monochromatic light) and the atomic transition energy were perfectly sharp (no uncertainty or spread in energy), there would be no stimulated emission (or absorption) unless the two happened to match up exactly, a rather rare occurance. In practice, there is always some width to both $h\nu$ and E_{21}, which allows stimulated emission to occur over a range of photon frequencies.

We will first consider the situation in which the photon spectrum is very broad compared with the width of the atomic transition. This approach was taken by Albert Einstein in 1917, and leads to a relationship between the spontaneous and stimulated emission rates. The opposite limit will then be considered, in which the photon spectrum is very narrow compared with the atomic transition width. This later situation is most relevant for computing the gain in a laser.

Broadband Radiation

The phenomenon of stimulated emission was first proposed by Einstein in 1917, in order to explain the interaction between atoms and electromagnetic radiation in thermal equilibrium. Since this is such an important concept, it is worthwhile presenting here the key

Photonics and Lasers: An Introduction. By Richard S. Quimby

steps in Einstein's derivation, in order to understand the historical development of the basis for laser action.

Consider two representative energy levels of an atom in a gas, as shown in Fig. 18-1. To be concrete, you can think of these as two of the energy states of the electron in a hydrogen atom. In thermal equilibrium, the number of atoms in each energy state remains constant in time. If there are processes such as absorption and emission that tend to change the number of atoms in a given level, then these processes must act in such a way that the number of atoms making a transition from state 1 to state 2 is equal to the number of atoms making a transition from state 2 to state 1. It was by studying the balancing of these processes in thermal equilibrium that Einstein was able to propose the new phenomenon of stimulated emission and show its relation to spontaneous emission.

To make the argument quantitative, we define the number of atoms in level 1 per unit volume to be N_1, and similarly for N_2. The probability per unit time that an atom in level 2 will relax by spontaneous emission to level 1 is given by A_{21}, the *Einstein A coefficient.* If there were N_2 atoms in level 2, then the total number of atoms per unit time making a transition from level 2 to level 1 by spontaneous emission would be N_2A_{21}. If there were no other types of transitions between the levels, then eventually all the atoms would end up in level 1. However, it was known by the early 1900s that in thermal equilibrium, there is a finite steady-state population of atoms in the excited state N_2, given by the *Boltzmann factor*

$$\frac{N_2}{N_1} = e^{-E_{21}/(k_B T)} \qquad \text{(Boltzmann factor)} \qquad (18\text{-}1)$$

where k_B is *Boltzmann's constant* and T is the absolute temperature (in degrees Kelvin). Therefore, there must be some other radiative process in addition to spontaneous emission.

Another process that was known to exist is absorption, in which an incoming photon promotes an atom from level 1 to level 2. Since this process depends on the presence of a photon, it makes sense to suppose that the absorption probability per unit time is proportional to the energy density of electromagnetic radiation. Each photon has a certain probability of being absorbed, and the more photons there are (the higher the energy density), the greater the probability that the atom will absorb one of them. The absorption probability also depends on how well the photon energy matches up with the energy difference between the levels, $E_{21} \equiv E_2 - E_1$. The agreement does not have to be perfect, because the energy difference E_{21} has a spread ΔE according to the uncertainty

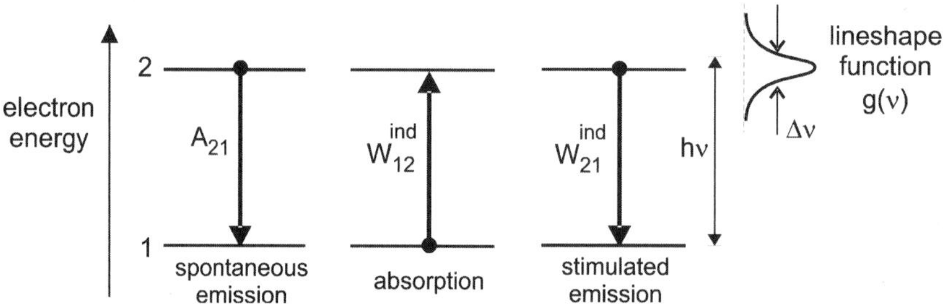

Figure 18-1 Transition rates for the three fundamental radiative processes in an atom. The photon energy $h\nu$ must be in the range $\Delta\nu$ about $(E_2 - E_1)/h$ for efficient emission or absorption.

principle (see Appendix B). Absorption will occur for those photons having energies $h\nu$ in a range ΔE around E_{21}.

Spectral Distribution and Lineshape Function

The number of photons having frequencies in the interval $\Delta\nu$ around ν can be described by the light's *spectral density,*

$$\rho_\nu(\nu) \equiv \frac{\text{energy in radiation}}{\text{(volume)(frequency interval)}} \tag{18-2}$$

which is a spectral *distribution function* (per unit frequency interval) as well as a density (per unit volume). The light-energy density that can interact with the atom is then $\rho = \rho_\nu(\nu)\Delta\nu$. Note that ρ and $\rho_\nu(\nu)$ have different units, J/m^3 and J s/m^3, respectively. Fig. 18-2 shows how $\rho_\nu(\nu)$ varies with frequency for blackbody (thermal) radiation, the so-called *blackbody spectrum.* It is very broad, varying as ν^2 at low frequency, and as $\exp(-h\nu/k_BT)$ at high frequency.

Also shown in Fig. 18-2 is the atomic *lineshape function,*

$$g(\nu) \equiv \frac{\text{probability of photon emission at } \nu}{\text{frequency interval}} \qquad \text{(lineshape function)} \tag{18-3}$$

which describes the relative probability that the atom will absorb or emit a photon of frequency ν. Like $\rho_\nu(\nu)$, this is a distribution function that gives a probability per frequency interval, so that $g(\nu)\, d\nu$ is the actual probability of emission in the frequency interval $d\nu$. The lineshape function is normalized to unit probability over all frequencies, so $\int g(\nu)\, d\nu = 1$.

Atomic lineshapes are much narrower than the blackbody spectrum, as indicated in Fig. 18-2. This means that $\rho_\nu(\nu)$ is approximately constant over the frequencies at which the atom absorbs or emits, and the probability of absorption should be proportional to the value of $\rho_\nu(\nu)$ evaluated at the lineshape center frequency ν_0. Thus, we can write the probability per unit time that an atom in level 1 is promoted to level 2 as

$$W_{12}^{\text{ind}} = \rho_\nu B_{12}$$

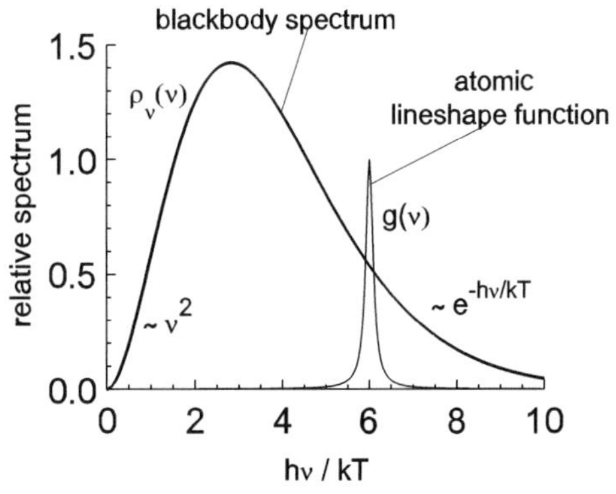

Figure 18-2 Blackbody radiation spectrum is broad compared with atomic transition.

where ρ_ν is evaluated at the atomic frequency, and the constant of proportionality B_{12} is the *Einstein B coefficient.* This absorption process is an *induced transition,* because the transition between states is "induced" by the incident photon.

Using this expression for the induced transition rate, the rate of change of the level 2 population N_2 is given by the *rate equation,*

$$\frac{dN_2}{dt} = N_1 B_{12} \rho_\nu(\nu) - N_2 A_{21}$$

where the first term gives the number of atoms (per unit volume) entering level 2 per unit time, and the second term gives the number leaving level 2 per unit time. In thermal equilibrium, $dN_2/dt = 0$, which implies that

$$\frac{N_2}{N_1} = \frac{B_{12}}{A_{21}} \rho_\nu(\nu) \qquad \text{(no stimulated emission)}$$

Putting this together with Eq. (18-1) would give $\rho_\nu(\nu) \propto \exp(-h\nu/k_B T)$, since at resonance $E_{12} = h\nu$. However, this does not agree with the blackbody spectrum, which was known from experiments to follow the so-called Planck distribution,

$$\rho_\nu(\nu) = \frac{8\pi h \nu^3}{c^3} \frac{1}{e^{h\nu/k_B T} - 1} \qquad \text{(Planck distribution)} \qquad (18\text{-}4)$$

This distribution, shown in Fig. 18-2, agrees with the $\exp(-h\nu/k_B t)$ dependence at higher ν, but not at lower ν. It is this disagreement between the calculated and measured blackbody spectrum that prompted Einstein to postulate a third radiative process, that of stimulated emission.

Stimulated Emission: Einstein Treatment

Einstein proposed that in addition to the usual absorption process, in which the atom goes from level 1 up to level 2, there is a corresponding downward-going process, termed *stimulated emission* (also called *induced emission*), which is induced by the incident light. In stimulated emission, the incident photon causes the atom to go from level 2 down to level 1, thereby emitting a new photon which joins the photon initially present. Like absorption, this is an induced transition, but the atom here is induced to go from a higher to a lower energy state, rather than from lower to higher. Since this process also depends on light being present, it is assumed to proceed at a rate proportional to $\rho_\nu(\nu)$, just as for absorption. The probability per unit time that an atom undergoes stimulated emission is written as

$$W_{21}^{\text{ind}} = \rho_\nu B_{21} \qquad \text{(stimulated emission rate)} \qquad (18\text{-}5)$$

where B_{21} is the second Einstein B coefficient. The three transition rates possible are summarized in Fig. 18-1.

When stimulated emission is included, the rate equation for level 2 becomes

$$\frac{dN_2}{dt} = N_1 B_{12} \rho_\nu(\nu) - N_2 B_{21} \rho_\nu(\nu) - N_2 A_{21} \qquad (18\text{-}6)$$

where $dN_2/dt = 0$ holds for thermal equilibrium, as before . Solving this for N_2/N_1 and using Eq. (18-1) gives

$$e^{-h\nu/k_BT} = \frac{B_{12}\rho_\nu(\nu)}{A_{21} + B_{21}\rho_\nu(\nu)} \tag{18-7}$$

which must be true at all temperatures T. The Einstein coefficients A_{21}, B_{21}, and B_{12} are independent of temperature, since they are properties of a single atom. The temperature dependence therefore comes from $\rho_\nu(\nu)$ on the right-hand side, and from the Boltzmann factor on the left-hand side.

At high temperature, it was known experimentally that $\rho_\nu(\nu) \propto k_BT$, which historically was referred to as the Rayleigh–Jeans law. Note that this is consistent with the Planck distribution of Eq. (18-4) in the limit $k_BT \gg h\nu$. In this limit the A_{21} in the denominator of Eq. (18-7) becomes small in comparison with $B_{21}\rho_\nu(\nu)$, since A_{21} and B_{21} are independent of temperature. Eq. (18-7) then becomes $1 \simeq B_{12}/B_{21}$, or

$$B_{21} = B_{12} \qquad \text{(Einstein } B \text{ coefficients)} \tag{18-8}$$

which means that the probability of stimulated emission is equal to the probability of absorption. This conclusion is valid at any temperature, and is a fundamental relationship between absorption and emission.* The equivalence of the Einstein B coefficients suggests that at a deep level, emission and absorption are fundamentally the same, one being an induced upward transition, and the other an induced downward transition. This equivalence was only fully appreciated much later with the development of quantum electrodynamics (QED), the quantum theory of light.

It should be noted that we have assumed nondegenerate energy levels, which means that there is only one quantum mechanical state having the energy of the given level. If the degeneracies of the lower and upper levels is g_1 and g_2, respectively, then the relation between the Einstein B coefficients becomes $g_2\, B_{21} = g_1 B_{12}$. For simplicity of notation, we will continue to assume nondegenerate levels throughout the following discussion, keeping in mind that the equations can always be generalized by multiplying by a ratio of degeneracy factors.

Using $B_{21} = B_{12}$ now in Eq. (18-7), and solving for $\rho_\nu(\nu)$ at arbitrary temperature, we have

$$\rho_\nu(\nu) = \frac{A_{21}}{B_{21}} \frac{1}{e^{-h\nu/(k_BT)} - 1} \tag{18-9}$$

This expression for $\rho_\nu(\nu)$ agrees with the blackbody spectrum of Eq. (18-4), provided that

$$\frac{A_{21}}{B_{21}} = \frac{8\pi h\nu^3}{c^3} \qquad \text{(Einstein relation)} \tag{18-10}$$

Eq. (18-10) is the principal result of Einstein's derivation, and plays a key role in understanding the physics of lasers. It relates the probability of stimulated emission to that of spontaneous emission, showing that one is proportional to the other. This means that,

*We did use the high-temperature limit to obtain this relation, but once we have it, it is valid in general since B_{21} and B_{12} are properties of a single atom and independent of T.

all other things being equal, a transition with a higher spontaneous decay rate will have a higher stimulated emission rate, and hence a stronger amplification of light. This has direct implications for the performance and characteristics of different gain media, as we will see in subsequent chapters.

Stimulated Emission: Quantum Viewpoint

Eq. (18-10) can be written in a different form that allows a simple physical interpretation. Recalling from Eq. (16-9) that $\beta_\nu(\nu) = 8\pi\nu^2/c^3$ is the number of cavity modes per unit volume per unit frequency interval, Eq. (18-10) becomes

$$\frac{A_{21}}{B_{21}} = \beta_\nu(\nu)\, h\nu \tag{18-11}$$

The spectral mode density $\beta_\nu(\nu)$ is related to the spectral energy density $\rho_\nu(\nu)$ by

$$\begin{aligned} \rho_\nu(\nu) &= \left[\frac{\text{modes}}{(\text{vol})\ \Delta\nu}\right]\left[\frac{\text{photons}}{\text{mode}}\right]\left[\frac{\text{energy}}{\text{photon}}\right] \\ &= \beta_\nu(\nu)\,\bar{n}h\nu \end{aligned} \tag{18-12}$$

where $\bar{n}$ is the average number of photons per cavity mode. Combining Eqs. (18-11) and (18-12) gives for the induced emission rate

$$W_{21}^{\text{ind}} = \rho_\nu(\nu)B_{21} = \bar{n}A_{21} \tag{18-13}$$

or

$$W_{21}^{\text{ind}} = \bar{n}W_{21}^{\text{spont}} \tag{18-14}$$

where the spontaneous decay rate is written as $W_{21}^{\text{spont}} \equiv A_{21}$.

Eq. (18-14) is a very fundamental and important result, the significance of which was not fully appreciated until the development of the quantum theory of radiation (quantum electrodynamics, or QED). It says quite simply that the probability of an induced transition is equal to the probability of a spontaneous transition times the number of photons per cavity mode. In the case of blackbody radiation, the average number of photons per mode is found, by combining Eqs. (18-4), (18-12), and (14-9), to be

$$\bar{n} = \frac{1}{e^{h\nu/k_BT} - 1} \qquad \text{(thermal equilibrium)} \tag{18-15}$$

Eq. (18-15) is actually a rather fundamental relation in itself, and anticipates certain ideas in quantum statistical mechanics. In the quantum mechanical view, each cavity mode is a quantum state, which may be either occupied or unocupied by one or more "quanta," which in this case are photons. The number of quanta per state is known as the *occupation number,* which can be greater than one for particles, like photons, that have integer spin. In statistical mechanics, such particles are called *bosons.* It turns out that the occupation number for bosons in a quantum state of energy $h\nu$ is precisely that of Eq. (18-15). For particles such as electrons, with half-integer spin (termed *fermions*), the occupation number is given by a similar expression, but with a plus sign in the denominator.

EXAMPLE 18-1

A He–Ne laser cavity has a cylindrical geometry with length 30 cm and diameter 0.5 cm. The laser transition is at 632.8 nm, with a frequency width of 1.5 GHz. Determine (a) The number of modes in the laser cavity that are within the laser transition bandwidth, (b) the average number of photons per cavity mode due to thermal radiation at room temperature (300 K), and (c) the average number of 300 K thermal photons that are in any cavity mode within the laser transition bandwidth.

Solution: (a) Designating the number of cavity modes within the laser transition bandwidth as p, we have

$$p \sim \frac{8\pi\nu^2}{c^3} V\,\Delta\nu = \frac{8\pi V}{c\lambda^2}\Delta\nu$$

$$= \frac{8\pi(7.5\times10^{-6})(1.5\times10^{9})}{(3\times10^{8})(633\times10^{-9})^2} = 2.4\times10^{9}$$

(b) The photon energy is

$$h\nu = hc/\lambda = \frac{(6.63\times10^{-34})(3\times10^{8})}{632.8\times10^{-9}} = 3.14\times10^{-19}\ \text{J}$$

and the thermal energy is

$$k_BT = (1.38\times10^{-23})(300) = 4.14\times10^{-21}\ \text{J}$$

The average number of photons per mode is then

$$\bar{n} = \frac{1}{e^{314/4.14}-1} = 1.15\times10^{-33}$$

(c) The average number of thermal photons in any mode within the gain profile is then

$$\bar{n}p = (1.15\times10^{-33})(2.4\times10^{9}) = 2.8\times10^{-24}$$

It is clear from these numbers that thermal photons play a negligible role in populating the cavity modes. We will see later that once lasing starts, the number of photons in a single mode can be much larger than 1.

Narrowband Radiation

Einstein's derivation of the A and B coefficients assumes that the blackbody radiation spectrum is very broad compared with the atomic absorption line shape. In this case, the number of photons per mode is nearly the same for all modes within the atomic line shape. For laser light, the opposite situation occurs, in which the light spectrum is very narrow compared with the atomic lineshape. Here, the number of photons per mode is very high for a few modes within the atomic lineshape, but much smaller for other modes, as illustrated in Fig. 18-3. The Einstein relation between A and B in Eq. (18-10) still applies in this case, but the relation of Eq. (18-5) between B_{21} and the induced transition rate W_{21}^{ind} must be generalized to

$$W_{21}^{\text{ind}} = \int B_{21}\rho_\nu(\nu)g(\nu)\,d\nu \tag{18-16}$$

This gives the relationship between the induced transition rate and the Einstein B coefficient for any combination of radiation spectral distribution $\rho_\nu(\nu)$ and atomic lineshape function $g(\nu)$. In the limiting case of blackbody radiation, where $\rho_\nu(\nu)$ is approximately constant over the width of $g(\nu)$, the function $\rho_\nu(\nu)$ can be brought outside the integral and evaluated at the lineshape center frequency ν_0. The induced transition rate then becomes $W_{21}^{\text{ind}} = B_{21}\rho_\nu(\nu_0)\int g(\nu)d\nu = B_{21}\rho_\nu(\nu_0)$, in agreement with the original definition of the Einstein B coefficient.

In the case of laser radiation that is close to monochromatic, the opposite limit generally applies, as shown in Fig. 18-4. Here it is the atomic lineshape function $g(\nu)$ that is approximately constant over the width of $\rho_\nu(\nu)$, so that $g(\nu)$ can be brought outside the integral, evaluated at the laser light center frequency ν'. Eq. (18-16) then becomes

$$\begin{aligned} W_{21}^{\text{ind}} &= B_{21}g(\nu')\int \rho_\nu(\nu)\,d\nu \\ &= B_{21}g(\nu')\,\rho \end{aligned} \tag{18-17}$$

where

$$\rho \equiv \frac{\text{light energy}}{\text{volume}} = \int \rho_\nu(\nu)\,d\nu \tag{18-18}$$

is the energy density, in MKS units of J/m^3. Note carefully the distinction between ρ and $\rho_\nu(\nu)$. The latter is a *spectral density,* the energy density per unit frequency interval, with units of J · s/m^3.

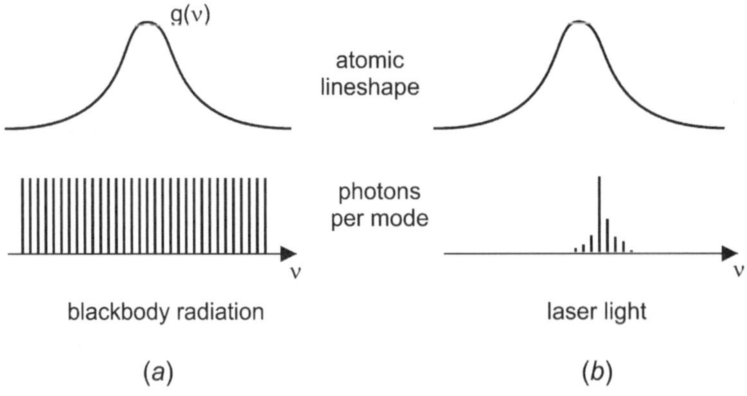

Figure 18-3 (a) Photon modes are equally populated under the atomic lineshape for blackbody radiation. (b) Only a few photon modes are highly populated in laser radiation.

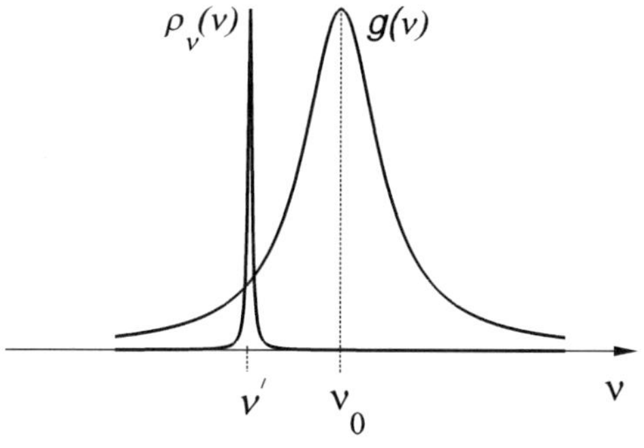

Figure 18-4 For laser light, the frequency distribution is much narrower than the lineshape of the atomic transition.

Eq. (18-17) can be written in terms of the light intensity I using $I = c\rho$ from Eq. (2-9). The induced rate then becomes

$$W_{21}^{\text{ind}} = B_{21} g(\nu')\, I/c \qquad \text{(stimulated emission rate)} \tag{18-19}$$

Since $B_{12} = B_{21}$ from Eq. (18-8), the induced transition rate from level 1 up to level 2 (absorption rate) is given by the same expression. Therefore,

$$W_{21}^{\text{ind}} = W_{12}^{\text{ind}} \equiv W^{\text{ind}} = B_{21} g(\nu')\, I/c \qquad \text{(induced transition rates)} \tag{18-20}$$

This will prove to be a useful relation in calculating optical gain.

Quantum Viewpoint: Photons Per Mode

We present here a more rigorous justification for the relation between B_{21} and W_{21}^{ind} given in Eq. (18-16). A fundamental principle of the quantum theory of light is that Eq. (18-14) applies not only to the total emission rate, but also to the emission rate into each mode separately. This can be expressed as

$$W_{21}^{\text{ind},i} = n_i W_{21}^{\text{spont},i} \tag{18-21}$$

where n_i is the number of photons in cavity mode i, $W_{21}^{\text{spont},i}$ is the probability per unit time that the atom emits a spontaneous photon into mode i, and $W_{21}^{\text{ind},i}$ is the probability per unit time that the atom emits an induced photon into mode i. The total induced rate from level 2 to level 1 can then be written as a sum over modes,

$$\begin{aligned} W_{21}^{\text{ind}} &= \sum_i W_{21}^{\text{ind},i} \\ &= \sum_i n_i W_{21}^{\text{spont},i} \end{aligned} \tag{18-22}$$

which is the most general relation between induced and spontaneous emission rates for an atomic transition. We will first consider some special cases, and then derive Eq. (18-16).

In the case of blackbody radiation previously considered, $n_i = \bar{n}$ for all modes within the atomic lineshape, and $W_{21}^{\text{ind}} = \bar{n}\Sigma W_{21}^{\text{spont},i} = \bar{n}W_{21}^{\text{spont}}$, consistent with the result in Eq. (18-14). In the opposite limit, where all the radiation is in a single cavity mode j,

$$W_{21}^{\text{ind}} = n_j\, W_{21}^{\text{spont},j}$$

The spontaneous emission rate $W_{21}^{\text{spont},j}$ into the single mode j can be estimated by assuming roughly equal emission into a total of p modes under the atomic lineshape, the fraction emitted into any single mode being $1/p$. We then have $W_{21}^{\text{spont},j} \simeq (1/p)\, W_{21}^{\text{spont}}$, which when combined with the above gives

$$W_{21}^{\text{ind}} \simeq \frac{n_j}{p} W_{21}^{\text{spont}} \qquad \text{(light in single cavity mode } j\text{)} \qquad (18\text{-}23)$$

This result shows that the total induced and spontaneous rates are related by the average number of photons per mode, even when the modes are unevenly populated. Equation (18-14) therefore applies quite generally, provided that $\bar{n}$ is interpreted as the average mode population.

An alternative form for Eq. (18-22) can be obtained by changing the sum over modes into an integral over mode frequencies,

$$W_{21}^{\text{ind}} = \int \left[\frac{\#\text{ modes}}{\text{frequency interval}} \right] n_i W_{21}^{\text{spont},i} d\nu \qquad (18\text{-}24)$$

where the # modes per frequency interval is given by Eq. (16-9) as $dN/d\nu = V\beta_\nu(\nu)$. The number of photons in mode i can be written as

$$\begin{aligned} n_i &= \left[\frac{\text{energy}}{\Delta\nu} \right] \left[\frac{1}{\text{energy/photon}} \right] \left[\frac{1}{\text{modes}/\Delta\nu} \right] \\ &= \frac{\rho_\nu(\nu)}{\beta_\nu(\nu) h\nu} \end{aligned} \qquad (18\text{-}25)$$

where $\Delta\nu$ is a small frequency interval and Eq. (18-2) has been used. The spontaneous emission rate into mode i can be written as

$$\begin{aligned} W_{21}^{\text{spont},i} &= \frac{\text{photons emitted}}{(\text{time})(\text{mode})} \\ &= \left[\frac{\text{photons emitted}}{(\text{time})(\Delta\nu)} \right] \left[\frac{1}{\text{modes}/\Delta\nu} \right] \\ &= W_{21}^{\text{spont}} g(\nu) \frac{1}{V\beta_\nu(\nu)} \end{aligned} \qquad (18\text{-}26)$$

where $g(\nu)$ is the lineshape factor defined in Eq. (18-3), normalized so that $\int g(\nu)\, d\nu = 1$. The lineshape factor gives the frequency distribution of photons emitted spontaneously from an atomic level, and plays an important role in characterizing a laser transition. The product $W_{21}^{\text{spont}} g(\nu)$ gives the photon emission rate per unit frequency interval, and integrating this over the entire lineshape gives the total spontaneous decay rate, $\int W_{21}^{\text{spont}} g(\nu)\, d\nu = W_{21}^{\text{spont}}$.

Putting Eqs. (18-24)–(18-26) together along with Eq. (18-11) gives the result

$$W_{21}^{\text{ind}} = \int B_{21}\rho_\nu(\nu)g(\nu)\, d\nu$$

which is the expression quoted earlier in Eq. (18-16).

18-2. OPTICAL GAIN

At the heart of a laser or optical amplifier is a gain medium that amplifies a light beam passing through it. We are now in a position to calculate this gain, using the expressions for induced transition rate developed in the previous section. In this section, we see how the degree of amplification can be characterized by either a gain coefficient or a gain cross section. We also discuss certain material properties that influence the gain on a laser transition.

Gain Coefficient

Let us assume that a nearly monochromatic light wave of frequency ν is incident on a collection of atoms having energy levels 1 and 2, as shown in Fig. 18-5. There will be other energy levels in the atoms, but we focus attention on the two particular levels involved in the laser transition. Light is added to the beam by stimulated emission at a rate $N_2 W_{21}^{\text{ind}}$, and is absorbed from the beam at a rate $N_1 W_{12}^{\text{ind}}$. Spontaneous emission also generates light, but this light is emitted in random directions, with only a small fraction in the direction of the original beam. The net number of photons added to the beam per unit volume per unit time is then given by

$$\begin{aligned}\frac{\text{photons added}}{\text{(volume)(time)}} &= N_2 W_{21}^{\text{ind}} - N_1 W_{12}^{\text{ind}} \\ &= \Delta N W^{\text{ind}}\end{aligned} \tag{18-27}$$

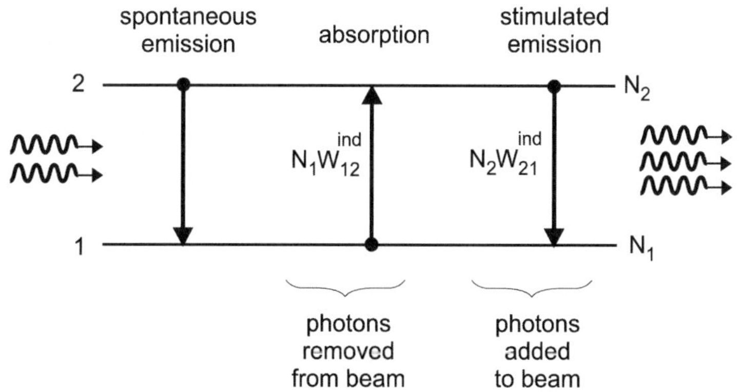

Figure 18-5 Induced emission and absorption add and subtract photons from the incident beam.

where $\Delta N \equiv (N_2 - N_1)$ is the *population difference,* and Eq. (18-20) has been used. Each added photon contributes an energy $h\nu$ to the beam, so the change in the beam's energy density with time is given by

$$\begin{aligned}\frac{\Delta\rho}{\Delta t} &= \frac{\text{energy generated}}{\text{(volume)(time)}} \\ &= \Delta N W^{\text{ind}} h\nu\end{aligned} \tag{18-28}$$

Assume now that the beam traverses a section of gain material with thickness Δz, as shown in Fig. 18-6. A slice of the beam with thickness $\ll \Delta z$ will take a time $\Delta t = \Delta z/c$ to pass through the section of gain material, and during this time the energy density in the slice will increase by

$$\begin{aligned}\Delta\rho &= \Delta N W^{\text{ind}} h\nu\Delta t \\ &= \Delta N W^{\text{ind}} h\nu\Delta z/c\end{aligned} \tag{18-29}$$

Using $\Delta I = c\Delta\rho$, Eq. (18-29) can be written as

$$\Delta I = \Delta N W^{\text{ind}} h\nu\Delta z \tag{18-30}$$

Writing W^{ind} in terms of I from Eq. (18-20), and using the Einstein relation between B and A in Eq. (18-10), this becomes

$$\Delta I = \frac{A_{21}c^2}{8\pi\nu^2} g(\nu) I \Delta N \Delta z \tag{18-31}$$

Dividing both sides by Δz and taking the limit as $\Delta z \to 0$ gives the differential equation,

$$\begin{aligned}\frac{dI}{dz} &= \left(A_{21}\frac{\lambda^2}{8\pi} g(\nu)\Delta N\right) I \\ &= \gamma(\nu) I\end{aligned} \tag{18-32}$$

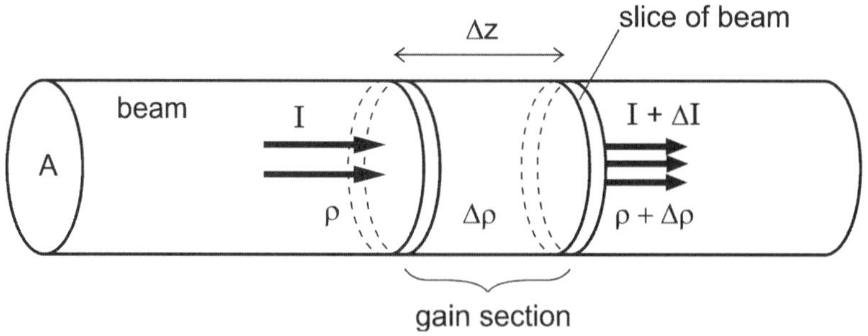

Figure 18-6 Change in beam intensity in traversing thickness Δz of gain medium is related to change in energy density $\Delta\rho$.

where $\gamma(\nu)$ is the *gain coefficient,* given by

$$\gamma(\nu) = A_{21} \frac{\lambda^2}{8\pi} g(\nu)\, \Delta N \qquad \text{(gain coefficient)} \tag{18-33}$$

In a medium of refractive index n, the wavelength to use in this equation is $\lambda = (c/n)/\nu$, the wavelength in the medium.

The gain coefficient $\gamma(\nu)$ is one of the most important parameters in the physics of lasers. It is defined as the fractional change in light intensity per unit propagation distance, $\gamma(\nu) = (\Delta I/I)/\Delta z$, and must be positive for light amplification to occur. For the gain coefficient to be positive, it is necessary that $\Delta N > 0$, which means that there are more atoms in the upper level 2 than in the lower level 1. For atoms in thermal equilibrium the opposite situation exists, since the populations are given by the Boltzmann factor of Eq. (18-1), and $\Delta N < 0$. To achieve amplification, then, more atoms must be put into the upper level than would normally be there in thermal equilibrium, a process termed *pumping* the upper level. A positive ΔN is termed a *population inversion,* since the relative sizes of N_1 and N_2 are inverted compared to thermal equilibrium. The process of pumping the upper level to obtain a population inversion will be considered in more detail in the next chapter.

If $\gamma(\nu)$ is positive and independent of z, the solution for $I(z)$ is easily obtained by dividing Eq. (18-32) by I and integrating over z,

$$\int_{I_0}^{I} \frac{dI}{I} = \int_0^z \gamma(\nu)\, dz$$

$$\ln I - \ln I_0 = \gamma(\nu)\, z$$

$$\ln\left(\frac{I}{I_0}\right) = \gamma(\nu)\, z$$

$$\frac{I}{I_0} = e^{\gamma(\nu) z}$$

or

$$I(z) = I_0 e^{\gamma(\nu) z} \tag{18-34}$$

where I_0 is the initial intensity at $z = 0$.

According to Eq. (18-34), the light intensity increases exponentially with propagation distance z, as shown in Fig. 18-7. The increase is gentle at first, becoming more pronounced as z increases. This qualitative behavior can be understood from the differential equation, which says that the rate of increase in I at a particular z is proportional to the intensity at that z. The higher the intensity, the more rapidly the intensity will increase with z. If the derivation leading to this result is traced backward, it is seen to arise directly from Einstein's assumption that the rate of stimulated emission is proportional to the energy density of the light beam.

The exponential increase in light intensity cannot continue indefinitely, of course, because the intensity would go to infinity, which is physically unrealistic. At some point, the gain will start to be limited because the population difference ΔN will decrease. One reason for a decrease in ΔN is the stimulated emission process itself, as illustrated in Fig. 18-8. Each photon created by stimulated emission decreases the upper-state population N_2 by 1, and increases the lower-state population N_1 by 1, for a change in ΔN of -2. The pump-

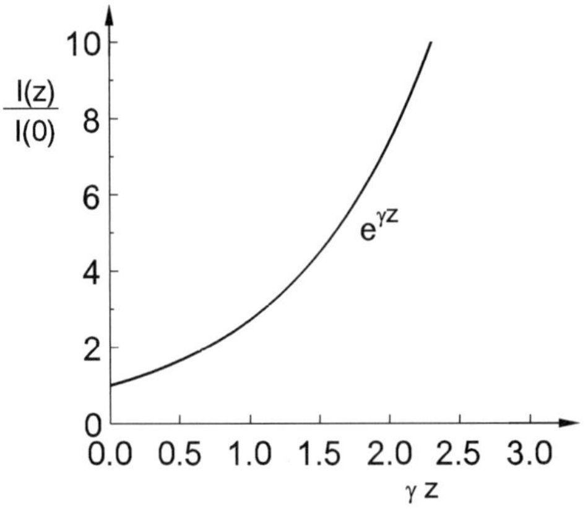

Figure 18-7 Exponential increase in light intensity when gain coefficient γ is independent of z.

ing mechanism can be used to maintain the population inversion, but only up to a point. When the decrease in ΔN from stimulated emission is greater than the increase in ΔN from the pump, the population inversion will decrease, thereby limiting the gain. This phenomenon of limited gain due to a decreased population inversion is referred to as *gain saturation,* and will be considered in more detail in the next chapter.

Gain Cross Section

The gain coefficient $\gamma(\nu)$ given in Eq. (18-33) depends on two types of parameters: those that characterize the properties of a single atom, and those that characterize how many atoms are in a particular energy level. It is useful to separate out all the single-atom properties into a single factor known as the *gain cross section* $\sigma(\nu)$, writing Eq. (18-33) as

$$\begin{aligned}\gamma(\nu) &= \Delta N\sigma(\nu) \\ &= N_2\sigma(\nu) - N_1\sigma(\nu)\end{aligned} \qquad (18\text{-}35)$$

where the cross section $\sigma(\nu)$ is given by

$$\sigma(\nu) = A_{21}\frac{\lambda^2}{8\pi}g(\nu) \qquad \text{(cross section)} \qquad (18\text{-}36)$$

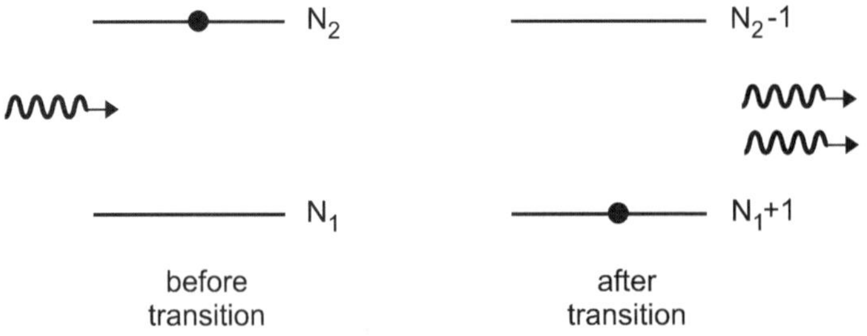

Figure 18-8 Stimulated emission causes the population difference to decrease.

Since the gain cross section is a property of a single atom, cross section values for atomic transitions of interest can be tabulated in handbooks. Cross section data is often presented in the form of a graph as a function of wavelength for a particular atomic transition. The gain coefficient cannot be similarly tabulated, because it depends on the number of atoms pumped into the higher energy level, which in turn depends on the pumping rate.

Eq. (18-35) assumes that the induced emission and absorption probabilities per atom are equal, as in Eq. (18-8). However, this is strictly true only for nondegenerate energy levels. In many cases of interest, the upper and lower energy states actually consist of a series of closely spaced sublevels, with transitions possible between any sublevel of the upper state and any sublevel of the lower state. This can be taken into account by generalizing Eq. (18-35) to

$$\gamma(\nu) = N_2\sigma_{\text{em}}(\nu) - N_1\sigma_{\text{abs}}(\nu) \tag{18-37}$$

where $\sigma_{\text{em}}(\nu)$ is the *emission cross section,* and $\sigma_{\text{abs}}(\nu)$ is the *absorption cross section.* The emission cross section is given by Eq. (18-36), with A_{21} now interpreted as the total spontaneous decay rate from the combined set of sublevels in the upper state. The corresponding absorption cross section is related to the emission cross section by the simple expression

$$\sigma_{\text{abs}}(\nu) = \sigma_{\text{em}}(\nu)e^{(h\nu-\varepsilon)/k_BT} \qquad \text{(McCumber relation)} \tag{18-38}$$

where the parameter ε is an effective energy difference between the two sets of sublevels, as illustrated in Fig. 18-9.

This relation between emission and absorption cross sections was first developed by McCumber (McCumber 1964), and constitutes a generalization of the Einstein A and B treatment for energy states with sublevels. It has proved useful in characterizing certain types of gain media for use as an amplifier or laser, such as rare earth ions doped into a solid. An example is given in Fig. 18-10, which shows $\sigma_{\text{abs}}(\nu)$ and $\sigma_{\text{em}}(\nu)$ for the lowest energy transition in Er^{3+}-doped silicate glass. Note that at higher photon energy (shorter wavelength), the absorption cross section is greater than the emission cross section,

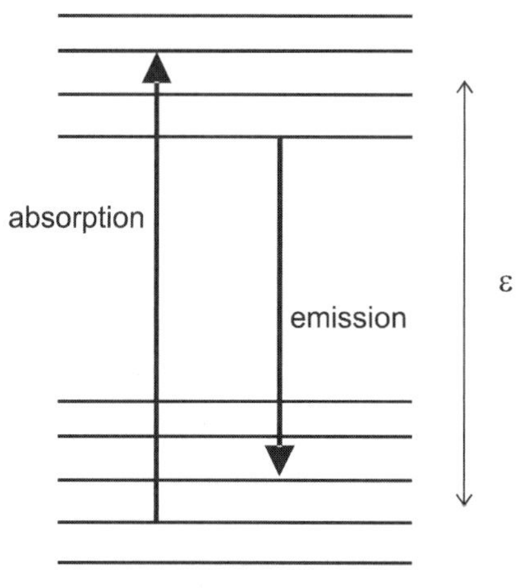

Figure 18-9 Absorption and emission between sublevels of the upper and lower states.

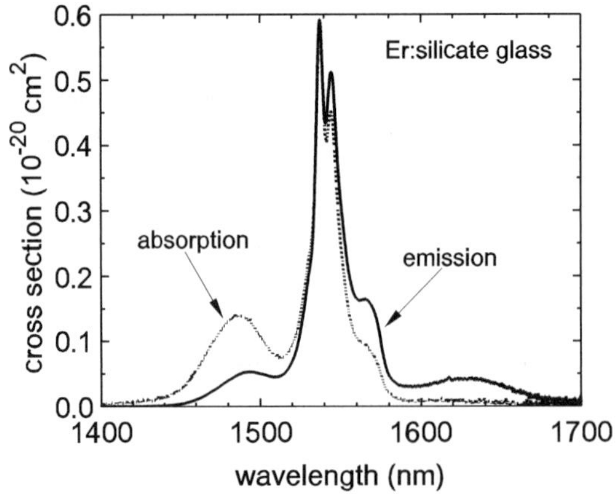

Figure 18-10 Absorption and emission cross sections versus wavelength for the rare earth ion Er^{3+} doped in silicate glass. (Data courtesy of Rodica Martin.)

whereas at lower photon energy the reverse is true. This agrees with Eq. (18-38), and can be understood by considering transitions between different pairs of sublevels in the upper and lower states, as shown in Fig. 18-9. Absorption transitions occur mostly from the lower sublevels of the bottom state, because the Boltzmann factor reduces the number of atoms in the higher sublevels. Similarly, emission occurs mostly from the lowest sublevels of the upper state. Therefore, there are more high photon energy transitions in absorption than in emission, and vice versa. This asymmetry between emission and absorption becomes more pronounced at low temperature.

For notational simplicity, we will assume that $\sigma_{abs}(\nu) = \sigma_{em}(\nu) \equiv \sigma(\nu)$ in much of our subsequent treatment of lasers. The analysis can always be generalized, however, by using Eq. (18-37) in place of Eq. (18-35), and where appropriate we will utilize the more general relation.

A cross section has units of area, since $\gamma(\nu)$ and ΔN in Eq. (18-35) have units of m^{-1} and m^{-3}, respectively. Cross section is also a property of a single atom, which suggests that it might correspond to an effective absorbing area for an atom. To explore this possibility, we consider a beam of light of area A_{beam} incident on a collection of atoms as in Fig. 18-11. The atoms will be assumed to be in the lowest energy state 1 (the *ground state*), with N_1 atoms per unit volume. The number of atoms in a slice of thickness Δz is then $N_1 V = N_1 A_{beam} \Delta z$, with the number of atoms per unit beam area in the slice Δz given by

$$\begin{aligned}\frac{\#\text{ atoms}}{\text{beam area}} &= \frac{N_1 V}{A_{beam}} \\ &= N_1 \Delta z\end{aligned}$$

We define an effective absorbing area A_{atom} for each atom, such that any light incident within this area will be absorbed, with any light outside this area transmitted. The combined absorbing area of the atoms per unit beam area is then

$$\frac{\text{area of atoms}}{\text{beam area}} = N_1 \Delta z A_{atom}$$

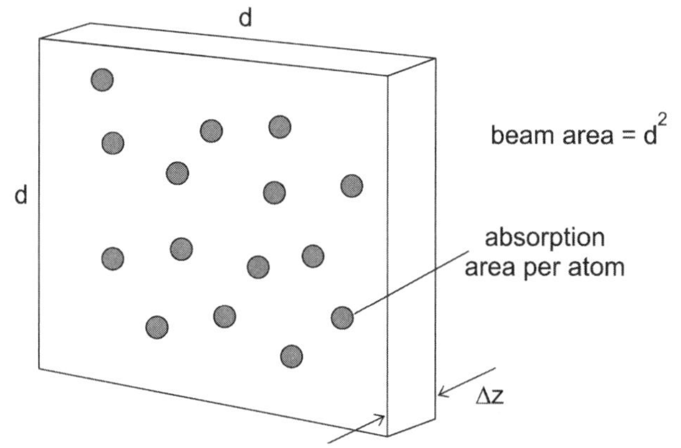

Figure 18-11 Each atom absorbs light from an area equal to its absorption cross section σ.

The above ratio of areas is the fraction of the total beam energy that is absorbed in thickness Δz. This ratio can also be expressed in terms of the *absorption coefficient* $\alpha(\nu)$, which is the fraction of the beam absorbed per unit length. From Eq. (18-37), $\alpha(\nu) = N_1\sigma_{abs}(\nu)$ when $N_2 = 0$, which gives for the fractional beam loss

$$N_1 \Delta z A_{atom} = N_1 \sigma_{abs}(\nu) \Delta z$$

Dividing by $N_1\Delta z$ gives $A_{atom} = \sigma_{abs}(\nu)$, which confirms that the absorption cross section can indeed be thought of as an effective absorption area for the atom. One must be careful with this analogy, however, because $\sigma(\nu)$ is not just a fixed quantity like the area of a shadow cast by a physical object. Instead, it represents the effective area over which an atom can "grab" a photon that is passing by. This grabbing ability is not a constant, but instead varies with the frequency or wavelength of the light. The cross section can be larger or smaller than the physical size of the atom.

As the beam propagates, its intensity varies with z according to Eq. (18-34), which in the case of $N_2 = 0$ becomes

$$I(z) = I(0)e^{-N_1\sigma_{abs}z} \tag{18-39}$$

The intensity decays exponentially with distance, as illustrated in Fig. 18-12, and corresponds to the Beer's law relation given in Eq. (5-1).

Fluorescence Lifetime

The Einstein A coefficient is a transition rate, giving the probability per unit time that an atom in the upper state 2 will make a transition to the lower state 1. An alternative way to characterize the transition probability is to specify the average time that an atom will remain in the upper state before decaying to the lower state. This average decay time is termed the *fluorescence lifetime,* since it describes the duration of spontaneously emitted light (fluorescence) by a collection of atoms that are initially placed in the upper state 2.

Considering only spontaneous emission processes, the time rate of change of the population of state 2 is given by Eq. (18-6) as

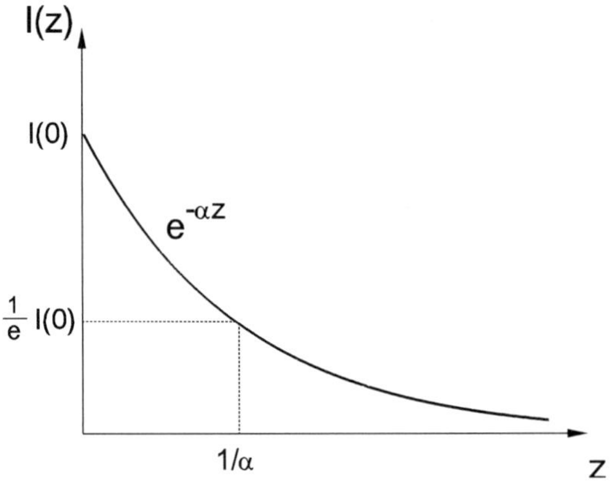

Figure 18-12 Exponential decrease of light intensity with propagation due to absorption.

$$\frac{dN_2}{dt} = -A_{21}N_2 \tag{18-40}$$

where N_2 is the number of atoms per unit volume in state 2. This simple first-order differential equation is in the same form as Eq. (18-32), and has the solution

$$\begin{aligned} N_2(t) &= N_2(0)e^{-A_{21}t} \\ &= N_2(0)\, e^{-t/\tau_{21}} \end{aligned} \tag{18-41}$$

where $N_2(0)$ is the initial population of the upper level, τ_{21} is the fluorescence lifetime or the *spontaneous lifetime* of the $2 \rightarrow 1$ transition, and we have defined

$$A_{21} \equiv \frac{1}{\tau_{21}} \tag{18-42}$$

The time dependence of $N_2(t)$ is that of an exponential decay, as shown in Fig. 18-13. The fluorescence lifetime τ_{21} is the time at which the number of atoms in the upper state has decreased to a fraction $1/e$ of the initial value. After each successive time interval of τ_{21}, the number remaining in the upper state decreases by an additional factor of $1/e$, eventually getting very small but never (in principle) going completely to zero.

Using Eq. (18-42), the cross section of Eq. (18-36) can be written as

$$\sigma(\nu) = \frac{1}{\tau_{21}} \frac{\lambda^2}{8\pi} g(\nu)$$

Integrating over ν and using the normalization condition $\int g(\nu)\, d\nu = 1$, the lifetime can be expressed as

$$\tau_{21} = \frac{\lambda^2}{8\pi \int \sigma(\nu)\, d\nu} \tag{18-43}$$

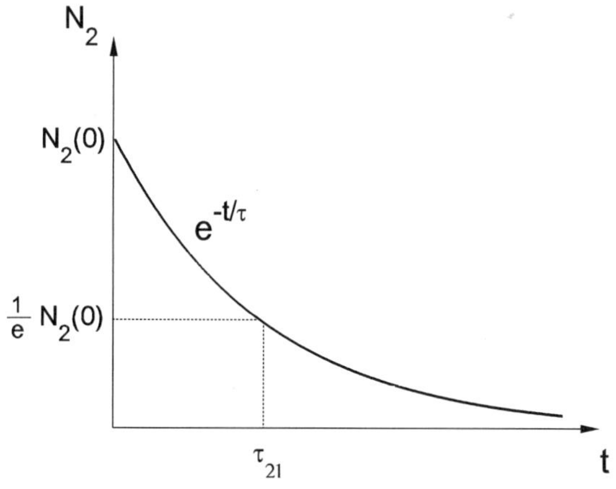

Figure 18-13 Population in excited state 2 decays exponentially with lifetime τ_{21}, due to spontaneous emission.

The lifetime therefore depends only on the wavelength and the integrated cross section. Sometimes, this integrated cross section is expressed in terms of a dimensionless parameter known as the *oscillator strength:*

$$f \equiv (4\pi\varepsilon_0)\,\frac{mc}{\pi e^2}\int \sigma(\nu)\,d\nu \qquad \text{(oscillator strength)} \qquad (18\text{-}44)$$

where m and e are the mass and charge of an electron, and ε_0 is the permittivity of free space. Transitions with $f \sim 1$ are termed *allowed transitions,* and have very short lifetimes, typically on the order of 10 ns for transitions in the visible and near IR regions. Some transitions (for example those of the rare earth ions) are partially forbidden by quantum mechanical selection rules, and have much lower oscillator strengths. For a typical rare earth ion transition, $f \sim 10^{-6}$ and $\tau \sim 10^{-3}$ s. The wide range of lifetimes and oscillator strengths has implications for various laser characteristics, as will be seen in subsequent chapters.

Quantum Yield

So far it has been assumed that level 2 decays only radiatively, by emitting a photon. For an isolated atom, this is the only decay process available for depopulating the excited state. However, if an atom in level 2 can interact with other nearby atoms, the energy in level 2 can be given to the surrounding atoms in a *nonradiative decay* process. In such a process, the excitation energy in level 2 is released from the atom without the emission of a photon. This can occur in a gas, for example, when energy is transferred between atoms during a collision. In a solid, energy can be transferred from level 2 to the vibrational modes (phonons) of the solid, which results in heating of the material. Another possible decay mechanism in a solid is *energy transfer,* in which the energy in an excited electronic state of one atom is transferred to an excited electronic state of a nearby atom. The probability per unit time that the atom in level 2 decays nonradiatively by any mechanism will be designated W_{nr}.

The two types of spontaneous decay processes for level 2 are summarized in Fig. 18-14. $W_r = A_{21}$ is the radiative decay rate (probability per unit time that atom decays radiatively), and W_{nr} is the sum of all possible nonradiative decay rates. The total decay rate out of level 2 is

$$\begin{aligned} W_{\text{tot}} &= W_r + W_{nr} \\ &\equiv \frac{1}{\tau} \end{aligned} \tag{18-45}$$

where τ is the fluorescence lifetime of level 2. In the rate equation of Eq. (18-40), A_{21} is now replaced by W_{tot}, and the population of level 2 decays in time as $N_2(0)\exp(-t/\tau)$.

The nonradiative decay process is generally detrimental to the operation of lasers, since it causes a certain fraction of the excitation energy to be wasted as heat. A quantitative figure of merit for a laser transition is the *quantum efficiency* (or quantum yield) ϕ, defined as the fraction of the time that an atom in state 2 will decay radiatively. This can be written as

$$\begin{aligned} \phi &= \frac{\text{(prob. for radiative decay)/(time)}}{\text{(prob. for any decay)/(time)}} \\ &= \frac{W_r}{W_{\text{tot}}} \end{aligned} \tag{18-46}$$

The radiative and nonradiative rates can also be written in terms of a lifetime according to

$$\begin{aligned} W_r &\equiv \frac{1}{\tau_r} \\ W_{nr} &\equiv \frac{1}{\tau_{nr}} \end{aligned} \tag{18-47}$$

so that ϕ can be written in the form

$$\phi = \frac{1/\tau_r}{1/\tau} = \frac{\tau}{\tau_r} \qquad \text{(quantum efficiency)} \tag{18-48}$$

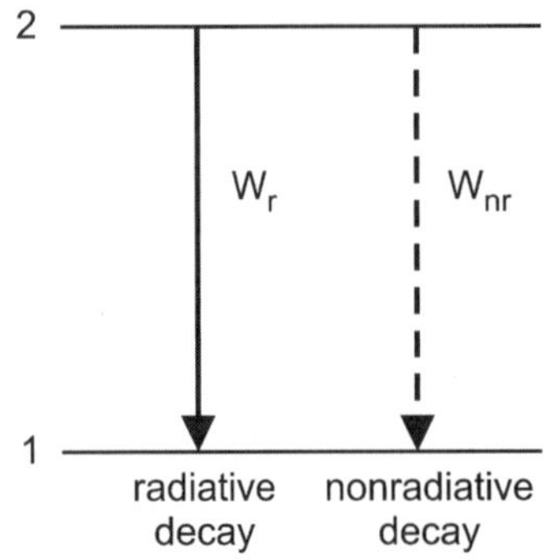

Figure 18-14 Excited state 2 can decay spontaneously either by radiative (solid line) or nonradiative (dashed line) processes.

Lineshape Function

We now consider in more detail the lineshape function $g(\nu)$, introduced in Eq. (18-3). A number of different types of lineshapes are encountered in the gain media for lasers. They may be classified as *homogeneously broadened* or *inhomogeneously broadened,* depending on whether the lineshape function is the same or different for each atom in the medium. A homogeneously broadened transition has the Lorentzian lineshape, given by

$$g(\nu) = \frac{1}{\pi} \frac{\Delta\nu/2}{(\nu - \nu_0)^2 + (\Delta\nu/2)^2} \qquad \text{(homogeneous lineshape)} \qquad (18\text{-}49)$$

where ν_0 is the center frequency of the atomic transition, and $\Delta\nu$ is the full width at half maximum (FWHM), as illustrated in Fig. 18-15. The maximum value of the lineshape function occurs at the line center frequency ν_0, and is given by

$$g_{\max} = g(\nu_0) = \frac{2}{\pi\Delta\nu} \qquad (18\text{-}50)$$

This reciprical relation between $g_{\max}$ and $\Delta\nu$ has important implications for the gain of a laser medium, since the gain coefficient is proportional to $g(\nu)$. Transitions with a narrow linewidth are, therefore, expected to have a higher gain.

The Lorentzian lineshape arises whenever the atomic state (described in quantum mechanics by the *wave function*) is perturbed by processes that occur randomly in time. If the average time between interuptions of the atomic state is Δt, the corresponding frequency width will be given by the uncertainty relation $\Delta\nu \sim 1/\Delta t$ (see Appendix B). For example, an atom in the excited state interacts with the electromagnetic field, and can emit a photon with a certain (constant) probability per unit time. This photon emission leads, as we have seen, to the exponential decay of the upper-state population, which is described by the fluorescence lifetime. The broadening of the lineshape due to photon emission is therefore termed *lifetime broadening*. The linewidth that results from lifetime broadening is referred to as the *natural linewidth,* since it occurs naturally for an isolated atom.

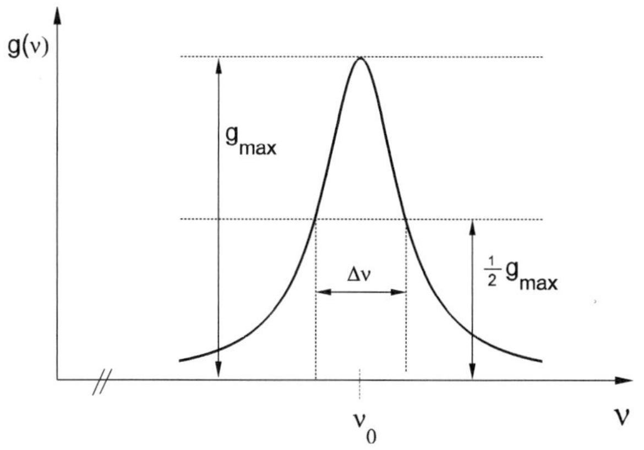

Figure 18-15 Lorentzian lineshape for homogeneously broadened transition, showing definition of full width at half maximum $\Delta\nu$.

Lifetime broadening is usually a minor contribution to the homogeneous linewidth, because there are other processes that generally interupt the atomic state more frequently than photon emission. In a gas, for example, collisions between atoms occur at a rate that is proportional to the gas pressure, resulting in a *pressure-broadening* contribution to the transition linewidth. In a solid, neighboring atoms do not "collide" directly, but they still interact via the vibrations of the solid. In quantum mechanics, this interaction is described by the absorption and emission of *phonons,* which are the quanta of lattice vibration. The resulting broadening of the linewidth is termed *phonon broadening,* which increases at higher temperatures where the amplitude of atomic vibrations (and hence the number of phonons) is greater.

In addition to the homogeneous broadening mechanisms discussed above, the transition lineshape may be inhomogeneously broadened as well. In this case, the lineshape is due to the superposition of many homogeneous lineshape components with different center frequencies, as illustrated in Fig. 18-16. For a gas medium, inhomogeneous broadening arises from the Doppler shift, in which atoms moving toward or away from the observer have their transition frequencies shifted up or down. For a solid, inhomogeneous broadening is due to the different local environment surrounding the various atoms in the medium, which causes a shift in the atoms' energy levels. This is especially pronounced in disordered media such as glasses or liquids, but it is present even in crystalline solids, where microscopic strains cause a spatially varying local environment.

The physical processes that give rise to inhomogeneous broadening are random in nature (velocity of atom in gas, site-to-site variation in local environment in solid), and the lineshape therefore represents the statistical distribution of a randomly varying center frequency. Mathematically, the relative probability of occurance for a random variable is given by the Gaussian distribution, which leads to the lineshape function

$$g(\nu) = ae^{-b(\nu-\nu_0)^2} \qquad \text{(inhomogeneous lineshape)} \qquad (18\text{-}51)$$

where ν_0 is the average center frequency. The constant b is determined by requiring $g(\nu_0 \pm \Delta\nu/2) = (1/2)\, g(\nu_0)$, whereas a is determined from the normalization condition $\int g(\nu)\, d\nu = 1$. The result is

$$\begin{aligned} a &= \frac{1}{\Delta\nu}\sqrt{\frac{4\ln 2}{\pi}} \\ b &= \frac{4\ln 2}{(\Delta\nu)^2} \end{aligned} \qquad (18\text{-}52)$$

Eq. (18-51) is valid when the homogeneous linewidth $\Delta\nu_h$ is much smaller than the inhomogeneous linewidth $\Delta\nu_{\text{inh}}$, and Eq. (18-49) is valid in the opposite limit, $\Delta\nu_h \gg \Delta\nu_{\text{inh}}$. In the intermediate case in which $\Delta\nu_h \sim \Delta\nu_{\text{inh}}$, the lineshape is given by the *Voight profile* (see, for example, DiBartolo 1968), which is the convolution of these two functions. This function is mathematically complex, and is not often used in practice.

Finally, it should be noted that the lineshape functions in Eqs. (18-49) and (18-51) only apply to transitions between a single (nondegenerate) pair of energy levels. Many atomic states actually consist of a series of closely spaced sublevels, as shown in Fig. 18-9. The many possible transitions between sublevels of the upper state and sublevels of the lower state lead to a rather complex lineshape function $g(\nu)$, as illustrated in Fig. 18-10 for the rare earth ion Er^{3+}. If the exact energies of the various sublevels are known, the lineshape

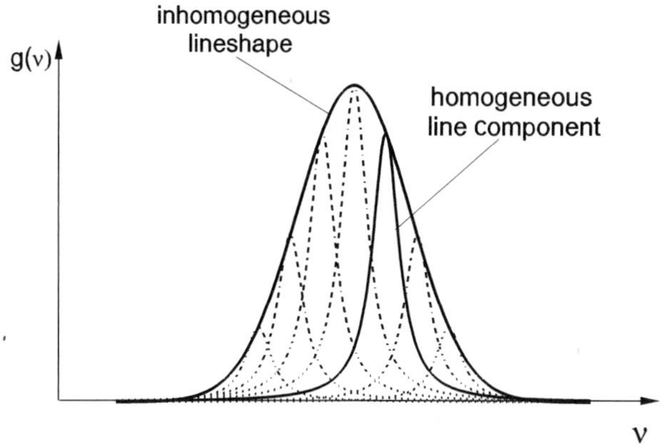

Figure 18-16 Inhomogeneously broadened lineshape is the superposition of many homogeneous lineshape components with different center frequencies.

can be written as a sum of Lorentzian or Gaussian lines with a finite number of center frequencies. In practice, the absorption or emission lineshapes are usually measured experimentally, and one is determined from the other using the McCumber relation of Eq. (18-38).

PROBLEMS

18.1 In an Er:glass laser, absorption from the ground state to the first excited state occurs at a wavelength of 1500 nm. At room temperature, determine the fraction of Er ions that are in the excited state if the material is in thermal equilibrium.

18.2 (a) Show that for a Lorentzian lineshape the peak cross section is

$$\sigma_{\text{peak}} = \frac{\lambda^2}{\tau_{21} 4\pi^2 n^2 \Delta\nu}$$

where $\Delta\nu$ is the full width at half maximum of the lineshape, τ_{21} is the radiative lifetime, λ is the free-space wavelength, and n is the material's index of refraction. (b) Derive a similar expression for a Gaussian lineshape function.

18.3 The active medium for a ruby laser consists of Cr^{3+} ions doped in an Al_2O_3 crystal. The spectroscopic parameters relevant for a ruby laser are given in Table 23-1. (a) Assuming that the emission spectrum has a Lorentzian lineshape, calculate the radiative lifetime of the transition (hint: see Problem 18.2). (b) Determine the quantum efficiency. (c) Determine the radiative and nonradiative decay rates for this transition.

18.4 For the ruby laser transition considered in Problem 18.3, calculate the absorption coefficient (probability of absorption per unit length) for fluorescence emitted at the peak wavelength. Assume the absorption and emission cross sections are equal, and take the Cr^{3+} ion density to be 1.6×10^{19} ions/cm^3. Use this to estimate the probability that an emitted photon will be reabsorbed as it passes through 0.5 cm of the ruby crystal.

18.5 (a) For the ruby laser transition considered in Problem 18.3, calculate the Einstein A and B coefficients. (b) During a Q-switched laser pulse, a typical intracavity peak intensity is 10^9 W/cm^2. Assuming that the laser pulse is spectrally narrow compared with the transition's Lorentzian linewidth, and occurs at the center of the lineshape, calculate the stimulated emission rate for a single Cr^{3+} ion in the excited state. How does this rate compare with the spontaneous emission rate from a single ion? (Note: in the relation between A and B, replace c by c/n for a medium with index of refraction n.)

18.6 An Er-doped glass fiber has a doping level of 8×10^{18} Er ions/cm^3. It is optically pumped with light of wavelength 1480 nm, and lasing occurs at a wavelength of 1560 nm. The absorption and emission cross sections are given in Fig. 18-10. (a) Determine the absorption coefficient at 1480 nm. (b) Determine the fraction of pump light absorbed in a fiber of length 2 m. (c) If the fiber is strongly pumped, so that most of the Er ions are in the excited state, determine the gain coefficient in the fiber at 1560 nm. (d) Calculate the net gain at 1560 nm in a 2 m length of this fiber (gain = power out/power in).

18.7 Obtain an expression for the radiative lifetime in terms of the transition oscillator strength. Calculate the lifetime for a transition with wavelength 500 nm and $f = 0.5$.

18.8 Determine the oscillator strength for the ruby laser and dye laser transitions, using the data from Table 23-1. Assume a Lorentzian lineshape. Which of these is an "allowed" transition?

18.9 A fiber amplifier uses a transition from the $^1G_4$ level of Pr^{3+} to provide amplification for 1300 nm light. The calculated radiative lifetime from the $^1G_4$ is 3.0 ms, whereas the measured fluorescence lifetime is 110 μs. Determine (a) the quantum efficiency, (b) the radiative decay rate, and (c) the nonradiative decay rate from this level.

18.10 In an absorption cell, 40% of the incident light is absorbed (60% is transmitted) in a path length of 10 cm. (a) Determine the absorption coefficient. (b) What fraction of the incident light would be transmitted in a similar absorption cell of length 25 cm?

18.11 Say that the lineshape function for a dye laser gain medium were modeled as a symmetrical triangle, rather than a Lorentzian curve. Assume the following parameters: spontaneous emission lifetime 8 ns, a center wavelength 590 nm, lineshape width 40 nm (FWHM), index of refraction 1.33, and population inversion $N_2 - N_1 = 1.0 \times 10^{18}$ cm^{-3}. Determine (a) the peak cross section for stimulated emission, (b) the peak gain coefficient, and (c) the thickness of this gain medium needed to amplify the light by a factor of 20.

Chapter 19

Optical Amplifiers

In the previous chapter, we found that optical amplification can occur when the population N_2 in the upper state is greater than the population N_1 in the lower state, a situation known as population inversion. The gain coefficient in the medium is proportional to the population difference $\Delta N = N_2 - N_1$, with the proportionality constant being the optical cross section $\sigma(\nu)$ for the transition. In this chapter, we show how to determine the population inversion using a rate equation approach. We use this to calculate the net gain of an optical amplifier, and discuss the efficiency with which the pump power is converted into amplified light power.

19-1. GAIN COEFFICIENT

In general, the gain coefficient may vary with position in the gain medium. We will therefore begin by focusing on a section of the gain medium small enough that the gain coefficient there can be considered to be approximately constant. The gain coefficient in this small section depends on the population inversion $N_2 - N_1$ in that section. To determine the level populations N_2 and N_1, we use a rate equation approach.

Rate Equation Approach

Consider two possible energy level schemes for obtaining a population inversion, as shown in Fig. 19-1. In both schemes, a higher state (level 3) is excited directly by the pumping mechanism, and this state decays quickly and nonradiatively to the upper laser level 2. Stimulated emission then occurs from level 2 to level 1, which is the optical gain transition. The difference between the two schemes is in the position of the lower laser level. For the *three-level system,* the lower laser level is the *ground state* (lowest energy level), whereas for the *four-level system* the lower laser level is an *excited state* of the system. It will be assumed that this lower laser level decays quickly back to the ground state. To achieve population inversion ($N_2 > N_1$) in the three-level system requires that at least half the atoms be pumped out of the ground state, which requires a good deal of pump energy. In contrast, the four-level system can achieve population inversion with only a small number of atoms raised out of the ground state.

Generally, it is easiest to obtain amplification and lasing with a four-level system because not as much pump energy must be wasted in removing atoms from the ground state. It is perhaps ironic, then, that the first laser, experimentally demonstrated in 1960, was based on the three-level energy scheme of ruby (Cr^{3+}:Al_2O_3). This laser was not efficient, however, and only operated in a pulsed mode. Soon, other lasers such as Nd:YAG were introduced, which were more efficient and operated continuously. These other lasers were

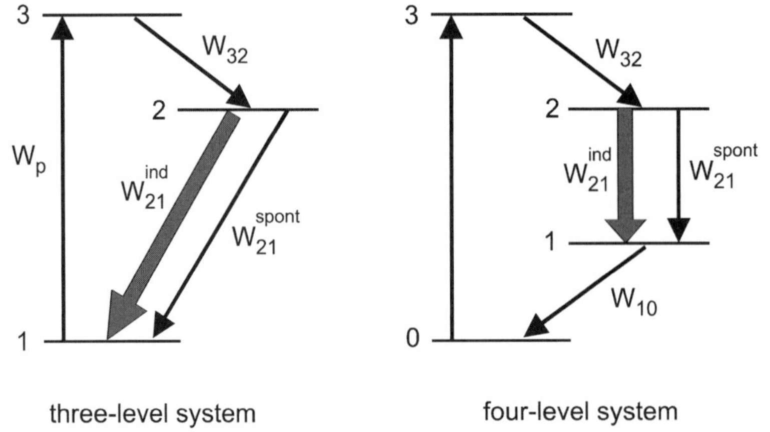

Figure 19-1 Two common energy level schemes for a laser or amplifier, showing relevant transition rates between levels. Thick arrows correspond to stimulated emission.

based on the more efficient four-level energy scheme. Since they are easier to treat mathematically, we will mostly confine our detailed analysis to four-level systems here.

The movement of population between levels is analogous to the flow of water between holding tanks in a recirculating system, as illustrated in Fig. 19-2 for a four-level system. The amount of water in a tank corresponds to the population N_i of that energy level. Water in the lowest holding tank (the "ground state") is "pumped up" by a water pump into the highest tank (#3), which drains quickly into the next-lowest tank (#2) through a large hole in the bottom. Tank #2 (the upper laser level) has a much smaller hole in the bottom, so water tends to build up there. The water that drips from tank #2 down into tank #1 is quickly drained from tank #1 by a large hole in its bottom, and returns to the lowest tank.

In our analogy, tanks with larger holes in the bottom correspond to levels with faster relaxation rates, which means that $W_{32} \gg W_{21}$ and $W_{10} \gg W_{21}$ for our system. It is clear that in the steady state there will be little water in tanks #1 and #3, which means that $N_1 \approx$

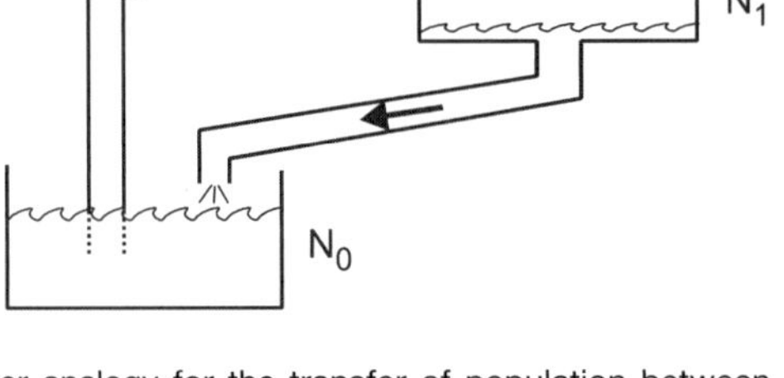

Figure 19-2 Flowing-water analogy for the transfer of population between energy levels in laser gain medium.

$N_3 \approx 0$. Population inversion is, therefore, readily obtained, since $\Delta N \approx N_2$, which is positive for any rate of pumping. It is also clear that level 2 is in effect being directly populated by the pump, since any water placed in tank #3 very quickly passes down to tank #2.

With the above approximations, only a single rate equation is required to describe the level populations in the four-level system. The rate of change of the upper laser state population can be written as

$$\begin{aligned} \frac{dN_2}{dt} &= N_0 W_p - N_2 W_{21}^{\text{ind}} + N_1 W_{12}^{\text{ind}} - \frac{N_2}{\tau_2} \\ &\simeq N_0 W_p - N_2 W_{21}^{\text{ind}} - \frac{N_2}{\tau_2} \end{aligned} \tag{19-1}$$

where τ_2 is the fluorescence lifetime of level 2 and the approximation $N_1 \ll N_2$ has been used. The rate W_p is the *pump rate,* defined as the probability per unit time that an atom is promoted by the pump from the ground state up to level 3. The rate W_{21}^{ind} is, as before, the probability per unit time for an induced transition.

The induced transition rate was found in the previous chapter to be proportional to the light intensity I that is resonant with the $2 \rightarrow 1$ transition. Using Eqs. (18-20), (18-10), and (18-36), it can be written in the form

$$\begin{aligned} W_{21}^{\text{ind}} &= \left(\frac{A_{21}}{8\pi\nu^2 (h\nu)/c^3} \right) \left(\frac{I}{c} \right) g(\nu) \\ &= \left(\frac{A_{21}\lambda^2 g(\nu)}{8\pi} \right) \frac{I}{h\nu} \\ &= \frac{I\sigma(\nu)}{h\nu} \end{aligned} \tag{19-2}$$

where $\sigma(\nu)$ is the gain cross section. Eq. (19-2) applies to induced rates for both absorption and emission,

$$\begin{aligned} W_{21}^{\text{ind}} &= \frac{I\sigma_{\text{em}}(\nu)}{h\nu} \\ W_{12}^{\text{ind}} &= \frac{I\sigma_{\text{abs}}(\nu)}{h\nu} \end{aligned} \tag{19-3}$$

where $\sigma_{\text{em}}(\nu)$ and $\sigma_{\text{abs}}(\nu)$ are the emission and absorption cross sections. In the case of optical pumping, the pump rate is given by $W_p = I\sigma_{\text{abs}}(\nu_p)/(h\nu_p)$, where ν_p is the frequency of the pump light.

Using Eq. (19-2) for W_{21}^{ind}, the rate equation of Eq. (19-1) can be written as

$$\frac{dN_2}{dt} = \mathcal{R} - N_2 \left(\frac{I\sigma}{h\nu} + \frac{1}{\tau_2} \right) \tag{19-4}$$

where $\mathcal{R} \equiv N_0 W_p$ is the total number of atoms pumped up to level 2 per unit volume per unit time. This equation relates the excited-state population N_2 to the light intensity I, and is one of the fundamental equations that will be used in this and subsequent chapters to understand the operation and behavior of lasers.

Gain Saturation

The gain coefficient $\gamma = N_2\sigma$ can be determined by solving Eq. (19-4) for N_2. This equation can be written in the more compact form

$$\frac{dN_2}{dt} = \mathcal{R} - \frac{N_2}{\tau_2'} \tag{19-5}$$

where τ_2' is an effective lifetime for the upper level, given by

$$\frac{1}{\tau_2'} \equiv \frac{1}{\tau_2} + \frac{I\sigma}{h\nu}$$

If I is constant in time, and if there is little depletion of the ground state ($N_2 \ll N_0$), then both τ_2' and $\mathcal{R}$ are constants. This makes Eq. (19-5) a linear, first-order differential equation, similar to that for a capacitor being charged through a resistor by a fixed voltage. The solution (which can be verified by substitution) is

$$N_2(t) = N_2(\infty)[1 - e^{-t/\tau_2'}] \tag{19-6}$$

where $N_2(\infty)$ is the limiting value of N_2 as $t \to \infty$, referred to as the steady-state value of N_2. The time dependence of Eq. (19-6) is that of an exponential rise to the steady state value, as illustrated in Fig. 19-3.

In the steady state, where $dN_2/dt = 0$, Eq. (19-5) becomes

$$\begin{aligned} N_2(\infty) &= \mathcal{R}\tau_2' \\ &= \frac{\mathcal{R}\tau_2}{1 + I\sigma\tau_2/(h\nu)} \end{aligned} \tag{19-7}$$

Note that a higher light intensity I gives rise to a smaller steady-state population $N_2(\infty)$. In our water system analogy, this corresponds to placing a second pump below tank #2,

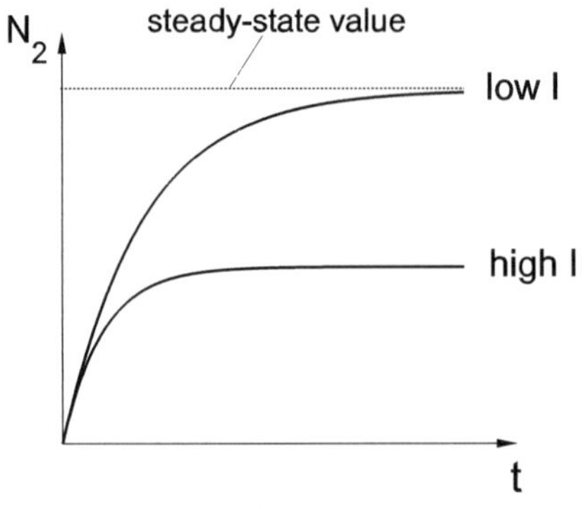

Figure 19-3 Excited state population N_2 versus time for constant pump rate and signal intensity I. Higher I decreases the population inversion.

which actively pulls water down from tank #2 to tank #1. It is intuitively clear that this will result is a smaller steady-state amount of water in tank #2. The steady state is also reached more quickly with a higher I, since the "time constant" τ_2' is smaller.

The gain coefficient in the steady state is then

$$\gamma(\nu) = N_2(\infty)\sigma(\nu) = \frac{\mathcal{R}\tau_2\sigma(\nu)}{1 + I\sigma\,\tau_2/(h\nu)} \tag{19-8}$$

which also decreases with higher I. The reduction in gain with increasing light intensity is termed *gain saturation,* and plays a key role in the operation of a laser or amplifier. It is convenient to characterize gain saturation by the *saturation intensity* I_s, defined as the intensity at which the gain is reduced by a factor of 2 from its low-intensity value. Setting the denominator of Eq. (19-8) equal to 2, we have

$$\frac{I_s\sigma\tau_2}{h\nu} = 1$$

or

$$I_s = \frac{h\nu}{\sigma\tau_2} \qquad \text{(saturation intensity)} \tag{19-9}$$

The gain coefficient in Eq. (19-8) can then be written as

$$\begin{aligned}\gamma(\nu) &= \frac{\mathcal{R}\tau_2\sigma(\nu)}{1 + I/I_s} \\ &= \frac{\gamma_0(\nu)}{1 + I/I_s}\end{aligned} \tag{19-10}$$

where

$$\gamma_0 \equiv \mathcal{R}\tau_2\sigma = \frac{\mathcal{R}h\nu}{I_s} \tag{19-11}$$

is the *unsaturated gain coefficient.*

The derivation leading to Eq. (19-8) assumes that all atoms in the gain medium are equivalent, which means that the lineshape is homogeneously broadened. In this case, a strong light intensity at frequency ν' will saturate each part of the gain curve to the same degree, as illustrated in Fig. 19-4a. If instead the lineshape is inhomogeneously broadened, then the light at ν' will only interact strongly with those atoms that have center frequencies within a homogeneous linewidth of ν'. The gain from these "spectrally nearby" atoms will be saturated, whereas the gain from atoms in other parts of the lineshape spectrum will remain at the unsaturated value. The result is a spectral "hole" in the gain spectrum, as illustrated in Fig. 19-4b, a phenomenon known as *spectral hole burning.* The gain coefficient at the saturating frequency ν' varies with I according to (Hawkes and Latimer 1995)

$$\gamma(\nu') = \frac{\gamma_0(\nu')}{\sqrt{1 + I/I_s}} \qquad \text{(inhomogeneous gain saturation)} \tag{19-12}$$

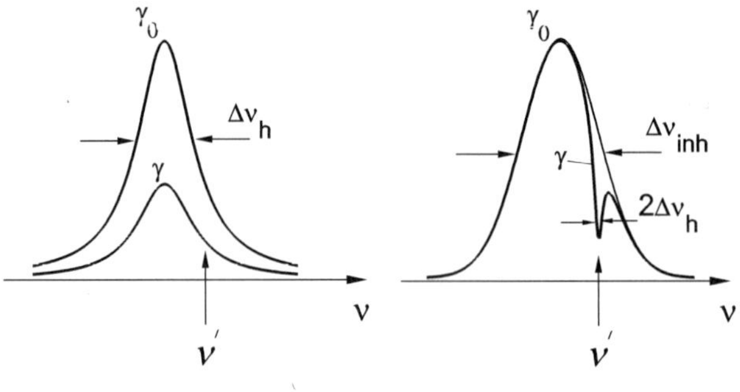

Figure 19-4 (a) Uniform saturation of gain for homogeneous lineshape. (b) Spectral hole burning in saturation of inhomogeneous lineshape. The frequency width of the hole is twice the homogeneous linewidth.

which is similar to the result for homogeneous broadening except for the square root in the denominator.

The weaker dependence of gain saturation on intensity I for inhomogeneous broadening can be understood qualitatively in the following way. As the intensity increases and saturates the atoms in the initial spectral hole, the gain from atoms farther away in the spectrum becomes more important in comparison with those nearby. Increasing the intensity still further causes even these farther-away atoms to be saturated, making the spectral hole wider. The gain saturates more slowly with intensity because the fraction of atoms that contribute to the gain increases as the intensity increases and the hole gets wider.

19-2. TOTAL GAIN OF AMPLIFIER

The gain coefficients obtained in the previous section can now be used to calculate the total gain of an optical amplifier of finite length L. The gain will be defined as $G \equiv I_2/I_1$, where I_1 and I_2 are the light intensities entering and leaving the amplifier, as shown in Fig. 19-5. The gain coefficient γ gives the fractional increase in intensity dI/I for light travers-

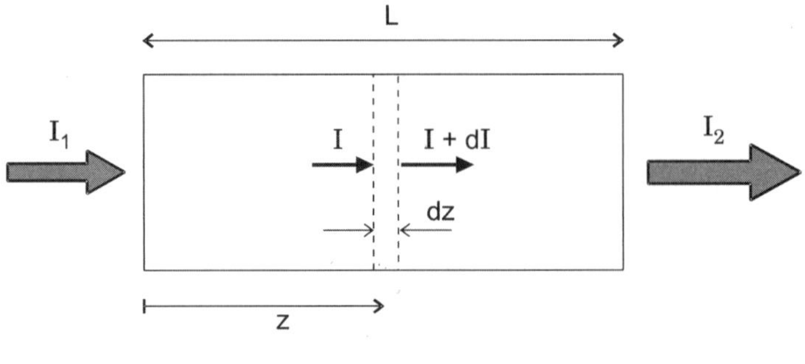

Figure 19-5 Input intensity I_1 and output intensity I_2 for optical amplifier of length L, showing intensity increase dI in thin slice dz.

ing a small slab of thickness dz, according to $dI = I\gamma\, dz$. Using Eq. (19-10) for γ with homogeneous gain saturation, this becomes

$$\frac{1}{I}\frac{dI}{dz} = \frac{\gamma_0}{1 + I/I_s} \tag{19-13}$$

To obtain the total gain G, Eq. (19-13) must be integrated over the entire length L of the amplifier. We consider first two simple special cases, and then the more general case.

Small Signal Gain

For signal intensities I small enough that $I \ll I_s$, Eq. (19-12) takes on the simple form

$$\frac{dI}{I} = \gamma_0\, dz$$

which can be directly integrated to give

$$I(z) = I(0)\, e^{\gamma_0 z} \tag{19-14}$$

This result, previously obtained in Eq. (18-34), gives the total gain as

$$G \equiv \frac{I_2}{I_1} = e^{\gamma_0 L} \qquad \text{(small signal gain)} \tag{19-15}$$

Since the small signal gain depends exponentially on the amplifier length, it is useful to describe this gain in decibel units:

$$\begin{aligned} \text{dB gain} &= 10 \log_{10} G \\ &= 10\, \gamma_0 L \log_{10} e \\ &= 4.34\, \gamma_0 L \end{aligned} \tag{19-16}$$

The dB gain is then linear with the amplifier length, and the unsaturated gain coefficient γ_0 can be given in units of dB/m. Thus, a value of $\gamma_0 = 1\ \text{m}^{-1}$ corresponds to 4.34 dB/m of gain. Note that each *addition* of 10 dB to the gain corresponds to a *factor of* 10 increase in the gain G.

The decibel concept also applies to the attenuation of light by absorption or scattering. The dB loss due to an attenuation coefficient α was found in Eq. (5-3) to be

$$\text{dB loss} = 4.34\, \alpha L$$

where α is the fractional loss per unit length. When both gain and attenuation are present in the gain medium, the net gain is

$$\text{dB net gain} = 4.34\, (\gamma_0 - \alpha)\, L \tag{19-17}$$

For a practical amplifier, it is necessary not only that $\gamma_0 > 0$, but also that $\gamma_0 > \alpha$.

EXAMPLE 19-1

A fiber amplifier has a net unsaturated gain of 25 dB in a length of 8 m. If the fiber loss is 2×10^{-4} cm^{-1}, determine the unsaturated gain coefficient in m^{-1}.

Solution: Writing the net gain in m^{-1},

$$\gamma_0 - \alpha = \frac{25}{(8)(4.34)} = 0.72 \text{ m}^{-1}$$

Therefore,

$$\gamma_0 = 0.72 + 0.02 = 0.74 \text{ m}^{-1}$$

Large Signal Gain

In the limit $I \gg I_s$, Eq. (19-13) simplifies to

$$\frac{dI}{dz} \simeq \gamma_0 I_s \tag{19-18}$$

where the right-hand side is a constant. Integrating this over z gives $\Delta I \simeq \gamma_0 I_s \Delta z$, or

$$I_2 - I_1 \simeq \gamma_0 I_s L \tag{19-19}$$

The output of the amplifier now increases linearly with amplifier length, rather than exponentially as in the small-signal case. The gain is found by dividing Eq. (19-19) by I_1,

$$G \simeq 1 + \frac{I_s}{I_1}\gamma_0 L \tag{19-20}$$

which is also linear with L.

In the large-signal limit, each additional amplification length *adds* a fixed amount of energy to the beam. This is in contrast to the small-signal limit, where each additional length *multiplies* the beam energy by a constant factor. To obtain some physical insight into this difference, it is instructive to consider the efficiency η with which absorbed pump power is converted into signal power. The increase in signal power P_{sig} in the large signal limit is

$$\begin{aligned} \Delta P_{\text{sig}} &= P_{\text{sig}}^{\text{out}} - P_{\text{sig}}^{\text{in}} \\ &= (I_2 - I_1)A \\ &= \gamma_0 I_s L A \\ &= \mathcal{R} h\nu L A \end{aligned} \tag{19-21}$$

where A is the cross-sectional area of the signal beam, and Eqs. (19-19) and (19-11) have been used. The power absorbed from the pump can be expressed as

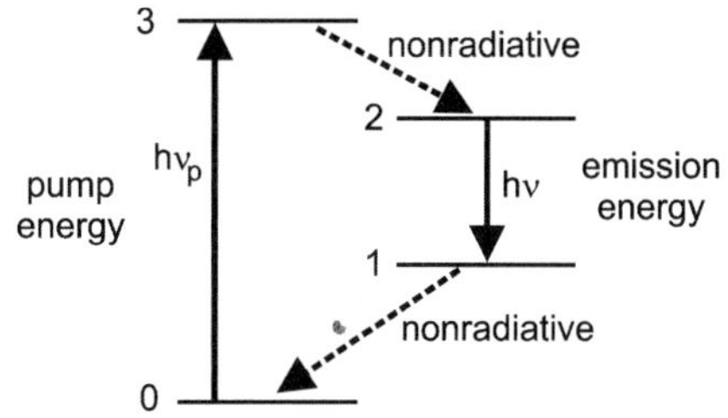

Figure 19-6 Emission energy is less than pump energy by the quantum defect.

$$P_{\text{pump}}^{\text{abs}} = \left[\frac{\text{atoms excited}}{(\text{time})(\text{vol})}\right]\left[\frac{\text{absorbed energy}}{\text{excited atom}}\right][\text{Vol}]$$
$$= \mathcal{R}h\nu_p LA \tag{19-22}$$

where LA is the volume of the pumped region of the gain medium, and $h\nu_p$ is the energy needed to put the atom in level 3 of the four-level system. In the case of optical pumping, this would be the energy of a pump photon. Equations (19-21) and (19-22) can be combined to give the desired energy conversion efficiency for the amplifier,

$$\eta \equiv \frac{\Delta P_{\text{sig}}}{P_{\text{pump}}^{\text{abs}}} = \frac{h\nu}{h\nu_p} \qquad \text{(large signal)} \tag{19-23}$$

Equation (19-23) gives a particularly simple result for the large-signal amplifier efficiency, which suggests a simple physical interpretation. Referring to the energy level scheme for the four-level system shown in Fig. 19-6, it is clear that the signal photon energy $h\nu$ is less than the pump photon energy $h\nu_p$, so that $\eta < 1$. Each absorbed pump photon can give rise to one generated signal photon, by the stimulated emission process. After the signal photon is generated, the atom returns rapidly to the ground state, where it may be excited again by the pump. Since the pump energy absorbed is $\propto h\nu_p$, and the generated signal energy is $\propto h\nu$ we expect that the maximum efficiency of converting absorbed pump energy into signal beam energy would be $h\nu/h\nu_p$, in agreement with Eq. (19-23). The difference between pump and signal photon energies is sometimes referred to as the quantum defect, and gives rise to heating of the gain material if the $3 \rightarrow 2$ and $1 \rightarrow 0$ transitions are mostly nonradiative. The quantum defect also limits the efficiency of lasers, as will be seen in Chapter 20.

In the small signal limit where $I_1 \ll I_s$, η becomes

$$\eta_{\text{small } I} = \frac{(I_2 - I_1)A}{\mathcal{R}h\nu_p LA}$$
$$= \frac{(e^{\gamma_0 L} - 1)\, I_1}{\mathcal{R}h\nu_p L} \tag{19-24}$$

where Eqs. (19-22) and (19-15) have been used. For small gain lengths such that $\gamma_0 L \ll 1$, this becomes

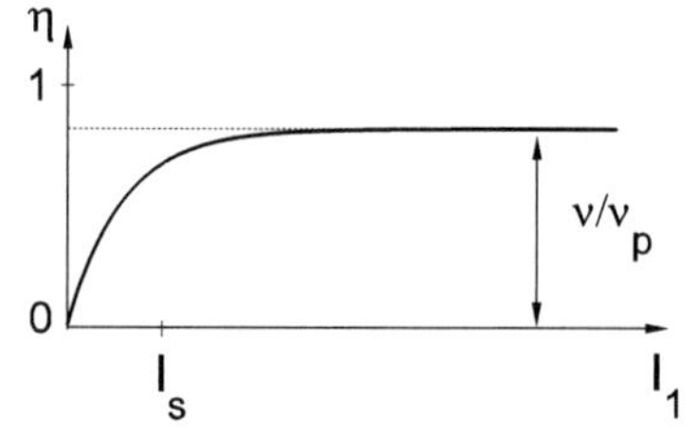

Figure 19-7 Energy conversion efficiency increases with input signal intensity I_1, reaching a limit due to quantum defect.

$$\eta_{\text{small } I} = \frac{\gamma_0 L I_1}{\mathcal{R} h\nu_p L} = \frac{I_1}{I_s}\frac{h\nu}{h\nu_p} \tag{19-25}$$

using Eq. (19-11). Note that the efficiency given in Eq. (19-25) is much smaller than the large signal limit $h\nu/h\nu_p$, since $I_1 \ll I_s$. As the signal intensity increases, the extraction efficiency increases proportionately, until the large signal limit is reached. The overall dependence of η on I_1 is shown schematically in Fig. 19-7.

The most notable feature of Fig. 19-7 is the low value for efficiency for small I_1. It is natural to ask: in what sense is the amplifier inefficient in this regime? The amplification is unsaturated here, so the signal growth is exponential, and it is "efficient" in that sense. The inefficiency indicated in Fig. 19-7 refers to the efficiency of converting absorbed pump energy into additional signal-beam energy. This is inefficient for $I_1 \ll I_s$ because much of the absorbed pump energy is emitted by the atoms as spontaneous fluorescence. The overall energy flow in the amplifier is depicted in Fig. 19-8. Pump energy absorbed in the material is converted into three different forms: additional signal energy, fluorescence energy, and heat energy. Symbolically, we can write

$$P_{\text{pump}}^{\text{abs}} = (P_{\text{sig}}^{\text{out}} - P_{\text{sig}}^{\text{in}}) + P_{\text{fl}} + P_{\text{heat}} \tag{19-26}$$

where P_{fl} is the emitted fluorescence power, and P_{heat} is the heating of the material due to the quantum defect. Both P_{fl} and P_{heat} are limited to some maximum value determined by the number of optically active ions in the material. However, the change in signal power $\Delta P_{\text{sig}} = P_{\text{sig}}^{\text{out}} - P_{\text{sig}}^{\text{in}}$ increases with increasing signal power because it depends on the stimulated emission rate. Therefore, in the strong signal limit stimulated emission dominates spontaneous emission, and $\Delta P_{\text{sig}} \gg P_{\text{fl}}$. In the small signal limit, however, $\Delta P_{\text{sig}} \ll P_{\text{fl}}$, and the ratio $\Delta P_{\text{sig}}/P_{\text{pump}}^{\text{abs}}$ becomes small.

Amplifier Gain: General Case

The general equation for amplifier gain including saturation is given by Eq. (19-13). This can be rewritten and integrated in the following way:

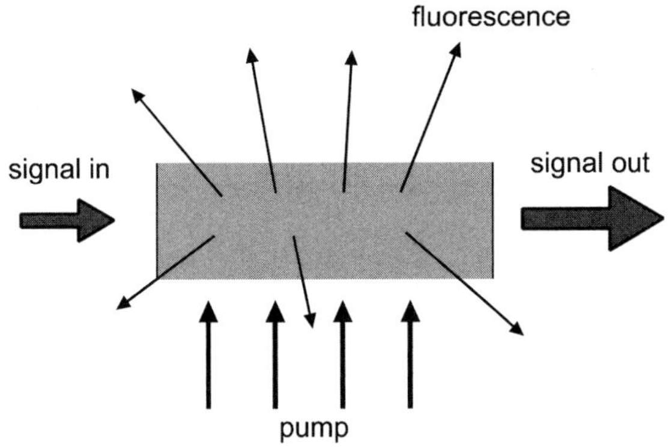

Figure 19-8 Pump energy is converted into fluorescence and heat, as well as an increase in signal beam energy. For a small input signal, most of the absorbed pump energy goes into fluorescence and heat.

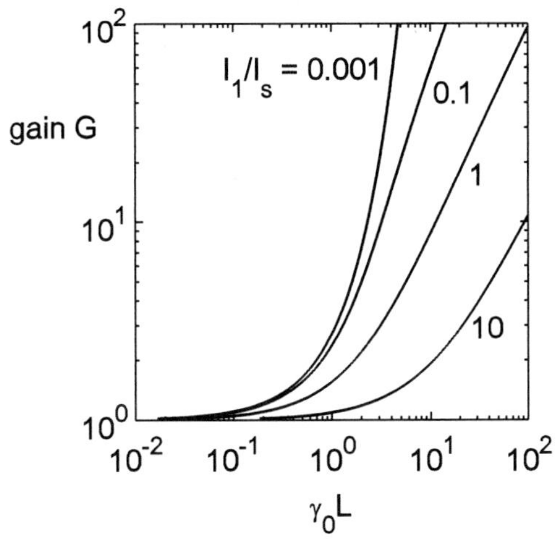

Figure 19-9 Total amplifier gain G versus unsaturated gain coefficient $\gamma_0 L$, on log-log scale. The gain G is reduced by saturation for higher input signal intensity I_1. Note that $\gamma_0 L \propto$ pump power.

$$\int_{I_1}^{I_2} \left(\frac{1}{I} + \frac{1}{I_s} \right) dI = \int_0^L \gamma_0 \, dz$$

$$\ln\left(\frac{I_2}{I_1}\right) + \frac{I_2 - I_1}{I_s} = \gamma_0 L \tag{19-27}$$

$$\ln G + \frac{I_1(G-1)}{I_s} = \gamma_0 L$$

which is an implicit equation for the gain G of the amplifier. Although this equation cannot be solved analytically for G, a plot of G versus L can be obtained by treating G as the independent variable and $\gamma_0 L$ as the dependent variable. Shown in Fig. 19-9 is a plot of G versus $\gamma_0 L$ obtained in this way for different values of I_1/I_s. For a fixed $\gamma_0 L$ (fixed pump rate), the gain G decreases with increasing I_1/I_s, due to saturation. If the gain G needs to be maintained under saturating conditions, a longer gain length L is required.

It should be noted that our discussion of amplifier saturation so far has assumed a four-level system with excitation rate $\mathcal{R}$ that is spatially uniform within the gain medium. Although these are sometimes good approximations, they are not valid for certain amplifier systems that are pumped from the end, such as fiber lasers and amplifiers. The modifications necessary to describe such systems will be discussed in Chapters 23 and 24.

PROBLEMS

19.1 (a) A dye molecule has emission cross section 4×10^{-16} cm^2 at $\lambda = 550$ nm, and fluorescence lifetime 3 ns. Assuming the transition is homogeneously broadened, calculate the signal intensity at which the gain is reduced by a factor of two. (b) The Nd:glass transition has peak emission cross section 4×10^{-20} cm^2 at $\lambda = 1054$ nm, and fluorescence lifetime 290 μs. If the transition is homogeneously broadened, determine the signal intensity at which the Nd:glass gain is reduced by a factor of two. Repeat if the transition is inhomogeneously broadened.

19.2 A Nd:glass amplifier material has peak emission cross section 4×10^{-20} cm^2 at the amplifier wavelength $\lambda = 1054$ nm, and fluorescence lifetime 290 μs. The Nd ion density is 3×10^{20} cm^{-3}, and the pump rate for one Nd ion is 170 s^{-1}. (a) Determine the number of Nd ions pumped to the excited state per unit volume per unit time. (b) If there is no signal light, determine the steady-state, upper-level population (number of excited ions per unit volume). (c) What fraction of Nd ions are in the excited state? (d) Determine the gain coefficient at 1054 nm.

19.3 For the Nd:glass amplifier of Problem 19.2, a signal beam of intensity 5×10^4 W/cm^2 is now introduced into the gain medium. Determine (a) the spontaneous and induced emission rates; (b) the new steady-state, upper-level population; (c) the new effective lifetime of the upper level; and (d) the new gain coefficient.

19.4 An optical amplifier boosts the intensity of a small signal by a factor of 200 over a path length of 15 cm. Determine the unsaturated gain coefficient, expressed in cm^{-1}. Also express the gain in dB/cm.

19.5 For the amplifier of Problem 19.4, assume that the saturation intensity is 2×10^4 W/cm^2, and that it is a four-level type system exhibiting homogeneous gain saturation. Determine the amplifier gain if the input signal intensity is (a) 4×10^5 W/cm^2, and (b) 2×10^4 W/cm^2.

19.6 If the optical amplifier in Problem 19.5 is in the form of a fiber with core diameter 50 μm, determine the amount of pump power that is converted into signal beam power for the two different input signal intensities.

19.7 An optical fiber amplifier of length 15 m has a small-signal gain of 1.5 dB/m at 1500 nm. When signal light of intensity 25 kW/cm^2 is coupled into the fiber, the measured signal output has intensity 500 kW/cm^2. Determine the saturation intensity of the gain medium assuming spatially uniform excitation and homogeneous gain saturation.

19.8 Consider the three-level system shown in Fig. 19-1. In this case, level 1 is the ground state, and the population N_1 can no longer be neglected in Eq. (19-1). Write the rate equations for levels 1 and 2 for the three-level system, making the assumption as before that N_3 is very small. (a) Derive an expression for the population difference $N_2 - N_1$ in terms of the signal intensity I, assuming that $\sigma_{em}(\nu) = \sigma_{abs}(\nu)$. (b) What is the minimum excitation rate $\mathcal{R}$ for a positive population inversion? (c) Write the expression for $N_2 - N_1$ in terms of a saturation intensity I_s, such that the gain decreases by a factor of two when $I = I_s$. Show how this expression for I_s compares with the one given in Eq. (19-9).

Chapter 20

Laser Oscillation

In the previous two chapters we have seen how light can be amplified by stimulated emission, which leads to a useful device: the optical amplifier. To make a laser, a feedback device such as a pair of mirrors must be added to the system. In this chapter, we consider the conditions under which the feedback provided by the mirrors will be sufficient to achieve laser oscillation. It will be seen that there is a minimum pumping power required to achieve lasing, termed the *threshold pump power.* We will then discuss the output properties of the laser above the lasing threshold, using a rate equation approach. Two different measures of the output efficiency of the laser will be discussed and contrasted. We will see how the mirror reflectivities can be chosen so as to optimize the laser output.

20-1. THRESHOLD CONDITION

The conditions under which lasing will occur can be determined by considering the simple laser cavity shown in Fig. 20-1. A uniform gain medium fills the region between two cavity mirrors, which are separated by distance L and have reflectivities R_1 and R_2. Let us say that there is a small amount of light at point A that happens to be moving in a direction toward mirror 2, as shown. As the light makes a round-trip through the cavity from point A to point B, it is amplified with a gain coefficient γ, while at the same time being attenuated by the loss coefficient α [see Eq. (19-17)]. After reflection from mirror 2, the light intensity is reduced by a factor R_2, and similarly for mirror 1. The intensity of the light arriving at point B (I_B) can then be written as

$$I_B = R_1 R_2 e^{(\gamma-\alpha)2L} I_A$$

where I_A is the intensity of light originating at point A. If $I_B < I_A$, then the light intensity will become progessively smaller with each round-trip, and lasing will not occur. The time dependence of light intensity in the cavity would be that of a decaying exponential, as in Fig. 16-6. For lasing to occur, it is necessary that $I_B > I_A$, so that the light intensity will grow exponentially in time. The *threshold condition* for laser oscillation then becomes

$$R_1 R_2 e^{(\gamma-\alpha)2L} \geq 1 \qquad \text{(threshold condition)} \qquad (20\text{-}1)$$

The smallest value of γ that satisfies this inequality is termed the *threshold gain coefficient,* and is denoted γ_{th}. Taking the log of Eq. (20-1) gives

$$\gamma_{\text{th}} = \alpha + \frac{1}{2L} \ln\left(\frac{1}{R_1 R_2}\right) \qquad \text{(threshold gain coefficient)} \qquad (20\text{-}2)$$

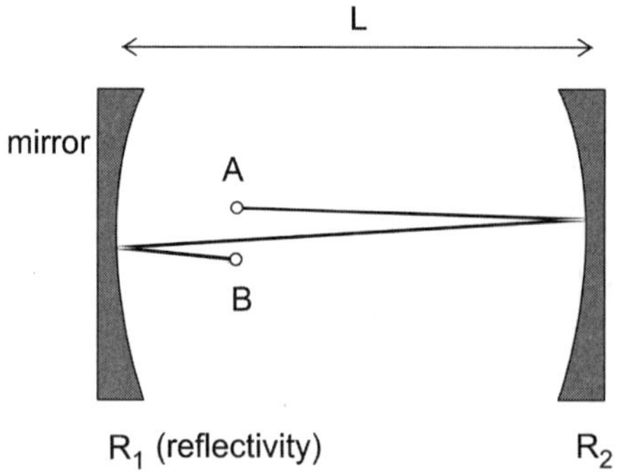

Figure 20-1 Lasing occurs when the round-trip gain from point A to point B exceeds the round-trip loss.

so that the condition for threshold can be written $\gamma > \gamma_{th}$. Note that in Eq. (19-17) it was found that an amplifier has net gain when $\gamma > \alpha$. A similar condition applies now to lasers, except that the losses include not just propagating losses, but also mirror reflection losses.

Writing the gain coefficient as in Eqs. (18-35) and (18-36), the threshold condition becomes

$$\gamma(\nu) \geq \gamma_{th}$$
$$\sigma(\nu)\,\Delta N \geq \gamma_{th} \tag{20-3}$$
$$A_{21}\frac{\lambda^2}{8\pi}\,g(\nu)\Delta N \geq \gamma_{th}$$

Solving this for ΔN gives

$$\Delta N \geq \frac{8\pi\gamma_{th}}{A_{21}\lambda^2 g(\nu)} \tag{20-4}$$

which is the population inversion required to achieve lasing. It takes a higher pump power to obtain a greater population inversion, so it is generally desirable that the required population inversion be as small as possible. Recalling from Fig. 18-15 that $g(\nu)$ takes on its maximum value g_{max} at the lineshape center frequency ν_0, and from Eqs. (18-50)–(18-52) that

$$g_{max} \simeq \frac{1}{\Delta\nu} \tag{20-5}$$

the minimum population inversion needed to achieve threshold is

$$\Delta N_{th} \simeq \frac{8\pi\gamma_{th}\Delta\nu}{A_{21}\lambda^2} \quad \text{(threshold population inversion)} \tag{20-6}$$

where $\Delta\nu$ is the full width at half maximum. Note that here $\lambda = (c/n)/\nu$ is the wavelength in the medium of refractive index n. The Einstein A coefficient in Eq. (20-6) is related to the oscillator strength of the transition f_{21} by

$$A_{21} = \frac{8\pi^2 e^2}{4\pi\varepsilon_0 mc\lambda^2} f_{21} \qquad (20\text{-}7)$$

using Eqs. (18-43) and (18-44). Substituting Eq. (20-7) into Eq. (20-6) gives an alternate form for the threshold population inversion,

$$\Delta N_{\text{th}} \simeq \frac{4\varepsilon_0 mc\Delta\nu\gamma_{\text{th}}}{e^2 f_{21}} \qquad (20\text{-}8)$$

The conditions that influence lasing threshold can be seen by inspection of Eqs. (20-8) and (20-2). The required population inversion will be lower (lasing more easily achieved) with:

1. Low loss coefficient α (low γ_{th})
2. High-reflectivity mirrors (low γ_{th})
3. Longer cavity (low γ_{th})
4. Narrow gain linewidth $\Delta\nu$
5. Large oscillator strength f_{21}

The first two of these lower the loss of the cavity, whereas the last three increase the gain. Although a longer cavity length results in a lower γ_{th}, this is not due to a reduction in cavity loss. Rather, the longer cavity allows the total round-trip gain to be larger for a given gain coefficient γ, and this allows γ to be smaller at the threshold of lasing.

It is important to note that the five conditions above, though leading to a smaller gain threshold, are not always desirable for a particular application. For example, a very narrow linewidth makes lasing easier, but restricts the ability to tune the laser over a range of wavelengths. Longer cavities make a laser less compact, which is sometimes undesirable. High oscillator strengths lead to short upper-state lifetimes, which can make pulsed operation difficult. Finally, very highly reflective mirrors allow little light to escape the laser cavity, giving a poor output efficiency. Considerations such as these are part of the engineering design trade-offs inherent in laser development.

EXAMPLE 20-1

A neodymium-doped glass laser is constructed by doping a phosphate glass rod of length 10 cm with Nd^{3+} ions, and placing mirrors at each end of the fiber. The mirror reflectivities are 1 for the end reflector, and 0.95 for the output coupler, and the attenuation coefficient in the rod is 0.2 m^{-1}. The lasing transition has a center wavelength of 1054 nm, a spectral width of 19 nm, and an oscillator strength of 7.5×10^{-6}. Determine the population inversion needed for lasing in this system.

Solution: The frequency width of the transition is

$$\Delta\nu = \Delta\left(\frac{c}{\lambda}\right) = \frac{c}{\lambda^2}\Delta\lambda = \frac{(3 \times 10^8)(19 \times 10^{-9})}{(1.054 \times 10^{-6})^2} = 5.1 \times 10^{12}\ \text{s}^{-1}$$

and the threshold gain coefficient is

$$\gamma_{th} = 0.2 + \frac{1}{(2)(0.1)} \ln\left[\frac{1}{(1)(0.95)}\right] = 0.456 \text{ m}^{-1}$$

The threshold inversion is then

$$\Delta N_{th} = \frac{(4)(8.85 \times 10^{-12})(9.1 \times 10^{-31})(3 \times 10^{8})(5.1 \times 10^{12})(0.456)}{(1.6 \times 10^{-19})^2(7.5 \times 10^{-6})}$$

$$\Delta N_{th} = 1.17 \times 10^{23} \text{ m}^{-3} = 1.17 \times 10^{17} \text{ cm}^{-3}$$

Since in a solid there are ~ 10^{23} atoms/cm^3, this requires only ~ 1 ppm (parts per million) of the atoms in the solid to be excited Nd^{3+} ions. Nd-doped phosphate glass lasers can be doped with Nd^{3+} concentrations in the range of 3×10^{20} cm^{-3}, so that less than 0.1 % of the Nd^{3+} ions need to be promoted to the excited state.

20-2. ABOVE LASING THRESHOLD

When the gain coefficient is greater than the value needed for lasing threshold, the light intensity will grow exponentially in time as the beam passes back and forth through the gain medium. Eventually, the light intensity will be high enough to cause saturation of the transition, as discussed in Section 19-1. In the steady state, the light intensity does not change in time, which means that there must be no net gain in a round-trip through the cavity. The maximum gain coefficient for steady state operation is, therefore, equal to the threshold value. We say that the gain coefficient becomes "pinned" at the threshold value for operation above threshold. To develop a quantitative understanding of laser operation above threshold, we turn next to a rate equation analysis of the excited-state population and light intensity.

Rate Equation Approach

In Section 19-1, we derived an expression for the rate of change of population in the excited state N_2. Eq. (19-4) relates dN_2/dt to both N_2 and the light intensity I. To solve this equation for N_2 and I, we need a second equation that relates these variables, since in general two equations are required to solve for two unknowns.

The additional equation can be obtained by considering how the light intensity varies as it propagates between the mirrors of the cavity. The net fractional increase in intensity after propagating a distance Δz is $(\gamma - \gamma_{th})\,\Delta z$, where γ_{th} is given by Eq. (20-2). This fractional increase occurs in a time $\Delta t = \Delta z/c$, since the beam is moving with speed c.* The fractional increase in intensity can then be written

$$\frac{\Delta I}{I} = (\gamma - \gamma_{th})\, c\Delta t \tag{20-9}$$

*For simplicity, we take the refractive index to be $n = 1$ in this discussion. When necessary, the equations can be generalized by making the replacement $c \rightarrow c/n$.

which is positive when $\gamma > \gamma_{th}$. Solving for $\Delta I/\Delta t$, we have

$$\begin{aligned}\frac{\Delta I}{\Delta t} &= c(\gamma - \gamma_{th})\, I \\ &= c\sigma N_2 I - \frac{I}{\tau_c}\end{aligned} \tag{20-10}$$

where Eq. (18-35) has been used with $\Delta N \simeq N_2$, and the cavity lifetime τ_c has been defined by

$$\begin{aligned}\frac{1}{\tau_c} &\equiv c\gamma_{th} \\ &= c\alpha + \frac{c}{2L}\ln\left(\frac{1}{R_1 R_2}\right)\end{aligned} \tag{20-11}$$

Eq. (20-11) generalizes the previous expression [Eq. (16-13)] for the cavity (or photon) lifetime, and is equivalent to it when $\alpha = 0$ and $R_1R_2 \approx 1$ (see Problem 20.3).

Taking the limit $\Delta t \rightarrow 0$ in Eq. (20-10) gives an equation for the rate of change of light intensity, which is the desired second equation relating N_2 and I. Writing this along with Eq. (19-4) yields the following set of two differential equations relating N_2 and I:

$$\frac{dN_2}{dt} = \mathcal{R} - N_2\left(\frac{I\sigma}{h\nu} + \frac{1}{\tau_2}\right) \tag{20-12}$$

$$\frac{dI}{dt} = c\sigma N_2 I - \frac{I}{\tau_c} \tag{20-13}$$

These are coupled equations, since both variables N_2 and I appear in each equation, and they are nonlinear equations because there are terms containing the product of the two variables. In the general case, these equations are difficult to solve, and numerical approaches are required. We will consider the time-dependent solutions to these equations in Chapter 22.

Although a complete solution to these equations is difficult, some general observations can be made about the time dependence of I. For example, if I is initially zero, then according to Eq. (20-13) the rate of change of I is also zero. This means that if there is initially zero intensity inside the cavity, the intensity will remain zero for all time. This would seem to preclude the possibility of lasing, unless we deliberately "seed" the cavity with a small amount of light.

Actually, there is one source of such "seed" light that we have not accounted for: spontaneous emission. Atoms in the excited state 2 will spontaneously emit light in random directions, and a very small fraction of this light will be directed into the laser cavity mode.

If there is a small amount of seed light in the cavity, then the light intensity will grow in time, provided that $dI/dt > 0$. According to Eq. (20-13) this will occur when $N_2 > N_{2,th}$, where

$$N_{2,th} = \frac{1}{c\sigma(\nu)\,\tau_c} \qquad \text{(threshold inversion)} \tag{20-14}$$

is the threshold population inversion. This expression for threshold is equivalent to that obtained earlier in Eq. (20-4), if Eqs. (18-36) and (20-11) are used. When N_2 exceeds the threshold value, the light intensity increases exponentially in time, until it becomes large enough to saturate the gain transition. Saturation decreases N_2, which decreases dI/dt, so that eventually the steady-state condition $dI/dt = 0$ is reached. To gain a further understanding of the laser's behavior above and below threshold, we turn next to a solution of Eqs. (20-12) and (20-13) in the steady state.

Steady-State Laser Output

In the steady state, Eqs. (20-12) and (20-13) become

$$0 = \mathcal{R} - N_2\left(\frac{I\sigma}{h\nu} + \frac{1}{\tau_2}\right) \tag{20-15}$$

$$0 = c\sigma N_2 I - \frac{I}{\tau_c} \tag{20-16}$$

These are still coupled, nonlinear equations, but they are algebraic rather than differential equations. The solutions for N_2 and I both above and below threshold can be found in the following way. Below threshold, I is very small, so $I\sigma/h\nu$ can be neglected compared to $1/\tau_2$ in Eq. (20-15). The excited state population is then

$$N_2 \simeq \mathcal{R}\tau_2 \qquad \text{(below threshold)} \tag{20-17}$$

which increases linearly with the excitation rate $\mathcal{R}$. When $\mathcal{R}$ reaches the threshold value

$$\mathcal{R}_{\text{th}} \equiv \frac{N_{2,\text{th}}}{\tau_2} = \frac{1}{c\sigma\tau_c\tau_2} \qquad \text{(threshold excitation rate)} \tag{20-18}$$

laser action begins, with Eq. (20-13) giving $dI/dt \geq 0$. The light intensity I builds up rapidly, but is prevented from going to infinity by saturation of the population inversion N_2. A further increase in $\mathcal{R}$ does not result in any additional increase in N_2, because the steady-state condition of Eq. (20-16) would then be violated. Instead, N_2 becomes pinned at the threshold value:

$$N_2 = N_{2,\text{th}} = \mathcal{R}_{\text{th}}\tau_2 \qquad \text{(above threshold)} \tag{20-19}$$

as illustrated in Fig. 20-2.

Although N_2 does not increase above threshold, the light intensity I does increase with increasing $\mathcal{R}$. This can be seen by combining Eqs. (20-19) and (20-15),

$$\mathcal{R} = \mathcal{R}_{\text{th}}\left(\frac{I\sigma\tau_2}{h\nu} + 1\right)$$

and solving for I to give

$$I = \frac{h\nu}{\sigma\tau_2}\left[\frac{\mathcal{R}}{\mathcal{R}_{\text{th}}} - 1\right] \tag{20-20}$$

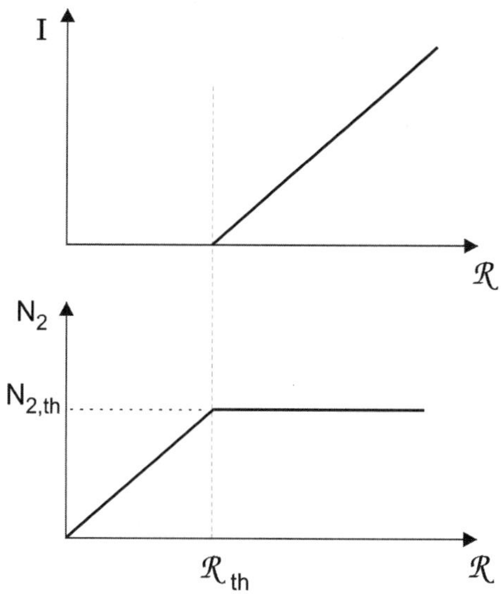

Figure 20-2 Light intensity I and upper-state population N_2 versus excitation rate $\mathcal{R}$. Above threshold, N_2 becomes pinned at $N_{2,th}$ and I increases linearly with $\mathcal{R}$.

The light intensity I is seen to increase linearly with excitation rate $\mathcal{R}$, as illustrated in Fig. 20-2. This linear increase in lasing intensity above threshold, along with the pinning of the population inversion, are key identifying features of laser action.

The dependence of N_2 and I on $\mathcal{R}$ can be understood physically by considering the flow of energy into and out of the laser. The pump energy absorbed by the laser material is converted into different forms, as indicated schematically in Fig. 20-3. Some is converted into *fluorescence,* which is light spontaneously emitted by atoms in the excited state. The power emitted by this fluorescence is proportional to N_2. Some of the absorbed pump energy is converted into heat by nonradiative processes, as discussed in Chapter 18. The remainder is converted into useful laser output energy, which is proportional to I.

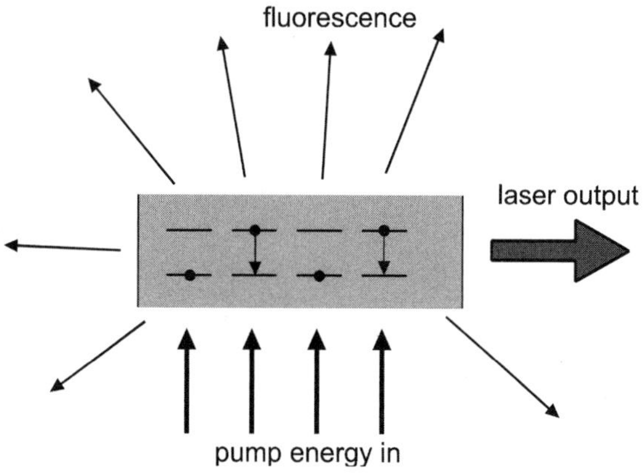

Figure 20-3 The absorbed pump power is converted into both fluorescence and laser light. The fluorescence power is $\propto N_2$, and becomes pinned above threshold.

The balance between fluorescence and laser output changes as the excitation rate $\mathcal{R}$ increases. Below threshold, there is no laser output, and any increase in absorbed energy from the pump leads to increased fluorescence. Above threshold, the fluorescence power becomes constant, since N_2 becomes pinned at the value $N_{2,\text{th}}$. Any increase in pump power above threshold, therefore, leads to an increased laser output. To make these arguments about energy flow in the laser more quantitative, we turn next to a calculation of output versus input power for a laser, and consider different measures for the laser output efficiency.

Laser Output Efficiency

The output power from the laser can be determined by referring to Fig. 20-4. The light wave inside the resonator has the form of a standing wave, which is equivalent to the superposition of two counterpropagating beams of intensities I_- and I_+ as shown. Each of these has half the intensity I of the light in the cavity, so $I_- = I_+ = I/2$. Assume for simplicity that one mirror (the left one, say) is perfectly reflecting with a reflection coefficient $R_1 = 1$, and that the other has a transmission T, so $R_2 = 1 - T$. Light will then leave the cavity only through the right mirror, with an intensity $TI_+ = TI/2$. If the cross-sectional area of the beam is A, the power exiting the laser becomes

$$P_{\text{out}} = \tfrac{1}{2} IAT \tag{20-21}$$

The input power to the laser can be taken as the absorbed pump power, which was evaluated in Eq. (19-22) to be

$$P_{\text{in}} = \mathcal{R} h\nu_p V \tag{20-22}$$

where $V = AL$ is the volume of the pumped region in the gain medium, L is the cavity length, and $h\nu_p$ is the energy required to put a single atom into the excited state. Combining Eqs. (20-21) and (20-22) with Eq. (20-20) yields

$$P_{\text{out}} = \frac{1}{2} ATI_s \left[\frac{P_{\text{in}}}{P_{\text{th}}} - 1 \right] \tag{20-23}$$

where $I_s = h\nu/(\sigma \tau_2)$ is the saturation intensity defined in Eq. (19-9), and

$$P_{\text{th}} \equiv \mathcal{R}_{\text{th}} V h\nu_p \tag{20-24}$$

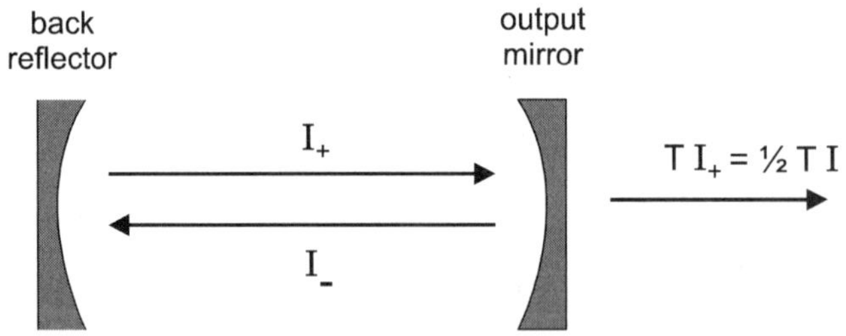

Figure 20-4 The standing wave in the laser cavity results from the superposition of two counterpropagating beams of intensities I_- and I_+. Only the I_+ beam is transmitted through the right mirror to give laser output.

is the threshold pumping power. Eq. (20-23) can be further manipulated into the simple form

$$P_{\text{out}} = \eta_s[P_{\text{in}} - P_{\text{th}}] \tag{20-25}$$

where

$$\eta_s \equiv \frac{\Delta P_{\text{out}}}{\Delta P_{\text{in}}} \qquad \text{(slope efficiency)} \tag{20-26}$$

is the *slope efficiency* of the laser. The slope efficiency is an important and widely used measure of a laser's efficiency, and gives the incremental change in output power for an incremental change in pumping power. It is defined as the slope of the output power versus input power curve for the laser, as illustrated in Fig. 20-5. Note that η_s is not simply the ratio of output to input power, but rather is the ratio of changes in those powers. By specifying both the threshold pump power and slope efficiency, the output power of the laser is determined for any given pumping power using Eq. (20-25).

The slope efficiency for the simple laser model developed here can be obtained from Eq. (20-23) using Eqs. (20-24), (20-18), and (19-9). The result after a few steps of algebra is

$$\begin{aligned} \eta_s &= \frac{1}{2} AT \frac{I_s}{P_{\text{th}}} \\ &= T \frac{h\nu}{h\nu_p} \frac{c\tau_c}{2L} \end{aligned} \tag{20-27}$$

The last factor above can be written in a more compact form by using Eq. (20-11) for τ_c,

$$\begin{aligned} \frac{2L}{c\tau_c} &= \alpha(2L) + \ln\left(\frac{1}{R_1R_2}\right) \\ &= \delta + \ln\left(\frac{1}{1-T}\right) \\ &\simeq \delta + T \end{aligned} \tag{20-28}$$

where $\delta = \alpha(2L)$ is the fraction of light lost in one round-trip due to absorption or scattering, and T is the fraction of light lost in one round-trip due to the finite mirror reflectivity.

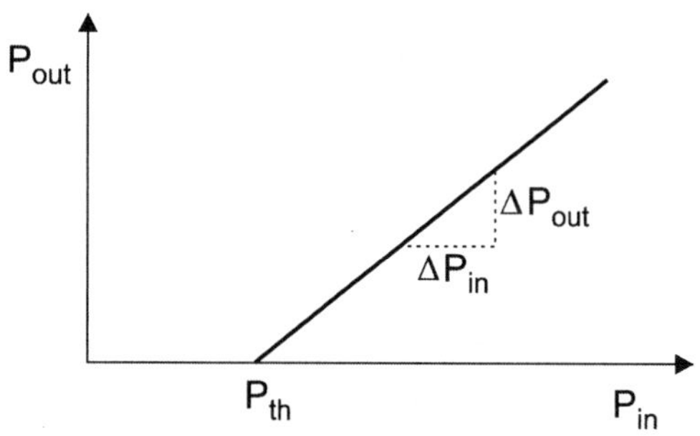

Figure 20-5 The slope efficiency is the slope $\Delta P_{\text{out}}/\Delta P_{\text{in}}$ of the P_{out} versus P_{in} curve.

We will refer to δ as the *internal loss* per round-trip, and T as the *coupling loss* per round-trip. It has been assumed in Eq. (20-28) that $\delta \ll 1$ and $T \ll 1$, so the total fractional loss per round-trip is small. Relaxing this assumption leads to much more complicated formulae, and does not provide much added insight. Therefore, we will maintain this small-loss assumption throughout our further discussion.

Combining Eqs. (20-28) and (20-27) yields the simplified result

$$\eta_s \simeq \frac{T}{\delta + T}\frac{h\nu}{h\nu_p} \qquad \text{(low loss slope efficiency)} \qquad (20\text{-}29)$$

The slope efficiency is seen to depend on two factors: the ratio of coupling loss to total loss, and the ratio of lasing photon energy to pump photon energy. When $T \gg \delta$, the slope efficiency is simply given by $\eta_s \approx (h\nu)/(h\nu_p)$, which is the quantum defect discussed earlier in connection with optical amplifiers (see Fig. 19-6). In physical terms, this means that every additional absorbed pump photon above threshold leads to an additional laser output photon, and the fact that the slope efficiency is < 1 is simply due to the smaller photon energy of the laser light compared with the pump light. A similar relation was found in Eq. (19-23) for the power conversion efficiency of an optical amplifier with a large signal. Although the mirror transmission T is considered a "loss" in the sense that it decreases the photon lifetime, it is a "useful loss" in the sense that it allows light to leave the laser cavity and become the output beam.

In contrast with the "good loss" of the mirror transmission T, the internal loss δ is always a "bad loss," because it reduces the useful output of the laser. The maximum slope efficiency occurs for $\delta = 0$, so it is desirable to reduce sources of internal loss such as scattering and absorption as much as possible. Any loss other than transmission through the output mirror is considered to be internal loss. For example, if the left cavity mirror #1 is partially transmitting, then δ includes the fractional loss $T_1 = 1 - R_1$ due to that mirror. Also, if there are any partially reflecting surfaces inside the laser cavity (for example filters or laser tube windows), then δ includes the round-trip fractional loss from these as well.

If $\delta > 0$, then according to Eq. (20-29) the slope efficiency is maximized by making T as large as possible. This may not maximize the total output power, however, because the laser threshold also varies with T. To see how P_{th} varies with T, Eq. (20-24) can be rewritten with Eqs. (20-18), (20-28), and (19-9), giving

$$\begin{aligned} P_{\text{th}} &= \frac{Vh\nu_p}{c\tau_c\sigma\tau_2} \\ &= (\delta + T)\left(\frac{V}{2L}\right)\frac{h\nu_p}{\sigma\tau_2} \\ &= \frac{1}{2}(\delta + T)\, I_s A\, \frac{h\nu_p}{h\nu} \end{aligned} \qquad (20\text{-}30)$$

The pump threshold is proportional to the total loss per round-trip, and increases with increasing T. The effect of this on laser output is shown in Fig. 20-6. At high pump power, the advantage of high slope efficiency outweighs the disadvantage of high pump threshold, and a larger T results in a higher output power. For smaller pump power, however, the reverse is true.

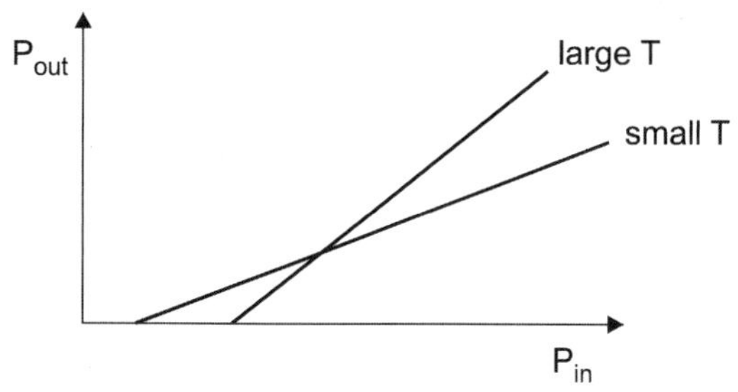

Figure 20-6 Laser output power versus pump power for two different values of mirror transmission T. The output power may increase or decrease with increasing T, depending on the pumping power.

It is clear that to describe the efficiency of a laser requires more than one parameter. The efficiency $\eta \equiv P_{out}/P_{in}$ is not a useful figure of merit for the laser, since it depends on the pump rate. The slope efficiency is independent of pump rate, but does not by itself give the output power for a given pump power. The combination of threshold pump power and slope efficiency, however, does completely describe the laser's efficiency at any pump power. These two parameters are often quoted as the figures of merit for various lasers.

At any given pumping power, there will be some value of T that maximizes the output power. To find this optimum value of T, we rewrite Eq. (20-23) to show explicitly the dependence on T. The ratio P_{in}/P_{th} in this equation can be written using Eqs. (20-22), (20-30), and (19-9) as

$$\begin{aligned} \frac{P_{in}}{P_{th}} &= \frac{2\mathcal{R}V\sigma\tau_2}{(\delta + T)\,A} \\ &= \frac{\gamma_0(2L)}{\delta + T} \end{aligned} \tag{20-31}$$

where $\gamma_0 = \mathcal{R}\sigma\,\tau_2$ is the *unsaturated gain coefficient* defined earlier in Eq. (19-11). Note that lasing threshold corresponds to the condition $\gamma_0\,(2L) = \delta + T$, which says that the unsaturated round-trip gain equals the total round-trip loss. Above threshold, the gain γ becomes pinned at the threshold value, but γ_0 continues to increase in proportion to the pump power.

Using the above, the expression for output power in Eq. (20-23) can be written as

$$P_{out} = \frac{1}{2}\,ATI_s\left[\frac{\gamma_0\,2L}{\delta + T} - 1\right] \tag{20-32}$$

The variation of power output with mirror transmission T is plotted in Fig. 20-7 for several values of the dimensionless ratio $K \equiv \gamma_0 2L/\delta$. For a given value of γ_0, there is some value of T that maximizes the output power, designated as T_{opt}. We can determine this optimum value of T by setting $dP_{out}/dT = 0$ in Eq. (20-32), and solving for T. The result is

$$T_{opt} = \sqrt{(\gamma_0 2L)\delta} - \delta \qquad \text{(optimum mirror transmission)} \tag{20-33}$$

which is valid when $\delta \ll 1$ and $T_{opt} \ll 1$. This result shows that the optimum value of T is larger for larger values of γ_0, in agreement with Fig. 20-7. When the unsaturated gain is

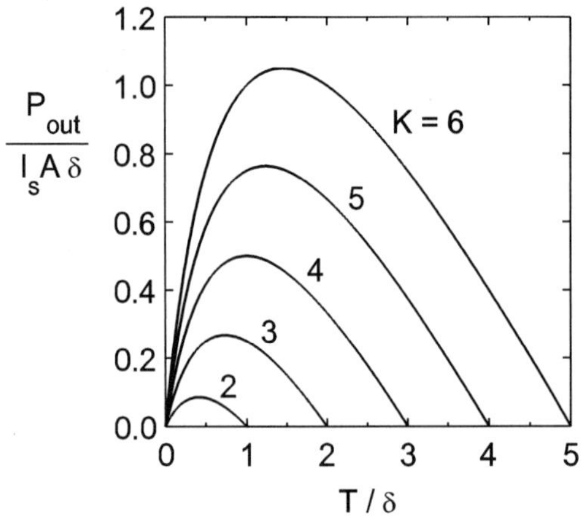

Figure 20-7 Laser output power versus mirror transmission T for several values of dimensionless parameter $K \equiv \gamma_0 2L/\delta$. P_{out} normalized to $I_s A\delta$, and T normalized to δ. The optimum value of T increases with increasing γ_0, and hence with increasing pump power.

just above threshold, it can be shown (Problem 20.12) that $T_{opt} \approx (\gamma_0 2L - \delta)/2$. There is no positive solution for T_{opt} below threshold, since lasing does not occur.

EXAMPLE 20-2

A Nd:YAG laser consists of a Nd:YAG rod of length 7.5 cm, situated between two mirrors with reflectivities $R_1 = 1$ and $R_2 = 0.85$, as shown in Fig. 20-8. The laser is optically pumped from the side with light of average wavelength 500 nm. The lasing transition in the Nd ion has the following characteristics: wavelength of 1064 nm, upper-level lifetime of 230 μs, and stimulated emission cross section $\sigma = 2.8 \times 10^{-19}$ cm^2. The beam area in the laser rod is 0.23 cm^2, and the attenuation coefficient is 5×10^{-3} cm^{-1}. Determine the threshold pump power for the laser.

Solution: The round-trip internal loss is $\delta = (5 \times 10^{-3})(15) = 0.075$, and the mirror transmission is $T = 1\text{-}0.85 = 0.15$. The saturation intensity is (using MKS units)

$$I_s = \frac{hc}{\lambda\sigma\tau_2} = \frac{(6.63 \times 10^{-34})(3 \times 10^8)}{(1.064 \times 10^{-6})(2.8 \times 10^{-23})(230 \times 10^{-6})} = 2.9 \times 10^7 \text{ W/m}^2$$

Using Eq. (20-30),

$$P_{th} \simeq \frac{1}{2}(0.15 + 0.075)(2.9 \times 10^7)(2.3 \times 10^{-5})\left(\frac{1064}{500}\right) \simeq 160 \text{ W}$$

This is the threshold for absorbed optical pump power. If the electrical power to the pump lamp is converted into absorbed pump power with efficiency $\eta_p = 0.05$, the threshold electrical power to the lamp is

$$P_{th}^{elec} \simeq \frac{160}{0.05} = 3200 \text{ W}$$

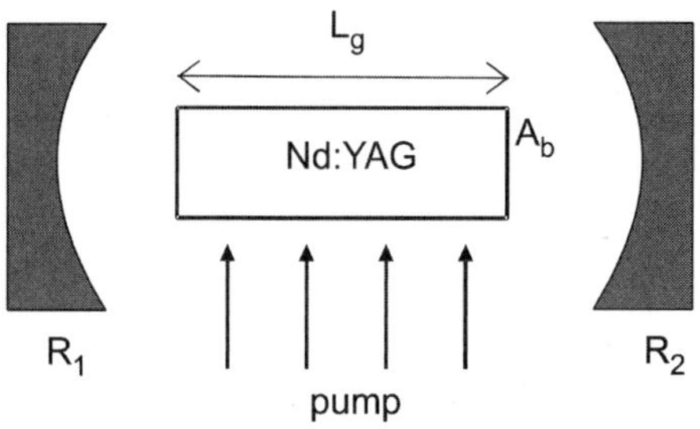

Figure 20-8 Diagram for Example 20-2.

EXAMPLE 20-3

For the Nd:YAG laser of Example 20-2, determine the slope efficiency.

Solution: Using Eq. (20-29) with $h\nu/h\nu_p = \lambda_p/\lambda$,

$$\eta_s \simeq \left(\frac{0.15}{0.075 + 0.15}\right)\frac{500}{1064} \simeq 0.31$$

EXAMPLE 20-4

For the laser of Example 20-2, determine the value of T that would maximize the output power if the pump power is twice the threshold value calculated in that example.

Solution: In Example 20-2, the threshold gain coefficient is determined by setting the round-trip gain equal to the round-trip loss,

$$\gamma_{th}(2L) = \delta + T = 0.225$$

When pumping at twice threshold, the unsaturated gain coefficient γ_0 is twice γ_{th}, so $\gamma_0(2L) = (2)(0.225) = 0.45$. The optimum mirror transmission is then found from Eq. (20-33) to be

$$T_{opt} \simeq \sqrt{(0.45)(0.075)} - 0.075 = 0.109$$

Note that the value originally chosen for R_2 was not quite optimum for this pumping rate.

PROBLEMS

20.1 The loss in a laser cavity of length $L = 1.5$ m is dominated by the 80% transmission of the output mirror. Determine the threshold gain coefficient for this laser cavity.

20.2 Assume that the transition linewidth is limited only by the lifetime of the upper state, according to the uncertainty relation (see Appendix B). In this case, derive a simplified expression for the threshold population inversion.

20.3 Show that the photon lifetime τ_c defined in Eq. (20-11) reduces to the previous definition in Eq. (16-13) in the limit $\alpha = 0$ and $R_1R_2 \approx 1$. (See also Problem 16.5.)

20.4 A Nd:YLF laser has the following parameters: stimulated emission cross section 1.8×10^{-19} cm^2, upper-state lifetime 480 μs, rod length 7.5 cm, mirror reflectivities of 100% and 95%, beam area 0.23 cm^2, and loss coefficient 8×10^{-3} cm^{-1}. If the laser is pumped optically with a lamp at an average wavelength of 500 nm, determine the minimum absorbed lamp power to achieve lasing threshold. Also, if the efficiency of converting the lamp electrical power into absorbed optical power is 5%, determine the minimum electrical power to the lamp for lasing.

20.5 A He–Ne laser has a tube length of 20 cm, a tube radius of 1 mm, and has mirror reflectivities of 0.99 (output coupler) and 0.998 (high reflector). Take other data for the He–Ne laser from Table 23-2, and assume that the lower laser level has negligible population. (a) Determine the threshold population inversion. What fraction of the available Ne atoms need to be in the upper laser state for lasing? (b) Determine the number of atoms that must be pumped into the upper laser level per unit time to achieve lasing. (c) The upper laser level is ~ 20 eV above the ground state. If each excitation of the upper laser level requires this much energy, what is the minimum pump power required to achieve lasing?

20.6 In the previous problem, the laser output power is 1.5 mW. Using the radiative lifetime of 30 ns, determine the ratio of stimulated emission rate to spontaneous emission rate for Ne atoms in the cavity.

20.7 An Er-doped optical fiber of length 1 m has the spectroscopic parameters given in Table 23-1. The fiber is strongly pumped so that the Er ions are nearly fully inverted (all in the excited state) along the entire fiber. (a) Determine the gain coefficient and the net gain of the fiber in dB. (b) Determine the mirror reflectivities needed (assume they are the same at each end) for lasing to occur. (c) Will the Fresnel reflection from the glass–air interface be enough to initiate lasing?

20.8 An optically pumped laser has a pump wavelength of 810 nm, a lasing wavelength of 1060 nm, and an output coupling per round-trip of 2%. When pumped with 2.5 W, the output of the laser is 180 mW, and when pumped with 3.5 W the output is 450 mW. (a) Determine the slope efficiency of this laser. (b) Determine the pump threshold for this laser. (c) It is desired to operate the laser at a pump power of 3.0 W. Determine the laser output at this pump power. (d) Determine the fractional internal loss per round-trip. (e) If the cavity length is 6 cm and the gain medium has refractive index 1.5, determine the photon lifetime.

20.9 A laser has an internal loss per round-trip of 1.5%, and when the output mirror transmission is 1% it has a pump threshold of 70 mW. (a) If this laser is pumped with 200 mW, what value of the output mirror transmission will maximize the laser output? (b) With the mirror transmission of part a, what is the new pump threshold? By what factor is the laser above theshold in this case?

20.10 A laser having an internal loss per round-trip of 0.015 is to be pumped at five times threshold. Determine the output mirror transmission that maximizes the output power of this laser.

20.11 The output power of a laser is 4 W when pumped at three times threshold. The optical mode in the laser has diameter 1 mm, and the output mirror transmission is 0.5%. Determine the saturation intensity of the laser transition.

20.12 Show that when the unsaturated gain γ_0 is just above threshold, the optimum mirror transmission is $T_{opt} \approx (\gamma_0 2L - \delta)/2$.

Chapter 21

CW Laser Characteristics

In the previous chapter, we considered the conditions under which lasing will occur. With steady-state pumping, there is a well-defined pump threshold, above which the intensity of laser light increases linearly with pump power. For some applications, it is not just the power of the laser light that is important but also its frequency spectrum. In this chapter, we consider the frequency distribution of laser light in the steady state. This is often referred to as *continuous wave* or CW operation. We explore the effect of gain saturation on the frequency spectrum, and consider ways in which the frequency can be stabilized or tuned over some range.

21-1. MODE SPECTRUM OF LASER LIGHT

In a laser cavity, the light intensity will build up to a high value only at the resonator mode frequencies, given by Eq. (16-3). Light in any of these modes will be amplified, provided that there is optical gain at the frequency of that mode. In order for lasing to occur in a given mode, the gain at that frequency must exceed or equal the threshold gain γ_{th}. The number of modes that will lase, therefore, depends on how many modes have a gain $\geq \gamma_{th}$. The number of modes with $\gamma \geq \gamma_{th}$ will in turn depend on the way in which the gain curve saturates with intensity.

Single-mode Lasing

In our previous discussion of gain saturation, we assumed that all atoms were equivalent, having the same center frequency ν_0. In this case, each atom saturates with intensity in the same way, and the gain curve is reduced by the same factor at each frequency ν. This is referred to as homogeneous saturation, and is illustrated in Fig. 19-4a.

For steady-state operation, the value of $\gamma(\nu)$ at any frequency ν cannot exceed the threshold value γ_{th}, because if it did the light intensity at that frequency would increase exponentially in time, and this would violate the assumption of steady-state operation. The gain curve $\gamma(\nu)$ must, therefore, saturate so that the peak value is just at threshold, as shown in Fig. 21-1. Only the mode closest to the line center will lase in this case, since other modes will have $\gamma < \gamma_{th}$. The result is *single-mode* lasing, with the output frequency spectrum as indicated in Fig. 21-1.

Multimode Lasing

Although single-mode operation is the easiest to describe mathematically, in practice, lasers often oscillate simultaneously in more than one mode. This is referred to as *multi-*

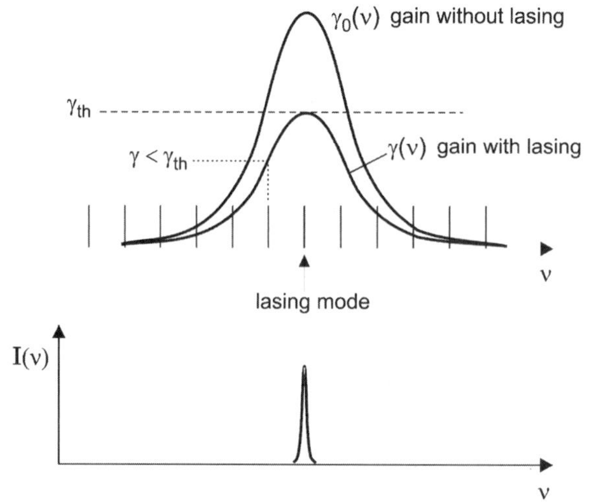

Figure 21-1 Homogeneous saturation of gain coefficient results in a single lasing mode.

mode lasing, and arises when the atoms in the gain medium are not all equivalent. Atoms can be inequivalent either because the spectrum is different for different atoms, or because different atoms are located in different spatial regions of the resonator mode pattern. In either case, the gain saturation is inhomogeneous, as shown in Fig. 19-4b. A strong light signal at one frequency creates a "hole" in the gain spectrum, and the gain is saturated only for a fraction of the atoms in the gain medium. The gain remains high at other frequencies, and lasing can occur at any frequency where $\gamma > \gamma_{th}$. The creation of such a hole in the gain spectrum is referred to as *spectral hole burning* or *spatial hole burning,* depending on the manner in which the atoms are inequivalent.

Spectral Hole Burning

If the atomic lineshape is inhomogeneously broadened, then the gain saturation is said to be due to *spectral hole burning.* In a gas, for example, the atomic center frequency varies randomly due to the Doppler shift. An atom with velocity component v_z along the laser cavity axis has its center frequency shifted from ν_0 to

$$\nu_0' = \nu_0\left(1 \pm \frac{v_z}{c}\right) \qquad \text{(Doppler shift)} \qquad (21\text{-}1)$$

where the + sign is taken when the atom and wave are moving in opposite directions, and the – sign when they are in the same direction. In a laser cavity, a mode with frequency ν_m corresponds to the superposition of two plane waves, one moving in the $+z$ direction, and the other in the $-z$ direction. As a result, a single mode actually saturates two distinct groups of atoms: those moving both left and right in the resonator with a speed $v = |v_z|$. This situation is depicted in Fig. 21-2, which shows atoms of group a moving to the left with speed v, atoms of group b moving to the right with the same speed v, and atoms of group c at rest. If the line center of the atomic transition coincides with the cavity mode m, then the mode of frequency ν_m will interact with atoms at rest, group c. The next-highest mode $m + 1$ will interact with both groups a and b, group a interacting with the travel-

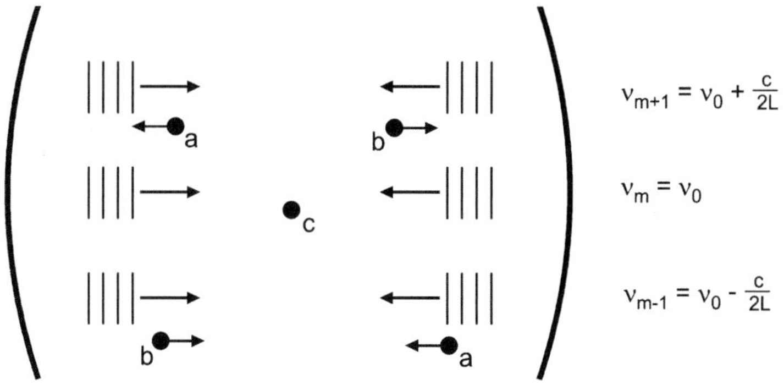

Figure 21-2 In a Doppler-broadened lineshape, modes below or above the atomic center frequency ν_0 interact with atoms moving left (group a) and atoms moving right (group b), whereas modes at ν_0 interact only with atoms at rest (group c).

ing wave component moving to the right, and group *b* interacting with the traveling wave component moving to the left. In a similar way, both groups *a* and *b* will interact with the next-lowest mode $m - 1$.

The gain saturation that results is illustrated in Fig. 21-3. The mode *m* at line center saturates independently of the modes $m - 1$ and $m + 1$, because it interacts with a different group of atoms. Modes $m - 1$ and $m + 1$ saturate in a similar manner, since they interact with the same group of atoms. However, this saturation results in simultaneous lasing at the distinct frequencies ν_{m-1} and ν_{m+1}, which contributes to the multimode character of the laser output. In general for a inhomogeneously broadened transition, the number of lasing modes will be equal to the number of modes for which $\gamma_0(\nu_m) > \gamma_{th}$. Therefore, a greater number of modes will lase when the unsaturated gain coefficient is increased (by increasing the pump power, for example), or when the cavity loss is decreased.

In a solid, inhomogeneous broadening results not from the Doppler shift, but rather from the variation of the atom's local environment from one location to another in the

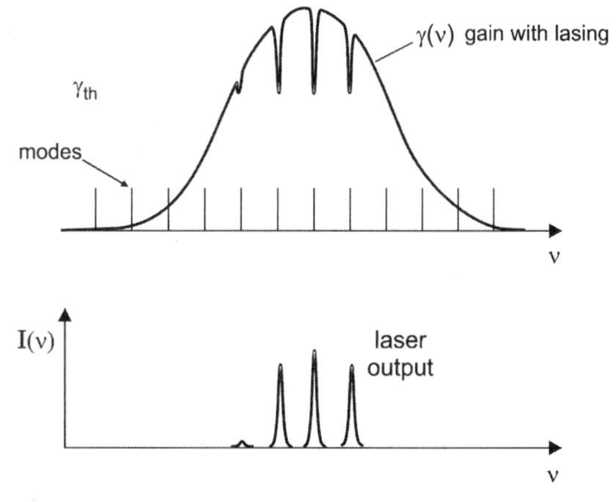

Figure 21-3 For an inhomogeneously broadened lineshape, multiple modes can lase simultaneously, since the different modes interact with and saturate the gain of different groups of atoms independently.

material. For example, the surrounding atoms in a glassy material or an imperfect crystal give rise to an electric field at the location of the active atom (i.e., the atom responsible for the gain transition), the magnitude of which varies randomly from one active atom to another. This electric field can change the energy level positions for the active atom, through what is known as the *Stark shift.*

As a result of the varying Stark shift, the center frequencies ν_0 vary randomly from one active atom to another, and the different cavity modes interact with different groups of atoms, just as for a gas. The gain curve, therefore, saturates independently at the various mode frequencies, and multimode laser output results, as in Fig. 21-3.

Although multimode lasing is similar conceptually for gas lasers and solid-state lasers, there is one important difference in how the different modes saturate the transition. In a gas laser, modes symmetrically located on either side of the gain peak interact with and saturate the same group of atoms, as discussed above. For a solid gain medium, however, each mode saturates independently, interacting with a distinct group of atoms. The reason for this difference is that in a gas, the frequency shift depends on the relative motion of the atom toward or away from one of the component traveling waves in the cavity, whereas in a solid, the frequency shift depends only on the position of an atom, not its velocity.

Spatial Hole Burning

If the atomic lineshape is homogeneously broadened, then each atom has the same center frequency, and there is no spectral hole burning. It might therefore be expected that the gain curve should always saturate homogeneously, as in Fig. 21-1, resulting in a single lasing mode. In fact, however, multimode lasing can still occur in this case, due to the phenomenon of *spatial hole burning.*

In spatial hole burning, different atoms interact preferentially with different modes because they are located in different spatial regions of the mode patterns. For example, higher-order transverse modes TEM_{lm} extend further from the cavity axis than do lower-order modes such as the Gaussian TEM_{00}, as indicated in Fig. 21-4a. When the TEM_{00} mode saturates the gain of atoms inside its spatial profile, there are still atoms outside this profile that are unsaturated. These unsaturated atoms can provide gain and lasing for the higher-order modes, which occur at different frequencies than the TEM_{00} mode. The result is simultaneous operation on more than one frequency, that is, multimode laser output. In this case, the "multiple modes" are different transverse modes.

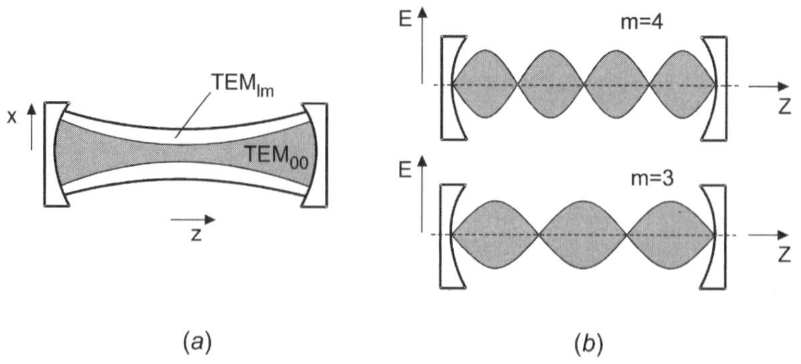

Figure 21-4 Spatial hole burning results from (a) different transverse modes or (b) different longitudinal modes. In both cases, atoms in different spatial locations saturate independently.

Spatial hole burning can also occur with only a single transverse mode. This can be understood by referring to Fig. 21-4b, which shows two longitudinal modes with different mode numbers m. The standing-wave pattern for each mode produces nodal points at which the intensity is zero at all times. The atoms at these nodal points will be unsaturated by laser light in that mode. Different modes, however, have their nodal points at different positions along the axis, and the nodal point for one mode may be at the same position as a maximum in intensity for another mode. Different modes can, therefore, interact with different groups of atoms, and can saturate independently. The laser output is then multimode, with different longitudinal modes lasing simultaneously.

The significance of spatial hole burning actually goes beyond simply the number of modes. It is clear from the standing wave patterns of Fig. 21-4b that the laser mode does not interact efficiently with all of the atoms in the gain medium. If pump energy is deposited uniformly into the atoms throughout the laser gain medium, this represents a loss of efficiency for the laser. In effect, some pump energy is being wasted because the atoms excited by that energy do not efficiently interact with the lasing light in the cavity. The same thing applies to the transverse modal pattern; atoms outside the spatial profile of the mode do not strongly interact with the lasing light, and the pump energy deposited there is wasted.

21-2. CONTROLLING THE LASER WAVELENGTH

In many applications, it is desirable to have some control over the laser wavelength. For example, holography and other interferometric applications require a long coherence time, which corresponds to a narrow frequency spectrum. The narrowest frequency spectrum is obtained when operating the laser in a single mode. Another application that benefits from single-mode operation is optical spectroscopy, in which the energy level structure of an atom, molecule, or solid is probed by laser light with a narrow frequency width. Optical spectroscopy is important not only for fundamental research, but also in applications such as remote atmospheric sensing, noninvasive medical diagnostics, optical communications, and many others. In all these applications, it is useful to be able to tune the laser light to different wavelengths, and also to stabilize the wavelength to be independent of environmental parameters such as temperature. In this section, we consider various methods for controlling the spectral output of the laser.

Achieving Single-mode Lasing

A single transverse mode can be obtained simply by placing a circular aperture inside the resonator, as illustrated in Fig. 21-5. This method takes advantage of the fact that higher-order transverse modes have a larger effective beam width than the fundamental Gaussian mode TEM_{00}. The aperture increases the loss coefficient α for these higher-order modes and prevents their lasing. The aperture will also increase α for the TEM_{00} mode to some extent, so care must be taken in optimizing the radius of the aperture. An adjustable iris is sometimes used for this purpose.

Operation in a single longitudinal mode can be obtained by placing a device inside the cavity that allows only selected frequencies to propagate with low loss. The Fabry–Perot filter, discussed in Chapter 16, is one such frequency-selective device. When the spacing between the parallel plates of the Fabry–Perot filter is fixed, the device is often referred to as an *etalon*. It can be formed by applying a highly reflective coating to the two surfaces

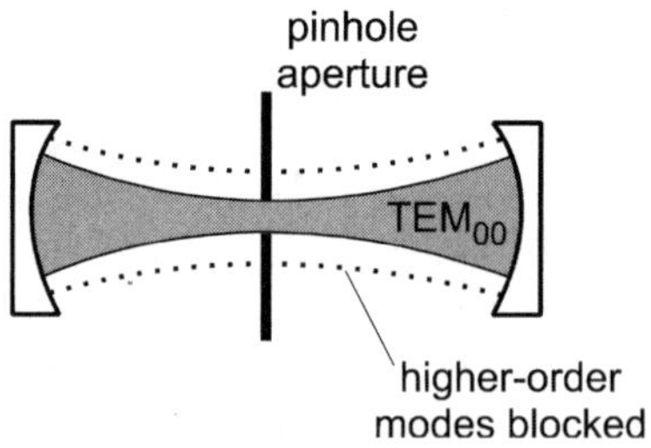

Figure 21-5 An intracavity aperture blocks the higher-order transverse modes.

of a thin glass slab of thickness d and refractive index n. The etalon's transmission spectrum for normal incidence, shown in Fig. 16-9, is a comb of narrow transmission peaks separated by the free spectral range $c/(2nd)$. When placed inside the laser cavity, the etalon provides high transmission (and hence low cavity loss) for only a small number of laser cavity modes, as illustrated in Fig. 21-6. A particular laser mode will oscillate only when it is within the gain bandwidth $\Delta\nu$ of the gain medium, and also within the spectral passband of the etalon mode. If the etalon modes are sufficiently narrow, single-mode operation can be achieved.

The passband frequency of the etalon can be tuned by tilting the etalon at an angle θ with respect to the cavity axis, as shown in Fig. 21-7. The frequency spacing between etalon modes can be shown (see Problem 21.1) to be

$$\delta\nu_{\text{etalon}} = \frac{c}{2nd\cos(\theta')} \qquad \text{(etalon mode spacing)} \tag{21-2}$$

where θ' is the angle of refraction inside the glass slab, related to the incident angle θ by $\sin\theta = n\sin\theta'$. As the angle θ is changed, the center frequency of each etalon mode

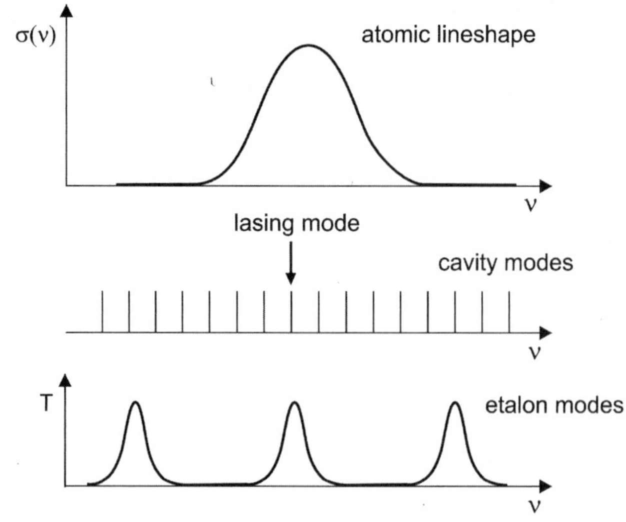

Figure 21-6 An etalon placed in the cavity allows lasing on only a single longitudinal mode.

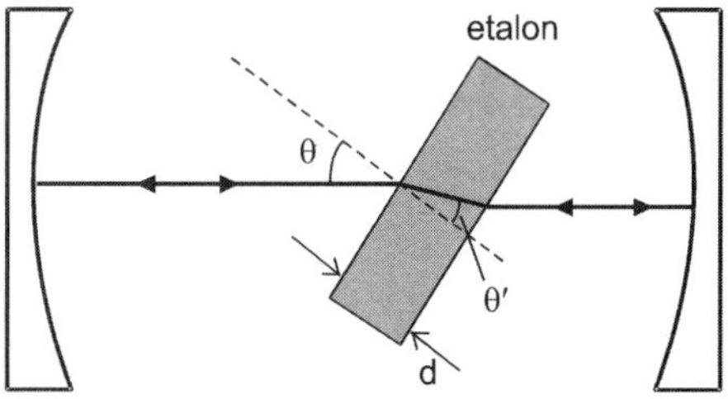

Figure 21-7 Rotating the etalon allows different cavity modes to be selected.

moves with respect to the fixed laser cavity modes. In this way, laser oscillation on a particular laser cavity mode can be selected.

Another method for selecting a single longitudinal mode, commonly used for semiconductor lasers, is Bragg reflection from a corrugated interface between semiconductor layers. This approach (known as a distributed-feedback laser) was described in Fig. (11-20), and gives rise to lasing at the (free space) wavelength $\lambda = 2n\Lambda$, where Λ is the corrugation period. A related method used for fiber lasers is the fiber Bragg grating, discussed in Chapter 8.

The goal of achieving single longitudinal mode operation can be obtained in a completely different way, by using the *ring laser* geometry shown in Fig. 21-8. In the ring laser, light is forced to circulate in one direction around the ring, by inserting an *optical diode* into the beam path. One such device utilizes *Faraday rotation,* in which a lightwave's polarization is rotated in a magnetic field. The direction of rotation depends on the propagation direction, and this allows a preferred direction to be selected using polarization filters. With the optical diode in place, the light in the cavity has the form of a traveling wave, rather than a standing wave. As a consequence, there is no spatial hole burning for the different longitudinal modes, and if the atomic transition is homogeneously broadened, the gain spectrum will saturate as in Fig. 21-1. This leads to single-mode operation without the need for intracavity filters or gratings. An added advantage of the ring geometry is that the pump energy is more efficiently utilized. The traveling waves in the ring laser lack the nodal points that are characteristic of standing waves (Fig. 21-4b), so that atoms at any point along the cavity axis interact equally with the light in the cavity mode.

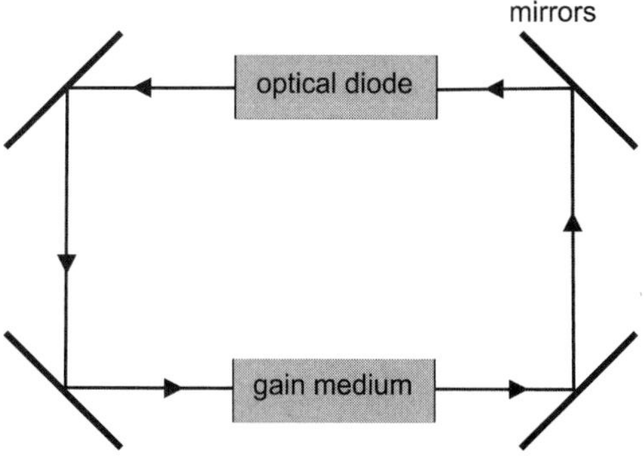

Figure 21-8 In a ring laser, the light travels in only one direction, so there are no standing waves to create spatial hole burning.

Frequency Stabilization

Operation in a single mode means that the laser has a well-defined frequency, but it does not guarantee that the frequency is constant in time. The mode frequencies given by Eq. (16-3) depend on the cavity length L and index of refraction n, both of which vary with temperature. As gas lasers such as the He–Ne warm up after being turned on, the cavity length increases due to thermal expansion, causing the mode frequencies to decrease. The modes then "sweep" across the atomic gain spectrum, in a phenomenon known as *mode sweeping.* The laser output power then oscillates slowly in time, as cavity modes enter and leave the region of maximum optical gain.

To stabilize the frequency against changes in mode frequency, it is necessary to prevent the cavity length from changing. A simple solution is to use materials in the laser cavity structure with a low thermal expansion coefficient. It is difficult to completely eliminate small drifts in the mode frequencies, however, and active stabilization is needed for precise frequency control.

One method for active frequency stabilization utilizes the gain saturation behavior of a gas laser. As a single lasing mode is tuned across the Doppler-broadened atomic lineshape, it might be expected that the laser power would be maximum at the atomic line center frequency ν_0, where the gain cross section is maximum. Instead, as shown in Fig. 21-9, there is a narrow dip in the power at ν_0, known as the *Lamb dip.* This narrow dip can be used to stabilize the laser frequency by providing a feedback signal to actively control the mirror separation via piezoelectric transducers.

The physical origin of the Lamb dip can be understood by considering the different groups of atoms that interact with the cavity mode, as illustrated in Fig. 21-2. Cavity modes either above or below ν_0 interact with two groups of atoms, those moving left (a) and those moving right (b) . A cavity mode right at ν_0, however, interacts with only one group of atoms, those at rest (c) . The gain coefficient for a given subset of the atoms saturates according to Eq. (19-12) for an inhomogeneous gain profile. Because of this saturation behavior, less intensity I is needed to pin the gain coefficient γ at γ_{th} when the light interacts with a smaller group of atoms, and the result is a dip in the output power as the mode is tuned across resonance (Siegman 1986).

Tuning the Laser Wavelength

For laser applications, it is often of interest to be able to tune the laser to a particular wavelength. There are several ways of doing this, depending on the degree of wavelength

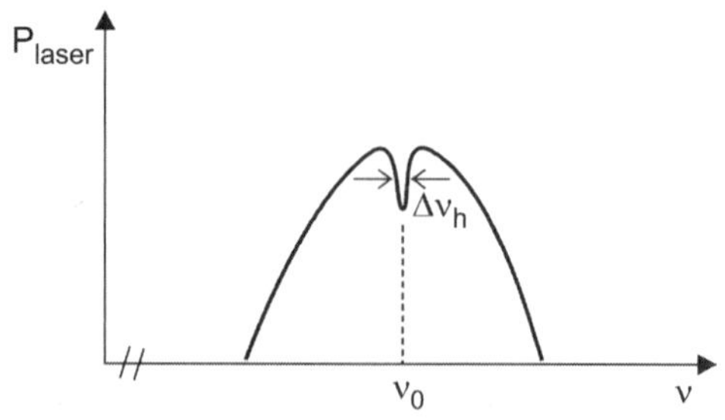

Figure 21-9 Laser output power versus frequency of the single lasing mode, as mode frequency is tuned across the inhomogeneous atomic lineshape. The decrease in power at line center is the Lamb dip.

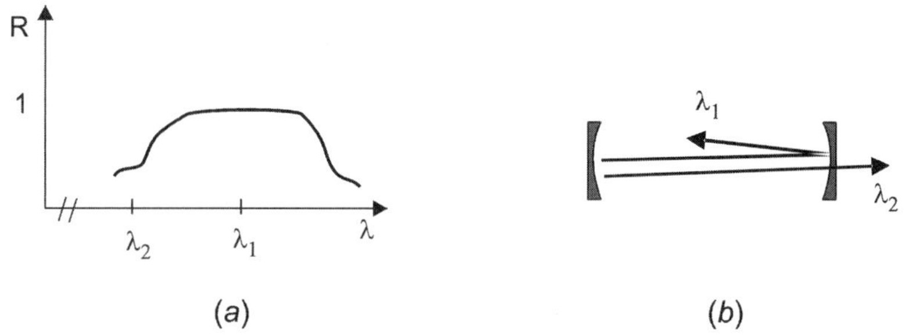

Figure 21-10 (a) Mirror reflectivity versus wavelength. (b) Only one wavelength is highly reflected by the mirrors.

selection required. For gain media that have a number of discrete laser transitions with widely separated wavelengths, it is sufficient to use mirrors that are highly reflecting in a wavelength range corresponding to only one of the transitions. This is illustrated in Fig. 21-10, which shows the laser transition at λ_1 being selected over the competing transition at λ_2. Mirrors that efficiently reflect light over a limited wavelength range are made by depositing multiple thin dielectric layers on a substrate (usually glass). The wavelength selectivity of the reflection results from constructive and destructive interference of light reflected from the various layers. As an example of this method of wavelength selection, the usual oscillation of the He–Ne laser at 632.8 nm can be suppressed in favor of a weaker transition at 543 nm by an appropriate choice of dielectric mirrors. Other He–Ne laser transitions can be selected as well, including 610 nm (orange color), 1.15 μm (near infrared), and 3.39 μm (mid infrared).

Another method for selecting individual discreet laser lines utilizes a prism placed inside the laser cavity, as shown in Fig. 21-11. The index of refraction of the glass in the prism varies with wavelength, a phenomenon known as *dispersion* (see also Fig. 6-2). As a result, only light of one particular wavelength will be refracted by the prism at the proper angle for retroreflection at the right mirror; other wavelengths will be deflected out of the laser cavity and will not lase. Rotating the mirror selects different wavelengths for retroreflection. This is useful for low-gain gas lasers such as the argon ion laser, in which there are discreet lasing transitions closely spaced in wavelength. The argon ion laser can be tuned in this way to a number of lines in the green-to-blue region between 514 and 454 nm.

A diffraction grating can also be used as the tuning element, as shown in Fig. 21-12. Light that satisfies the Bragg condition of Eq. (2-28) will be retroreflected from the grat-

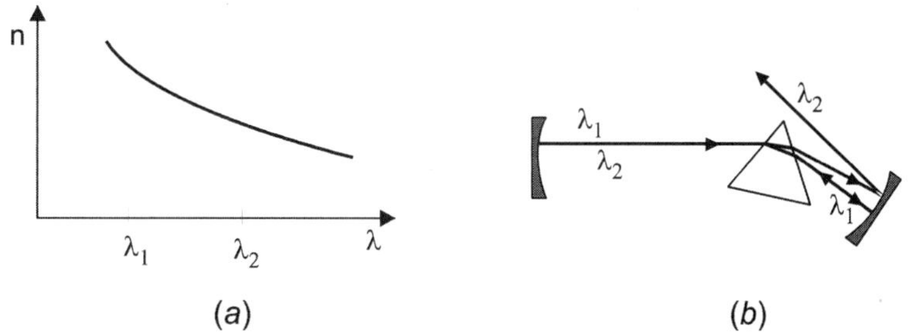

Figure 21-11 (a) Index of refraction n versus wavelength for glass in a prism. (b) Only one wavelength (λ_1) is retroreflected by the mirror through the prism.

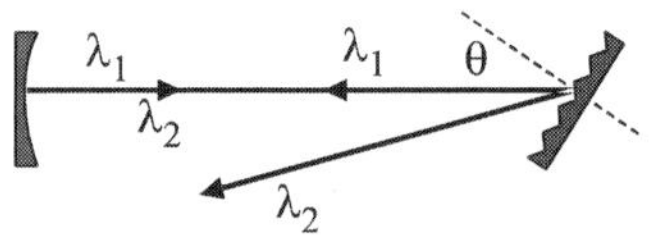

Figure 21-12 Tuning the wavelength with a diffraction grating. Only one wavelength is retroreflected by the grating.

ing, whereas other wavelengths will be diffracted out of the cavity. Different wavelengths can be selected by rotating the grating to change the incident angle θ. The efficiency of reflection from the grating is generally not as high as from a mirror, due to diffraction into other orders and other loss mechanisms. Therefore, gratings are used mostly with high-gain lasers such as CO_2 and pulsed-dye lasers.

PROBLEMS

21.1 Derive Eq. (21-2) for the etalon mode spacing. (Hint: consider two parallel rays incident on the etalon, and require constructive interference of these two rays.)

21.2 Eq. (21-2) applies to a solid etalon of refractive index n, as shown in Fig. 21-7. An alternative is the air-spaced etalon, which consists of two dielectric slabs such as this, coated on their facing surfaces with a high-reflectivity layer. These two surfaces are separated by a distance d with precision spacers. Derive an expression for the mode spacing in the air-spaced etalon when it is tilted at angle θ from the beam in the laser cavity.

21.3 A Nd microchip laser has index of refraction 1.8 and cavity length 3 mm. Mirrors of reflectivity 0.995 and 0.98 are mounted directly on the faces of the chip. The emission lineshape of the laser transition can be considered to be Lorentzian, with a linewidth of 5 THz. (a) Calculate the frequency spacing between modes of the laser. (b) Calculate the gain coefficient at lasing threshold. (c) Assume that the transition undergoes homogeneous gain saturation, with the lasing mode right at the center of the gain profile. Determine the gain coefficient for the mode just adjacent to the lasing mode, expressing your answer as a percentage difference. How realistic is it that only one mode will actually lase, in practice?

21.4 An argon ion laser has high reflectivity mirrors separated by 1.1 m, with a plasma of refractive index ≈ 1 in between. The inhomogeneously broadened emission lineshape is Gaussian with a full width at half maximum of 3.5 GHz. (a) Determine the frequency spacing between the modes of the laser cavity. (b) If the laser is pumped at twice threshold, how many modes can potentially lase? (c) Repeat part b if the laser is pumped at five times threshold.

21.5 To select one particular mode in the argon laser of the previous problem, a solid etalon is inserted into the cavity, at near-normal angle of incidence ($\theta \approx 0$). The etalon consists of a glass slab of thickness 5 mm and index 1.5, coated on both sides with a dielectric reflector having $R = 0.99$. (a) Calculate the frequency spacing of the etalon modes. (b) Calculate the finesse of the etalon, and the frequency width of the etalon modes. (c) Does the spacing and width of these etalon modes allow one particular argon laser mode to be selected? (d) Determine the frequency spacing of the etalon modes if the etalon is inserted at an angle $\theta = 40°$, where θ is defined in Fig. 21-7.

21.6 The emission lineshape in a He–Ne laser is Gaussian with a width (FWHM) of 1.5 GHz. The refractive index of the plasma in the discharge tube is ≈ 1. If the laser is pumped at twice threshold, what cavity length is required so that only one longitudinal mode can lase?

21.7 A He–Ne laser (wavelength 632.8 nm) has cavity length 25 cm, and is pumped at twice threshold. (a) How many modes can lase? (b) Thermal expansion causes the cavity length to slowly increase. If mode m is initially located at the center of the atomic transition lineshape, by how much must L change so that mode $m \pm 1$ moves into this position? Give both the absolute and fractional change in L. Will the mode number increase or decrease?

21.8 The wavelength of a laser is tuned with a diffraction grating, as shown in Fig. 21-12. Wavelength λ_1 is diffracted directly back in the same direction when incident at angle θ, whereas wavelength λ_2 is diffracted at a different angle $\theta + \Delta\theta$. The separation between mirror and grating is L, and the position of the lasing mode is defined by an aperture of radius a just in front of the mirror. (a) Using the grating equation [Eq. (2-28)], derive an expression for the range of wavelengths $\Delta\lambda$ for which the diffracted beam will pass through the aperture. Assume that $a \ll L$. (b) Evaluate $\Delta\lambda$ for a dye laser with $\lambda = 650$ nm, $\theta = 40°$, $L = 25$ cm, and $a = 1$ mm. (c) Compare this with the ~ 40 nm width of a laser dye's emission spectrum. Is this method suitable for tuning a dye laser? (d) What happens to $\Delta\lambda$ as $\theta \rightarrow 90°$? Gratings are sometimes used at "grazing incidence" to increase the spectral resolution.

Chapter 22

Pulsed Lasers

So far, our analysis of laser operation has considered only the steady state, in which the light intensity is constant in time. This type of operation, termed continuous wave (CW), is useful for applications requiring precise frequency control, such as optical spectroscopy and optical communications. However, there are applications for which pulses of light are more desirable. For example, in laser surgery, it is advantageous to deliver the laser energy in a short burst so the heat deposited in the tissue does not have time to spread out during the pulse. This results in a cleaner, more localized cut, with minimum damage to surrounding tissue. Short pulses can also be used for timing the propagation of light, in applications such as laser-based measurement of distance (laser ranging) and speed (for traffic enforcement), to mention just a few. In this chapter, we consider the theory behind pulsed laser operation, and also some practical methods for creating and controlling the laser pulses.

22-1. UNCONTROLLED PULSING

Although a pulsed laser output can be desirable for certain applications, laser pulsations sometimes occur even when an attempt is made to operate the laser in a continuous fashion. To understand how this arises, consider a laser in which the pump excitation rate $\mathcal{R}$ is turned on with a step-function time dependence, as shown in Fig. 22-1a.

The resulting time dependence of the laser light intensity I and atomic excited state population N_2 are determined by the rate equations Eq. (20-12) and Eq. (20-13), which are reproduced here for convenience:

$$\frac{dN_2}{dt} = \mathcal{R} - N_2\left(\frac{I\sigma}{h\nu} + \frac{1}{\tau_2}\right) \tag{20-12}$$

$$\frac{dI}{dt} = \left(c\sigma N_2 - \frac{1}{\tau_c}\right) I \tag{20-13}$$

Before N_2 reaches $N_{2,\text{th}}$, there is no lasing since $dI/dt < 0$. The term containing I in Eq. (20-12) is therefore negligible, and Eq. (20-12) then becomes

$$\frac{dN_2}{dt} \simeq \mathcal{R} - \frac{N_2}{\tau_2} \qquad \text{(below threshold)} \tag{22-1}$$

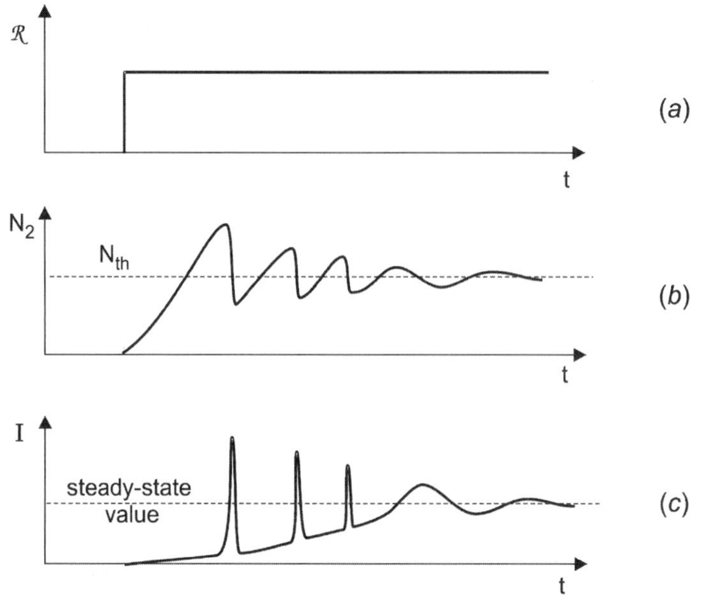

Figure 22-1. Excited state population N_2 and laser intensity *I* for a step increase in pump excitation rate $\mathcal{R}$, showing spiking behavior.

The solution for $N_2(t)$ is that of an exponential rise with time constant τ_2, as has been discussed previously in connection with Eq. (19-6) and Fig. 19-3. If the excitation rate $\mathcal{R}$ is sufficiently great, then at some point in time the excited state population will exceed the threshold value $N_{2,\text{th}}$. Equation (20-13) then gives $dI/dt > 0$, and the light intensity starts to increase exponentially in time. This is the beginning of the laser pulse.

After a small time delay, the light intensity has increased sufficiently that the stimulated emission rate $I\sigma/h\nu$ is greater than the spontaneous decay rate $1/\tau_2$, and this causes dN_2/dt in Eq. (20-12) to decrease and finally become negative. In physical terms, the excited state population decreases in time because stimulated emission is pulling the population down from the excited state faster than the pump is replacing it (think of the water pump analogy of Section 19-1). This decrease in N_2 will continue until N_2 goes below the threshold value again, making $dI/dt < 0$. After this time, the light intensity decreases, which brings an end to the laser pulse.

If the pump excitation rate $\mathcal{R}$ remains constant in time, then the excited state population N_2 will build up again, as indicated in Fig. 22-1b. The process just described now repeats itself, with a second laser pulse appearing a short time after N_2 once again exceeds $N_{2,\text{th}}$. The result of this interaction between light intensity and excited state population is a series of pulses, as shown in Fig. 22-1c.

Ideally, the solutions for $N_2(t)$ and $I(t)$ would tend toward a steady-state solution, each exhibiting oscillations with an amplitude that decreases in time. These oscillations are termed *relaxation oscillations,* and they decay in time at a rate that depends on the degree to which the laser is excited above threshold. In practice, perturbations such as temperature fluctuations, mechanical vibrations, and pump-light instabilities can lead to an irregular pattern of lasing pulses known as *spiking.* Spiking was observed in many early solid state lasers, including the first laser to be demonstrated experimentally, the ruby laser. A typical time separation between spikes is ~ 1–10 μs.

22-2. PULSED PUMP

The spiking behavior discussed above produces pulses which occur randomly in time. To be useful for applications, the timing of the pulses needs to be controlled. This is generally accomplished in one of three ways: using a pulsed pump excitation, Q-switching, or mode locking. The simplest method of producing a pulsed laser output is to pulse the pump excitation, as illustrated in Fig. 22-2. If the relaxation oscillations are sufficiently damped, the laser output will approximately follow the pump pulse, with the laser operating in a quasi-CW fashion during the pulse. Used by itself, this method is practical for creating pulses of long duration (much longer than the lifetime τ_2 of the upper laser level), in situations where the pump light is stable enough to avoid spiking. For shorter pulses, either Q-switching or mode locking must be used. Sometimes, the pulsed pump method is used in combination with Q-switching or mode locking to more efficiently utilize the pump energy.

22-3. THEORY OF *Q*-SWITCHING

To obtain intense laser pulses of width on the order of nanoseconds, the technique of Q-switching is often employed. The quality factor Q of a resonator was defined in Eq. (16-18) as the ratio of the frequency of a mode to its width. Using Eqs. (16-16), (20-11), and (20-14), the Q can also be written in the equivalent forms:

$$Q = 2\pi\nu\tau_c = \frac{2\pi\nu}{c\gamma_{\text{th}}} = \frac{2\pi\nu}{c\sigma N_{2,\text{th}}} \tag{22-2}$$

The first expression shows that Q can be interpreted physically as 2π times the number of oscillations of the light wave's electric field during a time equal to the cavity lifetime τ_c. A resonator with high Q "rings more freely" than one with lower Q. The second expression shows that the threshold gain coefficient is inversely proportional to Q. A high-Q cavity therefore has a lower threshold for lasing. The last expression shows that the threshold population inversion $N_{2,\text{th}}$ is also inversely proportional to Q. The population inversion required for lasing is lower for a higher-Q resonator.

The relationship between Q and $N_{2,\text{th}}$ allows us to understand the basic principle of Q-switching, which is illustrated in Fig. 22-3. The Q of the laser cavity is made to be ad-

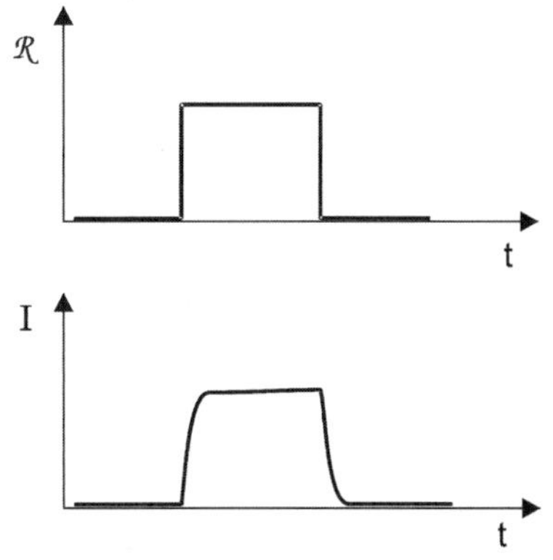

Figure 22-2. The laser intensity can be pulsed by pulsing the excitation rate $\mathcal{R}$.

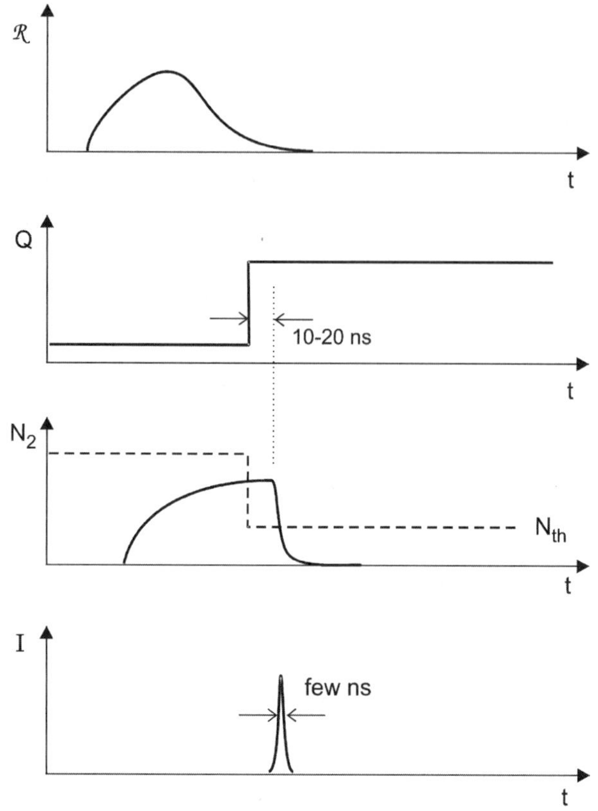

Figure 22-3. Laser intensity I, cavity Q, and excited-state population N_2 for Q-switching with a pulsed pump. The threshold inversion N_{th} shifts from high to low when the Q is switched.

justable, by mechanisms to be discussed in the next section. When the gain medium is first excited by the pump, the cavity is in its low Q state, with a very high threshold inversion $N_{2,th}$. During this initial time period, $N_2 < N_{2,th}$ and no lasing occurs. The excited state population is therefore allowed to build up to a high value, limited only by the pump energy.

At a time selected by the user, the Q is suddenly switched from a low to a high value. The inversion threshold correspondingly switches from a high to a low value, so that the population N_2 is now well above threshold. At this point, the light intensity in the cavity increases exponentially in time, and lasing ensues. There is a short delay (typically ~ 20 ns) between the switching of Q and the laser pulse, because it takes time for stray light in the cavity to be amplified sufficiently by the gain medium. The pulse intensity I reaches a maximum when $dI/dt = 0$. According to Eq. (20-13), this occurs when the excited state population N_2 has been brought down to the threshold value by stimulated emission. When N_2 goes below threshold, $dI/dt < 0$ and the light intensity decreases. Typical widths for Q-switched pulses are on the order of a few nanoseconds.

The Q-switching scheme presented in Fig. 22-3 generates only a single pulse, since the excitation rate $\mathcal{R}$ is small after the pulse, and the population N_2 is not allowed to build up a second time. Fig. 22-4 shows an alternative scheme, in which a series of Q-switched pulses is generated by continuous pumping and repetitive Q-switching. In this case, the excit-

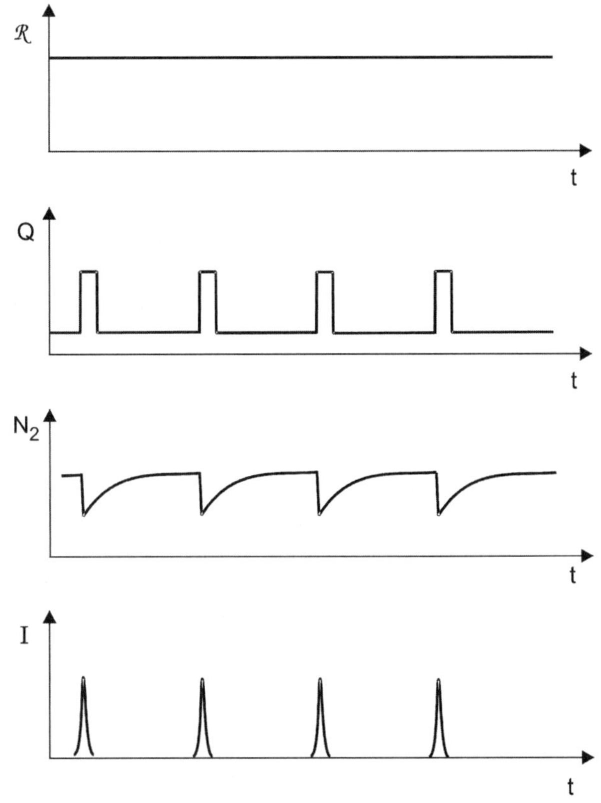

Figure 22-4. In repetitive Q-switching, the population N_2 recovers in between each pulse.

ed state population N_2 recovers after each pulse, rising exponentially with a time constant equal to the upper state lifetime τ_2.

The optimum time T_p between pulses in repetitive Q-switching can be determined by the following considerations. If $T_p \ll \tau_2$, then N_2 does not have time to fully recover from the previous pulse, and the pulse energy will be less than it could be. If $T_p \gg \tau_2$, on the other hand, then the atoms in the gain medium spend much of their time "idling" in the excited state, spontaneously emitting photons. These photons correspond to wasted pump energy, which decreases the overall energy efficiency of the laser. The pulse energy and laser efficiency can be optimized by choosing a pulse separation of $T_p \sim \tau_2$. For solid-state lasers such as Nd:YAG with upper-state lifetimes $\tau_2 \sim 10^{-3}$ s, this corresponds to an optimum Q-switching rate of ~ 10^3 pulses/second.

22-4. METHODS OF Q-SWITCHING

There are several practical methods for Q-switching a laser, each using a different approach to changing the Q of the resonator. A key requirement for all these methods is that the cavity Q be changed quickly enough that the population inversion N_2 remain nearly constant during the switching process. Generally, a switching time of ~ 10 ns is desirable. When the switching time is too long, multiple pulses may result as N_2 oscillates above and below threshold. (Svelto 1998)

Rotating Mirror

A simple mechanical method for Q-switching is illustrated in Fig. 22-5. One mirror in the laser cavity is fixed in place, and the other rotates at a high speed about a vertical axis. The Q of the cavity is then high only when the mirrors are parallel within some tolerance $\Delta\theta$. If the mirror is rotating at angular speed ω, then the cavity Q switches from high to low in a time ~ $\Delta\theta/\omega$. If the angular tolerance is $\Delta\theta \approx 10^{-3}$ rad, and the mirror speed is 10,000 rpm ($\approx 10^3$ rad/s), the switching time is ~ 10^{-6} s. Although this is longer than optimum, the simplicity of the method led to its use in early solid-state lasers such as the ruby laser. It is still used occasionally for solid-state lasers, and more recently, fiber lasers.

Electrooptic Shutter

Another way of changing the Q of the laser cavity is to insert a shutter inside the cavity. Mechanical shutters do not have a sufficiently fast response time, however, so a non-mechanical mechanism must be employed. One example is the *electrooptic shutter,* as illustrated in Fig. 22-6. This shutter is formed by placing two optical elements in the path of the beam inside the cavity. The first element is a polarizer, oriented to allow transmission of only one polarization of light (E field vertical, along the x axis, for example). The second element is a *Pockels cell* (see Fig. 9-22 and related discussion), a nonlinear crystal that rotates the polarization of the light when a high voltage is applied. With no applied voltage, vertically polarized light is efficiently transmitted through both elements, and the cavity Q is high. When an appropriate voltage is applied, vertically polarized light is rotated to horizontal polarization in two passes through the nonlinear crystal, and this light is then blocked by the polarizer. The net result is a low Q value when the voltage is applied. The Q-switching process proceeds by initially applying a high voltage to the Pockels cell, and then rapidly removing the voltage, which switches the Q from low to high.

The voltage required for the Pockels cell is $V_{\pi/2} = 0.5V_\pi$, where V_π is given in Eq. (9-49). The phase shift between the two polarization components only needs to be $\pi/2$ in one pass through the nonlinear crystal, because the accumulated phase shift in two passes is then π. In a laser cavity, the beam diameter is typically on the millimeter scale, and the electrode spacing d must be at least this large. According to Eq. (9-49), $V_\pi \propto d$, and so the required voltage for Q-switching is quite a bit higher than the value calculated for a waveguide in Example 9-3. Typical voltages required for a Pockels cell are a few kV, and switching times can be 20 ns or less. This type of Q-switching is in common use, but requires attention to safety because of the high voltages involved.

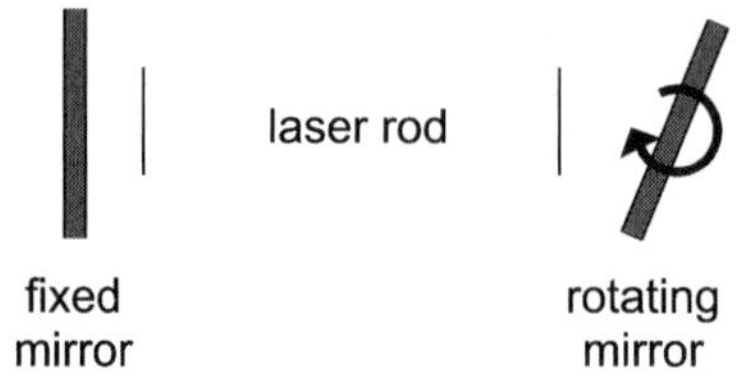

Figure 22-5. Q-switching can be achieved by rotating one of the laser mirrors at a high speed.

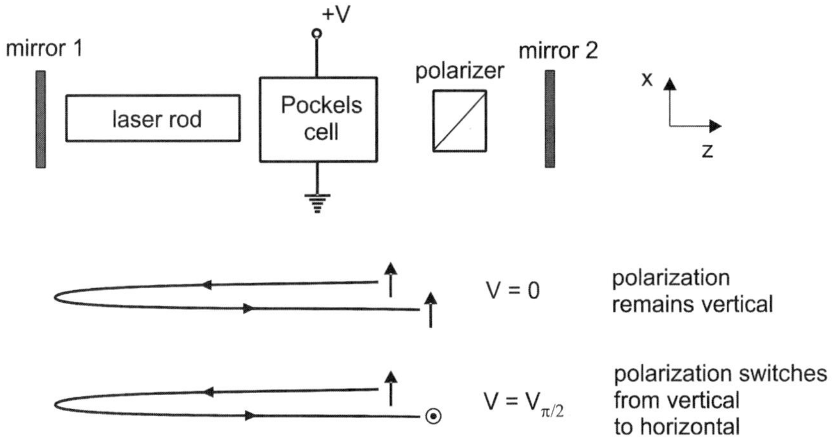

Figure 22-6. Electrooptic Q-switching is achieved by placing a Pockels cell and polarizer in the laser cavity. Vertical polarization is maintained when $V = 0$. When a voltage $V = V_{\pi/2}$ is applied, the polarization is rotated from vertical to horizontal after passing through the Pockels cell twice.

Acoustooptic Shutter

Another method for nonmechanical Q-switching is the *acoustooptic shutter,* depicted in Fig. 22-7. In this method, a transparent crystal is inserted into the laser cavity, and high intensity acoustic waves are generated in the crystal by an attached piezoelectric transducer (PZT). The acoustic waves create a periodic variation of the crystal's refractive index, which forms a volume-phase grating. Light that is diffracted from this grating increases the cavity loss and decreases Q. The Q-switching process starts with the acoustic waves turned on, such that Q is low enough to prevent lasing. The acoustic waves are then quickly turned off, which increases the Q and enables lasing.

A detailed view of the light passing through the crystal is presented in Fig. 22-8. Light enters the crystal with an angle of incidence θ_B', and is refracted inside the crystal at an an-

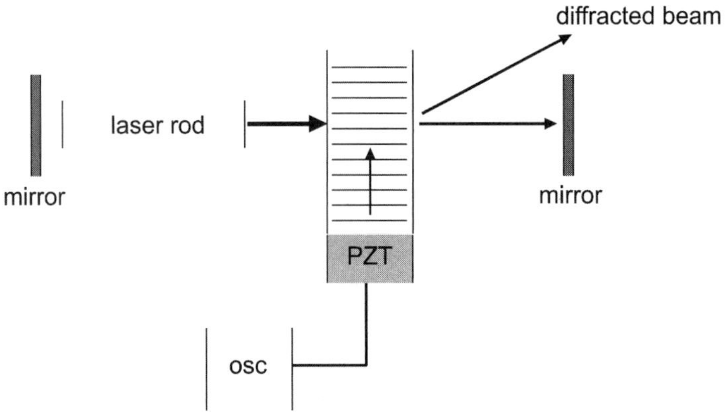

Figure 22-7. In acoustooptic Q-switching, sound waves create a refractive-index grating that diffracts part of the beam, spoiling the Q of the cavity.

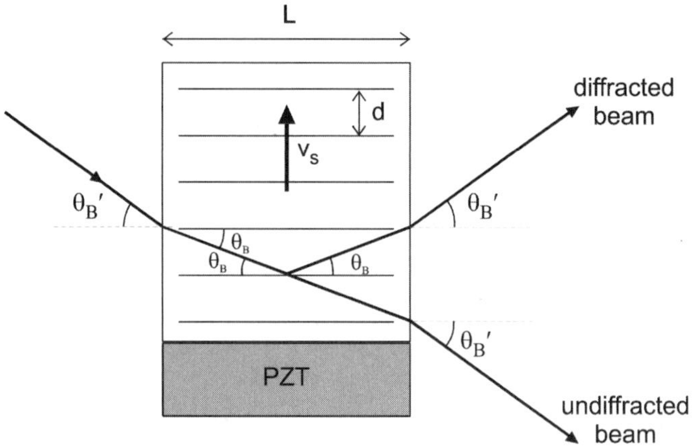

Figure 22-8. Acoustic waves of frequency f_a and wavelength d move with speed v_s in the crystal. Light with wavelength λ is efficiently scattered when it enters the crystal at an angle of incidence $\theta_B' = \lambda/2d$. The deflection of the beam is $2\theta_B'$.

gle θ_B. This light will be Bragg scattered from the acoustic waves if the thickness L of the crystal is large enough to form a thick grating. The distinction between thick and thin gratings can be seen by referring to Figs. 2-16 and 2-17. A grating is considered thick if light that is diffracted by the entrance side of the grating spreads out sufficiently in propagating a distance L that it interacts with several different grating planes at the exit side. Since light diffracted by an aperture of size d has an angular spread $\delta\theta \sim \lambda/d$, the light is spread out over a distance $L\delta\theta \sim L\lambda/d$ at the exit side of the grating, and the requirement is that this be much greater than d. The condition for a thick grating can then be written

$$L \gg \frac{d^2}{\lambda} \qquad \text{(thick grating condition)} \qquad (22\text{-}3)$$

For a thick grating of acoustic waves, light will be efficiently reflected when the Bragg condition is satisfied. Using Eq. (2-28) for first order ($m = 1$), and generalized to a medium with index of refraction n, this becomes

$$\theta_B \simeq \frac{\lambda/n}{2d}$$

where the small-angle approximation $\sin\theta \simeq \theta$ has been used. For small angles, Snell's law [Eq. (2-8)] becomes $\theta_B' \simeq n\theta_B$, so the exterior angle θ_B' for Bragg scattering is simply

$$\theta_B' \simeq \frac{\lambda}{2d} \qquad (22\text{-}4)$$

where λ is the free-space wavelength. The spacing d between the planes of constant refractive index is just the wavelength of the acoustic wave in the material, given by

$$d = \lambda_a = \frac{v_s}{f_a} \qquad (22\text{-}5)$$

where f_a is the acoustic frequency, and v_s is the velocity of sound in the material. Typical acoustic frequencies used in acoustooptic switching are in the radio frequency range of tens of megahertz.

EXAMPLE 22-1

An acoustooptic deflector is constructed using flint glass, which has a refractive index of 1.95 and a sound velocity 3×10^3 m/s. Acoustic waves with frequency 80 MHz are generated in the glass in a beam of width 2 cm. Determine the angle through which an optical beam of free-space wavelength 800 nm is deflected, and verify that the width of the acoustic waves is sufficient for Bragg scattering.

Solution: The acoustic wavelength is

$$d = \frac{3 \times 10^3 \text{ m/s}}{80 \times 10^6 \text{ s}^{-1}} = 37.5 \text{ } \mu\text{m}$$

and optical beam is deflected by

$$2\theta_B' = \frac{0.800 \text{ } \mu\text{m}}{37.5 \text{ } \mu\text{m}} = 2.3 \times 10^{-2} \text{ rad} = 1.22°$$

The acoustic wave width should be greater than

$$\frac{d^2}{\lambda} = \frac{(37.5 \times 10^{-6} \text{ m})^2}{800 \times 10^{-9} \text{ m}} = 1.75 \text{ mm}$$

Since $L = 2$ cm, this condition is satisfied.

Passive Q-Switching

So far we have considered *active Q-switching,* in which the time and duration of the change in Q are under active control. The voltage pulse applied to the Pockels cell, or the RF power sent to the acoustooptic deflector, occurs with a timing and repetition rate determined by the user. An alternative approach is to let the laser cavity Q-switch itself, independently of actions by the user. Such a method is termed *passive Q-switching.*

A laser can be passively Q-switched by placing a *saturable absorber* inside the cavity, as shown in Fig. 22-9. The saturable absorber has the property that its absorption coefficient decreases as the light intensity increases. This is the phenomenon of optical bleaching, discussed in Chapter 9 (see Fig. 9-6). As the absorption loss in the saturable absorber decreases, the Q of the cavity increases, and the threshold gain coefficient γ_{th} decreases. The gain coefficient γ in the laser gain medium is adjusted to be just a little above threshold for the low-Q condition, so that laser light starts building up slowly. As the light intensity increases, the difference $\gamma - \gamma_{th}$ increases, which causes the light intensity to build up faster. This higher light intensity makes $\gamma - \gamma_{th}$ increase still further, causing the light intensity to increase even faster still, in a positive feedback loop. In effect, the laser light "digs its own hole" through the saturable absorber. The result is an intense burst of light in a self-Q-switched pulse.

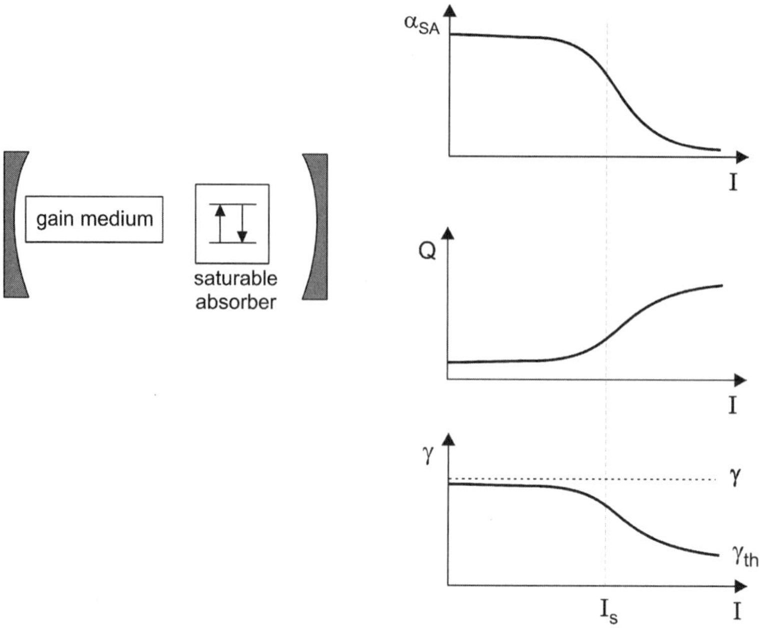

Figure 22-9. Placing a saturable absorber inside the laser cavity causes the laser to self-*Q*-switch.

The intensity at which the absorption coefficient is reduced by a factor of two is the saturation intensity I_s, given in Eq. (19-9) for a four-level system. It is generally desirable that I_s not be too large, so that *Q*-switching can occur at a reasonable power level. In practice, liquid dyes are often used for this purpose, since they have large absorption cross sections and correspondingly low values of I_s.

22-5. THEORY OF MODE LOCKING

The width of *Q*-switched pulses is generally limited to the order of a few nanoseconds, due to the finite cavity lifetime τ_c. To obtain shorter pulses, the technique of *mode locking* can be used. In this method, the laser is run continuously above threshold, with multiple modes lasing simultaneously. This contrasts sharply with the *Q*-switching technique, in which the laser makes a fast transition between nonlasing and lasing.

Two Lasing Modes

It may seem surprising at first that continuous-wave (CW) lasing can give rise to pulses of light. To see how this works, consider first the simple case of two modes lasing simultaneously. Let the two modes be designated 1 and 2, with time-dependent E fields given by

$$\begin{aligned} E_1(t) &= A\cos[(\omega_0 - \delta\omega/2)t] \\ E_2(t) &= A\cos[(\omega_0 + \delta\omega/2)t] \end{aligned} \qquad (22\text{-}6)$$

where ω_0 is the average (angular) frequency of the two modes, $\delta\omega$ is the mode separation, and A is the amplitude of each mode. We will assume that $\delta\omega \ll \omega_0$, a good approximation for laser cavities. The total electric field is the sum of the fields from the two modes, and with the trigonometric identity $\cos\alpha + \cos\beta = 2\cos\frac{1}{2}(\alpha+\beta)\cos\frac{1}{2}(\alpha-\beta)$ can be written

$$E(t) = E_1(t) + E_2(t) = 2A\cos(\omega_0 t)\cos\left(\frac{\delta\omega}{2}t\right) \tag{22-7}$$

The time-dependent E field given in Eq. (22-7) is illustrated in Fig. 22-10. It has the form of a fast oscillation at the average mode frequency ω_0, modulated by the slowly varying envelope function $\cos(\delta\omega/2)t$. The light intensity is $\propto |E|^2$, and will therefore pulsate with a repetition time T_{beat} given by

$$\begin{aligned} \frac{\delta\omega}{2}T_{\text{beat}} &= \pi \\ T_{\text{beat}} &= \frac{2\pi}{\delta\omega} = \frac{1}{\delta\nu} \end{aligned} \tag{22-8}$$

These intensity pulsations are the well-known beating phenomenon observed when combining waves of slightly different frequency, and the frequency $\nu_{\text{beat}} = 1/T_{\text{beat}}$ is the beat frequency.

We see from the above example of two oscillating modes that intensity pulsations can be obtained from two continuously lasing modes. The width of each pulse is still rather large, however, being on the order of half the beat period T_{beat}. For a laser with cavity length $L \sim 10$ cm, the spacing between adjacent modes is $c/2L \sim 1.5$ GHz, which corresponds to a beat period of $T_{\text{beat}} = 1/\delta\nu = 2L/c \sim 600$ ps. Although this is shorter than the pulse width of most Q-switched lasers, it is not sufficiently short for many applications. Also, the "pulses" are not well separated, and are more accurately described as smoothly varying oscillations in the light intensity.

N Lasing Modes

To obtain shorter pulses that are well separated, it is necessary to have a large number of modes oscillating simultaneously. If the gain bandwidth of the laser medium is inhomo-

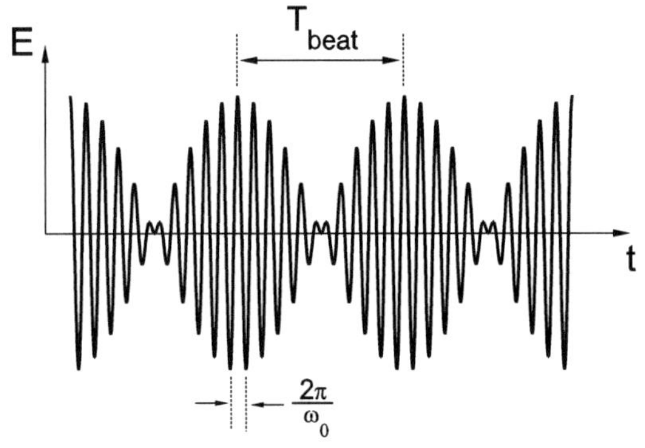

Figure 22-10. Electric field resulting from the addition of two sine-wave components with slightly different frequencies.

geneously broadened with width $\Delta\nu$, the number of simultaneously lasing modes will be roughly

$$N \simeq \Delta\nu/\delta\nu \tag{22-9}$$

where $\delta\nu$ is the mode spacing. The precise number of oscillating modes will depend on the ratio of γ_0 to γ_{th} for each mode, as shown in Fig. 22-11. Any mode with $\gamma_0/\gamma_{th} > 1$ will lase, and the rest will not. The number of lasing modes therefore depends on how far above threshold the laser is pumped.

When many modes are lasing, the general expression for the time-dependent E field can be written in complex exponential notation as

$$E(t) = \sum_m E_m \, e^{i[\omega_m t + \phi_m(t)]} \tag{22-10}$$

where E_m is the amplitude (taken as real), ω_m is the (angular) frequency, and $\phi_m(t)$ is the time-dependent phase of the mth lasing mode. The intensity of light varies in time as $I(t) \propto |E(t)|^2 = E(t)E^*(t)$, with $E^*(t)$ the complex conjugate of $E(t)$. Using the sum in Eq. (22-10) for $E(t)$ to evaluate $E(t)E^*(t)$ results in many terms, each with the general form

$$\text{term in sum} = E_m E_n e^{i(\omega_m - \omega_n)t} e^{i[\phi_m(t) - \phi_n(t)]} \tag{22-11}$$

If the phase of each mode is constant in time, then the time dependence of each term is a sinusoidal oscillation with (angular) frequency $\omega_m - \omega_n$. Since the modes are evenly spaced by $\delta\omega$ in frequency, the oscillation of each term is a harmonic (i.e., an integer multiple) of the fundamental frequency $\delta\omega$. Adding the various terms together, therefore, results in a superposition of a fundamental plus higher harmonics, which by the Fourier series principle should lead to a repeating pattern of short pulses separated in time by $2\pi/\delta\omega$. The width of each pulse is inversely related to the highest-frequency component in the Fourier series. A greater number of oscillating modes leads to higher harmonics, and therefore to shorter pulses.

The critical assumption made above is that the phase of each mode is constant in time. In general, this would not be true, since each mode is lasing independently, and

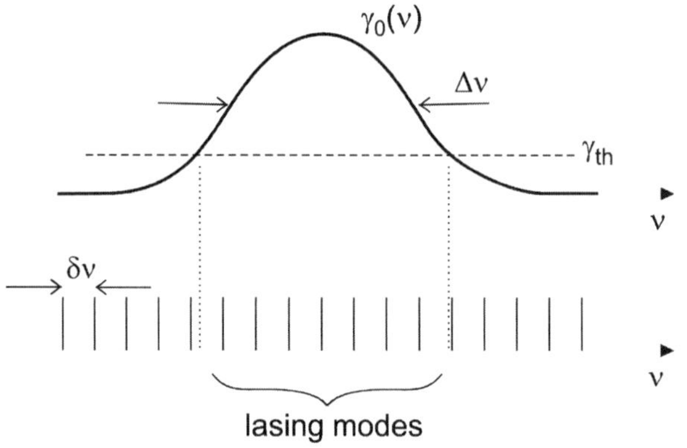

Figure 22-11. Unsaturated gain coefficient γ_0 versus frequency. Lasing will occur in those modes for which $\gamma_0 > \gamma_{th}$.

is subject to random perturbations due to vibrations, thermal effects, and other environmental variables. These perturbations cause the phase $\phi_m(t)$ of each mode to vary randomly in time, so that terms in Eq. (22-10) such as that of Eq. (22-11) average to zero for $m \neq n$. This leaves only the N terms of the form E_m^2 in the sum of Eq. (22-10), so that

$$\langle |E(t)|^2 \rangle = \sum_m E_m^2 \qquad \text{(no mode locking)} \tag{22-12}$$

where $\langle \ldots \rangle$ indicates a time average. For randomly varying phases, the mode intensities add to give the total intensity. If each mode has the same intensity, the total intensity for N lasing modes is therefore just N times the intensity in a single mode.

When the phase difference between modes is constant in time, we say that the modes are "locked in phase," and this leads to the phrase *mode-locking* to describe this method. Practical methods for causing the modes to lock in phase will be considered in the next section. For now, we assume that this is possible, and take the phase of all modes to be fixed at zero. Eq. (22-10) then becomes

$$E(t) = \sum_m E_m e^{i\omega_m t} \qquad \text{(modes locked)} \tag{22-13}$$

For simplicity, we will take each lasing mode as having the same amplitude $E_m = E_0$. Defining ω_0 as the average lasing mode frequency (at center of gain curve), and $\delta\omega$ as the mode spacing, Eq. (22-13) can be written as

$$E(t) = \sum_{\ell=-(N-1)/2}^{(N-1)/2} E_0 e^{i(\omega_0 + \ell\delta\omega)t} \tag{22-14}$$

where $\ell = 0, \pm 1, \pm 2, \ldots \pm(N-1)/2$ is an integer labeling the different lasing modes, and N is taken as odd. Since N is very large in practice, we will make the approximation $N - 1 \simeq N$ in what follows. Since E_0 and $e^{i\omega_0 t}$ are independent of ℓ, they can be factored out of the sum in Eq. (22-14), leaving

$$E(t) = E_0 e^{i\omega_0 t} \sum_{\ell=-N/2}^{N/2} e^{i\ell\delta\omega t} \tag{22-15}$$

The time dependence in Eq. (22-15) depends on the product of two factors, the first varying rapidly in time with average mode frequency ω_0, and the second varying much more slowly, with frequency $\delta\omega$. This is similar to the results obtained previously for two oscillating modes in Eq. (22-7) and Fig. 22-10. The second factor in Eq. (22-15) can be thought of as an envelope function that slowly modulates the amplitude of the rapid oscillations at frequency ω_0. The light intensity $I \propto |E(t)|^2$ depends on the magnitude of this sum of complex exponentials.

Pulse Width

To gain an intuitive understanding of the nature of mode-locked pulses, it is most instructive to perform the summation in Eq. (22-15) graphically rather than analytically. Each term in the sum is a vector in the complex plane, of unit magnitude and making an angle $\theta_\ell = \ell\delta\omega t$ with the real axis. At $t = 0$, all the vectors are along the real axis, and $E(0) = NE_0$. This time corresponds to the peak of a laser pulse. As time increases, the vectors fan out in the complex plane, as shown in Fig. 22-12, and the resultant amplitude decreases. When the vector with the highest value of $\ell = N/2$ reaches the negative real axis, as in Fig. 22-13, the vectors are uniformly distributed in angle, giving a resultant amplitude of zero. The point in time when this occurs corresponds to the end of the laser pulse.

The width Δt_p of the mode-locked pulse is, therefore, obtained by setting $\theta_\ell = \pi$ for $\ell = N/2$, which gives

$$\frac{N}{2}\,\delta\omega\Delta t_p = \pi$$

Solving for Δt_p yields

$$\Delta t_p \simeq \frac{2\pi}{N\delta\omega} = \frac{1}{N\delta\nu} = \frac{1}{\Delta\nu} \tag{22-16}$$

where Eq. (22-9) has been used. The last expression above shows that the pulse width is simply the reciprical of the gain bandwidth of the laser transition. This is another example of the uncertainty relation between time spread and frequency spread, given in Eq. (15-2). It is also equivalent to the Fourier transform principle (see Appendix B), which states that to construct a pulse of duration Δt from pure sine waves requires a distribution of frequencies of order $\Delta\nu \simeq 1/\Delta t$. The pulse duration $\Delta t_p \simeq 1/\Delta\nu$ is the minimum allowed for a bandwidth $\Delta\nu$, and a pulse with this minimum value is said to be *transform limited.*

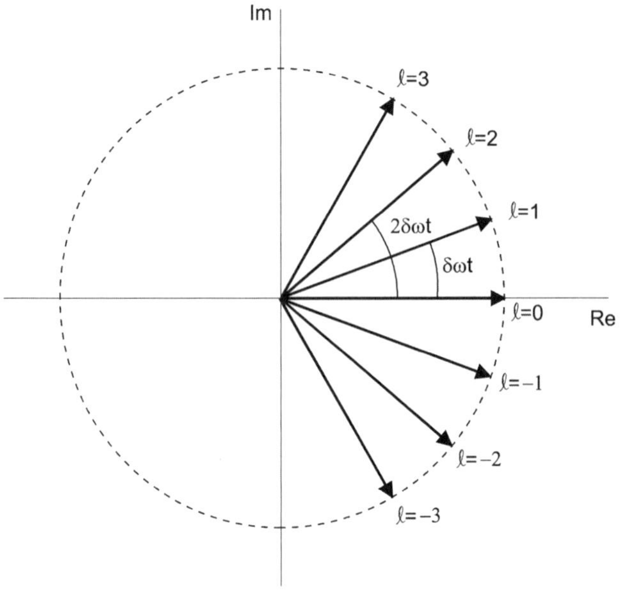

Figure 22-12. Phasors in the complex plane representing the terms of the form $\exp(i\ell\delta\omega t)$.

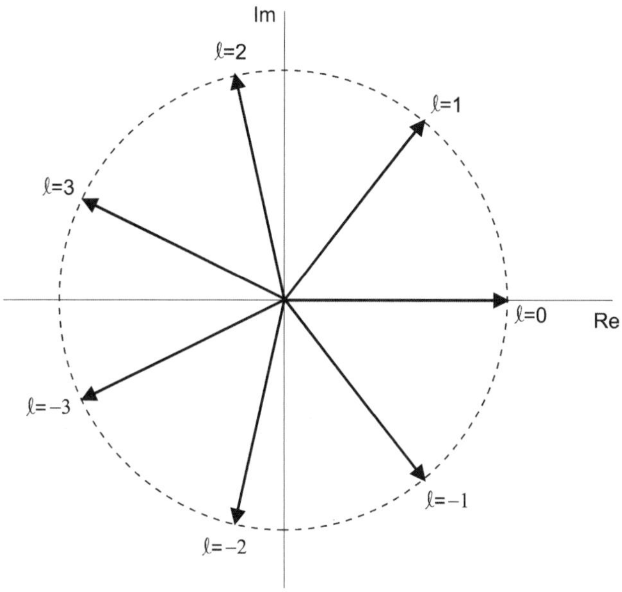

Figure 22-13. When the phasors span the complex plane, the pulse amplitude becomes very small.

Pulse Repetition Time

After the end of the laser pulse, the vectors in Fig. 22-13 continue to span the complex plane as they rotate, and the light intensity remains low. However, when the vector with $\ell = 1$ has rotated by 2π to again coincide with the real axis, the other vectors will have rotated by $\ell 2\pi$, and will also be along the real axis. The vectors will then be rephased, resulting in another laser pulse. The time T between rephasings (and hence between laser pulses) is given by $\delta\omega T = 2\pi$, or

$$T = \frac{2\pi}{\delta\omega} = \frac{1}{\delta\nu} = \frac{2L}{c} \tag{22-17}$$

The expression in Eq. (22-17) for the time between pulses has a simple physical interpretation. Since $2L$ is the distance that light must travel in making one round-trip through the cavity, T is the corresponding round-trip time. This is an intuitively satisfying result, and leads to the picture of a single pulse of duration Δt_p bouncing back and forth between the mirrors of the cavity. The pulse duration can be expressed in terms of the round-trip time as

$$\Delta t_p \simeq \frac{1}{N} T \tag{22-18}$$

using Eqs. (22-16) and (22-17). The "duty factor" for the light pulses (i.e., the fraction of time that the pulses are "on") is therefore $1/N$, as illustrated in Fig. 22-14.

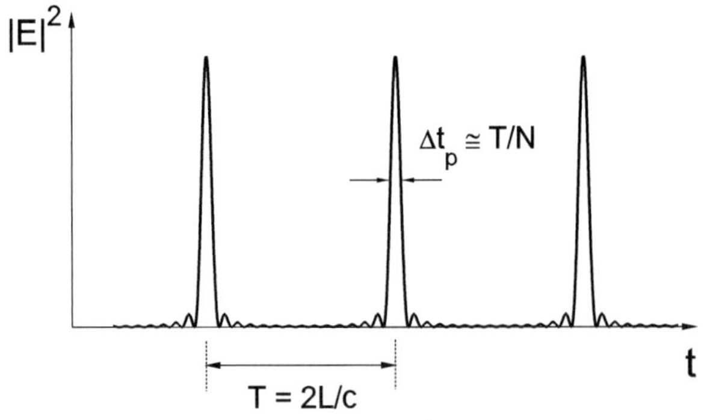

Figure 22-14. Time dependence of mode-locked pulses calculated for $N = 15$. Pulse width is Δt_p and pulse repetition time is T.

EXAMPLE 22-2

(a) A fiber laser of length 5 m operates at a free-space wavelength of 1530 nm with a gain bandwidth of 30 nm. Taking the refractive index of glass to be 1.5, determine the shortest possible mode-locked pulse, and the corresponding number of simultaneously lasing modes.

Solution:
The frequency bandwidth for the fiber laser is

$$\Delta\nu = \frac{c}{\lambda^2}\,\Delta\lambda = \frac{3\times 10^8}{(1.53\times 10^{-6})^2}(30\times 10^{-9}) = 3.84\times 10^{12}\ \mathrm{s}^{-1}$$

and the pulse width is

$$\Delta t_p = \frac{1}{3.84\times 10^{12}} = 2.6\times 10^{-13}\ \mathrm{s}^{-1} = 260\ \mathrm{fs}$$

The mode spacing (free spectral range) is

$$\frac{c}{2nL} = \frac{3\times 10^8}{(2)(1.5)(5)} = 20\ \mathrm{MHz}$$

which gives the number of lasing modes as

$$N = \frac{3.84\times 10^{12}}{20\times 10^6} = 1.92\times 10^5$$

(b) Repeat (a) for an argon ion laser of length 1 m, operating on a transition of wavelength 514.5 nm and frequency bandwidth 8 GHz.

Solution:
For the argon ion laser, the pulse width is

$$\Delta t_p = \frac{1}{8\times 10^9} = 1.25\times 10^{-10}\ \mathrm{s}^{-1} = 125\ \mathrm{ps}$$

and the mode spacing and number of lasing modes are

$$\frac{c}{2nL} = \frac{3 \times 10^8}{(2)(1)} = 1.5 \times 10^8 \text{ s}^{-1}$$

$$N = \frac{8 \times 10^9}{1.5 \times 10^8} = 53$$

It is clear from this example that solid-state lasers are capable of producing much shorter pulses than gas-phase lasers, due to the broader width of the gain transition.

Pulse Energy

The power contained in each lasing mode is the same with or without mode locking. The only difference is whether or not the E fields from the different modes periodically add together in phase to create pulses. When there is no mode locking, the laser output is continuous wave (CW), with an average power proportional to

$$\langle |E(t)|^2 \rangle = NE_0^2 \qquad \text{(no mode locking)} \tag{22-19}$$

Here we have used Eq. (22-12) with the simplifying assumption that each of the N modes has the same amplitude E_0. When mode locking occurs, the E field components in Eq. (22-13) add together in phase during the pulse to give a total E field $E_{\text{peak}} = NE_0$, for a peak pulse power proportional to

$$|E_{\text{peak}}|^2 = N^2 E_0^2 \qquad \text{(mode-locked pulse)} \tag{22-20}$$

Comparison of Eqs. (22-19) and (22-20) shows that the peak power P_{peak} in a mode-locked pulse is related to the average power $\langle P \rangle$ by

$$P_{\text{peak}} = N\langle P \rangle \tag{22-21}$$

A large number of oscillating modes can, therefore, greatly enhance the peak power of the mode-locked pulse.

The result in Eq. (22-21) has a simple physical interpretation based on Fig. 22-14. Mode locking does not change the average energy in the laser beam, but instead simply redistributes this energy in time. In a time interval equal to the pulse separation T, the energy delivered by the beam can be calculated in either of two ways: $\langle P \rangle T$ using the average power, or $P_{\text{peak}}\Delta t_p$ using the peak power. The relation between P_{peak} and $\langle P \rangle$ is therefore

$$\langle P \rangle T = P_{\text{peak}}\Delta t_p = P_{\text{peak}} \frac{T}{N}$$

which again gives Eq. (22-21).

22-6. METHODS OF MODE LOCKING

If the different modes can lase completely independently of each other, then the relative phase between modes will naturally drift due to environmental effects such as temperature, mechanical vibrations, and so on. In order to achieve mode locking, there must be some interaction, or coupling, between the modes. There are two general methods for accomplishing this.

Active Mode Locking

One technique for mode locking a laser is to introduce a variable loss in the cavity, as depicted in Fig. 22-15. This loss is modulated in time at a frequency ν_{mod} equal to the mode separation $\delta\nu$, typically by means of an acoustooptic or Pockels cell placed in the cavity. This modulation couples two adjacent cavity modes because, as we saw in Eqs. (22-6) and (22-7), an optical wave with time-dependent amplitude varying at frequency $\delta\nu$ corresponds to two component waves, with optical frequencies separated by $\delta\nu$. The modulation is said to create *sidebands* around the center frequency ν_0, which couples light energy from one mode into another as shown in Fig. 22-16. If the coupling is sufficiently strong, the phase of the different modes will be locked, resulting in mode locking.

The mode-locking process can also be understood in the time domain. Fig. 22-17 shows a sinusoidal time variation in cavity loss with modulation frequency $\delta\nu = c/(2L)$. The time between loss minima is $T = 2L/c$, the round-trip time for a pulse in the cavity. Light that arrives at the variable loss medium when the loss is a minimum has a lower round-trip loss than light arriving at other times. The process of mode locking becomes one of self-selection, the optical equivalent of Darwin's "survival of the fittest." Any linear combination of modes may spontaneously form through random variations in phase, but only the particular combination in which the modes are locked in phase has the lowest lasing threshold. The mode-locked laser output shown in Fig. 22-17 has the lowest round-trip loss of all possible solutions, and is therefore self-selected.

For a laser cavity with $L = 1$ m, typical for ion lasers, the mode spacing is $\delta\nu = c/2L = 150$ MHz, which is in the "radio frequency" (RF) region of the spectrum. To maintain efficient mode locking, an RF oscillator is needed that is very stable in frequency.

Passive Mode Locking

In the mode-locking process described above, the cavity loss is actively modulated to select for the mode-locked combination of modes. This method is therefore termed *active*

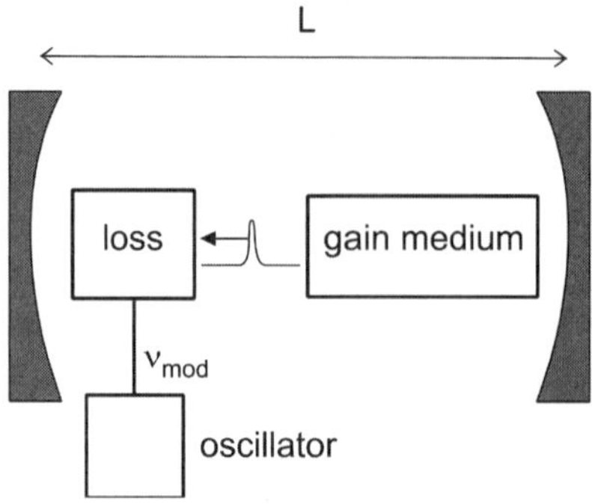

Figure 22-15. Active mode-locking scheme, in which the cavity loss is externally modulated at frequency ν_{mod}.

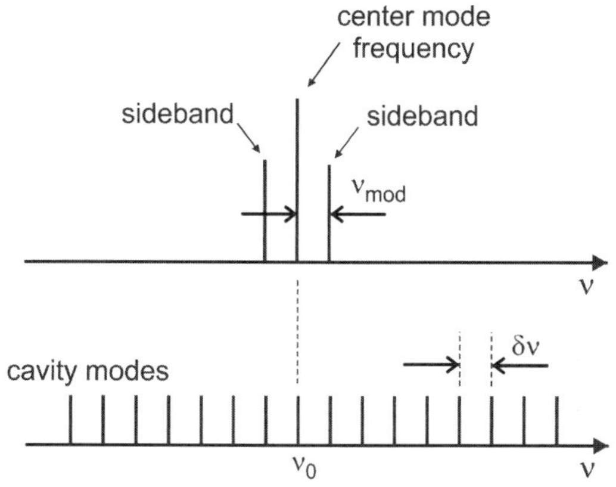

Figure 22-16. When light in a mode of frequency ν_0 is modulated at frequency $\nu_{mod} = \delta\nu$, sidebands are created at frequencies $\nu_0 \pm \delta\nu$ that inject light into the adjacent modes. In this way, neighboring modes become coupled.

mode locking. In contrast to this is *passive mode locking,* in which the mode-locked combination of modes occurs spontaneously in the cavity, without the need for a user-supplied modulation. This can be accomplished in a manner similar to that of passive *Q*-switching, by inserting a saturable absorber inside the cavity. As shown in Fig. 22-9, the absorption coefficient decreases with increasing light intensity, which favors the development of high intensity pulses.

The difference between the *Q*-switching and mode-locking arrangements is that for mode locking, the gain coefficient γ for CW lasing is set just *below* threshold, so that lasing will not be initiated on a single mode. In order for lasing to occur, the light intensity must be high enough to decrease the absorption loss in the saturable absorber. This can occur if the modes lock together in phase to create short pulses, since the peak power in each pulse is very high compared with the equivalent CW power. The process then again becomes one of self-selection, in which the mode-locked pulses themselves now create the conditions under which they can exist.

The time dependence of the gain threshold for passive mode locking is shown in Fig. 22-18. During the laser pulse, the gain is above threshold, and after the pulse it is below. To create short pulses, it is necessary to use a saturable absorber material that recovers its original absorption properties very quickly after the light intensity is reduced. Such a material is termed a *fast saturable absorber,* and is typically a dye molecule in solution or a

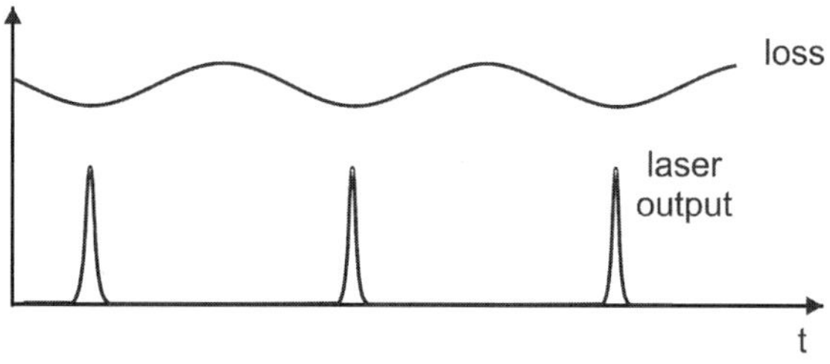

Figure 22-17. Time dependence of cavity loss in active mode locking. The lasing pulses pass through the loss medium when the loss is lowest.

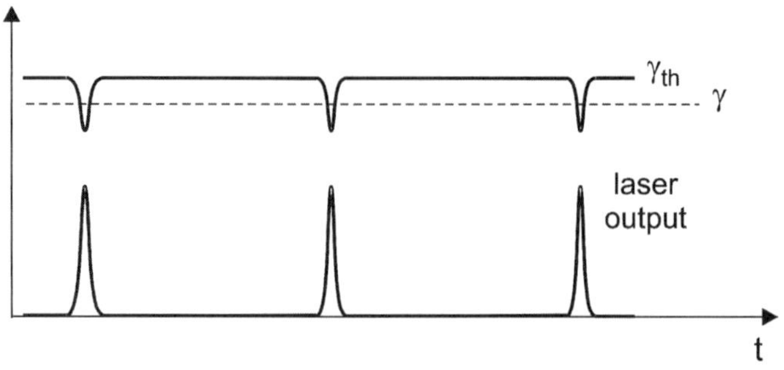

Figure 22-18. Time dependence of loss (gain threshold) in passive mode locking. The lasing pulse occurs when the gain threshold γ_{th} is reduced below the gain γ.

semiconductor. Passive mode locking has the advantage of simplicity, but gives less control over the pulses compared with active mode locking.

Kerr Lens Mode Locking

In the method described above, the duration of the mode-locked pulse is limited by the recovery time of the fast saturable absorber, usually ~ 1 ps. Pulses of much shorter duration can be obtained using the Kerr lens shutter effect, depicted in Fig. 9-19. This works on a femtosecond time scale, because it is based on the nonlinear refractive index that arises from distortions of the atom's electron cloud. It straightforward to implement, and simply requires the placement of an adjustable aperture in the laser cavity, as illustrated in Fig. 22-19. At low optical intensity, the beam is partially blocked by the aperture, preventing lasing. At sufficiently high intensity, however, self-focusing occurs, and this decreases the divergence of the beam and increases the amount of light getting through the aperture. This has the same effect on lasing as a saturable absorber, but on a much faster time scale. Using this method, ultrafast pulses have been generated in Ti:sapphire lasers, with durations down to about 5 fs.

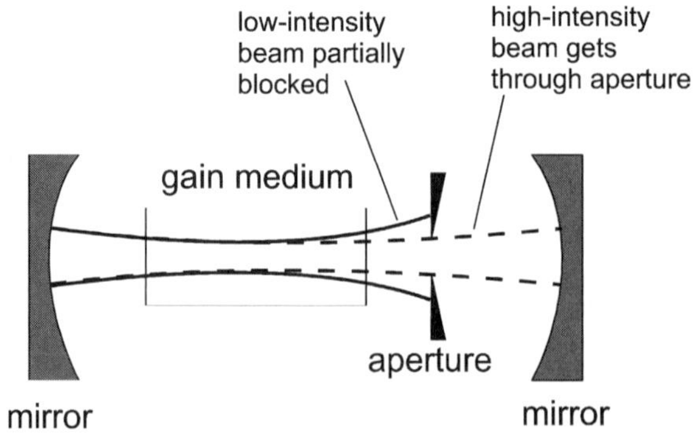

Figure 22-19. In Kerr lens mode locking, a CW beam of low intensity is partially blocked by an intracavity aperture, but a beam of high intensity is mostly transmitted, due to self-focusing. This naturally selects a combination of modes that corresponds to high-intensity pulses.

PROBLEMS

22.1 A Nd:YAG laser operates at 1064 nm, and puts out a Q-switched pulse of energy 100 mJ and duration 5 ns. Take the beam diameter inside the laser cavity to be 4 mm, and use the spectroscopic parameters for Nd:YAG given in Table 23-1. (a) Determine the peak laser output power. (b) Determine the number of photons in one laser pulse. (c) Assuming that the output mirror transmission is 70%, determine the stimulated emission rate at the peak of the laser pulse, and compare this with the spontaneous emission rate.

22.2 A laser cavity with $n = 1$ has mirrors of reflectivity 0.98 and 0.90 separated by 25 cm. After pumping and switching the Q, the population inversion in the laser medium is initially six times the threshold value. (a) Calculate the cavity lifetime. (b) Determine the time required for stray light in the cavity to increase by a factor of 10^5, assuming no gain saturation.

22.3 The pulse width of a Q-switched laser is generally somewhat greater (factor of two or three) than the cavity lifetime τ_c. Consider a fiber laser with index 1.5 and length 5 m operating at a wavelength of 1 μm. The mirror reflectivities are 0.98 and 0.95, and the loss coefficient in the fiber is 2 dB/km. (a) Determine the minimum duration of Q-switched pulses by calculating τ_c. (b) Repeat part a with the fiber length reduced to 1.5 m. (c) If this is the minimum length required for efficient absorption of the pump light, what else can be done to decrease the pulse duration?

22.4 The properties of an Er:glass laser are given in Table 23-1. (a) What is the optimum pulse repetition rate for repetitive Q-switching? (b) If the energy in each pulse is 5 mJ, what is the average laser output power?

22.5 A $LiNbO_3$ acoustooptic deflector is used to switch the direction of 1030 nm (free-space) light. $LiNbO_3$ has a sound velocity of 7.4×10^3 m/s, and refractive index of 2.3. (a) If the total deflection angle (external to the material) is required to be 5°, determine (a) the acoustic wave frequency that will accomplish this, (b) the internal total deflection angle, and (c) the minimum width of the $LiNbO_3$ chip such that the thick-grating condition applies.

22.6 The Ti:sapphire laser is a very wideband, tunable laser, with a center wavelength of 800 nm and operation range from about 700–1100 nm. Take the average output power to be 3 W, the effective (air-equivalent) cavity length as 90 cm, and assume all modes from 720–870 nm are lasing with equal amplitude. (a) How many modes are lasing? (b) What is the width of each pulse? (c) What is the time between pulses? (d) Calculate the energy and peak power of each pulse

22.7 A laser with an effective (air-equivalent) cavity length of 25 cm is actively mode locked with an intracavity modulator. The gain frequency spectrum has a Gaussian shape, with center wavelength 650 nm and width (FWHM) of 15 nm. (a) What modulation frequency is required? (b) If the laser is pumped at twice threshold, determine the number of lasing modes and the laser pulse width. (c) Repeat part b if the laser is pumped at five times threshold. How much does this decrease the pulse width?

22.8 A researcher wants to actively mode lock a laser using a modulation frequency that is either twice or one-half the usual value. Do you expect either of these methods to work? Give an explanation in both the time and frequency domains.

22.9 The sum in Eq. (22-15) was determined graphically by adding vectors in the complex plane. An alternative is to use the mathematical identity

$$1 + x + x^2 + \ldots + x^{N-1} = \frac{1 - x^N}{1 - x}$$

taking $x = \exp(i\delta\omega t)$. Show that this approach leads to a time-dependent total electric field of magnitude

$$|E(t)|^2 \propto \frac{\sin^2(N\pi t/T)}{\sin^2(\pi t/T)}$$

where $T = 2L/c$ is the cavity round-trip time. You can assume $N \gg 1$.

22.10 Using the results of the previous problem, (a) show that the time interval from the peak of a pulse to the first zero is given by Eq. (22-16), (b) show that the pulse repetition time is T, and (c) show that the full width at half maximum (FWHM) of the pulses is approximately given by Eq. (22-16)

22.11 An Er-doped fiber laser has cavity length 200 m, refractive index 1.5, and operates at wavelength 1550 nm. The laser is passively mode locked and produces pulses of energy 16 nJ and duration 1.3 ps. (a) Determine the pulse repetition time. (b) Calculate the number of oscillating modes. (c) Determine the wavelength range for the modes that are oscillating. (d) Calculate the peak power in each pulse. (e) Calculate the average output power.

22.12 The fiber laser in the previous problem is now actively mode locked by modulating the cavity loss at a frequency of 2 GHz. Assume that the average output power is the same as before. (a) How many pulses per second are produced now? Explain how this can be different than in the previous problem. (b) Assuming that the pulse duration is the same as before, determine the new pulse energy and peak power in each pulse.

22.13 An argon ion laser has cavity length 120 cm, loss coefficient in the gas discharge tube of 1.25×10^{-3} m^{-1}, and mirror reflectivities of 0.999 (high reflector) and 0.990 (output coupler). The Doppler-broadened gain linewidth (FWHM) is 3.5 GHz, and measurements using a high-resolution Fabry–Perot interferometer show that there are 36 modes lasing when the laser is excited with a CW drive current of 25 A. At this drive current, the CW output power (no mode locking) is 4.0 W. Calculate (a) the threshold gain coefficient, (b) the frequency separation between adjacent modes, (c) the expected pulse width with mode locking (assume equal-amplitude modes), (d) the modulation frequency ν_{mod} needed to achieve mode locking, (e) the peak power and pulse energy during the mode-locked pulse, and (f)the degree to which threshold is exceeded (i.e., the ratio of unsaturated gain coefficient to threshold gain coefficient).

22.14 In the previous problem, the drive current to the laser is increased to 50 A. Assuming that the excitation rate is proportional to the drive current, determine (a) the new pulse width, (b) the number of lasing modes, and (c) the new peak mode-locked power.

Chapter 23

Survey of Laser Types

In the preceding chapters, we have examined the fundamental principles that are common to all lasers. We turn now to the practical application of these principles in specific lasers. There are many different types of lasers, and they can be classified most fundamentally according to the pump mechanism (how energy is deposited in the upper laser level) and the nature of the energy levels. The most common pumping mechanisms are optical and electrical, although other energy sources such as chemical, nuclear, or particle-kinetic energy are possible as well. The energy levels can be electronic in nature (describing different energy states of an electron), vibrational (describing different energy states of the atomic vibrations in a solid), or some combination of these two. In this chapter, we give a brief survey of some of the more important laser types, organized primarily by pumping mechanism, and secondarily by the nature of the laser transition. This survey is not meant to be comprehensive, but is intended to give a sense of the different properties of various laser systems. More extensive treatments can be found in the Bibliography.

23-1. OPTICALLY PUMPED LASERS

In a laser that is optically pumped, the upper laser level is populated by absorption of a photon from some optical source. This light source can be a high-intensity lamp (lamp pumping) or another laser (laser pumping). The early lasers were mostly lamp-pumped, but the trend in recent years has been toward laser-pumped lasers. This may seem strange on the face of it; if you need a laser to pump another laser, why not just use the first laser instead? We will see, however, that there are distinct advantages to this approach.

Many optically pumped lasers have a gain medium consisting of rare earth or transition metal ions doped into an insulating dielectric solid. These are termed *solid-state lasers,* and include the historically important ruby laser, as well as the neodymium laser, a long-time industrial workhorse. In this section, we discuss the operation of these old but classic laser types, along with some newer ones based on a fiber geometry.

Electronic Transition

Ruby Laser

The first experimental demonstration of laser action (Maiman 1960) utilized a ruby rod as the gain medium. Ruby is a naturally occuring gem, but can also be produced artificially under carefully controlled conditions. It consists of Al_2O_3 (aluminum oxide crystal, also known as sapphire) doped with Cr^{3+} (chromium impurity ions). These triply ionized

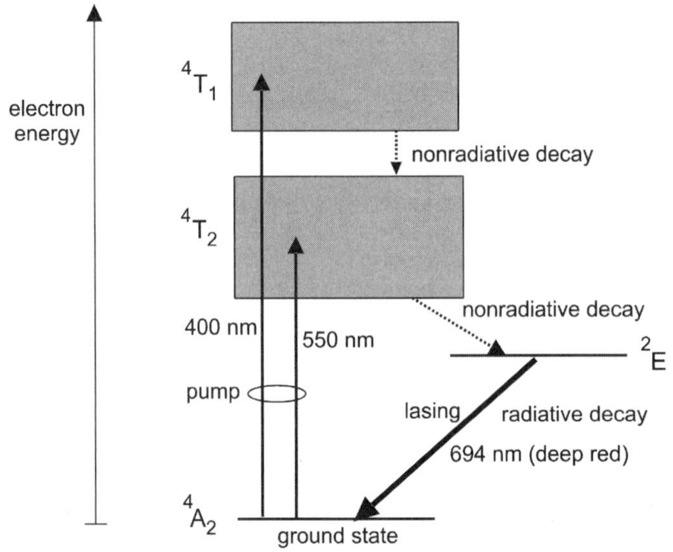

Figure 23-1 Energy levels in Cr^{3+}:Al_2O_3 (ruby). The lasing transition is from the first excited state ^2E to the ground state $^4A_2$, making this a three-level-type system.

chromium ions are responsible for the color of ruby (Al_2O_3 itself is colorless), and also for its laser action. The optically active electrons in Cr^{3+} can occupy states of different energies, as shown in Fig. 23-1. The origin of the notation ^2E, $^4T_2$, and so on, for the levels need not concern us here, and we can think of these as simply labels for the different electronic states. Electrons in the $^4T_1$ and $^4T_2$ levels* interact strongly with vibrational modes of the crystalline lattice, and these levels are significantly broadened in energy. Ruby absorbs visible light over a wide spectral range in the blue (centered at 400 nm) and the green (centered at 550 nm) regions. When it is illuminated with white light, the red light that remains gives ruby its characteristic color.

Electrons in the ^2E level interact less strongly with the lattice, and this level is relatively well defined in energy. The laser transition in ruby is from the ^2E to the ground state $^4A_2$, and occurs at a well-defined wavelength of 694 nm. To get electrons into the upper laser level, they are first promoted to the $^4T_1$ and $^4T_2$ levels by optical absorption, and from there they decay nonradiatively to the ^2E level. Because of the strong electron–lattice interaction for electrons in the $^4T_1$ and $^4T_2$ levels, this process occurs rapidly, on a ~ 100 ns time scale. The fluorescence lifetime for electrons in the ^2E level is comparatively quite long (3 ms), due to the weak electron–lattice interaction there. A level such as this with a relatively long lifetime is termed a *metastable state.*

Since the lower laser level is the ground state, ruby is an example of a three-level type laser system (see Chapter 19). To achieve the population inversion required for lasing, at least half the Cr^{3+} ions must be promoted from the ground state to the upper laser level ^2E. This requires a very high excitation rate, and it is rather ironic that the first example of lasing was in such a "difficult" system. In ruby's favor, however, is the broad absorption throughout the visible region, which allows excitation from a wide range of wavelengths to be funneled efficiently into the upper laser level. Also advantageous is the the relative-

*These are sometimes designated the $^4F_1$ and $^4F_2$ levels, respectively.

ly long lifetime of the upper state, which reduces somewhat the required excitation rate (see Eq. 19-7).

The high excitation rate required for population inversion is most easily obtained with a pulsed light source, and the ruby laser is, therefore, generally operated with a flashlamp pump. The first ruby laser utilized a spiral-shaped flashlamp that surrounded the laser rod, as depicted in Fig. 23-2. The flashlamp requires a high-voltage pulse of short duration, which can be obtained from the simple electrical circuit shown. An applied high voltage charges the capacitor, which is connected across a "spark gap" (two metal electrodes separated by a small space in air). When the capacitor voltage is sufficiently high, the air in the gap breaks down (becomes highly conducting), which causes the capacitor voltage to appear across the flashlamp tube. Other more sophisticated circuits are possible as well, but the basic requirement is that the voltage across the tube be switched quickly. The resulting optical pulse from the flashlamp has a typical duration on the order of 1 ms, which is shorter than the excited-state lifetime of ruby. This ensures that the upper-state population does not significantly decay during the excitation pulse, so the pump pulse energy is efficiently utilized in creating a population inversion.

When the ruby laser is excited with an appropriate flashlamp, the time dependence of the laser output exhibits spiking behavior (see Chapter 22). To obtain a more controlled output pulse, the laser can be *Q*-switched, for example by the rotating-mirror method. Typical *Q*-switched pulses may have an energy of ~ 0.1 J with duration ~ 10 ns, which corresponds to a peak power of ~ 10 MW. Because of the need for the power supply to recharge to a high voltage, the repetition rate of the pulses is limited to a few pulses per second, and so the average power is modest, on the order of a few watts.

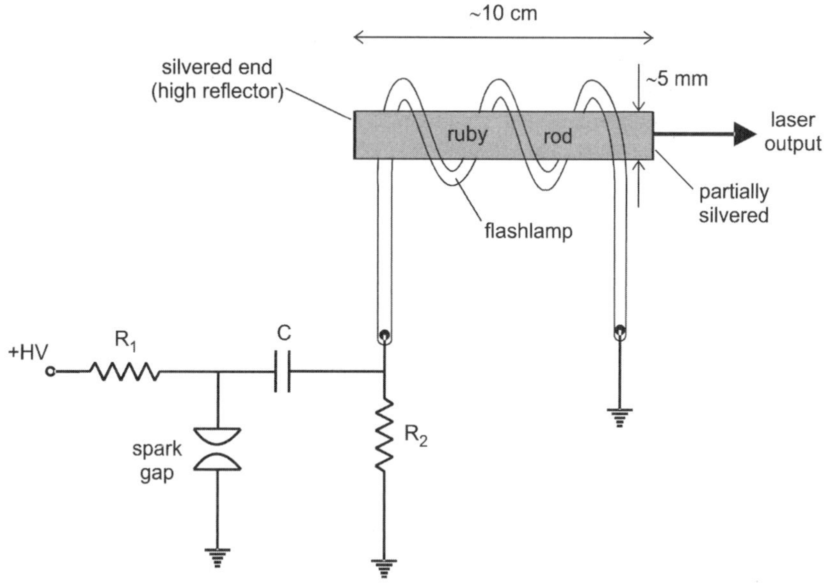

Figure 23-2 The early ruby lasers were pumped with a spiral shaped flashlamp, with one end silvered to give high reflection and the other end partially silvered to serve as the output coupler. The electrical circuit shown represents one simple way of providing a short-duration, high-voltage pulse to drive the flashlamp.

The ruby laser is important historically because it was the first to be demonstrated experimentally. It has found applications in rangefinding, holography, and medical therapy. However, it is less commonly used today, due to the development of more efficient and versatile lasers.

Neodymium Lasers

The first operating neodymium laser was developed in 1964 at Bell Labs, not long after the invention of the ruby laser. Unlike the ruby laser, however, the neodymium laser has continued to find new applications and to grow in importance, right up to the present day. The reason for this difference can be understood by considering the nature of the laser transition.

Neodymium (Nd) is one of the rare earths, the group of atoms with atomic number between 58 and 70. The triply ionized rare earths (Nd^{3+}, for example) have optical transitions in the visible and near infrared regions that are fairly well defined in energy, depending only slightly on the host solid into which the ion is doped. This insensitivity of the transition energy to the ion's environment comes about through a shielding effect unique to the rare earths. The optically active electron orbitals are designated 4f, which means principle quantum number $n = 4$ and angular momentum $l = 3$. The 4f orbitals happen to be located, on average, closer to the nucleus than the filled 5s and 5p orbitals. Electrons in these outer 5s and 5p orbitals, therefore, act like a spherical metallic shell in shielding the inner 4f electrons from the electric fields of neighboring atoms. The shielding is not perfect, but to a first approximation the energy of the various 4f levels is not affected by the environment surrounding the rare earth ion.

The energies of the lower-lying levels of Nd^{3+} are shown in Fig. 23-3. In principle, lasing can occur between any pair of levels, but the required population inversion is easily achieved only when the upper laser level has a long lifetime (see Eq. 20-18). The lifetime of most of the Nd^{3+} levels is rather short, due to efficient nonradiative relaxation to the next-lowest level. The $^4F_{3/2}$, however, has a large energy gap to the next-lowest state, and the probability of nonradiative decay is small. The lifetime of the $^4F_{3/2}$ is, therefore, reasonably long, making this the best choice for the upper laser level.

The most important laser transition in Nd^{3+} is from the $^4F_{3/2}$ (upper laser level 2) to the $^4I_{11/2}$ (lower laser level 1). Since the lower laser level here is not the ground state, this constitutes a four-level system (see Fig. 19-1). Achieving population inversion in a four-level system is much easier than in a three-level system, because it is not necessary to take half the ions out of the ground state. The required excitation rate is, therefore, much lower for the Nd^{3+} laser than for the ruby laser, and this is a primary reason for the Nd^{3+} laser's initial and continuing popularity. Other advantages of Nd^{3+} over ruby are an order-of-magnitude-higher peak cross section (for Nd^{3+} in a crystalline host), and the ability to use higher ion concentrations without significant lifetime quenching by ion–ion interactions. These both lead to a higher gain coefficient, which improves the lasing threshold and lasing efficiency.

The crystal host that has been most widely used for a Nd^{3+} laser is $Y_3Al_5O_{12}$ (yttrium aluminum garnet, or YAG), and the ion:host combination is referred to as Nd:YAG. In Nd:YAG, the laser transition occurs at 1064 nm, which is in the near infrared region. The origin of this particular wavelength can be understood by considering in more detail the nature of the upper and lower energy levels. Each "level" (the $^4F_{3/2}$ for example) is actually a *multiplet,* consisting of a number of sublevels. In general, a transition can occur from any sublevel of the $^4F_{3/2}$ down to any sublevel of the $^4I_{11/2}$, and the energy of the emitted

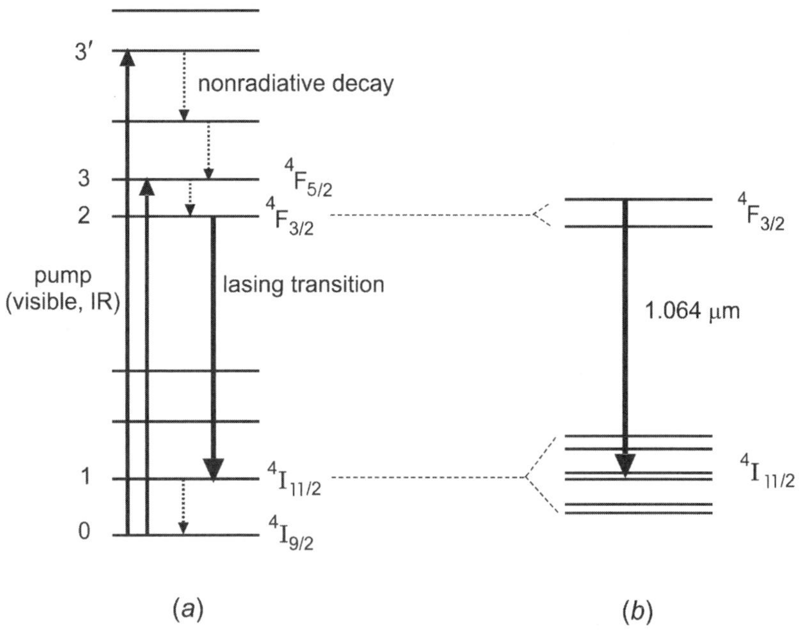

Figure 23-3 (a) Energy levels in Nd^{3+}. The lasing transition is from the metastable level $^4F_{3/2}$ to an excited state, the $^4I_{11/2}$, making this a four-level type system. The solid and dotted lines represent radiative and nonradiative transitions, respectively. (b) Each level consists of a number of sublevels, and laser transition is between one particular pair of sublevels.

photon will equal the energy difference between initial and final sublevels. The fluorescence spectrum for the $^4F_{3/2} \rightarrow {}^4I_{11/2}$ transition in Nd:YAG, shown in the lower-left panel of Fig. 23-4, gives an indication of the relative peak cross sections (and hence peak gain coefficients) for the various possible transitions. Lasing occurs at 1064 nm in Nd:YAG because this transition has the highest emission cross section. Once lasing starts on this transition, the gain becomes clamped, and gain on the other transitions remains below threshold (assuming homogeneous broadening).

Although the average energy of the sublevels in a manifold is fairly independent of the host material, the position of the various sublevels within the manifold varies considerably. The upper-left panel of Fig. 23-4 shows the emission spectrum of Nd^{3+} in a different crystal, $YAlO_3$. Although $Y_3Al_5O_{12}$ and $YAlO_3$ have the same atomic constituents, the crystal symmetries are different, and this makes the local environment of the Nd^{3+} ion different enough to significantly change the spectrum. When Nd^{3+} is doped in a glass, it can reside in any one of a great number of different "sites," each having a different local environment and symmetry. The emission spectrum in this case is an average over the many different sites, resulting in the rather smooth and broad spectra shown in the right panels of Fig. 23-4. The spectra are not entirely featureless, however, and differences in shape, width, and peak wavlength can be used to advantage for a particular application. This dependence on glass composition applies equally well to other rare earth ions doped into glass, and has implications for optical amplifiers as well as for lasers.

The Nd^{3+} laser is typically pumped with a lamp or with a diode laser. For lamp pumping, the lamp and laser rod are often placed at the foci of an elliptical reflector, as depicted in Fig. 23-5a. The law of reflection applied to an elliptical surface dictates that a light

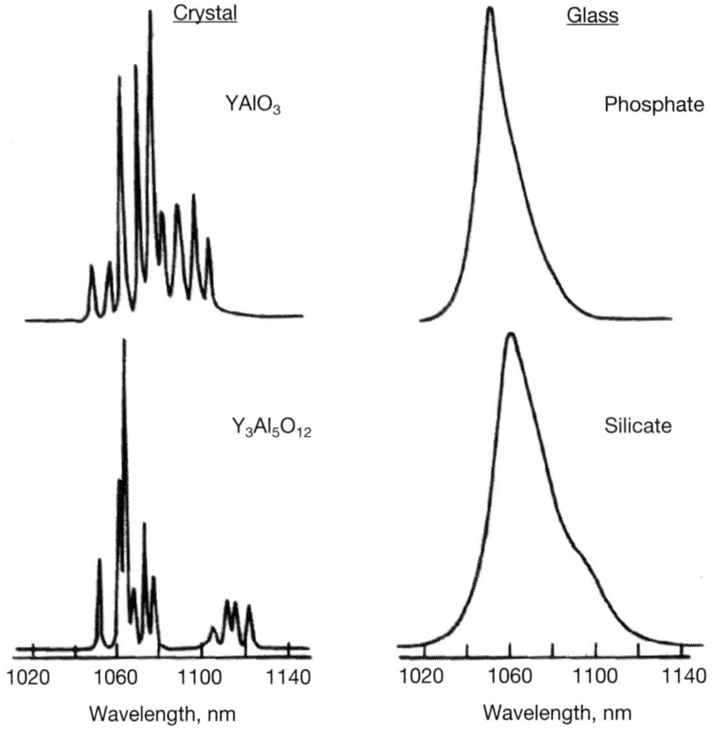

Figure 23-4 Emission spectrum for the ${}^4F_{3/2} \rightarrow {}^4I_{11/2}$ transition of Nd^{3+} in various hosts. In crystalline hosts (left two panels) the spectra consist of a series of narrow lines, with a distribution that depends strongly on the type of crystal. In glasses, the spectra are broad and smooth, with a weaker dependence on the type of glass. (After Weber 1979.)

ray emitted in any direction from one focus of the ellipse is reflected so that it passes through the other focus. This geometry ensures optimal coupling of the emitted lamp light into the laser rod. The pump light enters the laser rod from the side, and the laser is said to be "side pumped." In contrast to this, diode pumped lasers are often pumped from the end, or "end-pumped," as depicted in Fig. 23-5b. If the medium surrounding the laser rod is air, the pump light is trapped in the rod by total internal reflection, and the Nd^{3+} ions are efficiently excited by the pump.

For lamp pumping, the pump spectrum is very broad, and there are many levels above the upper laser level (for example, levels 3 and 3′ in Fig. 23-3a) that simultaneously absorb the pump light. Because of the close energy spacing of these levels, they all decay rapidly (nanosecond time scale) in a nonradiative cascade to the metastable ${}^4F_{3/2}$ ($\tau = 0.23$ ms for Nd:YAG). In this way, pump-light energy over a broad wavelength range is funneled into the upper laser level.

For diode laser pumping, in contrast, the excitation is at a single pump wavelength. For example, Nd:YAG has a strong absorption peak at a wavelength of 808 nm, which can be generated by an AlGaAs diode laser. Absorption of a photon at this wavelength promotes a Nd^{3+} ion from the ${}^4I_{9/2}$ ground state to the ${}^4F_{5/2}$ state (level 3). From here, the ion decays rapidly to the ${}^4F_{3/2}$ (level 2) in a single nonradiative step.

An important advantage of diode laser pumping is its efficiency. The energy difference between the pumped level (3) and the upper laser level (2) is smaller than the correspond-

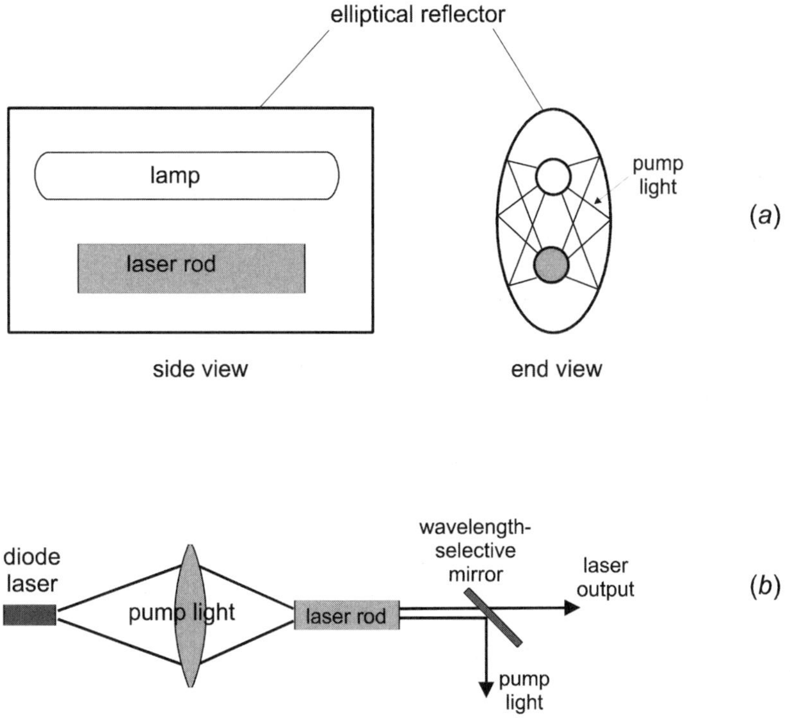

Figure 23-5 (a) For lamp pumping of a Nd:YAG laser, the lamp and laser rod are often placed at the foci of an elliptical reflector to maximize the coupling of pump light into the laser rod. (b) For diode-laser pumping, the pump light can be injected into the end of the rod as shown. Pump light that is not absorbed by the rod must then be separated from the laser beam.

ing difference in lamp pumping, so less pump energy is wasted in the nonradiative cascade to level 2. Stated another way, the quantum defect, defined earlier as the difference between pump and laser photon energies, is smaller in the case of diode laser pumping. The pump wavelength is closer to the lasing wavelength, and according to Eq. (20-29) this improves the laser slope efficiency.

The overall laser efficiency depends not only on how efficiently the laser medium converts absorbed pump power into laser output, but also on how efficiently the laser medium absorbs the pump light. This absorption efficiency is relatively low for lamp pumping, because the lamp spectrum contains many photons with an energy that falls in between the Nd^{3+} energy levels. However, the corresponding efficiency for diode laser pumping is high, since all of the diode laser power is concentrated at a wavelength at which the medium is highly absorbing. The overall efficiency of a laser is often expressed in terms of the "wall plug" efficiency, defined as the laser output power divided by electrical input power. Diode-pumped Nd:YAG lasers have a much higher wall plug efficiency (~ 30%) than their lamp-pumped counterparts (~ 3%), due largely to the difference in pump absorption efficiency.

Neodymium lasers have been industrial workhorses ever since their introduction. They can be operated efficiently in either continuous or pulsed mode, and have found application in cutting and drilling and other types of materials processing, as well as various medical applications (most of which involve cutting tissue). Although YAG has been the most commonly used crystalline host, other crystals such as YVO_4 and $YLiF_4$ have been used as well. Glass hosts have a much lower thermal conductivity than crystalline hosts, and heat dissipation becomes a problem for Nd:glass lasers operated at high average power. Also, the peak cross section for a glass host is smaller (see Table 23-1). For these reasons, Nd:glass lasers are mostly operated in pulsed mode. We will see an exception to this rule, however, when we discuss fiber lasers.

One of the more impressive applications of Nd:glass lasers is in the generation of power by nuclear fusion. In nuclear fusion, two hydrogen nuclei (or a nucleus of hydrogen and one of deuterium) are joined together to create a nucleus of helium, thereby releasing considerable energy. To get the nuclei to come together requires extraordinary conditions of temperature and compression that are quite difficult to achieve. One proposed scheme is to illuminate a small pellet of the hydrogen/deuterium mixture from all sides with a high-power laser pulse, which will then implode the pellet and create the necessary compression. The optical power needed for this is enormous, and the current proof-of-principle project is the National Ignition Facility (NIF), operated by Lawrence Livermore National Laboratory. Current plans call for 192 beamlines, each containing 16 rectangular slabs of Nd-doped phosphate glass oriented as shown in Fig. 23-6. The slabs are tilted at Brewster's angle to avoid reflection losses, and are pumped from the side by a series of flashlamps. The slabs act as optical amplifiers, increasing the energy of a seed pulse to over 15 kJ in a pulse duration of 3.5 ns. The total combined energy of the 192 beamlines would be some 3 MJ, and the peak power in the pulse would be $P = E/\Delta t = (3 \times 10^6)/(3.5 \times 10^{-9}) = 8.6 \times 10^{14}$ W! When constructed, this will be the world's most powerful laser.

Other Rare Earth Lasers

There are several other rare earth ions besides Nd^{3+} that can be used in an optically pumped solid-state laser. The lower-lying energy levels for the most commonly used rare earths are shown in Fig. 23-7, along with the most important laser transitions. Tm^{3+} and/or Ho^{3+} can be doped in a crystal such as YAG, to generate laser light in the 2 μm range. These transitions terminate on the ground state, and so this is a three-level type system that operates most efficiently in pulsed mode. Light in the 2 μm range is efficient-

Table 23-1 Typical parameters for selected solid-state or liquid lasers

Laser	Ion or dye	Host	Refractive index n	Emission λ (nm)	Peak emission σ (10^{-20} cm^2)	Ion density (10^{20} cm^{-3})	Linewidth $\Delta\nu$ (THz)	Excited state lifetime τ (ms)
Ruby	Cr^{3+}	Al_2O_3	1.76	694	2.5	0.16	0.33	3
Nd:YAG	Nd	$Y_3Al_5O_{12}$	1.82	1064	28	1.38	0.12	0.23
Nd:glass	Nd	phosphate glass	1.53	1054	4	3.2	5	0.29
Er:fiber	Er	silica glass	1.46	1530–1570	0.6	0.1	5	10
Yb:fiber	Yb	silica glass	1.46	980–1100	2.5	0.1	5	0.84
Ti:sapphire	Ti	Al_2O_3	1.76	660–1180	34	0.33	40	0.0038
Dye	R6G	ethylene glycol	1.43	570–640	2×10^4	0.1	35	5×10^{-6}

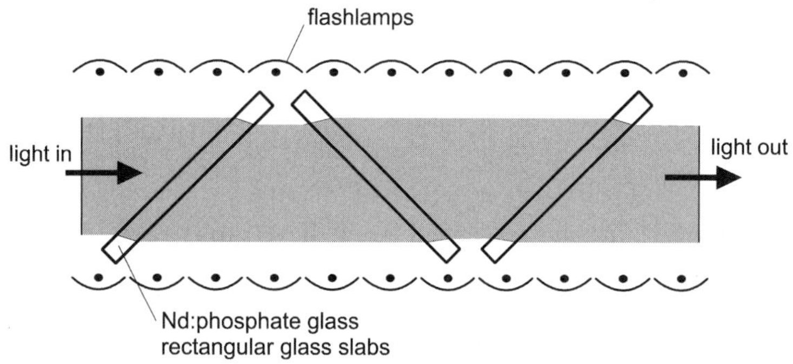

Figure 23-6 Light of wavelength 1.053 μm passes through a series of Nd-doped glass slabs in the amplifier section of the National Ignition Facility (NIF). The tilted geometry of the slabs not only reduces Fresnel reflection losses, but also allows convenient side pumping with a series of flashlamps. The slabs have dimensions 3.4 × 46 × 81 cm, and are oriented so as to present an approximately square cross section to the propagating beam.

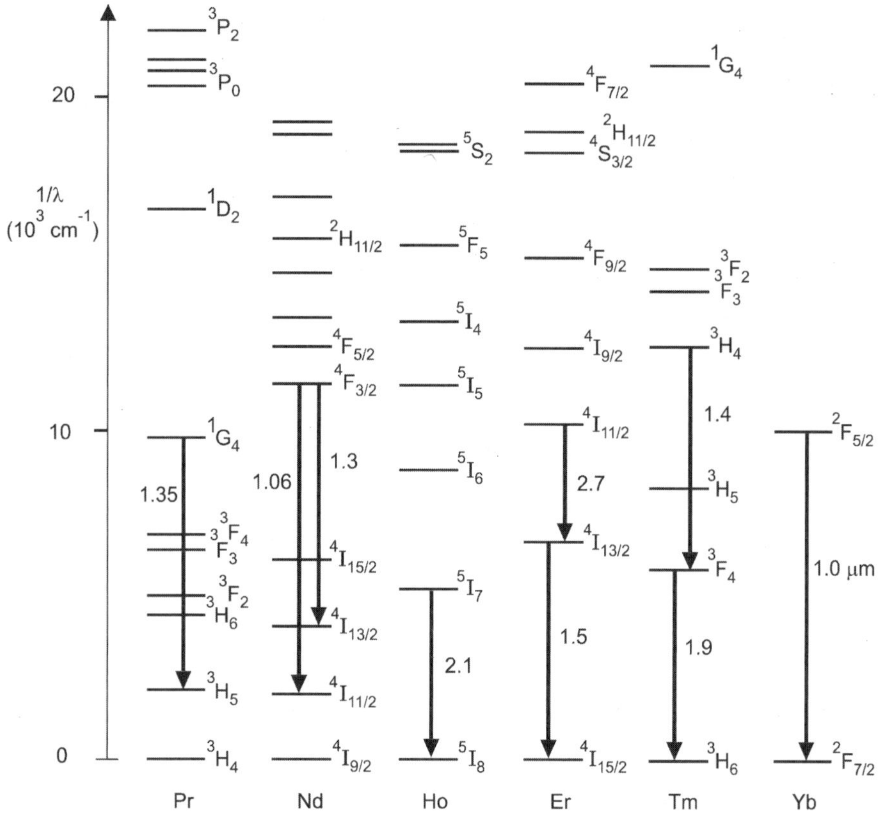

Figure 23-7 Lower-lying energy levels for selected rare earth ions (triply ionized), showing the important laser and amplifier transitions. The energy scale is written as an inverse transition wavelength, where $E = hc/\lambda$, so that 1 eV is equivalent to 8064 cm^{-1}. Transition wavelengths are also given in units of μm.

ly absorbed by the water contained in biological tissue, and these lasers have possible medical applications. The Er^{3+} ion generates light at 2.7 μm, and this wavelength is even more strongly absorbed by water, making it ideal for cutting tissue. The 2.7 μm transition of Er^{3+} is four-level in nature, and might be expected to be quite efficient. However, the lifetime of the lower level $^4I_{13/2}$ is much longer than that of the upper level $^4I_{11/2}$, because of the weaker nonradiative decay from the lower level. Steady-state population inversion is thus difficult to achieve, and the laser is mostly operated in the pulsed mode.

The Yb^{3+} ion has a particularly simple energy-level structure of just two levels, the ground state $^2F_{7/2}$ and the excited state $^2F_{5/2}$. It might seem at first that population inversion would never be attained here, since pump light resonant with this transition would at most simply equalize the population in the upper and lower levels (see Problem 23.7). However, the upper and lower states consist of closely spaced sublevels, as depicted in Fig. 23-8. The pump transition is, therefore, not the same as the lasing transition, and population inversion can indeed be achieved. For example, Yb:YAG can be pumped at 943 nm by a InGaAs/GaAs strained quantum-well laser, and lasing occurs at 1.03 μm. This laser wavelength is close to that of Nd:YAG, and Yb:YAG can, therefore, be used for many of the same applications. An advantage of Yb:YAG is that the quantum defect is smaller (the pump wavelength is closer to the lasing wavelength), so the slope efficiency can be higher. A disadvantage of Yb:YAG, however, is that it cannot be lamp pumped.

An interesting question is whether the Yb^{3+} laser should be classified as four-level or three-level. On the one hand, the lower laser level is not actually the lowest possible level, and so it might be considered four-level. But on the other hand, the lower laser level is thermally populated according to the Boltzmann factor $\exp(-\Delta/k_BT)$, and this can be non-negligible at room temperature. Some minimum level of excitation is, therefore, required to achieve population inversion, and this is characteristic of a three-level system. Systems like this are often referred to as *quasi-four-level,** and they have properties intermediate between true three- and four-level-type systems. Many of the rare earth transitions that terminate on the ground state are of this type, including those of Ho^{3+}, Er^{3+}, and Tm^{3+}. The performance of these lasers generally improves at low temperature, where the lower laser level is less thermally populated.

The other transitions indicated in Fig. 23-7 are in the wavelength range 1.3–1.6 μm,

*But equally often, they are referred to as quasi three-level.

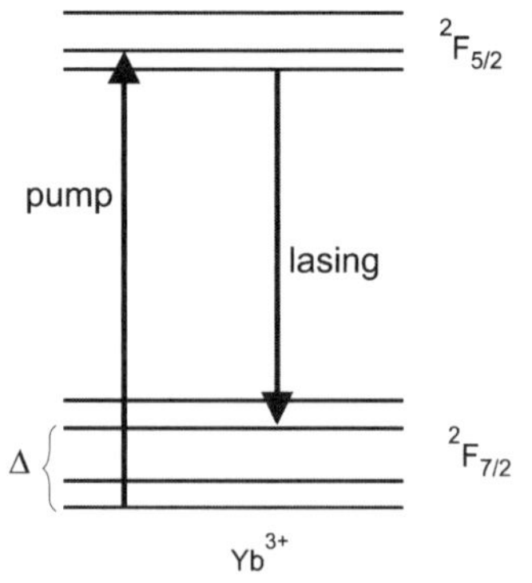

Figure 23-8 The upper and lower states of Yb^{3+} are split into a number of sublevels, and the pumping and lasing transitions are between different sets of these sublevels.

The other transitions indicated in Fig. 23-7 are in the wavelength range 1.3–1.6 μm, which overlaps the important second and third telecommunications windows. These transitions are particularly important for fiber amplifiers, as we will see when we discuss optical communications in Chapter 24. Lasers can also be constructed based on these transitions, but they are most commonly implemented in a fiber geometry, rather than in a conventional rod geometry of millimeter-to-centimeter scale. In the following section, we consider the many advantages of a laser with fiber geometry.

Fiber Lasers

Inside a conventional laser cavity (see Fig. 17-5), the spatial distribution of the optical mode is determined by the curvature and separation of the mirrors. Although there is no active confinement of the light perpendicular to the axis of the cavity (no "side mirrors"), the light is "self-confining" in this direction, according to the properties of a Gaussian beam. A cavity mode with large waist size w_0 has a low angular divergence, and remains nearly the same diameter everywhere between the mirrors. A mode with small w_0 diverges rapidly with position along the cavity axis, and the beam diameter becomes much larger at the mirrors. Because of this beam spreading, it is not possible to have an arbitrarily high intensity (small beam size) at all points along the axis of a conventional laser cavity.

An alternative approach, which overcomes this limitation, is to use an optical fiber for the laser cavity. Light in the core of an optical fiber is trapped by total internal reflection, and this provides a natural way of confining the lasing mode laterally. Light can be confined in the longitudinal direction by a mirror on each end of the fiber, as depicted in Fig. 23-9a. The mirrors in this case can be flat, because the optical mode is defined by the fiber rather than the mirrors. In this type of laser cavity, light can be confined to a small lateral size for an arbitrarily long cavity length, subject only to attenuation losses in the fiber. Gain in the cavity is provided by rare earth ions, which are doped into the fiber core. A device of this type is termed a *fiber laser*.

To obtain gain in the fiber laser, the rare earth ions need to be optically pumped with a suitable light source. The first fiber lasers, proposed and developed by Snitzer in the early 1960s, were lamped pumped from the side. However, this pump scheme does not take advantage of the long fiber lengths that are possible, and fiber lasers today are nearly always end pumped. One way to end pump the laser is to send pump light through a dichroic mirror at the fiber end, as illustrated in Fig. 23-9a. This mirror is highly transmitting at the pump wavelength λ_p, which allows the pump light to be coupled into the core of the fiber, but it is highly reflecting at the lasing (or "signal") wavelength λ_s, so it can serve as a high reflector in the laser cavity. This pumping arrangement is generally used for experimentation and setting up prototypes, since the cavity can be easily altered.

A more robust pumping arrangement is the all-fiber scheme shown in Fig. 23-9b. Pump light is injected into the lasing cavity using a fiber coupler (for example, the fused biconical taper coupler of Fig. 7-4), and the laser light in the output is separated from the unabsorbed pump light by another fiber coupler. Optical feedback in the laser cavity is provided by two fiber Bragg gratings written into the fiber, rather than end mirrors. The parameters for the couplers and fiber gratings can be designed for a particular application, and some limited degree of tunability is possible by stretching the fiber grating (changing Λ and therefore the lasing wavelength).

The fiber laser has a number of advantages over a conventional laser. It is compact, lightweight, and provides a very stable output beam. The spatial distribution of the output

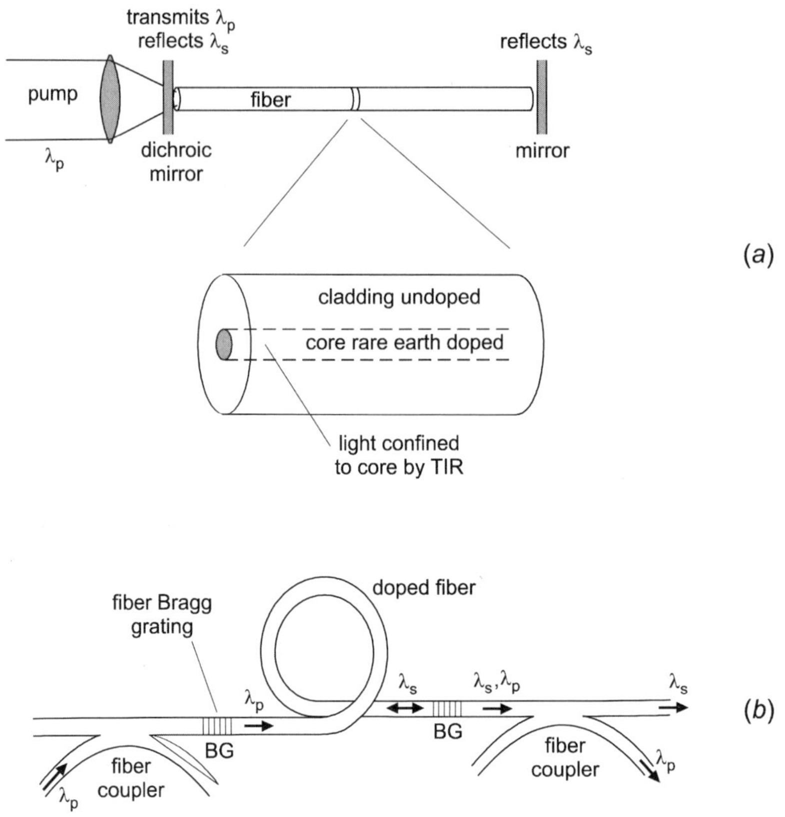

Figure 23-9 (a) A fiber laser can be constructed by doping the core of an optical fiber with a suitable rare earth ion, and placing two mirrors at either end for optical feedback. The laser is end-pumped through one of the mirrors. (b) In the all-fiber version, the mirrors are replaced by fiber couplers, and fiber Bragg gratings replace the end mirrors.

beam is nearly diffraction limited (Gaussian profile) when the fiber is single mode, and fiber lasers are ideal for creating a low-power "seed" beam for further amplification. An example of this is the NIF laser discussed previously, in which a low-power Yb fiber laser (pulse energy ~ nJ) provides the seed light that is eventually amplified into an enormous (~ MJ) pulse of energy. In applications of a more modest scale, a fiber laser's high beam quality allows the light to be efficiently focused onto a target for cutting or materials processing applications. For the same reason, fiber laser light can be efficiently coupled into a passive fiber for transporting the light to a distant target. This capability is especially useful, for example, in laser surgery.

The fiber geometry has an added benefit related to heat dissipation. The heat generated in an object is proportional to its volume (assuming uniform excitation), whereas the flow of heat away from that object is proportional to its surface area. In the steady state, therefore, a higher surface/volume ratio results in more efficient heat dissipation and a lower temperature rise. For a long rod of radius a and length L, the surface/volume ratio is $(2\pi aL)/(\pi a^2 L) = 2/a$. Since a fiber laser has a very small a, its core will rise in temperature by only a small amount, even with multiwatt power levels. In contrast, a conventional solid-state laser has a larger temperature rise that causes thermal lensing (see Fig. 9-8), limiting the output power and degrading the beam quality. As a result, fiber lasers generally require only air cooling, whereas conventional solid-state lasers must be cooled with a flowing liquid such as water. This is an important practical advantage of fiber lasers, allowing them to be compact, efficient, and reliable.

The above points are all important practical advantages of a fiber laser. The most fundamental advantage of the fiber geometry, however, is the ability to maintain a high optical intensity over an arbitrarily long path length. Because of the small core area (A_c), the intensity $I = P/A_c$ can be very high even for modest optical powers (P). These high intensities are sufficient to produce steady-state population inversion and lasing, even in three-level type systems such as Er, Tm, and Yb. The threshold pumping powers are much smaller in fiber lasers compared with their conventional laser counterparts, and the overall device efficiency is thereby improved. In the following, we consider in more detail the analysis of threshold and slope efficiency in fiber lasers.

Threshold in a Four-Level System

The condition for lasing threshold in a four-level system was developed in Section 20-1, under the assumption that the gain coefficient γ is constant. This assumption is not generally valid for fiber lasers, however, because the pump light is absorbed as it propagates down the fiber core, and this causes the pump intensity (and hence the population inversion and gain) to vary with position along the fiber. We must, therefore, generalize the results of Section 20-1 to include a spatially varying gain coefficient $\gamma(x)$.

Consider pump light of intensity I_{p0} and wavelength λ_p that is coupled into the core of a fiber of length L. The core has cross-sectional area A_c and is doped with N rare earth ions per unit volume. The pump light is attenuated with absorption coefficient

$$\alpha_p = N_0\sigma_p \qquad \text{(pump absorption coefficient)} \qquad (23\text{-}1)$$

where σ_p is the pump absorption cross section and N_0 is the number of ions per unit volume in the ground state. We will make the simplifying assumption that most of the rare earth ions remain in the ground state, so that $N_0 \approx N$, and the absorption coefficient $\alpha_p \approx N\sigma_p$ is approximately independent of position x along the fiber. In this case, the pump decays exponentially with x according to Beer's law (Eq. 5-1),

$$I_p(x) = I_{p0}e^{-\alpha_p x} \qquad (23\text{-}2)$$

as shown in Fig. 23-10.

To see how the gain coefficient varies with x, we first express it in terms of the level populations as

$$\gamma(x) = [N_2(x) - N_1(x)]\sigma_{se} \simeq N_2(x)\sigma_{se} \qquad (23\text{-}3)$$

where $N_1 \ll N_2$ has been assumed for the four-level transition. The excited-state population is $N_2 = \mathcal{R}\tau_2$ from Eq. (20-17), and the excitation rate $\mathcal{R}$ is

$$\mathcal{R}(x) = NW_p(x) = N\,\frac{I_p(x)\sigma_p}{h\nu_p} \qquad (23\text{-}4)$$

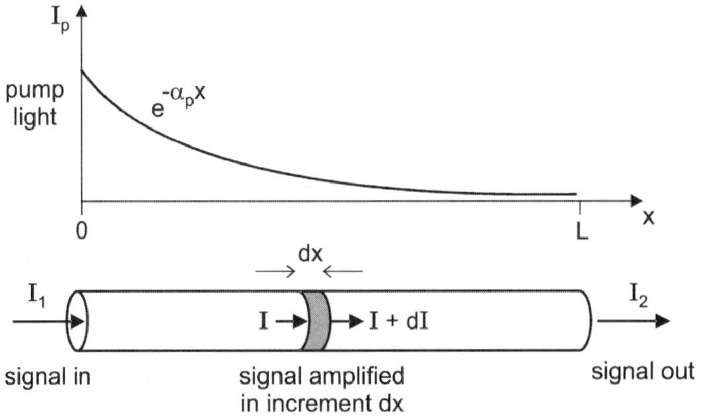

Figure 23-10 In a fiber laser, the pump light intensity varies along the fiber, giving rise to a position-dependent gain coefficient for the lasing (signal) light.

where W_p is the transition probability per unit time for a single ion [see Eq. (19-3)]. Putting these together we obtain

$$\begin{aligned}\gamma(x) &= N\frac{I_p(x)\sigma_p\sigma_{se}\tau_2}{h\nu_p} \\ &= \frac{\alpha_p\sigma_{se}\tau_2}{h\nu_p}I_p(x)\end{aligned} \tag{23-5}$$

where Eq. (23-1) has been used in the last step. This result shows that the gain coefficient $\gamma(x)$ varies with x in the same manner as the pump intensity $I_p(x)$.

The gain coefficient gives the fractional increase per unit length of light intensity at the lasing wavelength λ_s. For the small section dx of fiber depicted in Fig. 23-10, the increment in signal intensity is

$$dI = I\gamma(x)\,dx \tag{23-6}$$

Since $\gamma(x)$ is no longer a constant, the signal intensity now does not increase exponentially with x as it did in our previous analysis. However, it is still a simple matter to integrate this equation and obtain the net gain in a complete pass through the fiber. Dividing both sides of Eq. (23-6) by I and integrating over the entire fiber, we obtain

$$\begin{aligned}\int_{I_1}^{I_2}\frac{dI}{I} &= \int_0^L \gamma(x)\,dx \\ \ln\left(\frac{I_2}{I_1}\right) &= \frac{I_{p0}\sigma_{se}\tau_2}{h\nu_p}[1 - e^{-\alpha_p L}]\end{aligned} \tag{23-7}$$

where Eqs. (23-2) and (23-5) have been used. The single-pass gain in an optical amplifier was denoted previously as $G = I_2/I_1$ [Eq. (19-15)]. Using this definition, and taking the limit $\alpha_p L \gg 1$, Eq. (23-7) becomes

$$\ln G = \frac{I_{p0}\sigma_{se}\tau_2}{h\nu_p} \tag{23-8}$$

This is the single-pass logarithmic gain, assuming that all of the pump light is absorbed in the fiber (none is transmitted out the far end). Note that the gain does not depend on ion concentration, the pump cross section, or the fiber length. Only excited ions contribute to the gain, and it does not matter where those ions are located along the fiber; only the integrated excited-state population $\int N_2(x)\,dx$ is relevant for determining the net gain.

At lasing threshold, the total round-trip gain (treating losses as a gain less than unity) is 1. If the attenuation coefficient in the fiber is α and the mirror reflectivities are R_1 and R_2, this condition can be written as

$$R_1 R_2 G_{\text{th}}^2 e^{-2\alpha L} \qquad \text{(lasing threshold)} \qquad (23\text{-}9)$$

which is a generalization of the previous expression, Eq. (20-1). Taking the natural log of this equation gives

$$\ln (R_1 R_2) + 2 \ln G_{\text{th}} - 2\alpha L = 0$$

which can be written as

$$\ln G_{\text{th}} = \alpha L + \frac{1}{2} \ln \left(\frac{1}{R_1 R_2} \right) \qquad (23\text{-}10)$$

Comparing this with Eq. (20-2), we identify $\gamma_{\text{th}} L$ in the previous treatment with $\ln G_{\text{th}}$ in the present treatment. Other formulae in Chapter 20 can be generalized with this new notation as well. For example, the photon lifetime given in Eq. (20-11) becomes

$$\tau_c = \frac{L/c}{\ln G_{\text{th}}} \qquad \text{(photon lifetime)} \qquad (23\text{-}11)$$

As before, the replacement $c \to c/n$ should be made if the cavity medium has a refractive index n. In the case of the fiber laser, $n \approx 1.5$ for glass in the core.

The threshold pump power P_{th} can be evaluated from Eq. (23-8) using $I_{p0} = P_{\text{th}}/A_c$ and $G = G_{\text{th}}$. The result is

$$P_{\text{th}} = \frac{A_c h \nu_p \ln G_{\text{th}}}{\sigma_{se} \tau_2} \qquad \text{(four-level pump threshold)} \qquad (23\text{-}12)$$

with G_{th} given by Eq. (23-10). This expression is valid for complete absorption of the pump, and shows how the pump threshold depends on the properties of the fiber cavity (A_c and G_{th}) and the active ions (σ_{se} and τ_2). The product $\sigma_{se}\,\tau_2$ should be maximized for a low pump threshold, and can be considered to be a "figure of merit" for the laser material.

Slope Efficiency in a Four-Level System

When the fiber laser is pumped above threshold, the net round-trip gain must remain clamped at 1 in the steady state. If this were not the case, the light intensity would continue to increase exponentially in time, violating our assumption of steady-state conditions. This means that G remains clamped at G_{th}, in the same way that γ remained clamped at γ_{th} in the analysis of Chapter 20. The additional pumping power above threshold cannot go into increased fluorescence power, because the total excited state population $\int N_2(x)\,dx$ remains constant above threshold, and the amount of fluorescence is $\propto N_2$. Therefore, the additional pumping power goes into increasing the laser output power with a slope efficiency $\eta_s = \Delta P_{out}/\Delta P_{in}$ [Eq. (20-26)]. Using Eq. (20-27), along with Eq. (23-11), we obtain

$$\eta_s = T\left(\frac{h\nu}{h\nu_p}\right)\frac{1}{2\ln G_{th}} \qquad \text{(four-level slope efficiency)} \qquad (23\text{-}13)$$

Comparing Eqs. (23-12) and (23-13), we see that the threshold and slope efficiency depend in different ways on the cavity losses. For a fixed pump power, the output power can be optimized by varying the output mirror transmission T, as discussed in Section 20-2. Equations (20-31)–(20-33) apply as before, with the same asssumptions that $R_1 \approx 1$, and $T = 1 - R_2 \ll 1$.

EXAMPLE 23-1

A Nd:fluoride glass fiber laser is constructed with a fiber length of 50 cm, core diameter 40 μm, and mirror reflectivities $R_1 = 1$ and $R_2 = 0.9$. The fluorescence lifetime of the upper laser level ($^4F_{3/2}$) is 500 μs, and lasing occurs at 1050 nm. When pumped with 514.5 nm light from an argon laser, the pump threshold and slope efficiency are measured to be 35 mW and 0.2, respectively. Determine (a) the loss coefficient of the fiber, and (b) the stimulated emission cross section.

Solution: (a) Putting Eq. (23-13) in terms of wavelength, we have

$$2\ln G_{th} = \frac{T\lambda_p}{\eta_s\lambda} = \frac{(0.1)\,(514.5)}{(0.2)\,(1050)} = 0.245$$

Solving Eq. (23-10) for the round-trip internal loss δ gives

$$\delta = 2\alpha L = 2\ln G_{th} - \ln\left(\frac{1}{R_1R_2}\right) = 0.245 - \ln\left(\frac{1}{0.9}\right) = 0.139$$

and

$$\alpha = \frac{0.139}{2(0.5\text{ m})} = 0.139\text{ m}^{-1} = 1.39 \times 10^{-3}\text{ cm}^{-1} = 600\text{ dB/km}$$

where the conversion factor in Eq. (5-4) has been used.

(b) The core area is $A_c = \pi(20 \times 10^{-6})^2 = 1.26 \times 10^{-9}\text{ m}^2$, and the pump photon energy is $hc/\lambda_p = (6.63 \times 10^{-34})(3 \times 10^8)/(514.5 \times 10^{-9}) = 3.86 \times 10^{-19}$ J. Solving Eq. (23-12) for σ_{se} then gives

$$\sigma_{se} = \frac{A_c h\nu_p \ln G_{th}}{P_{th}\tau_2} = \frac{(1.26 \times 10^{-9})(3.86 \times 10^{-19})(0.1205)}{(35 \times 10^{-3})(5 \times 10^{-4})} = 3.3 \times 10^{-24}\text{ m}^2$$

$$\sigma_{se} = 3.3 \times 10^{-20}\text{ cm}^2$$

This cross section is in good agreement with reported values for the 1050 nm transition in Nd-doped glass.

Threshold in Three-level Systems

Several of the more important transitions used in fiber lasers are to the ground state, making the laser three-level in nature. Examples include Er at 1.5 μm, Tm at 1.9 μm, and Yb at 1 μm. We will consider Yb in some detail, to illustrate the principles involved and give typical numbers.

As discussed earlier, the upper and lower levels of Yb^{3+} are split into a series of closely spaced sublevels (see Fig. 23-8). Fig. 23-11 shows a typical absorption cross section spectrum $\sigma_{abs}(\lambda)$ and emission cross section spectrum $\sigma_{em}(\lambda)$ for Yb^{3+}-doped silica glass. The sharp peaks at 975 nm correspond to transitions between the lowest sublevel of the upper state and the lowest sublevel of the lower state, while the broader peaks correspond to transitions between other pairs of sublevels. Since there is both absorption and emission at the lasing wavelength λ, the gain coefficient is not simply $N_2\,\sigma_{se}$ as in an ideal four-level laser, but is instead given by the more general expression in Eq. (18-37). For lasing to occur, the net gain must be positive, and the criterion for lasing at wavelength λ becomes

$$\gamma(\lambda) = N_2\sigma_{em}(\lambda) - N_1\sigma_{abs}(\lambda) > 0 \tag{23-14}$$

where N_2 and N_1 are the populations of the upper ($^2F_{5/2}$) and lower ($^2F_{7/2}$) laser levels. The degree to which γ must be greater than zero depends on the cavity losses. For a perfectly lossless cavity, the condition $\gamma = 0$ would correspond to the lasing threshold, because the net gain for light circulating in a round-trip through the cavity would be unity. This is referred to as the *transparency* condition, because there is no net gain or loss in the laser medium (it is "transparent"). The transparency condition, therefore, sets a minimum requirement for lasing.

The population ratio at transparency is $N_2/N_1 = \sigma_{abs}(\lambda)/\sigma_{em}(\lambda)$ for a laser operating at wavelength λ. According to Fig. 23-11, $\sigma_{em}(\lambda) > \sigma_{abs}(\lambda)$ for $\lambda > 975$ nm, and so for these wavelengths $N_2/N_1 < 1$. This means that fewer than half of the Yb ions need to be raised to the excited state to achieve lasing, in contrast to the ideal three-level system which re-

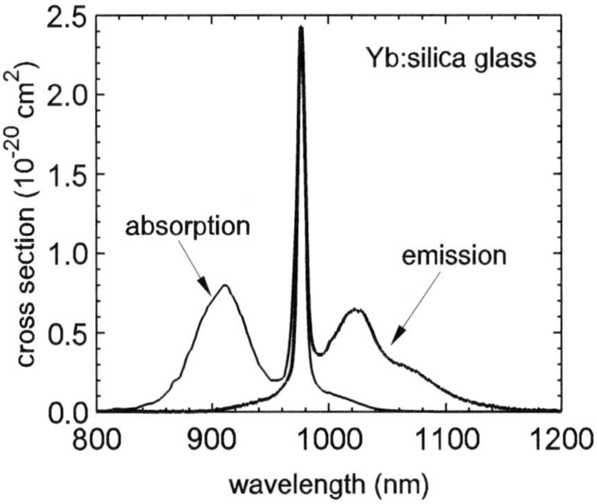

Figure 23-11 Typical absorption and emission cross-section spectra for Yb^{3+}-doped Al/P-silica glass. See Fig. 23-8 for the Yb^{3+} energy level diagram. (Data courtesy of Xiaojun Li.)

quires $N_2/N_1 \geq 1$. The lesser degree of inversion required in the Yb laser for $\lambda > 975$ nm is a consequence of the quasi-four-level nature of the lasing transition. The lower laser level is a thermally populated sublevel of the ground state, and absorption from this level is reduced by the Boltzmann factor $\exp(-\Delta/k_BT)$.

We can estimate the pump power required to achieve transparency by writing the rate equation for level 2,

$$\frac{dN_2}{dt} = N_1 W_p - N_2 W_{pe} - \frac{N_2}{\tau_2} \tag{23-15}$$

where

$$W_p = \frac{I_p \sigma_{\text{abs}}(\lambda_p)}{h\nu_p} \tag{23-16}$$

is the "pump rate" (probability per unit time that the ion makes an upward transition from $1 \rightarrow 2$ by absorbing a pump photon), and

$$W_{pe} = \frac{I_p \sigma_{\text{em}}(\lambda_p)}{h\nu_p} \tag{23-17}$$

is the "pump emission rate" (probability per unit time that the ion makes a downward transition from $2 \rightarrow 1$ by emitting a pump photon). This second process represents stimulated emission of the pump light, and the third term in Eq. (23-15) represents spontaneous emission from level 2. Setting $dN_2/dt = 0$ for the steady state, and using the constraint $N_1 + N_2 = N$ (the ions must be in either level 1 or level 2), we obtain after some algebra

$$N_2 = N \frac{W_p}{W_p + W_{pe} + 1/\tau_2} \tag{23-18}$$

$$N_1 = N \frac{W_{pe} + 1/\tau_2}{W_p + W_{pe} + 1/\tau_2} \tag{23-19}$$

These populations can be expressed in terms of the pump intensity I_p by substituting from Eqs. (23-16) and (23-17), with the result

$$N_2 = N \frac{I_p}{I_p(1 + \sigma_{pe}/\sigma_p) + I_{ps}} \tag{23-20}$$

$$N_1 = N \frac{I_p(\sigma_{pe}/\sigma_p) + I_{ps}}{I_p(1 + \sigma_{pe}/\sigma_p) + I_{ps}} \tag{23-21}$$

In the above, we have defined the *pump saturation intensity* as

$$I_{ps} \equiv \frac{h\nu_p}{\sigma_p \tau_2} \quad \text{(pump saturation intensity)} \tag{23-22}$$

and used a simplified notation $\sigma_p \equiv \sigma_{\text{abs}}(\lambda_p)$, $\sigma_{pe} \equiv \sigma_{\text{em}}(\lambda_p)$. Note the similarity of this definition to that of the signal saturation intensity in Eq. (19-9).

These expressions can be used to determine the level populations, and hence the gain coefficient, for any given pumping intensity. At low pump intensity where $I_p \ll I_{ps}$, $N_1 \approx N$ and $N_2 \approx 0$. Under these conditions, the gain coefficient is $\gamma(\lambda) \approx -N\sigma_{\text{abs}}(\lambda)$, which is negative for all wavelengths λ. In this case lasing will not occur because all signal wavelengths experience a net absorption. As I_p is increased, N_2 will increase and N_1 will decrease, making γ closer to zero. At some pump intensity the transparency condition $\gamma = 0$ will be achieved, and laser action is then possible.

We now derive an expression for the pump intensity required for transparency. From Eq. (23-14), the required population ratio is

$$\frac{N_2}{N_1} = \frac{\sigma_{\text{abs}}(\lambda)}{\sigma_{\text{em}}(\lambda)} \equiv \frac{\sigma_{sa}}{\sigma_{se}} \tag{23-23}$$

where $\sigma_{se} \equiv \sigma_{\text{em}}(\lambda)$ and $\sigma_{sa} \equiv \sigma_{\text{abs}}(\lambda)$. Substituting Eqs. (23-20)–(23-21) into the above and solving for I_p gives

$$I_p = I_{ps}\left(\frac{\sigma_{sa}/\sigma_{se}}{1 - \dfrac{\sigma_{sa}\sigma_{pe}}{\sigma_{se}\sigma_p}}\right) \tag{23-24}$$

This can be recast in a simplified form using the McCumber relation of Eq. (18-38), with the result (see Problem 23.8)

$$I_p = I_{ps}\left(\frac{\sigma_{sa}/\sigma_{se}}{1 - \exp[-(h\nu_p - h\nu)/k_BT]}\right) \qquad \text{(at transparency)} \tag{23-25}$$

An important feature of this equation is that the required pump intensity gets very large as $\nu_p \to \nu$, and becomes negative for $\nu_p < \nu$. This would be a nonphysical result (power can only be positive), which indicates that in a quasi-four-level scheme like this, lasing can occur only at wavelengths longer than the pump wavelength. In fact, this conclusion can be justified in a fundamental way based on thermodynamic arguments, and is true quite generally. Exceptions to this basic principle do occur, however, when nonlinear interactions become important (see Chapter 9). It is also useful to note that if the absorption at the signal wavelength becomes small ($\sigma_{sa} \to 0$), then $I_p \to 0$. This is to be expected, because in this case the material is already "transparent," and no excitation is needed to make it so.

The effect of pump intensity on the Yb gain spectrum is illustrated in Fig. 23-12. This shows the calculated gain coefficient in a Yb doped fiber that is pumped at 915 nm, for three different values of the pump intensity I_p. For each value of I_p, the gain is zero at some *transparency wavelength* λ_{tr}. The gain is positive at longer wavelengths, and negative at shorter wavelengths. This transparency point shifts to shorter wavelength as the pump intensity increases, so that the gain becomes positive over a wider wavelength range. As the pump intensity becomes arbitrarily large, the transparency wavelength approaches the pump wavelength. However, the condition $\lambda_{tr} > \lambda_p$ is always maintained, in accordance with Eq. (23-25).

EXAMPLE 23-2

A Yb-doped fiber with core radius of 2.3 μm is pumped at 915 nm, where the absorption cross section is 0.75×10^{-20} cm^2. At the lasing wavelength of 1025 nm, the emis-

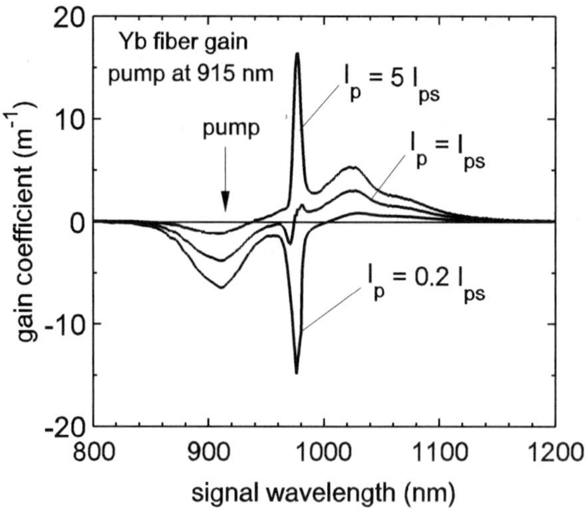

Figure 23-12 Calculated gain coefficient in Yb-doped silica fiber, for three different pump intensities. Cross-section spectra are the same as in previous figure, fiber core radius 2.3 μm, $N = 10^{19}$ Yb ions/cm^3 and $\tau_2 = 840$ μs. Pump light at 915 nm is assumed to be fully absorbed.

sion cross section is 0.64×10^{-20} cm^2, and the absorption cross section is 0.054×10^{-20} cm^2. Taking the excited-state lifetime to be 0.84 ms, determine the pump power needed to achieve transparency at 1025 nm.

Solution: The photon energies for the pump and signal are 2.17×10^{-19} and 1.94×10^{-19} J, respectively, using $h\nu = hc/\lambda$. At room temperature (20°C = 293 K), $k_BT = (1.38 \times 10^{-23})(293) = 4.04 \times 10^{-21}$ J. Using these, the exponent in the denominator of Eq. (23-25) is

$$-\frac{h\nu_p - h\nu}{k_BT} = -\frac{(2.17 - 1.94) \times 10^{-19}}{4.04 \times 10^{-21}} = -5.7$$

and the denominator becomes

$$1 - e^{-5.7} = 0.997$$

This value is ≈ 1 here because of the wide separation of pump and signal wavelengths. The pump saturation intensity is

$$I_{ps} = \frac{2.17 \times 10^{-19}}{(7.5 \times 10^{-25})(8.4 \times 10^{-4})} = 3.44 \times 10^8 \text{ W/m}^2$$

The pump intensity for transparency is then

$$I_p \approx I_{ps}\frac{\sigma_{sa}}{\sigma_{se}} = (3.44 \times 10^8)\left(\frac{5.4}{64}\right) = 2.9 \times 10^7 \text{ W/m}^2$$

The core area is $A_c = \pi(2.3 \times 10^{-6})^2 = 1.66 \times 10^{-11}$ m^2, and therefore the required pump power is

$$P_{tr} = I_p A_c = (2.9 \times 10^7)(1.66 \times 10^{-11}) = 4.8 \times 10^{-4}\ \text{W} = 0.48\ \text{mW}$$

This remarkably small value is a consequence of the small core area in a single-mode fiber. The threshold pump power for lasing will be somewhat higher, in order to overcome fiber and mirror losses. However, pump thresholds in the mW range are typical for rare earth doped fibers, even for transitions to the ground state.

Slope Efficiency in Three-level Systems

When a three-level laser system is pumped above threshold, the integrated gain $\int \gamma(x)\, dx$ remains clamped at the threshold value, as in a four-level system. The behavior of the three-level system above threshold is similar to that of a four-level system, with the additional pump power going into the laser output. The slope efficiency of the laser is again given by Eq. (23-13). Because of the low thresholds that are possible in fiber lasers, both four-level and three-level, the overall efficiency can be quite high. This capability of high efficiency holds true even for transitions that would operate only poorly in a conventional laser, and it is one of the hallmarks and distinct advantages of fiber lasers.

High Power

In principle, the output power of a fiber laser can be increased to any desired level simply by increasing the pump power. In practice, however, there is a limit to how much pump power can be coupled into the core of a single-mode fiber. The pump source for a fiber laser is typically a diode laser, and the coupling efficiency is fundamentally limited by the brightness of the diode laser light. Higher-power diode lasers require a larger emitting area, to avoid optical damage from the high intensity, and this larger area limits the brightness (see page 288). Output powers from diode-pumped, single-mode fiber lasers are generally less than 100 mW.

One solution to the brightness problem is to use a fiber with a double cladding structure, as shown in Fig. 23-13. The single-mode core is surrounded by an inner cladding with lower index, which in turn is surrounded by an outer cladding of still lower index. Pump light is coupled into the inner cladding, where it is trapped by total internal reflection at the boundary between the two claddings. The diameter of the inner cladding is made sufficiently large (~ 50 μm) so that it can accept a high pump power from a low brightness source. This pump light passes through the small core area as it zigzags down the fiber, exciting the rare earth ions that are doped in the core. Since the cladding is undoped, the fraction of pump light absorbed per unit length is rather small, and this requires long fiber lengths to efficiently absorb the pump. The result, however, is that the ions in the single-mode core interact with a much higher pump power than would otherwise be possible, and the maximum output power is correspondingly higher. This type of fiber is termed *double-clad fiber,* and the resulting laser is said to be *cladding pumped.*

Above threshold, the additional pump power is funneled into the laser output power, just as in a conventional laser. By conservation of energy, the output power must always be less than the incident pump power. However, since the core area is much smaller than the inner cladding area, the brightness of the laser output greatly exceeds that of the incident pump light. The device is essentially a "brightness converter" that transforms low-brightness pump light into high-brightness lasing light. This would seem to violate the brightness theorem (Appendix A). However, the brightness theorem only applies to pas-

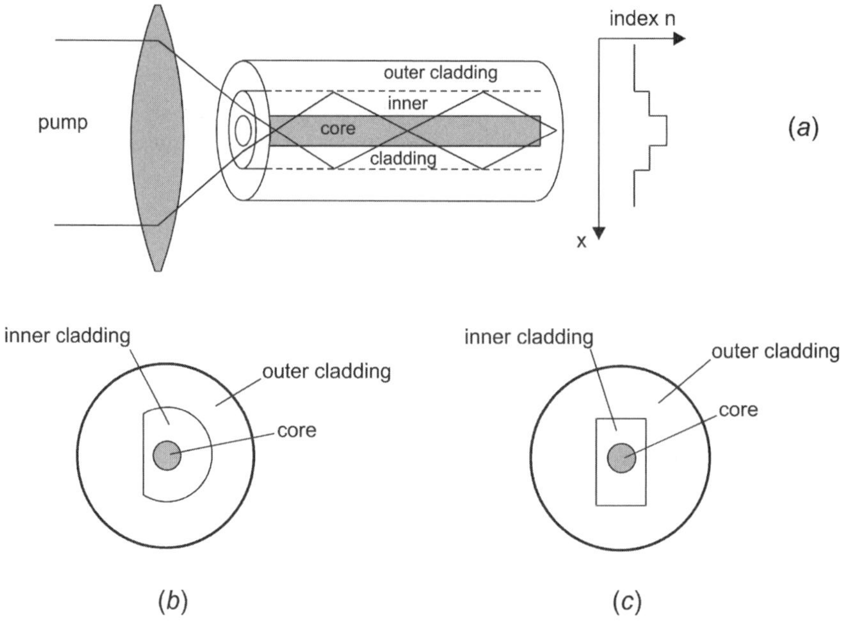

Figure 23-13 (a) A double-clad fiber has a core region that confines the signal light, and a larger inner cladding region that confines the pump light. To ensure that the pump uniformly fills the inner cladding, it often has a (b) "Dee" shape or (c) rectangular shape.

sive optical systems (lenses, mirrors, waveguides, etc.) that contain no energy sources. The laser is an active optical system that uses stimulated emission to create an output beam of high brightness. This is one answer to the question posed earlier: why would you use one laser to pump another, when you could just use the first laser? The reason is that the second laser may have much higher brightness and improved beam qualities, which makes it more useful for certain applications.

Cladding-pumped Yb fiber lasers have seen remarkable development in recent years. One key design feature that improves the performance is an asymmetrical inner cladding, depicted in Fig. 23-13b. The "Dee" or rectangular-shaped cladding suppresses helical inner cladding modes that would exhibit poor overlap with the core, and ensures that the pump beam uniformly fills the cladding area. With careful attention to such design details, CW powers of ~ 100 W can now be routinely obtained in a near-diffraction-limited Gaussian beam ($M^2 < 1.1$), and over 1 kW has been demonstrated in a beam with $M^2 \sim 3$.* These high power fiber lasers are now in a position to compete directly with more traditional solid-state lasers in applications such as laser cutting, drilling, and marking. With their higher efficiency, compactness, and reliability, they are destined to play an increasingly important role in this market niche.

Vibronic Transition

The rare earth transitions that we have considered in the previous sections are said to be electronic in nature, because they occur between two different electronic states of the ma-

*SPI (Southampton Photonics) press release, Jan. 2005.

terial. A different type of transition is also possible, in which the vibrational state of the material changes along with the electronic state. These are termed *vibronic transitions,* and they form the basis for widely tunable lasers.

The rare earth ions do not exhibit vibronic transitions to any significant degree, due to the efficient shielding of the 4f orbitals, and the resulting weak electron–lattice interaction. This is the exception to the general rule, however, and in most materials the electron–lattice interaction is strong enough to enable vibronic transitions. We consider here two examples that are representative of this type of laser system.

Dye Laser

In a dye laser, the optically active material is an organic dye molecule (the same type of molecule that is used to stain the fabric in your clothes). A typical dye molecule, depicted schematically in Fig. 23-14a, consists of a hydrocarbon backbone chain terminated by a more complicated structure on each end. In contrast to the solid-state lasers considered so far, the host for the dye molecules is a liquid, known in the language of chemistry as a *solvent.* Typical solvents for dye molecules might be water (H_2O) or ethylene glycol ($C_2H_6O_2$).

When an outermost electron in the dye molecule is optically excited, it becomes delocalized along the chain, and to a first approximation can be considered to be confined to a box of dimension L. The energy states for such an electron are given by the "particle in a box" model of quantum mechanics, and transitions between these energy states can give rise to laser action. One benefit of the great variety of organic dyes is that a molecule with the appropriate effective length L can be chosen so as to give a transition energy with the desired wavelength.

There are two details regarding these energy states that are important for the operation of dye lasers. The first is that each state actually consists of a number of sublevels, each corresponding to different vibrational and rotational motion of the molecule.* These sublevels are strongly coupled, with energy exchanged between them on a picosecond time scale, and this results in a quasithermal distribution within the sublevels of each state. The second detail is that each state is characterized by a quantity known as spin, which can be thought of as the "orientation" of the excited electron. In *singlet states,* the electrons in the molecule all pair up in opposite orientations, whereas in *triplet states,* one pair of electrons has the same orientation. Radiative transitions between two singlet states or between two triplet states can occur readily, but transitions between a singlet and a triplet state are strongly suppressed. Nonradiative transitions between these two types of states are allowed, however.

The resulting energy level structure for a typical dye molecule is shown in Fig. 23-14b. To achieve lasing, molecules originally in the ground state (the singlet S_0) are promoted to the first singlet state S_1 by absorption of a pump photon. The energy in the S_1 rapidly relaxes to the lower vibrational sublevels of this state, and laser action then occurs on transitions to the various vibrational sublevels of S_0. Because of the large vibrational width of the S_0 and S_1 states, light is emitted over a wide wavelength range. The dye laser is thus broadly tunable, a typical range being $\Delta\lambda \sim 40$ nm for a center wavelength $\lambda \sim 600$ nm. This tunability is the key advantage of dye lasers, which made them very popular after their introduction in the mid-1960s.

Because of the rapid nonradiative relaxation within the emitting state S_1, the average energy of light emission is lower than the average energy of light absorption. The spec-

*Rare earth levels also consist of sublevels, but these are different electronic states, not vibrational states.

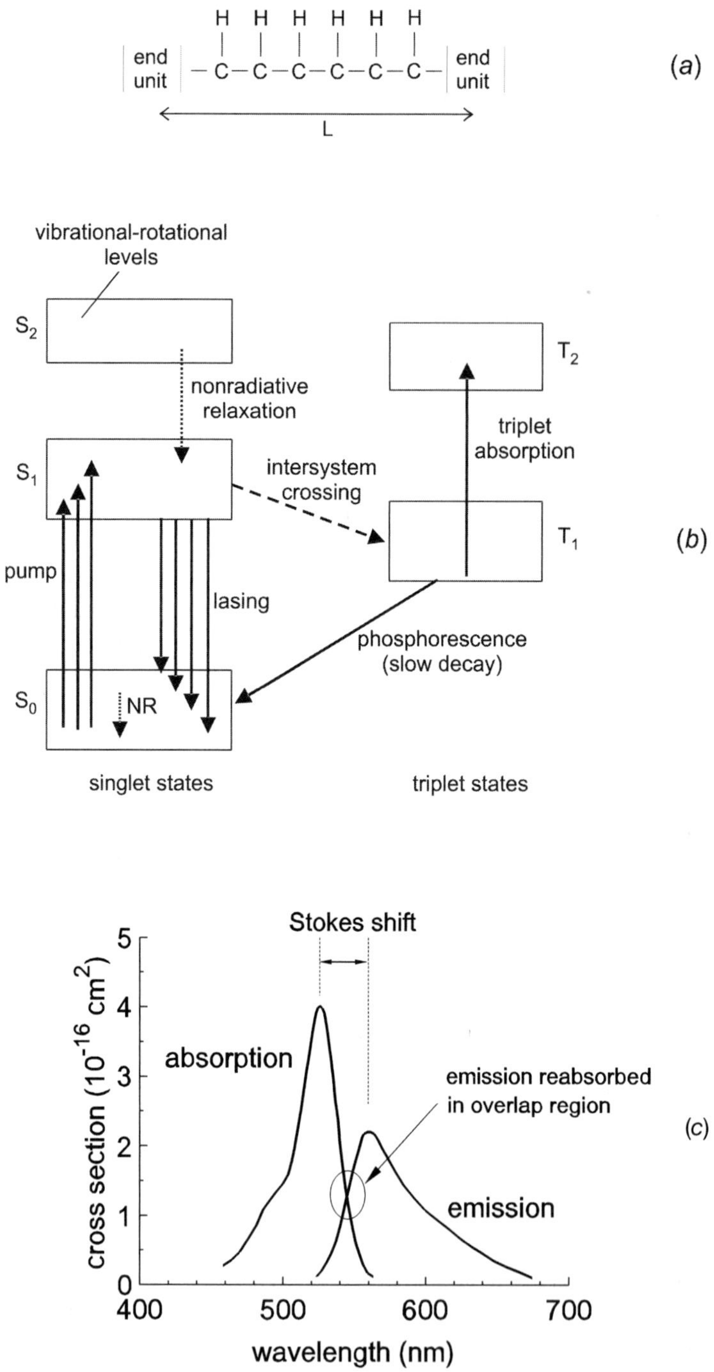

Figure 23-14 (a) Schematic view of typical dye molecule. (b) Energy states of dye molecule, showing singlet state on left and triplet states on right. Dotted and dashed lines represent nonradiative (NR) decay. (c) Absorption and emission spectra for a common laser dye, rhodamine 6 G.

trum of the emitted light is, therefore, shifted to longer wavelengths (lower energy) compared with the spectrum of absorbed light. This feature, illustrated in Fig. 23-14c, is known as the *Stokes shift,* and it is a universal feature of vibronic transitions. The Stokes shift is advantageous for laser action because it makes the system quasi-four-level. A larger Stokes shift leads to less overlap between absorption and emission spectra, and this reduces the undesirable reabsorption of emitted light.

The triplet states introduce a complication in this lasing scheme. The lowest triplet state T_1 is lower in energy than the lowest excited singlet state S_1, and energy can be transferred nonradiatively from $S_1 \rightarrow T_1$ (an *intersystem crossing*). Light emitted on the laser transition ($S_1 \rightarrow S_0$) can then be absorbed by the triplet states ($T_1 \rightarrow T_2$), which tends to quench the lasing. This is referred to as *triplet quenching,* and was a significant problem in the early development of dye lasers. One solution is to operate only in the pulsed mode, so the T_1 level never has time to build up a significant population. The short lifetime of the S_1 state (typically 2–5 ns) requires a very short pump pulse for efficient operation.

Another solution to the triplet quenching problem is to flow the liquid dye at a high velocity through the pumped region, so that fresh dye (with no triplet population) is continually introduced into the laser cavity. A typical arrangement suitable for CW lasing is illustrated in Fig. 23-15, which shows a three-mirror folded cavity with space for a tuning element. The dye flows through a flattened nozzle, creating a thin sheet of liquid flowing at high speed without turbulence (laminar flow). The flowing dye stream has the further advantage that heat deposited in the pumped region is rapidly carried away, thereby greatly reducing thermal distortions in the liquid.

Dye lasers are typically pumped in the blue or UV regions, where organic dyes are highly absorbing. Energy deposited in the higher singlet states (such as S_2) relaxes nonradiatively to the upper laser level S_1. Possible pump sources include the Ar ion laser (for CW operation), and the nitrogen or excimer laser (for pulsed operation). Dyes can be chosen to cover the entire visible and near-infrared regions, making this the ideal choice when a wide spectral coverage in the visible region is required. They can be highly efficient as well, with conversion efficiencies of 20% (absorbed pump to laser output) not uncommon. They can

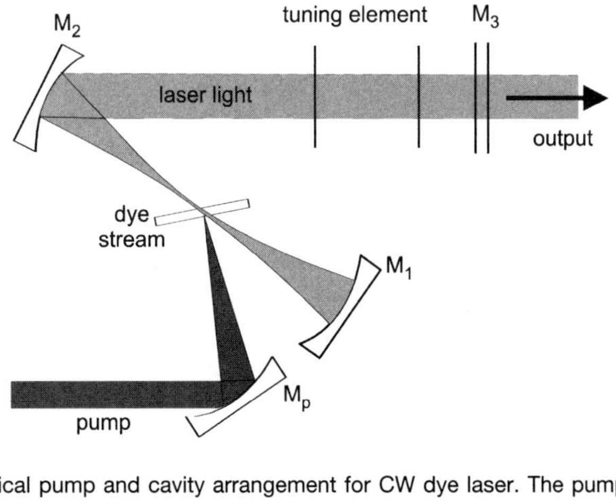

Figure 23-15 Typical pump and cavity arrangement for CW dye laser. The pump is focused with mirror M_p onto a thin stream of dye flowing perpendicular to the page. The laser cavity mode (light shaded area) has a beam waist at the pumped spot. A tuning element such as a birefringent filter can be inserted to select the lasing wavelength.

also be mode locked, with extremely short pulse durations (femtosecond range) made possible by the broad spectral width. Dye lasers have played (and continue to play) an important role in scientific research. However, due to the need for handling messy dye solutions and additives, which pose a health hazard and tend to degrade over time, they have not been as successful in commercial applications. There are now alternative choices for a widely tunable laser in the near infrared, as we will see in the next example.

Ti:Sapphire Laser

Earlier in this chapter, we saw examples of lasers based on rare-earth-doped solids. The transition metals also give rise to optical absorption and emission when doped in a transparent solid, and can serve as the gain medium for a laser. In the transition metal ions, it is the 3d orbitals (quantum numbers $n = 3$, $l = 2$) that are optically active, and these are less well shielded from the surroundings than are the 4f orbitals of the rare earths. As a consequence, the electron–lattice interaction is stronger, and the transitions become vibronic, rather than purely electronic as in the rare earths. This results in broad optical transitions that can be used to make a widely tunable solid state laser. There are now many such lasers with different ion/host combinations, the Ti:sapphire laser having the distinction of the widest tuning range, from 660–1180 nm. Other tunable solid state lasers include alexandrite (Cr:$BeAl_2O_4$), tunable from 700–820 nm; Cr:LiSAF (Cr:$LiSrAlF_6$), tunable from 780–1010 nm; and Co:MgF_2, tunable from 1.8–2.5 μm. We focus on the Ti:sapphire laser here because it is the most commonly used.

The energy level structure of Ti^{3+} is particularly simple, since there is just one 3d electron. There is a ground state (^2T) and excited state (^2E), each broadened into a series of vibrational sublevels, as shown in Fig. 23-16a. The absorption is strong in the wavelength range 450–580 nm, which permits pumping with an Ar ion laser (514.5 and 488 nm) or frequency-doubled Nd:YAG (532 nm) laser. Lasing can occur from the lower vibrational levels of the ^2E to any of the vibrational levels of the ^2T. The potential tuning range is so large that ordinary laser mirrors (typically multilayer dielectric stacks) are not sufficiently reflective over the entire range. It is therefore common to have a number of mirror sets for the laser, as indicated in Fig. 23-16b.

The Ti:sapphire laser has many of the advantages of the dye laser, without the problem of messy laser dyes. It is not only widely tunable (690–1080 nm in commercial lasers), but also highly efficient, with 20–30% conversion efficiency of pump light into laser output typical around 800 nm. When pumped with a frequency-doubled Nd:YAG, it makes an all-solid-state laser system that is efficient and highly reliable. Furthermore, the wide emission bandwidth of Ti^{3+} allows generation of extremely short pulses in mode-locked operation (as short as 5.5 fs, $\approx$ 35 fs typical in commercial lasers). The principle disadvantage compared with dye lasers is that the tuning range is limited to ~ 670 nm on the short-wavelength side. However, frequency doubling the Ti:sapphire laser output can provide continuously tunable laser light in the range 400–550 nm.

23-2. ELECTRICALLY PUMPED LASERS

We turn now to the large class of lasers that achieve population inversion in the gain medium via electrical excitation. One important example is the semiconductor laser, but since this has been treated in Chapter 11 we do not consider it further here. The other lasers that are electrically pumped are mostly lasers with an active medium in the gas phase. The reason that these lasers are not optically pumped is that the optical transitions

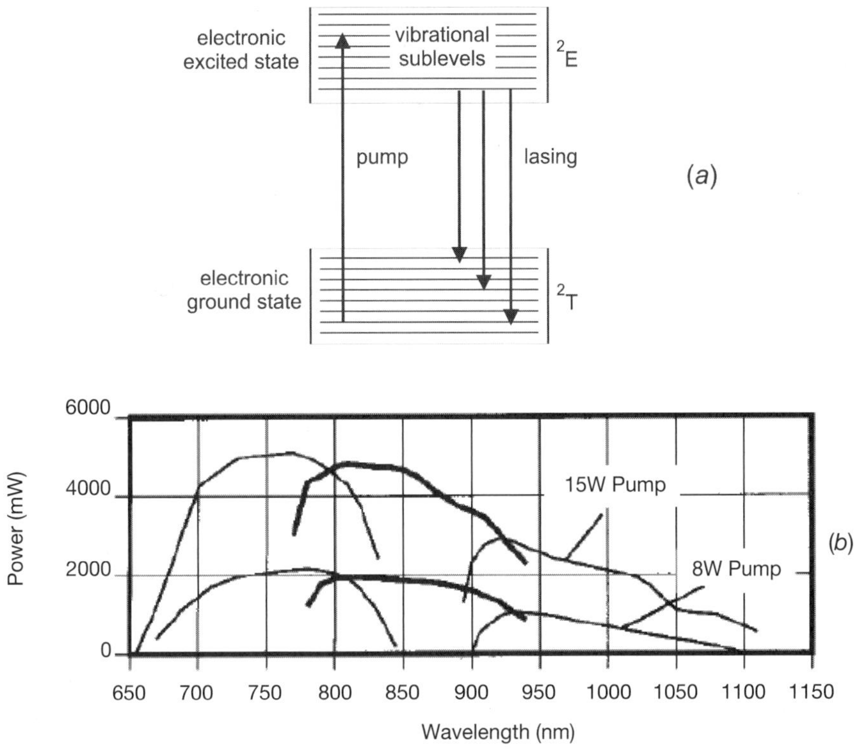

Figure 23-16 (a) Energy levels for Ti:sapphire. (b) Typical tuning range for a Ti:sapphire laser (courtesy of Coherent Inc.). Generally, different mirror sets are required to cover the entire range.

have a much narrower spectral width than those in solids, due to the weaker interaction between adjacent atoms. Optical pumping in this case would be quite inefficient due to poor spectral overlap.

Gas-phase lasers have certain advantages compared with solid-state lasers. For example, the gain medium is resistant to optical damage and distortion, and is easily replenished. Also, the laser wavelengths are sharply defined, a feature that can be useful in optical spectroscopy and metrology. There are some disdvantages, however. The low atomic density that is characteristic of a gas leads to a small gain coefficient, and this necessitates high-reflectivity mirrors for achieving lasing threshold. The narrow spectral widths at well-defined wavelengths can be a mixed blessing as well, since this restricts operation to one of several fixed wavelengths. These lasers can be "tuned" in the sense of choosing between certain fixed wavelengths, but cannot be continuously tuned over a large wavelength range as can solid-state lasers.

In this section, we illustrate the variety of electrically pumped lasers by describing three lasers with an electronic transition, and one with a vibrational transition.

Electronic Transition

In the visible, UV, and near IR regions, most lasers operate on an electronic transition. We will consider here the He–Ne, argon ion, and excimer lasers, which represent electrically pumped lasers of this type.

He–Ne Laser

The He–Ne (helium–neon) laser is one of the most familiar lasers, often used for classroom demonstrations of optical phenomena. Although it can operate on a number of transitions in the visible and near infrared regions, the most commonly used by far is the red line at 632.8 nm. A typical construction is shown in Fig. 23-17. A glass tube with two metal electrodes inside is evacuated and then backfilled with a low-pressure (a few torr) mixture of helium and neon gas. High voltage (a few kilovolts) is applied between the anode (positive electrode) and cathode (negative electrode), and this results in an electrical discharge, with electrons accelerated to a high kinetic energy. When these fast-moving electrons collide with gas atoms, they transfer some of their kinetic energy, raising the atoms to an excited electronic state. The excited atoms then emit light on various transitions, in much the same way that light is emitted by a neon sign. Laser oscillation occurs when this emitted light is efficiently reflected back by the mirrors at either end. One mirror (the end reflector) has a very high reflectivity at the lasing wavelength, while the other (output mirror) has a mirror transmission T that optimizes the laser output [see Eq. (20-33)]. Only one of the mirrors need be curved to form a stable cavity (see Fig. 17-7).

The energy states involved in the He–Ne laser are depicted in Fig. 23-18. The lasing transitions are all within the excited states of Ne, and the purpose of the He atoms is simply to facilitate the excitation of Ne. In the excitation scheme shown, He atoms are promoted from their ground state (both electrons in the 1s orbital) to the first two excited states (one electron promoted to the 2s orbital) by collisions with fast-moving electrons, a process termed *electron-impact excitation.* The two states 2^3S and 2^1S correspond to the two He electrons having the same or different spin "orientation," respectively. The excited He atoms then collide with and transfer their energy to Ne atoms, promoting them from their ground state (1s, 2s, and 2p shells filled) to an excited state (one Ne electron in the 4s or 5s orbital).* The important 632.8 nm transition is from the 5s state to the 3p state, an "allowed" transition with $\Delta l = \pm 1$. Spectroscopic parameters for this transition are given in Table 23-2. To achieve population inversion, the 3p lower laser level must be depleted at a sufficiently high rate. Fortunately, this is the case because the 3p $\rightarrow$ 3s radiative decay is fast (again, an allowed transition). However, the 3s level does not decay as rapidly, and electron impact excitation can repump the Ne atom from the 3s back up to the 3p state, thereby destroying the population inversion. This problem is partially mitigated by collisions of Ne atoms with the walls of the laser tube, which depopulate the 3s nonradiatively. The result is a practical limit on tube diameter for a given pressure. Optimum performance in the He–Ne laser is found to occur when the product of tube diameter and total gas pressure is $D \times P \simeq 3.6\text{-}4$ torr $\times$ mm.

Attempts to scale the He–Ne laser up to high output power are not successful, because as the discharge current increases, the lower laser level (3p) becomes fed more efficiently than the upper laser level (5s). Beyond some optimum current density, the laser output power then starts to decrease with increasing current. The only way to make a higher-power He–Ne laser is to increase the tube length, but this requires a proportionate increase in the high voltage, which at a certain point becomes impractical. He–Ne lasers are thus inherently low power, with typical output powers of a few mW for a laser 10–20 cm long. The efficiency is quite low, with electrical-to-optical conversion efficiencies of

*These excited Ne states are often labelled 2s and 3s, the so-called Paschen notation. We follow here the more intuitive notation that labels the states according to the principal and angular momentum quantum numbers n and l.

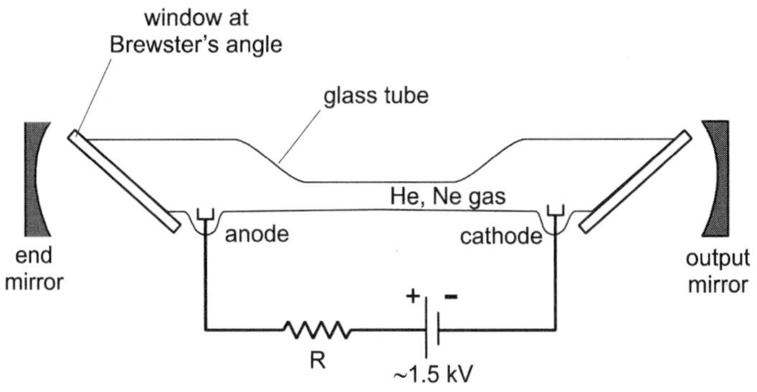

Figure 23-17 Schematic view of He–Ne laser with external mirrors. Windows at end of tube are at Brewster's angle to minimize reflection losses. This results in vertically polarized laser light.

$\eta \sim 0.02\%$. In comparison, a diode laser in the same wavelength range can be quite efficient ($\eta > 10\%$), and requires only a low-voltage power supply.

The reason that He–Ne lasers are still widely used, in spite of the above limitations, is that they have an inherently circular beam, with a beam divergence that is close to being diffraction limited. In contrast, the diode laser beam is inherently asymmetrical, and must be circularized with specialized optics. Another advantage of the He–Ne laser is that the coherence length is naturally long—typically 10–20 cm for a laser with multiple longitudinal modes, and hundreds of meters for a laser with a single longitudinal mode. This makes the He–Ne laser ideal for holography and interferometry when the optical power required is not too high.

The lasing transition at 632.8 nm is the most well known and commonly used, but other wavelengths are possible as well. In fact, the first He–Ne laser (made by Javan in 1960)

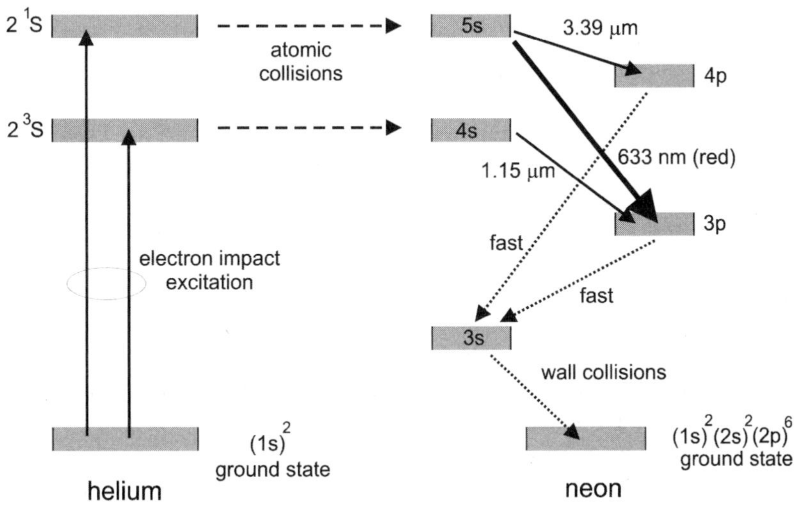

Figure 23-18 Energy levels of helium and neon that are relevant for laser operation. The thick solid arrow is the important 632.8 nm transition, thin solid arrows are other lasing transitions, and dotted arrows are other radiative and nonradiative decays. Level positions are not drawn to scale (the helium 2^1S and neon 3s states are $\approx$ 20.6 eV and $\approx$ 16.7 eV above the ground state, respectively).

Table 23-2 Typical parameters for selected electrically pumped lasers.

Laser	Gas partial pressures (torr)	Concentration of active gas molecules (10^{14} cm^{-3})	Emission wavelength (μm)	Peak emission σ (10^{-14} cm^2)	Excited state lifetime τ (ns)	Transition linewidth $\Delta\nu$ (GHz)
He–Ne	2.5 He 0.5 Ne	120	0.6328	30	~100[a]	1.5
Ar ion	0.1 Ar	24	0.5145	25	6	3.5
KrF	90 Kr 5 F 1800 He	~1[b]	0.248	0.05	~10	3000
CO_2	1 CO_2 1 N_2 8 He	240	10.6	0.018	$\approx 6 \times 10^5$	0.06

[a]Longer than the 30 ns radiative lifetime due to radiation trapping.
[b]Since there are no "ground state" KrF molecules, this is typical excimer concentration.

operated at 1.15 μm, and is important historically as the first gas laser, as well as the first CW laser. The infrared transition at 3.39 μm is also quite efficient, and must be suppressed in order to obtain lasing at 632.8 nm. Lasing on the 1.15 and 3.39 μm transitions can be prevented by using mirrors that are highly reflective at 632.8 nm, but transmitting at those longer wavelengths. Another lasing wavelength that is possible in the He–Ne laser is 543 nm, which corresponds to a transition between the 5s and 3p states, the same pair of states involved in 632.8 nm emission. The reason that different wavelengths can be emitted on the same transition is that the 5s and 3p "states" are really manifolds of closely spaced sublevels, and the two emission wavelengths are generated by transitions between different pairs of sublevels. To achieve lasing at 543 nm, the 632.8 nm lasing must be suppressed by the use of wavelength-selective mirrors.

Argon Ion Laser

The argon ion laser is similar in many ways to the He–Ne laser just discussed. The gain medium is a noble gas (Ar) that is confined at low pressure (0.1 torr) inside an evacuated tube, and the gas is excited by energetic electrons in an electrical discharge. The primary difference is that lasing occurs in an ion, rather than a neutral atom. The krypton ion laser operates in a similar manner, using another noble gas (Kr). The Kr ion laser can operate on a wide range of discreet wavelengths across the visible spectrum, whereas the Ar ion laser operates only in the green, blue, and UV regions. Because of their similarity, we discuss only the argon ion laser in detail here.

Fig. 23-19 depicts the laser excitation process on an energy level diagram. First, a neutral argon atom is ionized by electron impact excitation, creating the singly charged ion Ar^+, a process requiring $\approx$ 16 eV of energy. Next, one of the 3p electrons in the Ar^+ ground state is excited by electron impact excitation into the upper laser level (4p), which requires an additional $\approx$ 19.5 eV of energy. The 4p and 4s states consist of a number of sublevels, and transitions can occur between the various sublevels in the 4p to the various sublevels in the 4s. This results in a number of discrete laser transitions in the blue-green portion of the spectrum, the most prominant of which are at 514.5 and 488.0 nm. When

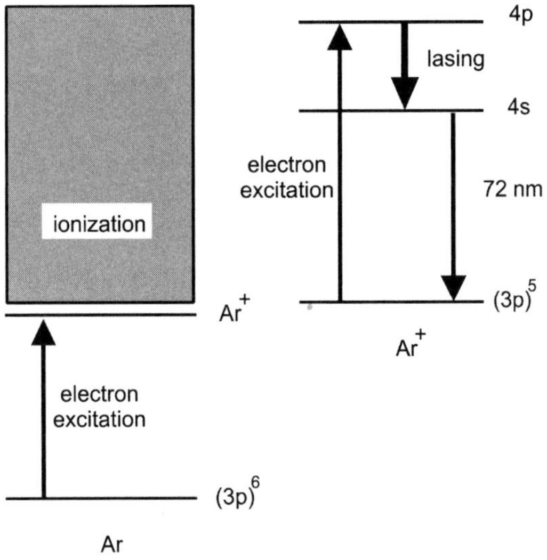

Figure 23-19 Energy levels (not drawn to scale) relevant for an argon ion laser. The lasing transitions (thick arrow) occur between various sublevels of the 4p and 4s states of the Ar^+ ion.

maximum output power is desired, simultaneous lasing on several lines can be achieved by using a mirror that is highly reflective at all the relevant wavelengths (a "broadband reflector"). Alternatively, individual lines can be selected for lasing using the prism tuning method (Fig. 21-12).

Unlike the He–Ne laser, the Ar ion laser can be scaled up to high power by increasing the drive current. Since it takes two excitation steps to populate the upper laser level (first to ionize Ar, then to raise Ar^+ to an excited state), the effective pumping rate goes as the square of the drive current, and very high current densities (~ 10^3 A/cm^2) are required to achieve lasing threshold. A typical medium-power Ar laser, shown schematically in Fig. 23-20, has an optical output on all lines of ~ 5 W when driven by a 400 V, 30 A power supply. The low wall-plug efficiency (η = 5 W/12 kW = 0.04%) means that most of the electrical power is converted into heat, which must be removed from the tube to prevent thermal damage. Lasers with output powers > 1 W are generally cooled by flowing water around the tube. The output power can be scaled up to $\approx$ 100 W by increasing the electrical pump power to $\geq$ 60 kW.

Tube lengths range from $\approx$ 10 cm for low-power (100 mW) lasers, to $\approx$ 2 m for the highest-power lasers. There is often a magnet surrounding the tube, creating a B field along the laser axis that serves to confine the electrons to the center of the discharge region. A filament inside the tube is heated to generate the free electrons needed for the discharge. To initiate the discharge, a high-voltage starting pulse is applied to the tube from an inductively coupled starter circuit. After the discharge starts, it can be maintained with a lower voltage (a few hundred volts), and the current is controlled with a current regulator. Typical currents are ~ 30 A for a 5 W laser, and ~ 60 A for a 25 W laser.

The argon ion laser has applications in ophthalmology, and has been used for laser light shows (along with the similar krypton ion laser) because of the visible wavelengths. Its most important application has perhaps been as a pump for dye and Ti:sapphire lasers, since the 514.5 and 488 nm lines are efficiently absorbed by both. In recent years, however, this role has been somewhat supplanted by the frequency-doubled Nd:YAG laser,

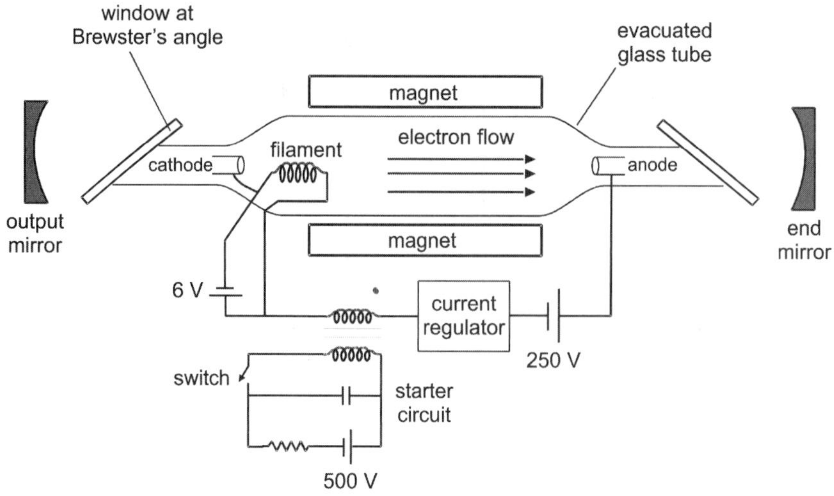

Figure 23-20 Schematic view of a typical 5 W argon ion laser. Windows at the tube ends are oriented at Brewster's angle to minimize reflection loss. External mirrors define the laser cavity modes. To start the laser, a capacitor is charged to high voltage, and this is then switched to an inductor that couples a larger voltage spike into the tube, initiating the plasma discharge.

which operates at 532 nm. The all-solid-state nature of the doubled Nd:YAG/Ti:sapphire combination has important advantages in efficiency, compactness, and reliability. The argon ion laser still has its niche, however, because like the He–Ne laser, it naturally produces a near-Gaussian beam of low divergence and high spectral purity. It continues to find new applications such as mastering of video disks, detecting latent fingerprints, stereolithography (making 3-D images), as well as in optical spectroscopy research.

Excimer Laser

The excimer laser is a pulsed laser that operates in the UV region. As in the He–Ne and argon ion lasers, the laser transition is between electronic energy states, and it is excited by an electrical discharge. In contrast to those lasers, though, the gain medium in the excimer laser consists of molecules rather than isolated atoms. The molecules involved in excimer laser operation are rather peculiar, however, since they exist as bound molecules only when the molecule is in the excited electronic state. When the molecule returns to the ground state, it dissociates, breaking up into two separate atoms. Such a molecule is termed an *excimer,* which is a contraction for "excited dimer."

Laser action in the excimer molecule can be understood by referring to the energy level diagram of Fig. 23-21. This shows the electronic energy of a KrF "molecule" as a function of the separation of the two nuclei. When an atom of Kr and F are brought together, each in their ground state, the energy of the system increases monotonically due to Coulomb repulsion. Since systems always tend toward the point of lowest energy, the atoms naturally tend to separate, and the "molecule" is unbound. This is characteristic of the noble gases (He, Ne, Ar, Kr, and Xe), which are generally unreactive and not inclined to form stable molecules.

If the KrF molecule is promoted to the next-highest electronic state, however (by electron bombardment in an electrical discharge, for example), then the energy curve has a

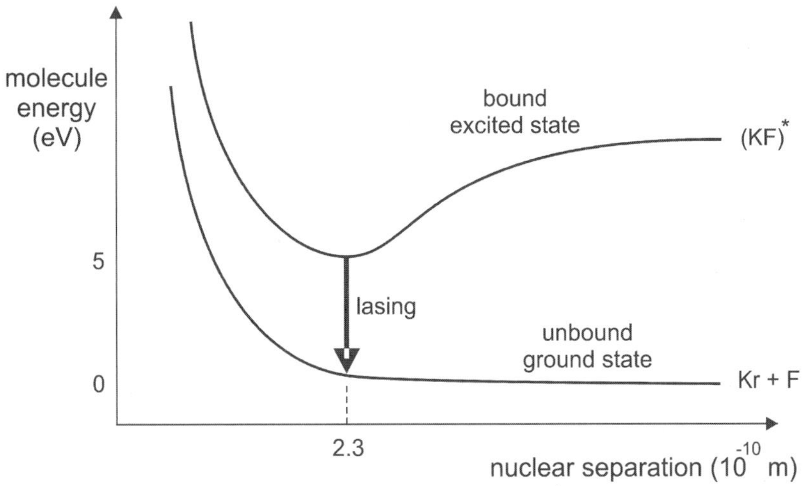

Figure 23-21 Energy versus Kr–F nuclear separation for ground and excited electronic states of KrF. The excited state is bound, and known as an excimer state. The lower state is unbound, so the KrF dissociates into two separate atoms of Kr and F.

minimum at an internuclear separation $R = 0.23$ nm. This gives rise to a bound state at this value of R, which remains stable as long as the molecule remains in the excited state ($\approx$ 10 ns for KrF). During this excited-state lifetime, lasing can occur on the transition from the bound excited state to the unbound ground state. Since there is no stable population in the lower laser level, this is an example of a perfect four-level system. In KrF, lasing occurs at a wavelength of 248 nm. Other noble-gas–halide pairs operate in the same fashion, but with different transition wavelengths. Commonly used excimer lasers include XeCl at 309 nm, XeF at 351 nm, and ArF at 193 nm.

Excimer lasers have moderate efficiencies (~ 1%), short pulse duration (~ 10 ns), high pulse energies (0.1–1 J), and can be scaled up to high average power (100 W). They have numerous applications, including pumping dye lasers, photolithography, materials processing, laser surgery, and ophthalmology. The short wavelengths are strongly absorbed by most materials, and this is a distinct advantage for precision cutting. The principal disadvantage of these lasers is the need to work with highly reactive gases such as fluorine, which requires proper ventilation and safety precautions. Typically, the gases are circulated through the electrical discharge chamber to maintain the purity of the reactants. A high concentration of helium (about 2 atmospheres of pressure) is added as a buffer gas to facilitate the reactions between noble and halide gases.

Vibrational Transition

In all the laser types discussed so far, the laser transition has involved a change in electronic state. We now consider lasers in which the vibrational state changes, but the electronic state remains the same. These are termed *vibrational transitions,* and should not be confused with the vibronic transitions discussed earlier.* Vibrational energy levels in a molecule have energy separations on the order of ~ 0.1 eV, rather than the ~ 1 eV separa-

*In vibronic transitions, both the electronic and vibrational states change.

tion typical of electronic levels, and this means that lasers operating on a vibrational transition will generate wavelengths in the infrared region. The most important laser of this type is the carbon dioxide (CO_2) laser, operating in the range 9–11 μm, with principal lines at 9.6 and 10.6 μm. Other vibrational lasers include the carbon monoxide (CO) laser operating at 5–6.5 μm, and the nitrous oxide (NO_2) laser operating at 10–11 μm, although these are much less commonly used. Some vibrational lasers utilize a chemical reaction to achieve an excited vibrational state, and are termed *chemical lasers.* Examples are the hydrogen fluoride (HF) laser operating at 2.6–3.3 μm, and the deuterium fluoride (DF) laser operating at 3.5–4.2 μm. We will focus our attention on the CO_2 laser here, since it illustrates well the characteristics of vibrational lasers, and is the most common.

CO_2 is a linear molecule, with the carbon and two oxygen atoms lying along a common axis. There are three fundamental ways that such a molecule can vibrate, as illustrated in Fig. 23-22a. Each of these fundamental vibrational patterns is referred to as a *vibrational mode,* and has a characteristic frequency f_v [see Eq. (5-8) for the simpler case of two atoms]. In general, the vibrations of the molecule can be described by a linear combi-

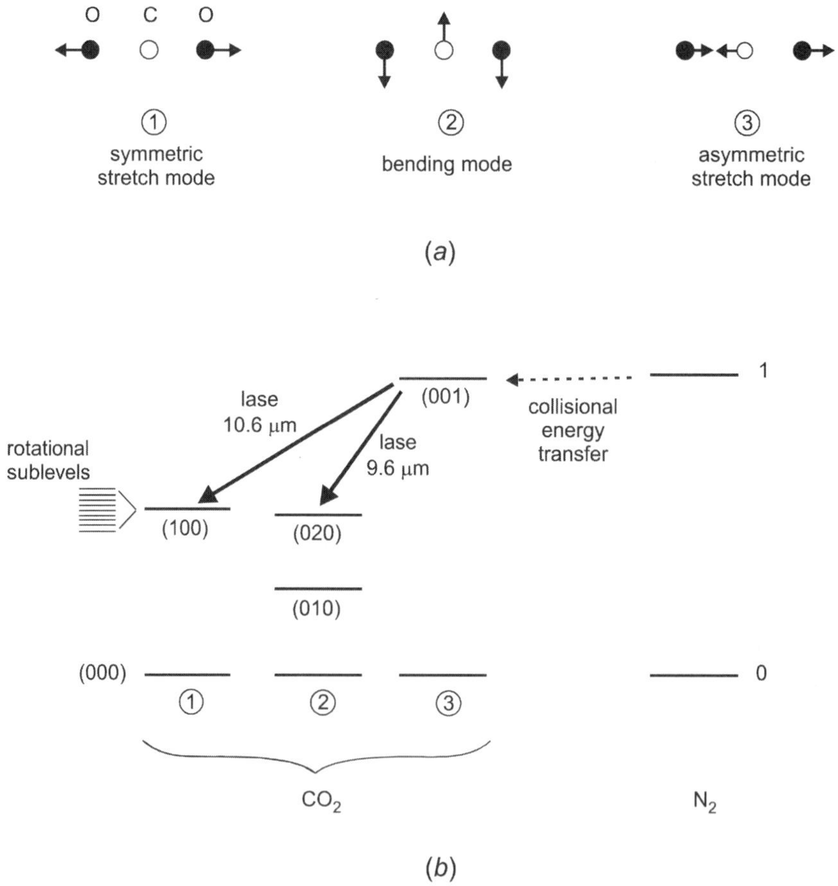

Figure 23-22 (a) Vibrational modes of a CO_2 molecule. (b) Energies of CO_2 vibrational modes, showing two possible laser transitions. Also shown is the molecular nitrogen vibrational level, which is nearly coincident with (001) vibration of CO_2. Each vibrational state contains a number of closely spaced rotational sublevels.

nation of these fundamental modes. In CO_2, the frequencies of the modes are $f_{v1} \approx 40$ THz, $f_{v2} \approx 20$ THz, and $f_{v3} \approx 70$ THz, where the subscripts refer to the modes as labelled in Fig. 23-22a. According to quantum mechanics, the energy contained in each vibrational mode can be any integer multiple of the energy quantum hf_v, so the spectrum of allowed energy states is a ladder of equally spaced levels. When there are n_1 quanta in mode 1, n_2 quanta in mode 2, and n_3 quanta in mode 3, we use the notation $(n_1n_2n_3)$ to describe the complete vibrational state of the molecule.

Fig. 23-22b shows the vibrational levels of CO_2 that are relevant for laser action. The upper laser level corresponds to one quantum in the asymmetric stretch mode, with no quanta in the other two modes. This is denoted (001) in the notation described above. The lower laser level has either one quantum in the symmetric stretch mode (100), or two quanta in the bending mode (020). Both of these states have approximately the same energy, since $2f_{v2} \approx f_{v1}$, and, therefore, $2hf_{v2} \approx hf_{v1}$. The laser photon energy will be equal to the difference in energy between the upper and lower levels, which in this case is $hf_{v3} - hf_{v1}$. It is easily shown (see Problem 23.15) that this leads to a laser wavelength ≈ 10 μm. The transition to the (020) level occurs at a slightly shorter wavelength than the transition to the (100) level, because the (020) level is slightly lower in energy and the corresponding photon energy is greater.

Treating the vibrational energy levels of the CO_2 molecule as single states gives the basic idea of the CO_2 laser, but it leaves out one important detail. The molecule can rotate, as well as vibrate, and in quantum mechanics this rotational motion is also quantized. Each level, such as the one labeled (100), actually consists of a number of discrete sublevels, spaced by ~ 10^{-4}–10^{-3} eV. This is more than two orders of magnitude smaller than the spacing between vibrational levels, and would not be noticeable when the energy level diagram is drawn to scale. However, it has important implications for the CO_2 laser, because it means that the laser can be tuned over a limited range, to any one of the individual vibrational–rotational transitions. Tuning in the CO_2 laser is usually performed with a grating, as shown in Fig. 21-13.

A typical configuration for the CO_2 laser is shown in Fig. 23-23. An electrical discharge excites the gas, and lasing occurs between two mirrors (often made of copper, a good reflector for $\lambda = 10$ μm). In addition to CO_2, the gas mixture includes nitrogen (N_2) to help populate the upper laser level, and helium (He) to help depopulate the lower laser level. The relative concentration of CO_2, N_2, and He varies from laser to laser, but they are usually comparable.

The added gases N_2 and He function in the following way. The N_2 molecule has a single vibrational mode of frequency $f_v \approx 70$ THz, which is excited by electron bombardment in the discharge. When an excited N_2 molecule collides with a ground state CO_2 molecule, the close energy match between the excited N_2 and CO_2 (001) levels allows the molecules to exchange energy, with the N_2 going down to the ground state and the CO_2 up to the excited state (001). This process is called *energy transfer,* and it is efficient because the N_2 remains in the excited state a long time (it is metastable). This long lifetime is due to the symmetry of the N_2 molecule, which inhibits radiative transitions between the vibrational states. The He gas depopulates the lower vibrational levels by a different kind of energy transfer process, in which the vibrational energy is converted into translational kinetic energy of the He atoms. He atoms serve best for this purpose, because they have a small mass, and can take up a greater amount of energy in an elastic collision.

CO_2 lasers are very efficient, with wall-plug efficiencies of $\approx 30\%$ possible. This is due in part to the rather direct excitation of the upper laser level, which does not require excitation of higher-lying levels. These lasers can also be scaled up in power quite readi-

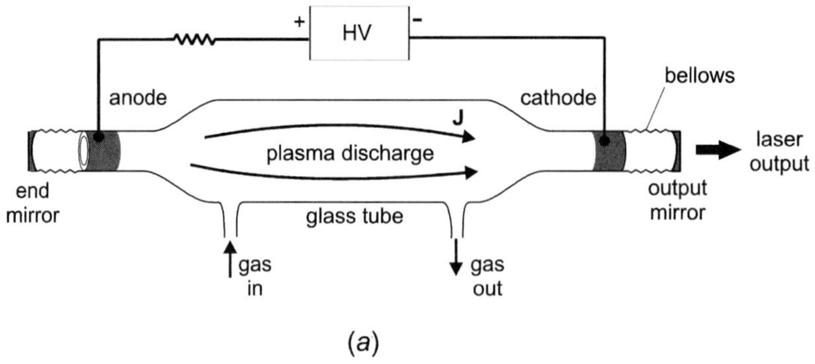

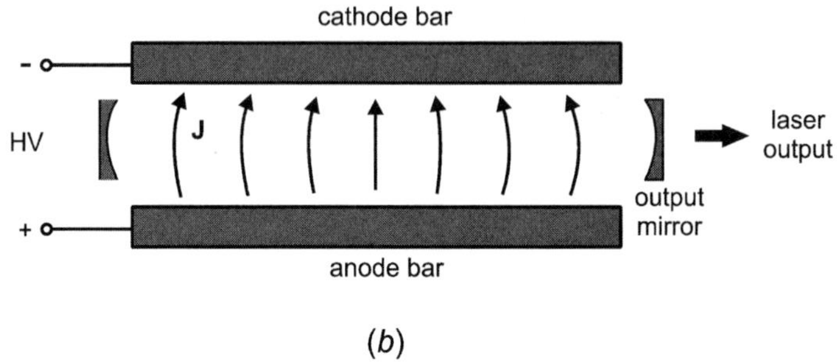

Figure 23-23 Excitation schemes for CO_2 laser. (a) Longitudinal discharge, with current in same direction as optical beam. (b) Transverse discharge, with current perpendicular to direction of optical beam.

ly, by increasing the gas pressure and length of the laser. However, for high gas pressures and long cavity lengths, it becomes impractical to use a longitudinal discharge such as that in Fig. 23-23a. At 10 torr of pressure, the discharge requires an electric field of ~ 8 kV/m, and this value scales in proportion to the pressure. The required voltage for high-power lasers would then be prohibitively high.

An alternative scheme is the transverse discharge geometry, depicted in Fig. 23-23b. Here, the required voltage depends on the lateral separation between electrodes, not on the cavity length. By placing the electrodes close together, lasers with this type of configuration can achieve the required electric field with reasonable voltages. The gas is often flowed transversely through the discharge region as well, for more efficient cooling. The power can be scaled up as high as desired by increasing the gas pressure to ≈ 100 torr, and increasing the cavity length. CW powers as high as a few kW per meter of gain length are possible in this way. At pressures >100 torr, instabilities in the discharge make it necessary to operate the laser in the pulsed mode. When operated at atmospheric pressure or higher, these are often called TEA lasers (for Transverse Excitation Atmospheric).

Since its invention in 1964, the CO_2 laser has been an industrial workhorse. Although the 10 μm wavelength is not efficiently absorbed by metals, the very high power of these lasers overcomes this drawback and makes them useful for a variety of materials processing applications such as cutting, drilling, welding, surface heat treatment, and so on. They have also been used for laser surgery, because biological tissue strongly absorbs the 10 μm light. Pulsed lasers are particularly suited for this application because less average power is required, and this minimizes the damage to surrounding tissue.

PROBLEMS

23.1 (a) The sides of a Nd:YAG rod are surrounded by flowing water for cooling. What fraction of the fluorescence emitted from within the rod will be trapped by TIR inside the cylindrical rod? Repeat for a glass rod of index 1.5. (b) The Nd:YAG rod in part a is end pumped. What is the maximum angle an incident ray can make with the rod axis if it is to be trapped in the rod by TIR? Repeat for a glass rod.

23.2 Carry through the steps leading to Eq. (23-7)

23.3 A Pr-doped fiber laser is pumped at 1015 nm, at which the absorption cross section is 4×10^{-22} cm^2, and it lases at 1310 nm, at which the stimulated emission cross section is 4×10^{-21} cm^2. The laser operates as a four-level system, with an upper-state lifetime of 110 μs. The fiber has length 3 m, core diameter 40 μm, attenuation coefficient 10 dB/km, and mirrors of reflectivity 0.99 (high reflector) and 0.97 (output coupler). The refractive index of the core glass is 1.5, and it is doped with 3×10^{19} Pr ions per cm^3. (a) Determine the cavity lifetime. (b) Verify that most of the pump light is absorbed in the fiber. (c) Calculate the threshold pump power. (d) Determine the slope efficiency. (e) If the laser is pumped with 250 mW, determine the laser output power

23.4 In the previous problem, the fiber length is shortened to 30 cm. (a) Determine the fraction of pump light absorbed in the fiber. (b) Determine the new pump theshold power. (c) Determine the new slope efficiency. (d) Which of the quantities calculated in parts b and c is most affected by the shorter fiber length? Explain.

23.5 The ruby rod in a ruby laser has length 6 cm and diameter 6 mm, and is pumped with a short lamp pulse that excites nearly every Cr^{3+} ion into the excited state. Once the medium is fully inverted, the cavity Q is switched and a single laser pulse is produced. Additional data for ruby are given in Table 23-1. (a) Considering this to be an ideal three-level system, determine the maximum possible output pulse energy. (b) If the pulse duration is 10 ns, calculate the peak power during the pulse.

23.6 Show that Eq. (23-12) for the four-level pump threshold can be obtained using Eqs. (20-24) and (20-18).

23.7 Consider an optically pumped atomic system consisting of just two energy levels, with populations N_1 and N_2 in the lower and upper states, respectively. Write the rate equation for the upper level, and show that no matter how high the pump intensity, at most one-half the atoms can be pumped to the excited state.

23.8 Show that Eq. (23-25) follows from Eq. (23-24), using the McCumber relation of Eq. (18-38).

23.9 Use the McCumber relation to show that the gain coefficient is always negative when the pump wavelength is longer than the signal wavelength.

23.10 Consider the Yb-doped fiber of Example 23-2, with the pump wavelength changed to 975 nm. At this wavelength, the absorption and emission cross sections are both

2.4×10^{-20} cm^2. Assume the signal wavelength is still 1025 nm. (a) Repeat the calculation of the pump power required for transparency. (b) Obtain an analytical expression for the gain coefficient as a function of pump power. Sketch this graph, and compare with the results of part a.

23.11 In Fig. 23-12, each gain curve is seen to cross the zero point only once, with positive gain at longer wavelength and negative gain at shorter wavelength. Show that this is a universal feature of such gain spectra, valid whenever the absorption and emission cross-section spectra are connected by the McCumber relation of Eq. (18-38).

23.12 Using the data in Fig. 23-14, estimate the shortest wavelength that can lase in the R6G dye laser, assuming that two-thirds of the dye molecules are pumped into the excited singlet state S_1. Repeat this if only 10% of the molecules are in the excited state.

23.13 The 2 ^{1}S level in He is $\simeq$ 20.6 eV above the He ground state. Using this and other data given in Fig. 23-18, determine the energies of the 4p, 4s, and 3p levels in Ne with respect to the Ne ground state. Assume that the He 2 ^{1}S level has the same energy as the Ne 5s level.

23.14 An argon ion laser operates with an output power of 2 W at 488 nm, and the tube is excited with a voltage of 250 V and drive current 30 A. (a) If the distance between electrodes is 60 cm, calculate the electric field in the tube, and the tube resistance (assume Ohm's law applies when there is a plasma discharge in the tube). (b) Assuming that the electrons in the tube give up all their energy in steps, sequentially exciting the 4p level of a number of Ar^+ ions (see Fig. 23-19), calculate the maximum number of Ar^+ ions that could be excited to the 4p level per unit time. (c) Assuming steady-state operation, calculate the number of stimulated emission decay processes occuring per unit time. (d) If the laser is far enough above threshold that the stimulated emission rate dominates the spontaneous emission rate, estimate the actual number of Ar^+ ions excited to the 4p level per unit time. (e) Compare the answers to parts b and d, and comment on the efficiency with which the Ar^+ 4p level is excited.

23.15 Take the frequencies of the three CO_2 vibrational modes (see Fig. 23-22) as $f_{v1} \approx$ 40 THz, $f_{v2} \approx$ 20 THz, and $f_{v3} \approx$ 70 THz. (a) Show that the laser wavelength for a (001) $\rightarrow$ (100) transition is $\approx$ 10 μm. (b) If the CO_2 laser were made to lase on the (001) $\rightarrow$ (010) transition, what would be the lasing wavelength?

Chapter 24

Optical Communications

We have seen in the preceding chapters how light is generated for photonics applications, how that light propagates, both in fibers and in free space, and how the light can be detected and transformed into an electrical signal. In this final chapter, we show how these different parts all work together in one important application, that of optical communications. This is by no means the only important application, but it nicely illustrates the elements of design that are common in photonic systems. Some of these design issues have already been mentioned earlier in the text. We will now take more of a systems point of view, and consider how the behavior of the various components affects the overall performance of a communications system.

24-1. FIBER OPTIC COMMUNICATIONS SYSTEMS

In the introductory chapter, we overviewed the topics to be covered by relating them to the simplified optical communications scheme of Fig. 1-2. This overall scheme is repeated in Fig. 24-1, but now with more detail about the choices for the various components. The data can be in analog or digital form, and can modulate the light source either directly, by varying the source's drive current, or indirectly by passing the light through an external modulator. An example of an external modulator is the electrooptic Mach–Zehnder device (Fig. 9-21). The light source can be an LED or a laser diode, and the laser diode can be either multiple longitudinal mode (MLM) or single longitudinal mode (SLM). The MLM laser has no frequency-selective element other than the natural Fabry–Perot resonances between end facets, and is sometimes referred to as a *Fabry–Perot laser diode* (FP). The SLM laser has an additional frequency-selective element, such as the Bragg grating in a distributed feedback (DFB) or distributed Bragg reflector (DBR) laser. Light can be coupled into the fiber using an evanescent wave device such as the fused biconical taper coupler (Fig. 7-4) or a fiber grating (Fig. 8-5). The fiber can be step-index or graded index, and the step index fiber can be single-mode or multimode. Finally, light exiting the far end of the fiber can be detected with a PIN photodiode or avalanche photodiode (APD).

The combination of these photonic elements that is appropriate for a particular communications system depends on the length scale over which data is transmitted, and the maximum required data rate. Fiber optic networks generally fall into one of the three broad categories listed in Table 24-1. The smallest in scale is the *local area network* (LAN), in which different rooms and/or buildings in a campus setting are connected over a distance of ~ 0.1 – 2 km. This often requires only modest data rates, and the short fiber lengths mean that fiber losses do not have to be as low as possible. For the LAN, connector losses can dominate fiber losses.

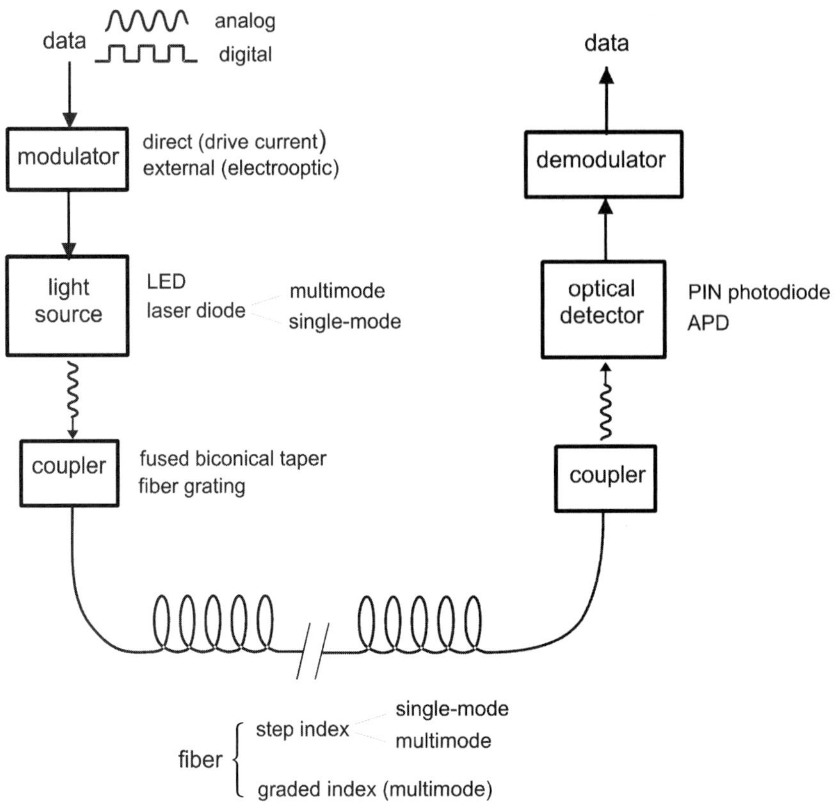

Figure 24-1 Overview of fiber optic communications system, showing different choices for the various components.

The *metropolitan area network* (MAN, or simply "metro") is wider in geographic scope, encompassing large sections of a city, or regions in a rural or suburban area. Transmission distances can be as great as 100–200 km, so fiber losses here become much more important. Optical amplifiers are generally not needed for fiber links less than 100 km in length, provided that data rates are not too high, and that the low-loss second (1300 nm) or third (1550 nm) telecommunications windows are used. Although the boundary be-

Table 24-1 Characteristics of different types of fiber optic systems

	LAN	MAN	Long-haul/WAN
length (km)	0.1–2	2–200	> 200
light source	LED/MM laser diode	MM/SM laser diode	SM laser diode
modulation	direct	direct	external
fiber type	graded index	single-mode	single-mode
detector	PIN photodiode	PIN photodiode	APD
wavelength (nm)	850/1310	1310	1550
data rate (Mb/s)	10–1000	500–2500	2500–40,000
geographical scope	building or campus	city or metropolitan area	state/nation or global

tween these categories is fuzzy, the lack of need for amplification can be thought of as a defining characteristic of the MAN.

The third category is the *wide area network* (WAN), which extends over a scale greater than 200 km. Included here is the long-haul global telecommunications backbone that carries voice and internet data between cities, countries, and continents. The distances are too great for unamplified transmission, and electronic repeaters or optical amplifiers must be inserted periodically, with a typical separation of 40–120 km. To minimize the number of required amplifiers, fiber loss must be kept as low as possible, and this makes 1550 nm the band of choice. The receiver is often an APD for improved sensitivity. Chromatic dispersion limits the maximum data transmission rate in the single-mode fiber, and can be minimized using a narrow-linewidth SLM diode laser, with external modulation. Direct modulation is not used at the highest data rates because it causes frequency chirp, a time-varying optical frequency in the laser-diode output.

The three categories of fiber optic networks can be likened to the hierarchy of roads in the physical transportation network. The long-haul WAN backbone is analogous to the interstate highway system, with limited access but very high speed and high volume traffic possible. Branching off the long-haul backbone are the metro networks, which are analogous to the numbered state highways. These have a higher level of access and connectivity, but still with moderately high speed and capacity. Branching off the metro networks are the LANs, which in our analogy are equivalent to the local roadways of villages and towns. These are lower-speed roads with more limited capacity, but they have the advantage of a much higher connectivity that is more easily modified. Although the boundaries between these three categories can sometimes be blurred, their functioning is usually different enough to make the separation useful.

24-2. SIGNAL MULTIPLEXING

The data-rate capacity of a single optical fiber is far greater than what is required for a single data source, such as a phone conversation or a computer connecting to the Internet. In order to take advantage of a fiber's large capacity, it is necessary to combine data from many different sources, a process termed *multiplexing*. There are two basic types: *time-division multiplexing,* in which data trains from different sources are interleaved in time, and *wavelength-division multiplexing* (WDM), in which data signals at different optical wavelengths are sent simultaneously down the same fiber. Before discussing each of these in detail, we first consider the format of the data that needs to be multiplexed.

Data Format

Data can be represented in analog format, with a voltage varying continuously in time, or in digital format, with a series of "ones" and "zeros" that correspond to a binary coding. Voice and video signals start off as analog signals, and can be transmitted in that format through an optical fiber. To do this, the light intensity in the fiber is continuously modulated in proportion to the signal voltage, and this faithfully reproduces the original data after photodetection. Any interuption in this data stream will distort the signal, and these analog transmission schemes, therefore, require a dedicated line for one signal.

In contrast to this, computer data is inherently digital, and is sent through a network in a very different manner. The computer data is broken up into small bunches called *pack-*

ets, each packet containing directions for the final destination (a *header*), along with the actual data. These packets are directed to their destination by a series of network hubs called *routers,* in a process known as *packet switching.* Different packets may take very different physical routes in getting from their origin to their destination, even if they are part of the same "message." This method of data transfer is efficient for computers because for much of the time, any given computer is just idling away, waiting to send or receive data. It sends and receives data in short spurts of activity, and is then idle again for some time. When many computers are sending information over the same network, packet switching turns out to be a very efficient way to utilize the network's capacity.

From the above, it is clear that analog voice and digital computer data are rather incompatible, and therefore not easily multiplexed. However, if the voice signal were in digital rather than analog format, then multiplexing could be readily accomplished. In fact, telephone signals were converted to digital format long before the advent of optical communications, largely for this reason. This process is known as *analog to digital conversion,* and it has other benefits as well, as we will see in the following.

Consider the representative analog waveform shown in Fig. 24-2, which could correspond to the voltage waveform for a telephone conversation. The waveform's voltage is sampled periodically with a time T_{phone} between samples, and the voltage level at the time of each sample is allocated to one of 256 "bins" along the voltage axis. The bin number is expressed in binary notation, so that, for example, the bin 178 becomes

$$178 = (1 \times 2^7) + (0 \times 2^6) + (1 \times 2^5) + (1 \times 2^4) + (0 \times 2^3) + (0 \times 2^2) + (1 \times 2^1) + (0 \times 2^0)$$

which is the binary number 10110010. In a digital data stream, this 8-bit number would be transmitted as indicated in Fig. 24-3, as a series of high and low voltage levels.* A high voltage level corresponds to a digital "one," and a low voltage level corresponds to a digital "zero." A separate 8-bit number such as this would be transmitted for each sample, so the total bit rate for digitizing the analog signal is (8 bits/sample) × (sampling rate). According to the *Nyquist criterion,* the sampling rate must be at least twice the signal bandwidth in order to digitize the signal without any loss of information. Taking 4 kHz as a sufficient bandwidth for a telephone connection, the required bit rate for one phone conversation becomes

$$BR_{\text{phone}} = \left(8\ \frac{\text{bits}}{\text{sample}}\right)\left(2 \times 4 \times 10^3\ \frac{\text{samples}}{\text{s}}\right) = 64\ \text{kb/s} \tag{24-1}$$

where the unit kb/s stands for kilo (10^3) bits per second.

A key advantage in digitizing the data is that it becomes more immune from degradation due to noise. When noise is added to a digital signal, the receiver compares the average signal level during a bit period to a preset *decision level,* as depicted in Fig. 24-4. If the average signal is higher than the decision level, the bit is interpreted as a digital "one," and if it is lower, the bit is interpreted as a digital "zero." Even in the presence of significant noise, it is still possible to reconstruct the original waveform almost perfectly. This high fidelity in signal transmission comes at a cost, however. Processing the 64 kb/s digital signal electronically requires a receiver bandwidth of ~ 64 kHz, which is much greater than the original 4 kHz bandwidth of the analog signal. Fortunately, this increased bandwidth require-

*This is the so-called NRZ (non-return-to-zero) format, in which there is no "dead space" between bits. In the alternative RZ format, the voltage always returns to zero for a full period T between the data bits.

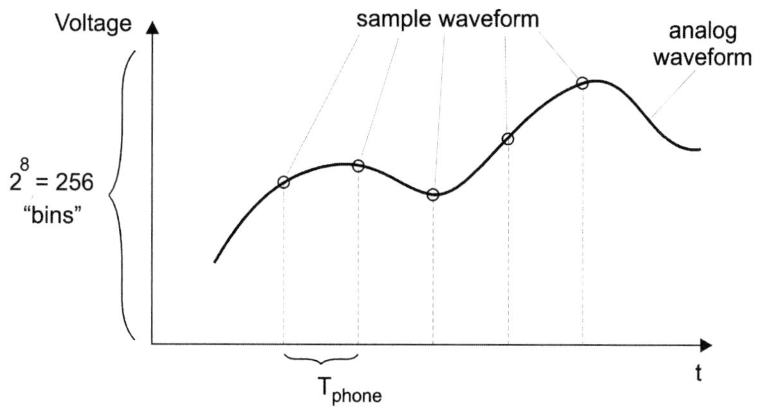

Figure 24-2 An analog waveform can be sampled periodically to generate a digital representation.

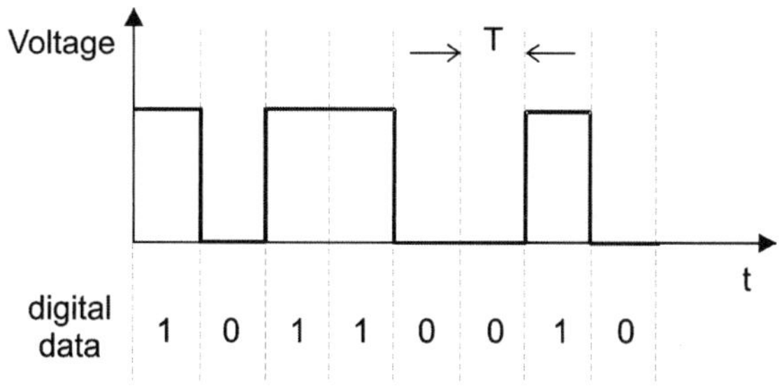

Figure 24-3 Digital waveform representing the decimal number 178, which is 10110010 in binary. An 8-bit number such as this would be sent for each of the sampled voltages in the waveform of Fig. 24-2.

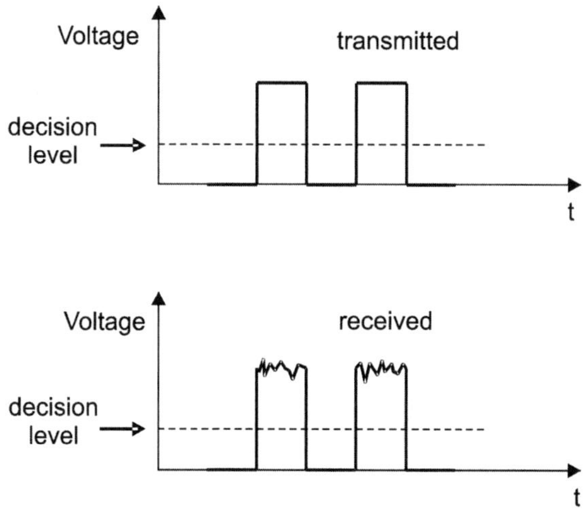

Figure 24-4 A digital signal is interpreted as a "one" or "zero" depending on whether the average signal level during a bit period is higher or lower than the decision level. Moderate amounts of noise do not result in any significant loss of data.

ment is not really a problem. As we will see in the following section, there is enough available bandwidth in a fiber for thousands of simultaneous phone conversations.

Time-Division Multiplexing

The basic idea of time-division multiplexing is illustrated in Fig. 24-5 for three representative data sources labeled A, B, and C. In this discussion, we will consider the data to be digitized phone conversations, but the same concept applies to computer data and other forms of digital data as well. The multiplexer combines the three data streams by interleaving them in time, as shown. If the time between bits in the original phone data is designated T_{phone}, then the time between bits in the combined data stream is $T = T_{phone}/3$. Generalizing this to N phone conversations, the relation becomes $T = T_{phone}/N$. Since the bit rate is the reciprical of the time between bits, the number of phone channels that can be multiplexed into a single bit stream is

$$\#\text{ phone channels} = \frac{T_{phone}}{T} = \frac{BR}{BR_{phone}} = \frac{BR}{64\text{ kb/s}} \tag{24-2}$$

The channel capacity for an optical fiber is, therefore, proportional to the bit rate in the multiplexed data stream.

To promote the compatability of data exchanged between different users, a number of telecommunications standards have evolved that specify data format and standard bit rates. Table 24-2 lists the most commonly used standard rates in North America, and Table 24-3 lists the corresponding European/international rates. The North American rates designated T1, T2, and T3 were developed for the long-distance transmission of telephone traffic through so-called *trunk lines,* which connect different telephone switching centers. Well before the advent of fiber optics, long-distance telephone data was being multiplexed in a hierarchic fashion, with four T1 lines multiplexed into one T2 line, and seven T2 lines multiplexed into one T3 line. In fact, the first use of fiber optics in a public phone system was as a replacement for a trunk line operating at the T3 rate of 45 Mb/s. Today, these standard rates are still in common use, especially the T1 and T3 rates, and they are used not only for digital telephone, but also for other types of data. For example, a "high-speed" connection to the internet for a business or university is often one or more T1 connections.

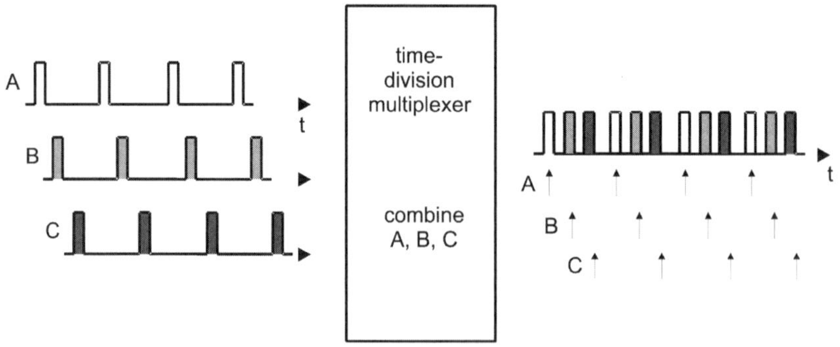

Figure 24-5 A time-division multiplexer interleaves a number of original data streams into a single data stream with a time T between bits.

Table 24-2 Data rate standards for North America

Rate name	Type	Data rate	# voice channels
T1	digital telephone	1.544 Mb/s	24
T2	digital telephone	6.312 Mb/s	96
T3	digital telephone	44.736 Mb/s	672
OC-1	SONET	51.84 Mb/s	672 (1 T3)
OC-3	SONET	155.52 Mb/s	2016
OC-12	SONET	622.08 Mb/s	8064
OC-48	SONET	2.5 Gb/s	32,256
OC-192	SONET	10 Gb/s	129,024
OC-768	SONET	40 Gb/s	516,096

The higher-speed OC-1 to OC-768 rates have been developed more recently, in connection with the *Synchronous Optical Network* (SONET) standard. The European/international counterpart to this is the *Synchronous Digital Hierarchy* (SDH) standard, with rates designated STM-1 to STM-256. The purpose of these new standards is to facilitate the transmission of mixed data types over fiber. They are designed to be compatible with the older data standards, so that, for example, the OC-1 rate can contain data from a T3 line. Although the SONET and SDH standards are somewhat different, especially in their compatibility with the older digital telephone standards, they are very similar in the higher-speed realm, where compatibility across international borders is especially important.

Today, the OC-192 rate is in common use for long-haul applications, and the OC-768 rate is starting to be used. There is a limit to this upward march in data rates, however. For one thing, it becomes increasingly difficult to design light modulators, detectors, and associated electronics that work at such high speed. More fundamentally, however, the maximum transmission rate is limited by dispersion in the fiber, as we saw in Chapter 6. To pack even more information into a fiber requires a different kind of multiplexing, which we consider in the following section.

Wavelength-Division Multiplexing (WDM)

The transmission capacity of a single optical fiber can be greatly increased by sending separate optical signals at different wavelengths down the fiber simultaneously. This is

Table 24-3 Data rate standards for Europe and international

Rate name	Type	Data rate	# voice channels
Level 1	digital telephone	2.048 Mb/s	30
Level 2	digital telephone	8.448 Mb/s	120
Level 3	digital telephone	34.3 Mb/s	480
Level 4	digital telephone	139 Mb/s	1920
STM-1	SDH	155.52 Mb/s	(1 Level 4)
STM-4	SDH	622.08 Mb/s	8064
STM-16	SDH	2.5 Gb/s	32,256
STM-64	SDH	10 Gb/s	129,024
STM-256	SDH	40 Gb/s	516,096

known as *wavelength-division multiplexing* (WDM), and is depicted schematically in Fig. 24-6. A set of laser diodes with wavelengths $\lambda_1, \lambda_2, \lambda_3, \ldots, \lambda_N$ are modulated individually with separate data signals, and the wavelengths are then combined in a multiplexer and sent through the fiber. After leaving the far end of the fiber link, the individual wavelengths are separated in a demultiplexer, and the data stream carried by each wavelength is detected in a separate receiver. These different wavelengths constitute a set of channels for optical data transmission, in much the same way that different frequencies in the electromagnetic spectrum are assigned to the various TV channels and radio stations. The main difference is that optical frequencies are in the 10^{14} Hz range, whereas radio and TV channels are in the 10^6–10^9 Hz range.

The optical channels in WDM can be represented in terms of a frequency spectrum, as shown in Fig. 24-7. This shows the optical power in the fiber as a function of optical frequency $\nu = c/\lambda$. It is customary in WDM to space the channels evenly in frequency, rather than wavelength. The ITU (International Telecommunications Union) has set up a series of standard frequencies for WDM channels, spaced by 100 GHz or 50 GHz. For example, two adjacent frequencies with 100 GHz spacing would be at precisely 195.90 THz and 195.80 THz. Using the value $c = 2.99792458 \times 10^8$ m/s, this corresponds to wavelengths 1530.33 and 1531.12 nm, respectively, which is a wavelength spacing between channels of $\approx$ 0.8 nm. Note, however, that this wavelength spacing is not a constant.

To maximize the number of channels, and hence the data carrying capacity of the fiber, the channels should be spaced as closely as possible. This is termed *dense wavelength division multiplexing* (DWDM), which is roughly defined by a channel spacing $\delta\nu < 200$ GHz. The limit on how closely the channels can be spaced is determined by the spectral width $\Delta\nu_{ch}$ of each channel. When $\Delta\nu_{ch} \geq \delta\nu$, the light from one channel can cross over and be detected on another channel, resulting in "cross-talk." Therefore, a small value of $\Delta\nu_{ch}$ is desirable for DWDM. There are many sources of linewidth for an optical source, but at a minimum, the width is determined by the Fourier transform of the time-dependent waveform. Since the waveform varies on a time scale of T for bit rate $BR = 1/T$, the Fourier transform-limited linewidth is

$$\Delta\nu_{ch} \sim 1/T = BR \qquad \text{(transform-limited channel width)} \qquad (24\text{-}3)$$

If the channel spacing is reduced to the minimum value $\delta\nu \sim \Delta\nu_{ch}$, then the number of channels in an available optical bandwidth $\Delta\nu_{\text{band}}$ is

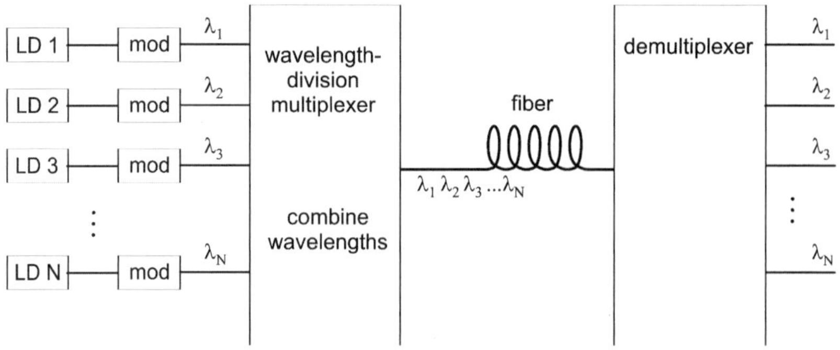

Figure 24-6 In wavelength-division multiplexing, signals at several different wavelengths are combined and sent through the same fiber.

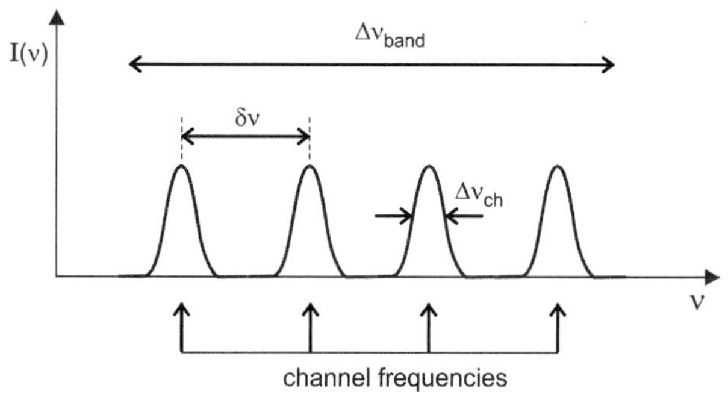

Figure 24-7 Optical power per unit frequency in wavelength-division multiplexing. Channels of width $\Delta\nu_{ch}$ are spaced evenly in frequency with a separation $\delta\nu$.

$$N_{ch} = \frac{\Delta\nu_{\text{band}}}{\Delta\nu_{ch}} \sim \frac{\Delta\nu_{\text{band}}}{BR} \qquad \text{(number of channels in bandwidth)} \qquad (24\text{-}4)$$

Since each of these N_{ch} channels carries BR bits/s, the combined bit rate in all channels is

$$BR_{\text{tot}} = N_{ch}\,BR \sim \Delta\nu_{\text{band}} \qquad \text{(combined bit rate, all channels)} \qquad (24\text{-}5)$$

This last equation gives the remarkably simply and important result that the combined maximum bit rate in all channels is on the order of the available bandwidth. Note that BR_{tot} does not depend on BR, the data rate in one channel. The reason for this is that as BR goes up, the maximum number of channels N_{ch} goes down, so that their product remains a constant. The result is an engineering trade-off, in which the choice of bit rate and channel spacing depends on the practical difficulties of high-speed electronics on the one hand, and the demultiplexing of closely spaced wavelengths on the other hand.

Spectral Efficiency

The exact numerical proportionality factor to be used in Eq. (24-5) depends on the type of modulation and the coding scheme, among other things. It can be written as a *spectral efficiency,* defined as

$$\eta_{sp} \equiv \frac{\text{total bit rate}}{\text{total bandwidth}} \qquad \text{(spectral efficiency)} \qquad (24\text{-}6)$$

In practice, it has been possible to achieve spectral efficiencies of about $\eta_{sp} \approx 0.4$ (bits/s)/Hz for well-designed systems. The following example illustrates the bit rate capacity made possible by this optimized multiplexing.

EXAMPLE 24-1

An optical fiber communications system uses WDM to obtain an optimized spectral efficiency of 0.4 (bits/s)/Hz. Determine the combined maximum bit rate capacity in the wavelength interval 1530–1560 nm (this corresponds to one of the bands of the Erbium-doped fiber amplifier).

Solution: Taking $\lambda_1 = 1530$ nm, and $\lambda_2 = 1560$ nm, the two corresponding optical frequencies are

$$\nu_1 = \frac{2.99792458 \times 10^8}{1530 \times 10^{-9}} = 1.9594 \times 10^{14}\ \text{Hz} = 195.94\ \text{THz}$$

and

$$\nu_2 = \frac{2.99792458 \times 10^8}{1560 \times 10^{-9}} = 1.9217 \times 10^{14}\ \text{Hz} = 192.17\ \text{THz}$$

The frequency bandwidth is then $\Delta\nu_{\text{band}} = \nu_1 - \nu_2 = 3.77$ THz, which gives a maximum combined bit rate of

$$BR_{\text{tot}} = \eta_{sp}\, \Delta\nu_{\text{band}} = \left(0.4 \frac{\text{b/s}}{\text{Hz}}\right)(3.77\ \text{THz}) = 1.5\ \text{Tb/s}$$

Bit rate capacities in the Tb/s (10^{12} bits/s) range are truly enormous. To put it in perspective, we can divide this by the required bit rate for one phone conversation, 64 kb/s, to obtain an estimated capacity of $\approx 2.3 \times 10^7$ phone channels. That's a lot of talking on one fiber! By way of comparison, the maximum bandwidth in a microwave or wireless link is limited to a few GHz, some three orders of magnitude less than an optimized fiber link. Although wireless has its conveniences, it can never approach the raw data capacity of fiber.

WDM Multiplexers

There are a number of methods for combining and separating wavelengths in a WDM system. One approach using a fiber Bragg grating is shown in Fig. 24-8a. For separating out more than one wavelength, several fiber gratings can be cascaded, and this is a practical method when the number of wavelengths is not too large. Another approach uses a bulk-optic diffraction grating to separate the wavelengths, as shown in Fig. 24-8b. Light signals at different wavelengths are diffracted at different angles from the grating, and are then focused by a graded index lens into one of a series of fibers attached to the lens. The spectral resolution for this type of device is not generally good enough for DWDM, but it can be used in *coarse wavelength division multiplexing* (CWDM), which is characterized by a channel separation $\delta\nu \geq 1$ THz.

A third approach to wavelength separation in WDM is the arrayed-waveguide grating (AWG), depicted in Fig. 24-8c. This is a planar waveguide device that separates the wavelengths by multiple-path optical interference. Light containing several wavelengths enters the input slab waveguide coupler, where it spreads out and is uniformly distributed among a number of individual waveguides in a waveguide array. The waveguides in this array are designed so that each one is longer or shorter than its neighbor by a fixed amount ΔL. There is then a fixed optical path length difference $n\Delta L$ between adjacent waveguides, where n is the index of refraction. When the light reaches the output slab waveguide coupler, it again spreads out, and is coupled into another series of waveguides.

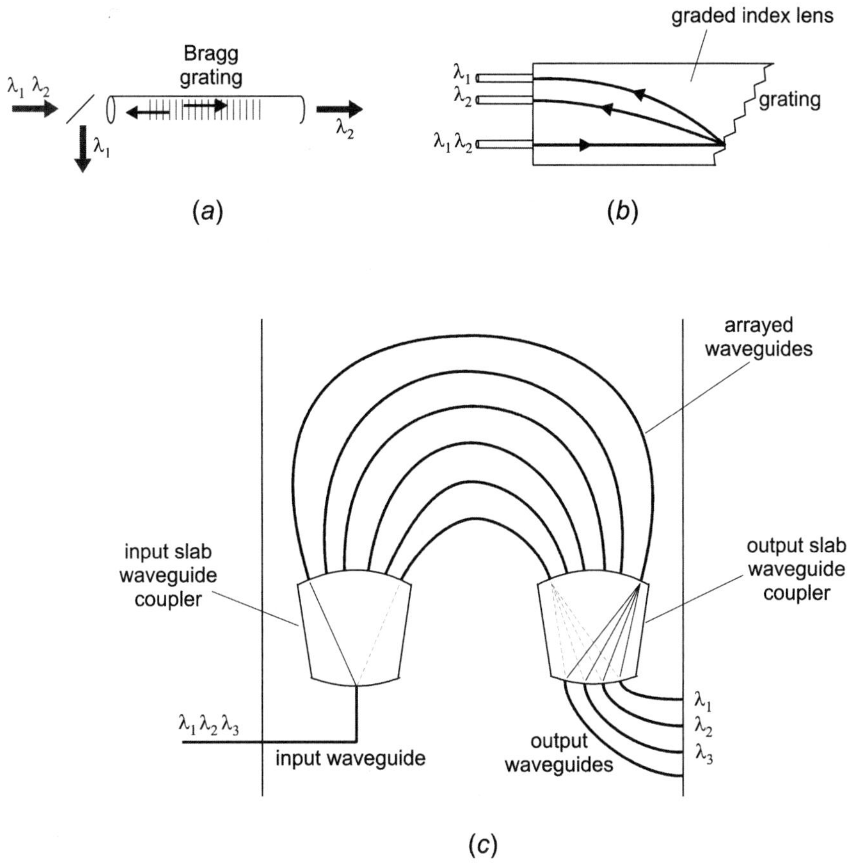

Figure 24-8 Wavelengths can be combined and separated in WDM systems using (a) fiber Bragg gratings, (b) bulk-optic gratings, or (c) arrayed-waveguide gratings.

The amount of light that is coupled into any one of these output waveguides depends on the linear superposition of the lightwave fields from the various arrayed waveguides. Because of the path length difference ΔL between waveguides, the lightwave field coming from each waveguide in the array is shifted in phase by $(2\pi/\lambda)\ (n\Delta L)$, which depends on wavelength. Constructive interference will then occur in different directions (at different output waveguides), depending on the wavelength. The operating principle here is much the same as in a diffraction grating (Fig. 2-16), in which phase shifts due to path length differences cause the direction of the diffracted beam to depend on wavelength. This similarity to a diffraction grating gives the arrayed waveguide "grating" its name, even though there is no physical grating in the device. The AWG can be designed to separate channels as closely spaced as 50 GHz, and can therefore be used for DWDM.

Nonlinear Channel Mixing in WDM

In DWDM it is important to minimize coupling (cross talk) between adjacent channels. One source of cross talk that we have already discussed is the finite spectral width of each channel in comparison to the channel separation. Another source of cross talk arises from nonlinear interactions between light at different frequencies, as depicted in Fig. 24-9. Light in three adjacent frequency channels ν_1, ν_2, and ν_3 can couple together in a four-wave mixing process (see Section 9-2) to create light at a fourth frequency ν_4. Energy conservation requires that ν_4 be at some combination of sums and differences of the input frequencies. In the example shown, $\nu_4 = \nu_1 - (\nu_3 - \nu_2)$. Since $\nu_3 - \nu_2$ is the channel spacing $\delta\nu$, the new frequency is $\nu_4 = \nu_1 - \delta\nu$, exactly on resonance with the frequency channel one step lower than ν_1. The even spacing of the frequency channels makes this coupling process quite efficient for high powers and long path lengths.

One way to minimize this problem is to make the different frequencies travel at slightly different velocities down the fiber. The phase matching that is necessary for frequency conversion then occurs over a limited path length, and this reduces the overall efficiency of the process. But we have seen in Chapter 6 that different frequencies do in fact travel at different velocities, due to dispersion. From this point of view, dispersion is not something to be eliminated entirely, but rather to be managed for optimum performance. The dispersion should be great enough to reduce the efficiency of four-wave mixing between channels, but small enough to allow a high bit rate. This has led to the development of dispersion-flattened fiber, as discussed in connection with Fig. 6-4.

24-3. POWER BUDGET IN FIBER OPTIC LINK

In the previous section we examined several techniques that can be used to maximize the rate of data transmission in an optical fiber. These high data rate capacities are certainly an important benefit of fiber optic communications. Equally important, however, is the ability to transmit this data over great distances. The transmission distance can be limited by either dispersion or by optical losses, which reduce the optical signal below the detectable level. A convenient way to analyze the effect of losses is to set up an optical *power budget,* which systematically accounts for the different losses and gains in the fiber link. Losses and gains are fundamentally multiplicative (signal power is decreased or increased by a certain factor), but when expressed in decibel units, they are additive. The overall power budget can therefore be written as

$$\text{injected power} - \text{losses} + \text{gain} = \text{received power}$$

with all quantities expressed in dB. It is always a good idea to design a fiber link in such a way that the expected (nominal) level of received power is somewhat greater than the minimum detectable level. This difference, expressed in dB, is known as the *system mar-*

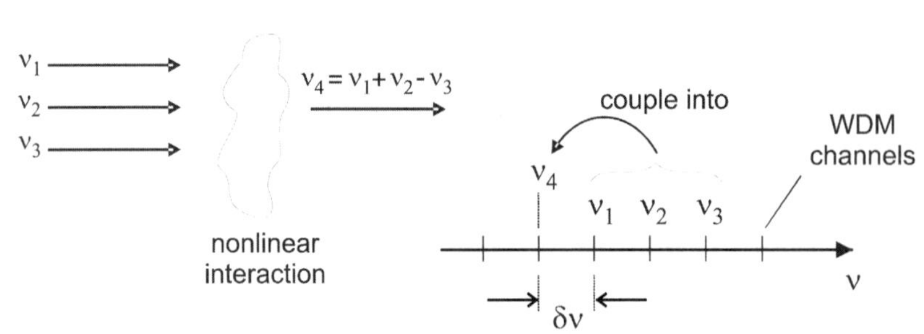

Figure 24-9 At high signal powers, four-wave mixing in the fiber can cause nonlinear coupling between WDM channels.

gin. Denoting the signal power in dBm as $\mathcal{P}$, using the definition in Eq. (1-7), the power budget can be expressed symbolically as

$$\mathcal{P}_T - \alpha L - K + G_{dB} = M + \mathcal{P}_R \tag{24-7}$$

where $\mathcal{P}_T$ is the signal power (in dBm) injected into the fiber by the transmitter, α is the fiber loss coefficient in dB/km, L is the fiber length in km, K is the dB loss due to fiber connections and splicing, G_{dB} is the dB gain (if any) due to amplifiers inserted into the link, M is the system margin, and $\mathcal{P}_R$ is the minimum receiver power (in dBm) that can detected reliably. The power level $\mathcal{P}_R$ is known as the *receiver sensitivity.*

Typical values for these parameters are as follows. Injected powers are ~ 1 mW (0 dBm) for laser diodes, but only ~50 μW (–13 dBm) for LED's coupled into multimode fiber. The lower value for an LED comes from the source–fiber coupling efficiency, which was found in Eq. (12-11) to be NA^2. LEDs are seldom used with single-mode fiber, because the area mismatch makes the coupling efficiency much lower still. Fiber connection losses are usually ~0.1 dB per splice, and 0.2–1 dB per connector. Fiber attenuation depends on the wavelength, typically ~0.25 dB/km for 1550 nm, and 2.5 dB/km for 850 nm. System margins in the range 3–10 dB are generally desirable.

It is not as easy to specify the receiver sensitivity $\mathcal{P}_R$ as a single number. It depends not only on the type of photodetector used, but also on the bit rate. To understand the reason for this, we turn next to a detailed examination of receiver sensitivity.

Receiver Sensitivity

The signal to be detected can be represented by a plot of received power versus time, as depicted in Fig. 24-10. The power is P_1 when the pulse is on (digital "one"), and P_0 when the pulse is off (digital "zero"). Usually $P_0 = 0$, so the average power P_R is one–half the peak power P_1. Fundamentally, the job of the receiver is to determine whether each bit in the digital data stream is a "one" or a "zero." To be considered a digital "one," a bit must have at least one photon of energy, since this is the minimum amount that will give rise to photoexcitation of an electron–hole pair. However, more photons per bit than this are required for reliable detection, because photons do not come evenly distributed in time. Photon generation and detection is a statistical process, with the number n_1 of photons detected in one bit varying according to the Poisson probability distribution:

$$P(n_1) = \frac{(\overline{n_1})^{n_1}\, e^{-\overline{n_1}}}{n_1!} \qquad \text{(Poisson distribution)} \tag{24-8}$$

We previously encountered this distribution function in connection with electrical current shot noise (Section 13-3). Here, $\overline{n_1}$ is the average number of photons in a digital "one" bit, taken over many such bits. If there were only a single photon on average per "one" bit ($\overline{n_1} = 1$), the probability that no photons would be present in any particular "one" bit would be

$$P(0) = \frac{1^0}{0!}\, e^{-1} = e^{-1} = 0.368$$

This means that every third "one" would be read as a "zero," an unacceptably large error rate. For the general case of $\overline{n_1}$ photons on average per "one" bit, the probability of an error (i.e., getting $n_1 = 0$) is

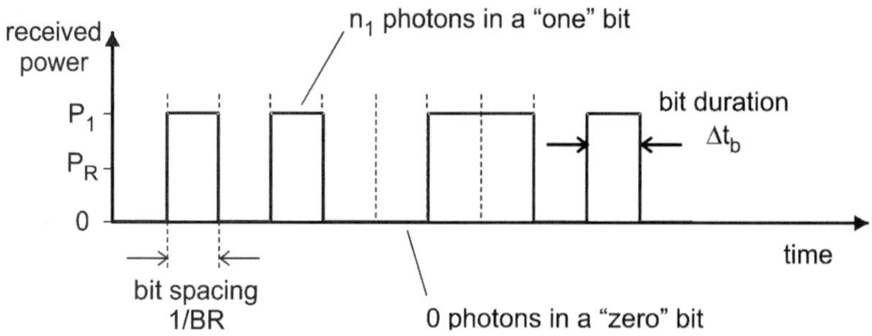

Figure 24-10 Time dependence of optical power incident on receiver in an optical communications system. Illustrated here is the NRZ scheme, in which the pulse duration of each bit (Δt_b) is equal to the time spacing between bits, $T = 1/BR$. In the RZ scheme, $\Delta t_b = T/2$.

$$P(0) = \frac{(\overline{n_1})^0 e^{-\overline{n_1}}}{0!} = e^{-\overline{n_1}} \tag{24-9}$$

Clearly, there are fewer errors when there are more photons per bit.

The reliability of digital data is often expressed in terms of the *bit error rate,* or BER. This is defined as the probability that a digital bit will be read incorrectly—either a "one" will be read as a "zero," or vice versa. However, a "zero" cannot be mistakenly read as a "one," because the light beam is switched off during a digital "zero," and there are no fluctuations about an average of zero. Therefore, the overall BER is P(0)/2, assuming an equal number of digital one's and zero's overall. There is no hard and fast rule about what error rate is acceptable, but for purposes of comparison a maximum BER = 10^{-9} is often assumed. Using this criterion, the condition on $\overline{n_1}$ becomes

$$10^{-9} = \tfrac{1}{2} e^{-\overline{n_1}}$$

or

$$(\overline{n_1})_{\text{min}} = -\ln(2 \times 10^{-9}) = 20 \tag{24-10}$$

This is the *quantum limit* for the number of photons in a logical "one" bit, and sets a lower limit on any receiver's sensitivity. It is customary to express this limit as $\overline{n}$, the average over all bits (in contrast to $\overline{n_1}$, which is an average over just the "one" bits). The "zero" bits contain no photons, and so if half the bits are "zero" on average, the quantum limit becomes

$$\overline{n}_{\text{min}} = \tfrac{1}{2}(\overline{n_1})_{\text{min}} = 10 \qquad \text{(quantum limit)} \tag{24-11}$$

We therefore conclude that, on average, there must be at least 10 photons per bit in a digital signal for an acceptably low error rate.

For a power-budget analysis, the receiver sensitivity must be specified as a power level in dBm, rather than as a number of photons per bit. To relate the two, we write

$$\text{power} = \frac{\text{energy}}{\text{time}} = \left(\frac{\text{energy}}{\text{photon}}\right)\left(\frac{\text{photons}}{\text{bit}}\right)\left(\frac{\text{bits}}{\text{time}}\right)$$

which translates symbolically into

$$P_R = h\nu\,\bar{n}_{\min} BR \qquad \text{(receiver sensitivity)} \qquad (24\text{-}12)$$

This equation gives the average power required at the receiver in terms of the minimum average number of photons per bit. It can also be written in terms of the peak power, as $P_1 = h\nu(\overline{n_1})_{\min} BR$. The interesting (and important) feature of this result is that the required receiver power increases in proportion to the bit rate. This has implications for a power-budget analysis, because it means that the receiver sensitivity is not a fixed number; rather, it must be specified at a particular bit rate. We will pursue this point further, but first we need to consider how these results apply to real detectors.

Real photodetectors never do quite as well as the quantum limit, because they are subject to thermal noise and electronic shot noise. Well-designed APD receivers can respond at the quantum level, but there is added noise due to the statistics of avalanche multiplication. They typically require ~ 500 photons per bit, rather than the quantum limit of 10. PIN photodiode receivers are invariably thermal-noise limited, and in this case detectability is better understood in terms of a signal-to-noise ratio, as discussed in Section 14-5. Under small signal conditions, SNR $\propto P^2/B$, where P is the incident optical power and B is the detector electronic bandwidth. The optical power required to maintain a constant minimum SNR is then expected to be $P \propto \sqrt{B}$. The electronic bandwidth required to detect a waveform with bit rate BR is $B \sim BR$ and, therefore, one would predict that the receiver sensitivity for thermal-noise-limited amplifiers would vary as $\sqrt{BR}$, rather than linearly with BR. This scaling assumes that all other amplifier parameters remain the same, whereas, in practice, amplifier circuits are optimized in different ways for different frequency regions. For practical purposes, therefore, the relation $P_R \propto BR$ can be taken to hold approximately for PIN receivers, as well as for APD and quantum-limited receivers. A typical $\bar{n}_{\min}$ value for PIN receivers is ~ 5000 photons/bit. This is a factor of 500 (27 dB) higher than the quantum limit.

EXAMPLE 24-2

A digital receiver operates at a wavelength of 1.3 μm and a bit rate of 400 Mb/s. Determine the receiver sensitivity in dBm for (a) the quantum limit, and (b) a PIN photodiode circuit requiring 5000 photons per bit.

Solution: (a) The photon energy is

$$h\nu = \frac{hc}{\lambda} = \frac{(6.63 \times 10^{-34})(3 \times 10^{8})}{1.3 \times 10^{-6}} = 1.53 \times 10^{-19}\ \text{J}$$

For the quantum limit, the required receiver power is

$$P_R = (10)(1.53 \times 10^{-19}\ \text{J})(4 \times 10^{8}\ \text{s}^{-1}) = 6.12 \times 10^{-10}\ \text{W}$$

which corresponds to

$$\mathcal{P}_R = 10 \log_{10}\left(\frac{6.12 \times 10^{-10}}{1 \times 10^{-3}}\right) = -62.1\ \text{dBm}$$

(b) For the PIN photodiode receiver,

$$P_R = (5000)(1.53 \times 10^{-19}\text{ J})(4 \times 10^{8}\text{ s}^{-1}) = 3.06 \times 10^{-7}\text{ W}$$

which corresponds to

$$\mathcal{P}_R = 10\log_{10}\left(\frac{3.06 \times 10^{-7}}{1 \times 10^{-3}}\right) = -35.1\text{ dBm}$$

We now use the receiver sensitivity given in Eq. (24-12) to analyze the power budget. Taking the log of this equation to obtain the receiver sensitivity in dBm, we obtain a bit rate dependence of the form

$$\mathcal{P}_R(BR) = \mathcal{P}_R(BR_0) + 10\log_{10}\left(\frac{BR}{BR_0}\right) \tag{24-13}$$

where BR_0 is a reference bit rate. The receiver sensitivity measured in dBm is seen to vary logarithmically with bit rate. For example, if the bit rate increases from $BR_0 = 3$ Mb/s to $BR = 300$ Mb/s, this equation predicts an increase of $10\log_{10}(300/3) = 20$ dB in the required receiver power. This corresponds to a factor of 100 increase, which is consistent with the scaling prediction of Eq. (24-12). The effect of bit rate on power budget calculations is illustrated in Fig. 24-11, which shows typical variations of $\mathcal{P}_R$ with bit rate on a semilog scale. In this graph, the separation between transmitter power $\mathcal{P}_T$ and receiver sensitivity $\mathcal{P}_R$ gives the maximum total loss from all sources, including system margin. As the bit rate increases, less and less total loss is allowable. One of the sources of loss is

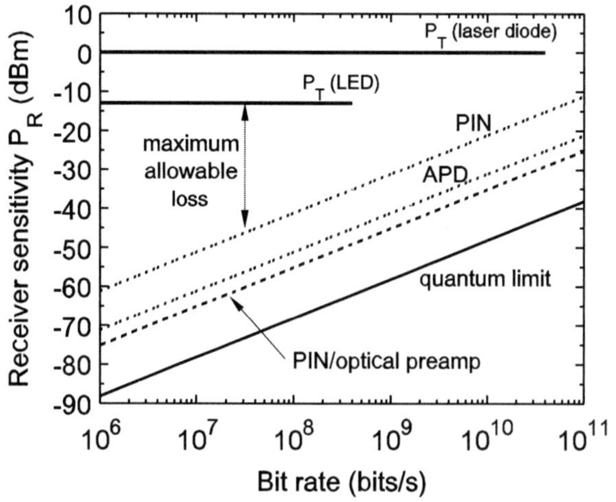

Figure 24-11 Receiver sensitivities in dBm versus bit rate, calculated from Eq. (24-12) for a wavelength of 1.3 μm. Values of $\bar{n}_{min}$ used for the plot are 10 for the quantum limit, 200 for a PIN photodiode with an optical preamplifier, 500 for an APD, and 5000 for a PIN photodiode without an optical preamplifier. Also shown are typical injected powers for a laser diode (1 mW) and an LED (50 μW).

fiber attenuation, and the limits on total loss lead to limits on fiber length. This is considered in more detail in the following section.

Maximum Fiber Length

The maximum allowable fiber length in a optical communications system is found by solving Eq. (24-7) for L:

$$L_{\max} = \frac{1}{\alpha}[\mathcal{P}_T - K + G_{dB} - M - \mathcal{P}_R] \tag{24-14}$$

with α in dB/km, losses in dB, and powers in dBm. The use of this equation is best made clear by considering some examples.

EXAMPLE 24-3

An 850 nm LED couples 0.1 mW into a multimode fiber that has an attenuation coefficient of 2.5 dB/km. The LED is modulated at 100 Mb/s, and a PIN photodiode receiver is used that requires 5000 photons per bit. Losses due to splices and connectors are 3 dB, and the desired system margin is 3 dB. Determine the maximum fiber length.

Solution: The photon energy is

$$h\nu = \frac{hc}{\lambda} = \frac{(6.63 \times 10^{-34})(3 \times 10^{8})}{850 \times 10^{-9}} = 2.34 \times 10^{-19}\ \text{J}$$

and the receiver sensitivity is

$$P_R = (5000)(2.34 \times 10^{-19})(10^{8}) = 1.17 \times 10^{-7}\ \text{W}$$

which corresponds to $\mathcal{P}_R = -39.3$ dBm.

The maximum length is, therefore,

$$L_{\max} = \frac{-10 - 3 - 3 - (-39.3)\ \text{dB}}{2.5\ \text{dB/km}} = 9.3\ \text{km}$$

The above system utilizes relatively inexpensive components in the original 850 nm telecommunications window. It may be suitable for LAN networks or short-distance metro applications. In contrast, consider the following example using higher-end components.

EXAMPLE 24-4

A 1550 nm laser diode couples 1 mW into a single-mode fiber that has an attenuation coefficient of 0.25 dB/km. The laser is modulated at 100 Mb/s, and an APD receiver is used that requires 500 photons per bit. There are 20 splices, each with a loss of 0.1 dB, and two connectors, each with a loss of 0.8 dB. The desired system margin is 6 dB. Determine the maximum fiber length.

Solution: The photon energy is

$$h\nu = \frac{hc}{\lambda} = \frac{(6.63 \times 10^{-34})(3 \times 10^8)}{1550 \times 10^{-9}} = 1.28 \times 10^{-19}\ \text{J}$$

and the receiver sensitivity is

$$P_R = (500)(1.28 \times 10^{-19})(10^8) = 6.4 \times 10^{-9}\ \text{W}$$

which corresponds to $\mathcal{P}_R = -51.9$ dBm. The combined loss from splices and connectors is

$$K = (20 \times 0.1) + (2 \times 0.8) = 3.6\ \text{dB}$$

The maximum length is, therefore,

$$L_{\text{max}} = \frac{0 - 3.6 - 6 - (-51.9)\ \text{dB}}{0.25\ \text{dB/km}} = 169\ \text{km}$$

The more expensive components used in the above example might be employed in a long-haul system, where separating repeater stations by the greatest distance is a high priority. This minimizes overall system expense and simplifies maintenance.

Dependence on Bit Rate

We have seen in these examples that the maximum fiber length depends, among other things, on receiver sensitivity. But the receiver sensitivity depends on the bit rate, and, therefore, the maximum fiber length depends on bit rate. This dependence can be made explicit by combining Eqs. (24-7) and (24-13), giving

$$L_{\text{max}} = \frac{1}{\alpha}\left[\mathcal{P}_T - K + G_{dB} - M - \mathcal{P}_R(BR_0) - 10\log_{10}\left(\frac{BR}{BR_0}\right)\right] \tag{24-15}$$

According to this result, the maximum fiber length decreases logarithmically with increasing bit rate. To illustrate this dependence graphically, we plot L_{max} versus BR in Fig. 24-12, calculated using Eq. (24-15) with the system parameters given in Example 24-4. It can be seen that up to ~ 1 Gb/s, there is only a modest (less than factor of two) reduction in maximum length, and only at the higher rates ($BR > 10$ Gb/s) does the fiber length becomes significantly limited by attenuation.

There is another limit to the fiber length, however, due to dispersion. We saw in Chapter 3 that when dispersion spreads a pulse in time by Δt, the bit rate is limited to $BR_{\text{max}} = 1/(2\Delta t)$ [Eq. (3-37)]. The time spread in a multimode fiber is $\Delta t \approx Ln\Delta/c$ [Eq. (3-35)], whereas in single-mode fiber it is $\Delta t = LD_c\Delta\lambda$ [Eq. (6-11)]. In both cases, $\Delta t \propto L$, making

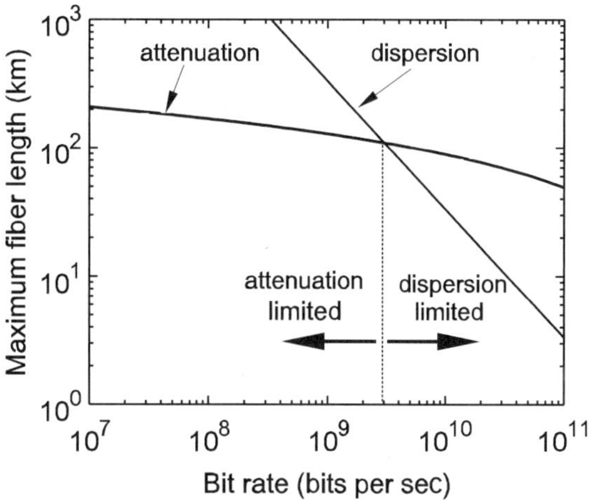

Figure 24-12 Limit on fiber length in a communications system due to fiber attenuation and dispersion. Attenuation parameters are taken from Example 24-4, and dispersion parameters are taken from Example 24-5b. The operating point must lie below both the attenuation and dispersion curves.

$BR_{max} \propto L^{-1}$. There is thus a maximum product of bit rate and fiber length for each case, given by

$$(BR \times L)_{max} = \frac{c}{2n\Delta} \qquad \text{(multimode)} \tag{24-16}$$

and

$$(BR \times L)_{max} = \frac{1}{2D_c\Delta\lambda} \qquad \text{(single-mode)} \tag{24-17}$$

When limited by dispersion, bit rate can be traded off against fiber length. One system might have a higher bit rate over a shorter length, while another might have a slower bit rate over a longer length. However, subject to the restrictions given above, a system cannot have both a high bit rate and long length.

EXAMPLE 24-5

(a) Calculate the product of maximum bit rate and length for a system using multimode fiber with fractional index difference $\Delta = 0.01$. (b) Repeat part a for a system using single-mode fiber and a laser diode at 1550 nm with spectral linewidth 0.1 nm. (c) For the system of part b, determine the dispersion-limited fiber length when operating at bit rate 100 Mb/s.

Solution: (a) Using Eq. (24-16),

$$(BR \times L)_{max} = \frac{3 \times 10^8}{2(1.5)(0.01)} = 10^{10} \text{ m/s}$$

To put this in more useful units we write it as

$$10^{10}\ \frac{\text{m}}{\text{s}}\left[\frac{1\ \text{Mb/s}}{10^6\ \text{b/s}}\right]\left[\frac{1\ \text{km}}{10^3\ \text{m}}\right] = 10\ \text{km}\cdot\text{Mb/s}$$

(b) From Fig. 6-3, we estimate the chromatic dispersion at 1550 nm to be $D_c \approx 15$ ps/(nm · km). Eq. (24-17) then gives

$$(BR \times L)_{\text{max}} = \frac{1}{2\,(15\,\dfrac{\text{ps}}{\text{nm}\cdot\text{km}})(0.1\ \text{nm})} = 0.333\ \frac{\text{km}}{\text{ps}}$$

Converting units this becomes

$$0.333\ \frac{\text{km}}{\text{ps}} = 333\ \frac{\text{km}}{\text{ns}} = 333\ \text{km}\cdot\text{Gb/s}$$

This is over four orders of magnitude higher than the $BR \times L$ product found in part a.
(c) Writing the bit rate as 0.1 Gb/s, we have

$$L_{\text{max}} = \frac{333\ \text{km}\cdot\text{Gb/s}}{0.1\ \text{Gb/s}} = 3330\ \text{km}$$

Comparing this last result with that of of Example 24-4 (which assumed a similar fiber and bit rate), we see that the fiber length is actually limited by attenuation rather than dispersion. This conclusion only applies to the particular fiber and bit rate assumed, however. To see more generally what limits the fiber length, we show in Fig. 24-12 the maximum length versus bit rate for both attenuation and dispersion in the same fiber. The attenuation data is taken from Example 24-4, and the dispersion data is taken from Example 24-5b. It can be seen that there are two regions, separated by the point at which the curves cross. At bit rates below the crossover point, the fiber length is limited by attenuation, whereas at bit rates above this point, the length is limited by dispersion. The bit rate at which the curves cross depends on many parameters, including the fiber loss coefficient, the light source wavelength and spectral width, and the type of receiver used.

It should be noted that in practice, the bit rate is limited not only by fiber dispersion and attenuation, but also by the response time of the transmitter and receiver. Designating these by Δt_{xmtr} and Δt_{rcvr}, respectively, the total response time that limits the bit rate is

$$\Delta t = \sqrt{\Delta t^2_{\text{xmtr}} + \Delta t^2_{\text{fiber}} + \Delta t^2_{\text{rcvr}}} \qquad \text{(system response time)} \qquad (24\text{-}18)$$

where Δt_{fiber} is the total fiber dispersion, given in Eq. (6-15). The maximum bit rate is then $BR_{\text{max}} = 1/(2\ \Delta t)$, with Δt calculated from Eq. (24-18).

24-4. OPTICAL AMPLIFIERS

We have seen that fiber attenuation is often the factor that limits the length of an optical fiber link. For very long fiber spans, it is necessary to periodically amplify the light signal, either with an electronic repeater, or with an optical amplifier. Optical amplifiers have the advantage that the data format does not need to be known in order for amplification to occur. What comes out of the optical amplifier is simply a higher-power replica of what came in. In contrast, an electronic repeater must be designed to decode and retransmit digital data with a specified rate and format. The optical amplifier has the related advantage that it can simultaneously amplify signals at several different wavelengths, a feat not possible with traditional repeaters. This has made the optical amplifier an enabling technology for wavelength-division multiplexing in long-haul telecommunications.

The principle drawback of optical amplifiers is that they do not reconstitute a signal that has been degraded by dispersion. At certain points in a long-distance system, it is still occasionally necessary to reshape and retime the pulses using electronic amplifiers. The need for this can be minimized, however, by using various *dispersion-compensation* techniques. An example of this would be to pass light that has been dispersed in one fiber through a second fiber, which has a dispersion coefficient D_c of sign opposite that of the first. The pulse dispersion is then "undone" in the second fiber, and the original pulse shapes are restored.

There are two basic types of optical fiber amplifiers. The first type utilizes stimulated emission in a rare-earth-doped fiber for the amplification, and is essentially a fiber laser without the end mirrors. The second type utilizes stimulated Raman scattering, a nonlinear optical process that becomes efficient at very high optical intensities. The optical amplifier that has so far had the greatest impact on fiber optic communications is of the first type, the *erbium-doped fiber amplifier* or EDFA. In this section we first consider the EDFA in some detail, and then follow this with a look at other kinds of doped amplifiers, as well as Raman amplifiers.

Erbium-doped Fiber Amplifier (EDFA)

The elements and operation of an EDFA are shown schematically in Fig. 24-13. There are many similarities to a fiber laser (Fig. 23-9), with pump light injected into the doped fiber core through a fiber coupler. However, in this case there are no Bragg reflectors in the fiber, and no laser oscillation occurs. Instead, weak-signal light that is coupled in through the fiber coupler is amplified by stimulated emission, drawing energy from the pump, and exits as a higher-power version of the input signal.

The Er^{3+} ion has a number of energy levels (see Fig. 23-7), but only the lower three of these, shown in Fig. 24-14, are normally relevant for EDFA operation. Amplification occurs on the ${}^4I_{13/2} \rightarrow {}^4I_{15/2}$ transition, which occurs at ~ 1550 nm. This coincides nicely with the wavelength of minimum attenuation in silica fiber (the third telecommunications window), making the EDFA a good fit with the needs for amplification in long-haul telecommunications. The quantum efficiency of the transition is high as well, because nonradiative decay processes are weak over the large energy gap between the ${}^4I_{13/2}$ and ${}^4I_{15/2}$ levels. These two features of high efficiency and optimum wavelength range have made the EDFA an important component of fiber optic communications systems. The lower level in the Er^{3+} transition is the ground state, which makes this a three-level-type system. One way to excite the upper level is to pump on the short wavelength (high energy) side of the ${}^4I_{15/2} \rightarrow {}^4I_{13/2}$ absorption transition, around 1480 nm. This is very similar to the pumping of Yb^{3+} in a Yb-doped fiber laser, and much of the analysis and discussion relating to the Yb fiber laser in Section 23-1 applies to Er^{3+} as well. One difference is that Er^{3+} has several excited states, whereas Yb^{3+} has just one. This allows Er^{3+} to be pumped at other wavelengths, for example at $\approx$ 980 nm. Absorption of a 980 nm photon promotes

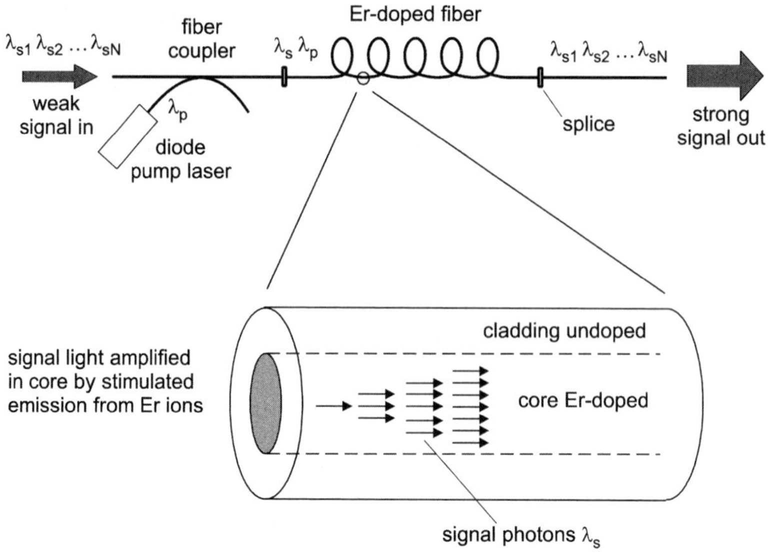

Figure 24-13 Schematic representation of erbium-doped fiber amplifier (EDFA).

Er^{3+} to the $^4I_{11/2}$, which decays rapidly to the $^4I_{13/2}$ by nonradiative relaxation. One advantage of 980 nm pumping over 1480 nm pumping is a higher possible population inversion, leading to higher gain (see Problem 24.12).

Gain Transparency

The gain coefficient for a transition from level 2 to level 1 is given in Eq. (18-37) as

$$\gamma(\lambda) = N_2\sigma_{em}(\lambda) - N_1\sigma_{abs}(\lambda) \qquad \text{(gain coefficient)}$$

written here in terms of wavelength rather than frequency. Since N_1 is the population of the ground state, it is much larger than N_2 under conditions of weak pumping, and the

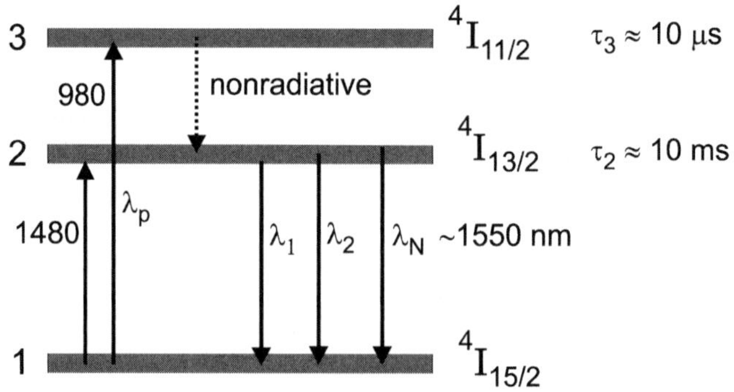

Figure 24-14 Lower three energy levels in Er^{3+}, showing pump transitions at 1480 and 980 nm, and signal transitions at ~1550 nm. Several wavelengths can be amplified simultaneously without interference.

gain is negative. A minimum pump intensity is, therefore, required to achieve the transparency condition $\gamma = 0$, just as for Yb^{3+}. In fact, the analysis in Eqs. (23-15)–(23-24) developed for Yb^{3+} applies equally well to Er^{3+}, provided that σ_{pe} is set to zero for 980 nm pumping.*

The dependence of gain coefficient on pump power is illustrated in Fig. 24-15, for a pump wavelength of 1480 nm and signal wavelength of 1550 nm. Eqs. (23-20) and (23-21) were used to calculate the level populations N_2 and N_1 for this plot, with cross-section data taken from Fig. 18-10. Fiber parameters assumed were core radius 2.5 μm, ion density 10^{19} cm^{-3}, and upper-state lifetime 10 ms. The transparency condition occurs in this fiber at a pump power slightly over 2 mW, a remarkably small value for a three-level-type transition. This ease of achieving transparency is a direct result of the small core size, and is one of the primary benefits of the fiber geometry for fiber lasers and amplifiers.

Above the transparency point, the gain coefficient is positive and increases with increasing pump power. However, it saturates at some maximum gain coefficient γ_{max}, because the upper-state population N_2 is limited by N, the total number of Er ions per unit volume. Under conditions of complete population inversion, where $N_2 \approx N$ and $N_1 \approx 0$, the maximum gain coefficient is

$$\gamma_{max}(\lambda) = N\sigma_{em}(\lambda) \qquad \text{(maximum gain coefficient)} \qquad (24\text{-}20)$$

With 1480 nm pumping the maximum gain will be somewhat less than this, due to the incomplete inversion.

Gain Spectrum

We saw in Fig. 24-15 how the gain varies with pump power at a fixed signal wavelength. Also of interest is how the gain varies with signal wavelength at a fixed pump power. This is referred to as the *gain spectrum,* and is calculated in Fig. 24-16 assuming four different pump powers. The fiber parameters are taken to be the same as those in Fig. 24-15. An important characteristic of these curves is that each one crosses the zero gain line only once, with positive gain at longer wavelengths and negative gain at shorter wavelengths. A similar behavior was seen in Fig. 23-12 for the gain spectrum of Yb^{3+}. This is a universal feature of such gain spectra, valid whenever the absorption and emission cross-section spectra are connected by the McCumber relation, Eq. (18-38). As the pump power increases, the zero-crossing point moves to shorter wavelength, and approaches the pump wavelength. However, as mentioned in connection with Yb^{3+}, it never becomes shorter than the pump wavelength. As a result, positive gain can only occur for signal wavelengths longer than the pump wavelength.

An interesting aspect of the gain spectrum curves is how they change shape and grow in magnitude as the pump power is increased. For low pump powers, the curves appear "distorted," because near the transparency point, the gain spectrum is a nearly equally weighted combination of $\sigma_{em}(\lambda)$ and $\sigma_{abs}(\lambda)$. At high pump powers, however, far above the transparency point, the gain spectrum is mostly due to $\sigma_{em}(\lambda)$, since $N_2 \gg N_1$. Note also that the magnitude of the peak gain coefficient increases only slightly when the pump power is increased from 5 mW to 100 mW. This is in accord with the saturation of gain with pump power shown in Fig. 24-15.

*This is due to the short lifetime of level 3, which makes N_3 small and emission from this level insignificant.

Integrated Gain

The gain coefficient gives the fractional increase in signal intensity per unit length, $\Delta I/I = \gamma\Delta x$. To determine the total gain over a fiber of length L we must integrate,

$$\int_{I_1}^{I_2} \frac{dI}{I} = \int_0^L \gamma(x)\, dx \tag{24-21}$$

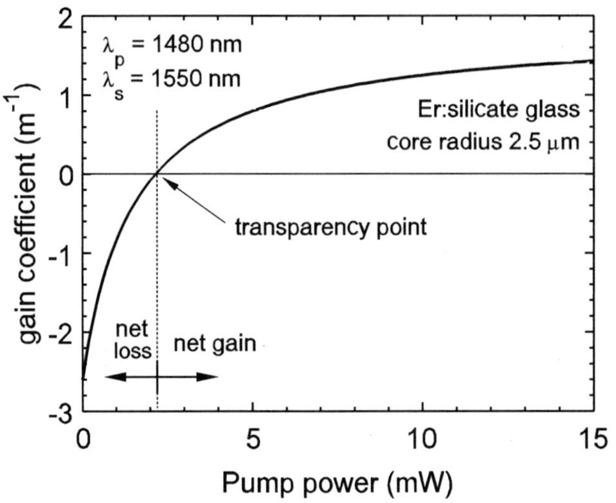

Figure 24-15 Gain coefficient γ versus pump power at 1480 nm, for signal wavelength 1550 nm. Parameters assumed are core radius 2.5 μm, Er density 10^{19} ions/cm^3, τ_2 = 10 ms, and cross-section values from Fig. 18-10.

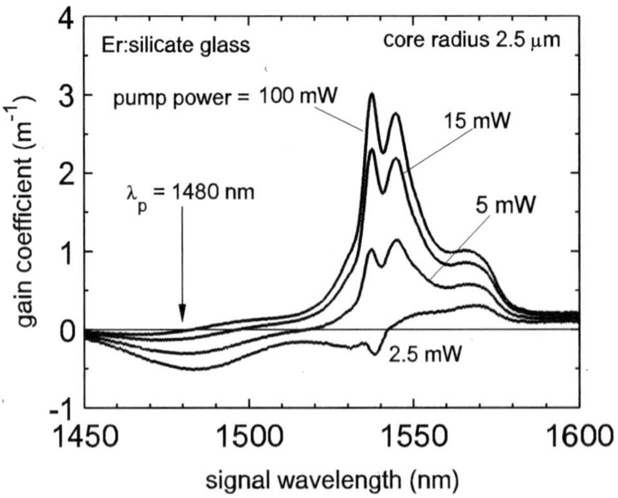

Figure 24-16 Gain spectrum for different values of pump power, calculated using the same parameters as in Fig. 24-15. Pump wavelength position is indicated by the arrow. Note that positive gain only occurs for $\lambda_s > \lambda_p$.

where I_1 and I_2 are the signal intensities entering and leaving the fiber, respectively. If the gain coefficient $\gamma(x)$ were independent of x, this would lead to the simple result

$$G = \frac{I_2}{I_1} = e^{\gamma L} \qquad \text{(constant gain coefficient)} \tag{24-22}$$

However, the gain coefficient depends on pump power, as we have just seen, and the pump power decreases along the fiber length due to absorption. This makes the integration fairly complicated in general, and numerical integration is often needed in practice. We showed in Eq. (23-7) how this integration can be performed in a four-level type system, when the pump power decreases according to Beer's law. However, Beer's law applies only when the majority of ions are in the ground state, and this will not be true for three-level systems because the ground state must be significantly depleted to achieve transparency. It would seem, then, that any simple analytical expression for gain in a three-level system is out of the question.

Fortunately, there is a considerable simplification at high pump power. The variation of pump power with position along the fiber can be determined by applying energy conservation to a small length Δx of fiber, as depicted in Fig. 24-17. The pump power entering this small section is $P_p(x)$, and the pump power leaving it is $P_p(x + \Delta x)$. Also leaving this small section is fluorescence, emitted at a rate

$$\frac{\text{fluorescence photons emitted}}{\text{unit time}} = N_2(A_c \Delta x)\frac{1}{\tau_2}$$

where A_c is the area of the fiber core. For each of these fluorescence photons, a pump photon of energy $h\nu_p$ must be absorbed. The energy balance equation then becomes

$$P_p(x) - N_2(A_c \Delta x)\frac{h\nu_p}{\tau_2} = P_p(x + \Delta x)$$

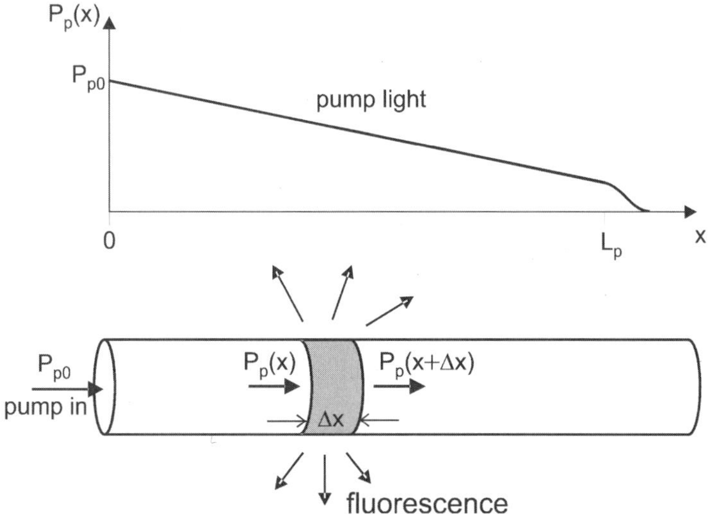

Figure 24-17 By applying energy conservation to the small fiber section shown, it is found that under strong pumping conditions, the pump power $P_p(x)$ decays linearly with position x along the fiber.

which can be written

$$\Delta P_p \equiv P_p(x + \Delta x) - P_p(x) = -\left(\frac{N_2 A_c h\nu_p}{\tau_2}\right)\Delta x \tag{24-23}$$

where ΔP_p is the change in pump power in distance Δx. The quantity in brackets is approximately a constant for large pump power, because in that case N_2 approaches a limiting value independent of P_p. For simplicity, we will consider 980 nm pumping, in which case $N_2 \to N$ at large P_p. Eq. (24-23) therefore predicts a linear decrease in pump power with position, as indicated in Fig. 24-17. This is quite different from a Beer's law dependence, which is exponential. In Beer's law, the pump loses a fixed *fraction* of its energy per unit distance, whereas in a strongly pumped fiber, the pump loses a fixed *amount* of energy per unit distance.

This linear decrease in the pump power makes it a simple matter to estimate the fiber length L_p required to fully absorb an incident pump power P_{p0}. Setting $\Delta x = L_p$ and $\Delta P_p = 0 - P_{p0}$ in Eq. (24-23), we obtain

$$L_p \approx \frac{P_{p0}\tau_2}{NA_c h\nu_p} \quad \text{(absorption length for strong pump)} \tag{24-24}$$

The required fiber length is thus proportional to the incident pump power, and can be many times longer than the Beer's law attenuation length $1/\alpha_p = 1/(N\sigma_p)$. Over most of this distance, the gain coefficient γ has a constant value, given by Eq. (24-20). Combining this with Eq. (24-24), the integrated gain over a fiber of length L_p becomes

$$\ln G \approx \frac{P_{p0}\sigma_{em}\tau_2}{A_c h\nu_p} \quad \text{(small signal gain, strong pump)} \tag{24-25}$$

It is interesting to note that this result agrees precisely with the relation derived in Eq. (23-8) for a four-level system, where the pump light was assumed to decay according to Beer's law. The agreement between these two expressions points to the fact that the gain depends only on $\int \gamma(x)\, dx$, and not on the way that $\gamma(x)$ is distributed along the fiber.

EXAMPLE 24-6

An EDFA is pumped at 980 nm with a power of 10 mW, and amplifies signals at 1550 nm. It contains fiber with core radius 2.5 μm, Er concentration 10^{19} cm^{-3}, and Er upper-state lifetime 10 ms. (a) Estimate the fiber length required to absorb the pump light, and (b) estimate the gain of the amplifier in dB for this fiber length.

Solution: (a) The photon energy of the pump light is

$$h\nu_p = \frac{hc}{\lambda_p} = \frac{(6.63 \times 10^{-34})(3 \times 10^8)}{980 \times 10^{-9}} = 2.03 \times 10^{-19}\ \text{J}$$

The core area is $A_c = \pi(2.5 \times 10^{-6})^2 = 1.963 \times 10^{-11}$ m^2, and the emission cross section is estimated from Fig. 18-10 to be $\sigma_{em}(1550\ \text{nm}) \simeq 3.2 \times 10^{-25}\ m^2$. Putting these into Eq. (24-24) yields

$$L_p \approx \frac{(10^{-2}\text{ W})(10^{-2}\text{ s})}{(10^{25}\text{ m}^{-3})(1.963 \times 10^{-11}\text{ m}^2)(2.03 \times 10^{-19}\text{ J})} = 2.5\text{ m}$$

(b) The maximum gain coefficient is $\gamma_{max} = (10^{25}\text{ m}^{-3})(3.2 \times 10^{-25}\text{ m}^2) = 3.2\text{ m}^{-1}$, and the gain is, therefore,

$$G \approx e^{(3.2\text{ m}^{-1})(2.5\text{ m})} \simeq 3 \times 10^3$$

or

$$10 \log_{10} G \approx 35\text{ dB}$$

The efficiency of an optical amplifier is often given in terms of the dB gain per mW of absorbed pump power. In this example, the gain efficiency is 35/10 = 3.5 dB/mW, a typical value for an EDFA. With special attention to design, such as a small core and Er confinement to the center of the core where the pump intensity is highest, values at high as ≈ 10 dB/mW can be achieved.

Gain Saturation

So far, we have assumed that the gain is determined by the pump intensity I_p, but not by the signal intensity I. This amounts to the small-signal limit, where the stimulated emission rates due to I are much smaller than the pump and spontaneous emission rates. As I becomes larger, stimulated emission reduces the upper-state population, which decreases the gain. This is termed *gain saturation,* and was discussed in Section 19-2 for spatially uniform pumping. Fig. 19-9 shows how the gain G is reduced as the signal intensity increases in relation to the signal saturation intensity I_s.

A similar calculation for the general case of nonuniform pumping is beyond the scope of this book, but qualitatively the same behavior is observed. Figure 24-18 shows gain measurements on an erbium-doped fiber with core diameter of 5.5 μm, pumped at 975 nm with powers ranging from 18 to 33 mW. At low signal levels, the gain is constant (independent of signal power), but at higher signal levels the gain saturates. The degree of saturation for a given signal level depends on the pump power. As the pump power increases, the signal output power at which saturation starts to occur also increases.

This behavior can be easily understood in terms of energy conservation. The optical amplifier can be thought of, essentially, as a device for converting pump power into signal power. The conversion efficiency must be less than the quantum limit $\eta = h\nu/h\nu_p$ given in Eq. (19-23). The data in Fig. 24-18 are consistent with this, with the increase in signal power always less than the incident pump power.

Gain Flattening

An optical amplifier is often used in combination with wavelength division multiplexing (WDM), to increase the span length of high-data-density telecommunications links. If there is a significant gain difference between wavelength channels, then after several amplifying steps there will be much more optical power in some channels than in others. This complicates system design and results in poor system performance. It is therefore de-

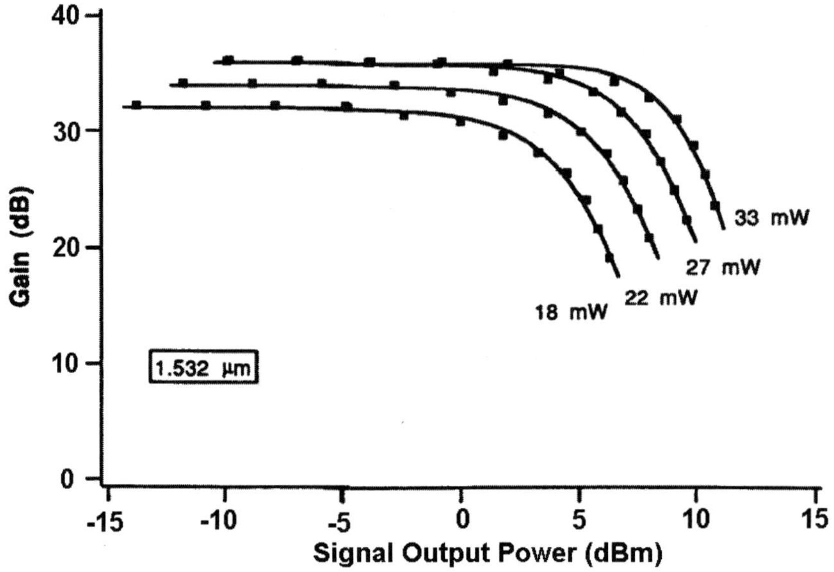

Figure 24-18 Measured gain at 1532 nm in erbium-doped fiber amplifier with 5.5 μm diameter core, pumped at 975 nm with powers ranging from 18 to 33 mW. Gain saturation occurs at higher signal power when higher pump power is used. (After Lindgard et al. 1990.)

sirable that the gain be as "flat" as possible, "flat" here meaning independent of wavelength. The gain spectrum of Er in a typical silicate glass is not at all flat, however, as can be seen from Fig. 18-10. Therefore, special materials or methods need to be employed to make the gain curve more uniform, a process referred to as *gain flattening.*

One way to make the EDFA gain spectrum flatter is to use a different glass host. High-aluminum-content silica glass and fluoride glass have both been found to have a fairly smooth gain profile. The Al-silica glass is more practical, since it is more compatible with conventional silica glass fiber. Another approach is to use an optical filter that is designed to attenuate the spectral regions of high gain. This works quite well, although it lowers the overall amplifier efficiency. One type of spectral filter is the *long-period fiber grating,* which is similar to the fiber Bragg grating except that the grating period Λ is about 100 times longer. This type of fiber grating couples light at certain wavelengths from guided core modes into cladding modes, resulting in selective spectral attenuation. Still another method for spectral flattening is to operate the amplifier at less than full population inversion. As seen in Fig. 24-16, the gain curve is flatter at lower pump power, where the inversion ratio N_2/N_1 is lower.

In practice, a number of these approaches are often combined. For example, Fig. 24-19 shows the gain spectrum of a two-stage EDFA using high-aluminum Al-silica fiber. One stage is pumped at 1480 nm and the other at 980 nm, so the degree of inversion can be controlled. Spectral filtering with a long-period fiber grating is added between the stages, which results in a gain that is flat to within 1 dB over a spectral range of 40 nm.

Other Optical Amplifiers

We saw in Fig. 5-4 that the wavelength range for low fiber loss ($\alpha < 1$ dB/km) extends from about 1200 to 1700 nm. The EDFA normally operates over only a portion of this

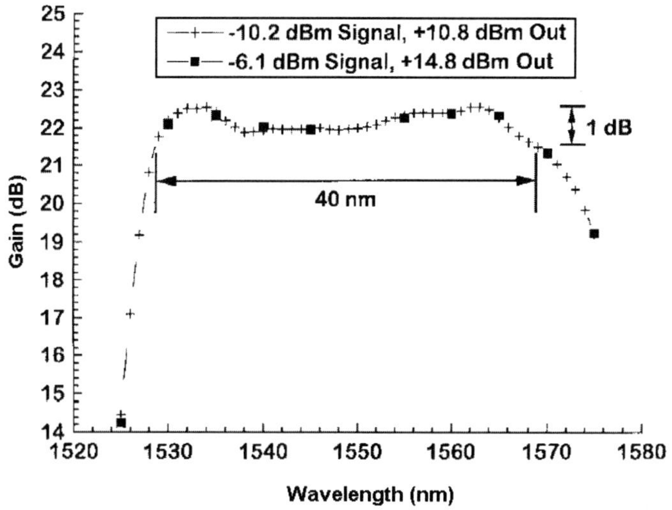

Figure 24-19 Measured gain spectrum in a two-stage EDFA using high-aluminum silica fiber, with spectral filtering by a long-period fiber grating to flatten the gain. (After Wysocki et al. 1997, © 1997 IEEE.)

range, from 1530 to 1565 nm, and this has come to be called the "C band" (C for conventional). The gain of Er-doped glass extends out to longer wavelengths as well (see Fig. 18-10), so the EDFA can also be used in the range 1565–1625 nm, the so-called "L band" (L for long). The gain of Er^{3+} is smaller here, however, and it is difficult to design a single amplifier that works well in both bands. If both C and *L* bands are needed for WDM, the wavelengths are separated with a coarse wavelength filter into two groups, one with C band wavelengths and one with *L* band wavelengths. Each group is then amplified separately with an EDFA optimized for that band.

To promote consistency in nomenclature, the ITU (International Telecommunications Union) has proposed a naming scheme for other bands in the 1200–1700 nm range. Figure 24-20 shows the spectral range for each band, with the "O band" (O for original)* from 1260 to 1360 nm, the "E band" (E for extended) from 1360 to 1460 nm, the "S band" (S for short) from 1460 to 1530 nm, the "C band" from 1530 to 1565 nm, the "L band" from 1565 to 1625 nm, and the "U band" (U for ultralong) from 1625 to 1675 nm. Amplifiers for the bands other than C and *L* are not currently in widespread use, although they are under development. In this section, we take a look at possible amplifiers for these other bands, as well as some competing technologies for amplification in the C and *L* bands.

Other Rare Earth Dopants

Although the third telecommunications window at 1.5 μm is now the clear choice for the lowest-possible attenuation loss, much of the fiber that is currently installed is so-called "legacy fiber," designed with operation at 1.3 μm in mind. In fact, the 1.3 μm ("O") band is still in common use, and it has been a long-standing goal to find a doped-fiber amplifier for this band that works as well as the EDFA does for the C band.

*This was the "original" band for long-haul telecommunications, but it was actually the second telecommunications band, the first being at 850 nm.

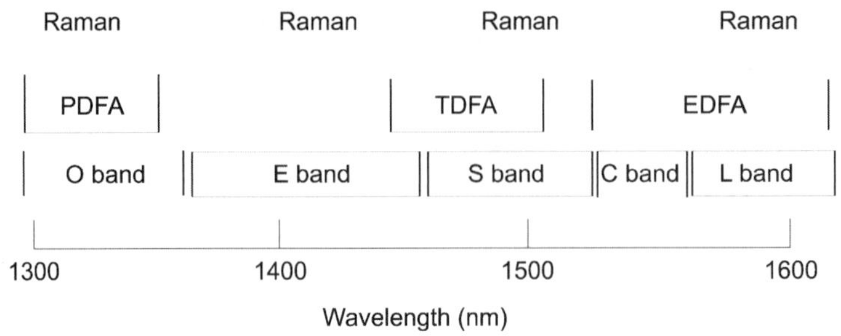

Figure 24-20 Proposed ITU nomenclature for the telecommunications bands. Also shown are the gain regions for three doped fiber amplifiers. Fiber Raman amplifiers can operate over the entire range.

Unfortunately, nature has not been as kind to us here. The energy levels of some relevant rare earth dompants are shown in Fig. 23-7. The $^4F_{3/2} \rightarrow {}^4I_{13/2}$ transition of Nd^{3+} was considered at first, but it has the problem that signal photons are absorbed from the $^4F_{3/2}$ level, promoting the Nd^{3+} ion to a higher level. This process is termed *excited-state absorption,* or ESA, and it prevents the gain at certain wavelengths from ever being positive. The ESA can be reduced somewhat by using a fluoride glass host, but there is still another problem, that of competing strong emission to the $^4I_{11/2}$ level (the well-known 1.06 μm Nd^{3+} transition). As a result, Nd^{3+} has lost favor as a 1.3 μm amplifier.

A more promising approach is the praseodymium-doped fiber amplifier (PDFA), which operates on the $^1G_4 \rightarrow {}^3H_5$ transition of Pr^{3+} at ~ 1.3 μm. Unlike Nd^{3+}, there are no stronger competing emissions from the upper level, and no significant ESA. However, Pr^{3+} does have the problem that in a silica glass host, the upper level $^1G_4$ is strongly quenched by nonradiative relaxation to the next-lowest level. The nonradiative quenching can be reduced by using a fluoride glass host, but the resulting amplifier efficiencies (typically ~ 0.15 dB/mW) are still much lower than for an EDFA. Commercial PDFAs are available, but have not found widespread use.

As the need for additional bandwidth in the 1.5 μm low-loss window has grown, there has been increasing interest in optical amplifiers for the S band. Most prominant among these is the thulium-doped fiber amplifier (TDFA), which operates on the $^3H_4 \rightarrow {}^3F_4$ transition of Tm^{3+}. Nonradiative quenching of the upper level by nonradiative relaxation is a problem for Tm^{3+}, as it is for Pr^{3+}, and this requires the use of a nonsilica glass host such as fluoride glass or heavy-metal oxide glass. The amplifying transition has the additional complication that the lower-level lifetime is significantly longer than the upper-level lifetime. This is not a good feature for any laser or amplifier transition, because it tends to prevent a steady-state population inversion. One solution for the TDFA is to pump simultaneously with two different pump wavelengths (the "dual-pumping" scheme). The first pump puts population into the $^3H_4$ by ground-state absorption, while the second pump takes population out of the $^3F_4$ by excited-state absorption. Fiber amplifiers with efficiencies of ~ 0.1–0.2 dB/mW have been demonstrated in this way, covering the spectral range 1460–1520 nm. The TDFA has promise, but is still in the developmental stage.

There has not been much interest to date in developing optical amplifiers for the E band, because of the strong absorption at ~1400 nm in conventional optical fibers. This absorption is due to vibrations of the OH molecule, a ubiquitous impurity in glass that comes from water contamination during the fiber manufacturing process. Methods are now available, however, to significantly reduce this "water peak" in telecommunications-grade fiber. For newly installed fiber systems, it might, therefore, be advantageous to have an amplifier in the E band. But the number of fiber links that contain this new fiber is likely to remain small for some time, and in practice the E band will likely continue to be avoided for long-haul systems.

Semiconductor Amplifiers

Since any laser can be turned into an amplifier by removing the optical feedback elements, it is natural to consider the use of semiconductor laser diodes, which are already quite well developed. Indeed, there are several advantages to this approach. The gain coefficient is very high, which makes these amplifiers compact and compatible with integrated optics. In fact, the gain is so high that lasing can occur due to just the Fresnel reflections from the end facets of the semiconductor chip. For an optical amplifier, this feedback must be reduced by antireflection coating or beveling the ends. Another advantage is that the amplifier gain can be modulated or switched electrically, by varying the drive current. This again makes them highly compatible with integrated optics. Finally, and perhaps most importantly, the wavelength of operation depends on the bandgap of the semiconductor, and this can be chosen for a particular application. By varying the semiconductor composition, virtually the entire range from 1250 to 1675 nm can be covered. This is in distinct contrast to doped fiber amplifiers, which rely on a few rare-earth transitions at particular wavelengths.

There are several drawbacks to the use of semiconductor optical amplifiers, however. Some problems arise from their planar waveguide geometry, which is quite different from the cylindrical geometry of fibers. For example, the coupling of light from fiber into waveguide is inefficient, due to the small (usually <1 μm) thickness of the semiconductor waveguide. Another consequence of the semiconductor's rectangular geometry is a strong polarization sensitivity, which makes the gain different for different polarizations. Optical fiber amplifiers, in contrast, are mostly polarization insensitive. Semiconductor amplifiers also tend to introduce more noise into the signal beam than optical fiber amplifiers.

There is one feature of the semiconductor amplifier that is both a blessing and a curse. Because the upper-state lifetime can be short (subnanosecond), it is possible to rapidly switch the gain, a distinct advantage in optical processing applications. However, this fast time response becomes a disadvantage in WDM applications, where more than one wavelength is being amplified in the same device. A strong signal in one wavelength channel causes the gain to saturate, and this changes the amplification in a second wavelength channel. Because of the fast response time, the time-dependent intensity (signal information) of one channel becomes impressed on the signal in another channel. This is known as *cross-gain modulation,* and is generally undesirable.

A similar interaction between signals occurs in an EDFA, but the Er^{3+} ion response time is on the order of milliseconds, rather than nanoseconds. This slow response time is a great benefit, because it significantly attenuates gain fluctuations on the time scale of the signal bit rate. Even though the average gain may change slowly, as the amount of signal power changes, there is little cross talk between channels.

The semiconductor optical amplifier is no match for the EDFA as a power booster in long-haul systems, but it does have applications in optical processing and switching. One example is frequency conversion, in which a digital waveform at one wavelength is used to generate a duplicate digital waveform at another wavelength. This is essentially what happens in the cross-gain modulation process described above, and is one case in which it can be considered an advantage. Frequency conversion can also occur via nonlinear four-wave mixing processes in the semiconductor, which are especially efficient because of the large nonlinear χ_3 coefficients in semiconductors.

Fiber Raman Amplifiers

All the optical amplifiers and lasers that we have discussed so far are based on stimulated emission, in which an atom in an excited state is stimulated to emit a fluorescence photon into a light mode that already contains one or more photons. The enhancement of the emission rate into this mode is proportional to the number of photons in the mode, according to Eq. (18-14). We now consider an entirely different type of amplifier, in which it is a scattering process that is stimulated, rather than a fluorescence process.

Fig. 24-21 illustrates how this stimulated scattering works for the case of Raman scattering, which was discussed previously in Section 5-2.* In Raman (Stokes) scattering, light of frequency ν interacts with molecules having vibrational frequency f_v, and this creates scattered light of a lower frequency ν'. The difference between incident and scattered light frequencies, $\nu - \nu' \equiv \Delta\nu_R$, is known as the *Raman frequency shift.* Spontaneous Raman scattering occurs under weak pumping conditions, when the scattered light intensity is small enough that the number of scattered photons per mode is $\ll 1$, on average. When the incident pump intensity becomes sufficiently large, there will be more than one scattered photon per mode, and the scattering process will be enhanced. This is termed *stimulated Raman scattering* (SRS), and is the physical basis for the *fiber Raman amplifier.*

The gain coefficient in a traditional optical amplifier is proportional to the Einstein A coefficient, which is the rate at which fluorescence is spontaneously emitted from an excited state. In a similar way, the gain coefficient in a Raman amplifier is proportional to the rate at which pump photons are spontaneously scattered. This scattering rate is proportional to the pump intensity I_p, because when more photons come in (higher intensity), more photons are scattered. It also depends on the frequency shift $\Delta\nu_R$, because energy conservation requires $\Delta\nu_R = f_v$, and there is only a limited range of vibrational frequencies available in a given material. The gain coefficient for SRS can therefore be expressed as

$$\gamma = g_R I_p \qquad \text{(Raman gain coefficient)} \qquad (24\text{-}26)$$

where g_R is the *Raman gain function.*

The dependence of g_R on the frequency shift $\Delta\nu_R$ is shown in Fig. 24-22 for silica glass. The peak gain occurs at $\Delta\nu_R \approx 13$ THz, and the width of the main peak (FWHM) is ≈ 8 THz. At a wavelength of 1500 nm, this corresponds to a gain bandwidth of 60 nm, larger than that of the EDFA C band. A key advantage of the Raman amplifier is that the range of signal wavelengths λ_s that can be amplified is not restricted by any dopant energy levels, as in the EDFA, but instead depends only on the pump wavelength λ_p. To provide amplification in a different spectral region, it is only necessary to select a different pump wavelength, and the entire 1.2–1.7 μm range can be covered in this way. The fiber Raman laser can also be combined with an EDFA to create a single hybrid amplifier with a very large gain bandwidth. Such a device can be used to increase the maximum number of channels in WDM applications.

*Brillouin scattering can also be stimulated, but we do not discuss this here.

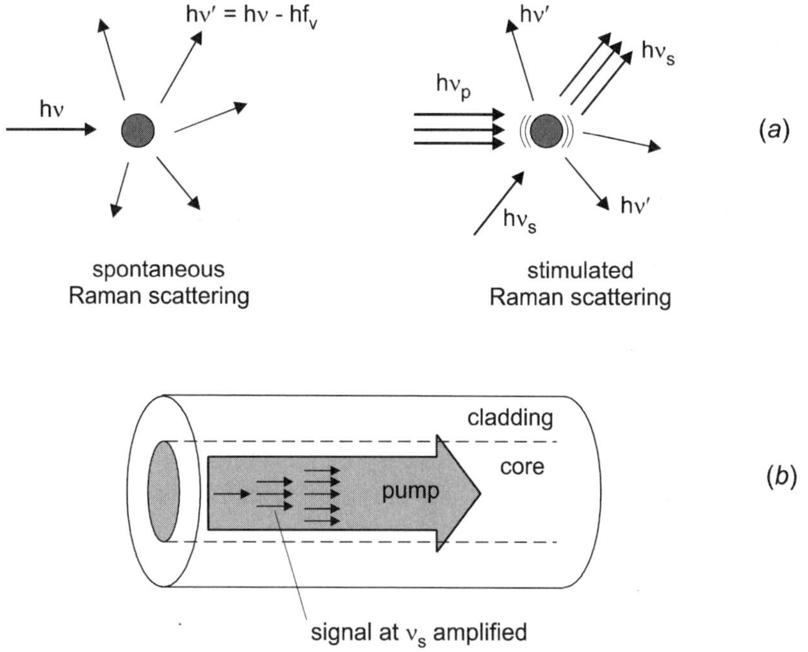

Figure 24-21 (a) In stimulated Raman scattering, a signal beam of frequency ν_s propagating along a particular direction is amplified by the large number of pump photons that are spontaneously scattered in this same direction. (b) In a fiber, the pump and signal beams propagate in the same direction, and the signal is amplified while extracting energy from the pump.

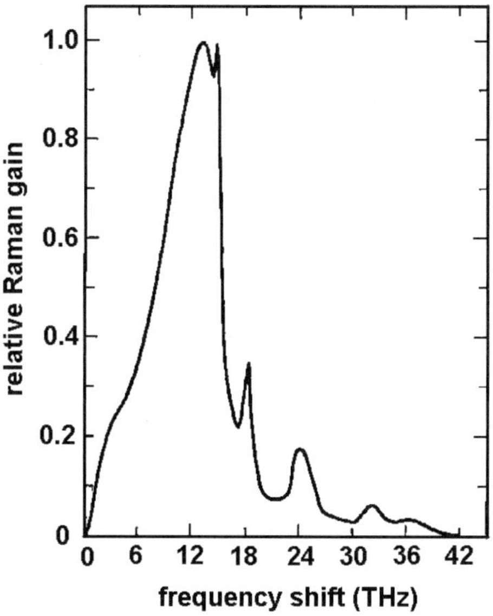

Figure 24-22 Raman gain function g_R in silica glass versus Raman frequency shift $\Delta\nu_R$, normalized to unity at the gain peak. (After Stolen et al. 1989.)

The value of the Raman gain function in silica glass, at the peak of the gain curve, is given by

$$g_{R,\max} \simeq \frac{10^{-13}}{\lambda_p}\left[\text{units of } \frac{\text{m}}{\text{W}}\right] \tag{24-27}$$

where the pump wavelength λ_p is expressed in units of μm. If the pump intensity is approximately uniform over a length L of fiber, the total gain can be written as

$$G \equiv \frac{P_{\text{sig}}(\text{out})}{P_{\text{sig}}(\text{in})} = e^{\gamma L} = e^{g_R I_p L} \qquad \text{(Raman gain)} \tag{24-28}$$

This equation, along with Eq. (24-27) and Fig. 24-22, allows us to make an estimate of the pump power required to achieve a particular value of gain in a fiber Raman amplifier.

EXAMPLE 24-7

A fiber Raman amplifier is to be designed that will amplify light at 1550 nm with 30 dB of gain. It will use 1 km of single-mode silica fiber with a mode field diameter of 2w = 10 μm. Determine (a) the required pump wavelength, and (b) the required pump power.

Solution: (a) The optical frequency of the signal light to be amplified is

$$\nu_s = \frac{c}{\lambda_s} = \frac{3 \times 10^8}{1550 \times 10^{-9}} = 1.935 \times 10^{14} \text{ Hz} = 193.5 \text{ THz}$$

The Raman frequency shift for maximum gain is $\Delta\nu_R = 13$ THz, and therefore the required pump frequency is

$$\nu_p = \nu_s + \Delta\nu_R = 193.5 + 13 = 206.5 \text{ THz}$$

This corresponds to a pump wavelength

$$\lambda_p = \frac{3 \times 10^8}{2.065 \times 10^{14}} = 1.452 \times 10^{-6} \text{ m} = 1452 \text{ nm}$$

(b) A gain of 30 dB corresponds to $G = 10^3$, so

$$10^3 = e^{g_R I_p L} = e^{g_R P_p L/A_p}$$

where P_p is the pump power and A_p is the effective area of the pump, taken to be $A_p = \pi(5 \times 10^{-6})^2 = 7.85 \times 10^{-11}\ \text{m}^2$. Solving this for pump power gives

$$P_p = \frac{A_p \ln(10^3)}{g_R L}$$

The peak value of g_R for the 1452 nm pump wavelength is

$$g_{R,\max} = \frac{10^{-13}}{1.452} = 6.89 \times 10^{-14} \text{ m/W}$$

and the pump power is then evaluated to be

$$P_p = \frac{(7.85 \times 10^{-11})(6.9)}{(6.89 \times 10^{-14})(10^3)} = 7.9 \text{ W}$$

The large value of pump power calculated in this example illustrates the primary disadvantage of the fiber Raman amplifier. The gain efficiency is 30/7900 = 0.0038 dB/mW, three orders of magnitude lower than the efficiency of an EDFA. The required pump power can be decreased by using specially developed fiber with small mode field diameter, but it is preferable to utilize existing fiber. A longer fiber length will also decrease the required power, but this is ultimately limited by fiber attenuation. In practice, pump powers of ~ 1 W are required. This is not quite the obstacle that it once was, due to recent developments in high-power diode and fiber laser sources.

24-5. FREE-SPACE OPTICS

In this book, we have considered optical communications to be practically synonymous with optical *fiber* communications. However, an optical signal can be transmitted through free space as well as through a fiber, and, in fact, Alexander Graham Bell's original proposal for the "photophone" envisioned just such a system. Optical technology was not sufficiently advanced in Bell's day, however, and his system never became practical. Today, the same photonic devices that are used for optical fiber communications can be adapted for use in what has come to be called *free-space optics* (FSO) or *optical wireless.* FSO has long been of interest to the military, because the directional nature of a collimated optical beam makes it practically invulnerable to interception. Radio waves, in contrast, spread out rapidly by diffraction, and are easier to intercept.

More recently, there has been growing interest in civilian applications as well, but primarily for another reason. The motivation comes from a very practical issue: how to connect a local LAN in a dense urban environment with the high-speed MAN or WAN systems that pass close by (but not into) the building containing the LAN. Typically the "high-speed" connection to a building's LAN is a T1 line operating at ≈ 1.5 Mb/s, and this does not allow the LAN to fully exploit the much higher capacity of metro and wide-area networks. The connection could be made with fiber, of course, but digging up sidewalks or city streets to lay new fiber is expensive.

One solution to this so-called "last mile" bottleneck problem is the use of FSO, which does not require the laying of any new fiber. Installation is easy, and the main requirement is that there be a direct line of sight between the optical transmitter and receiver. In an urban environment, the ideal location for these would be on upper floors or the roof of a high-rise building. There are added benefits of FSO, as well. In contrast to conventional wireless, the connection is quite secure, as noted above. Data rates can be as high as for fiber, in the Gb/s range. Furthermore, there is no licensing requirement since the FCC does not regulate the optical spectrum.

In spite of these advantages, there are some drawbacks to FSO. One obvious issue is reduced atmospheric transmission in bad weather. Rain and snow are not as much of a factor as might be expected, but fog can be a real problem. The optical signal can also be temporarily lost due to obstructions, such as a bird flying through the beam. Even in good weather with no obstructions, there are fluctuations in the beam direction (beam "wander") due to random variations in the atmosphere's refractive index. The same phenomenon is responsible for the twinkling of stars in the nighttime sky. In FSO this is referred to as *scintillation,* and it adds noise to the detected signal. A further problem is that the transmitter and receiver are not really precisely fixed in location, because the top of a tall building moves slightly under wind loading, a phenomenon termed *building sway.* The lateral motion of the top of a building is typically in the range $H/200$ to $H/800$, where H is the building height. Taking $H = 100$ m for a 25 story building, we can estimate the lateral motion to be $\approx H/400 = 100/400 = 0.25$ m, or 25 cm.

Problems such as building sway, scintillation, and beam obstruction can be made less severe by expanding the optical beam diameter. This is accomplished using a lens to collimate the beam, as shown in Fig. 24-23. Typical initial beam diameters are in the ~ 20 cm range. If the beam were diffraction limited, we could use Eqs. (17-4) and (17-5) to estimate the Rayleigh range and beam divergence. Taking a beam waist of $w_0 = 10$ cm, we find

$$z_0 = \frac{\pi(0.1)^2}{1.55 \times 10^{-6}} \simeq 20 \text{ km}$$

and

$$\theta = \frac{1.55 \times 10^{-6}}{\pi(0.1)} \simeq 5 \times 10^{-6}$$

This suggests that the FSO beam could propagate some 20 km before spreading out significantly. However, maximum transmission distances in FSO systems are more on the order of ~ 2 km, and such a highly coherent beam is generally not necessary. Typical beam divergence in practice is $\geq 10^{-4}$ rad, with the beam diameter expanding to ~ 1 m at a distance of 2 km. Figure 24-24 shows a rooftop FSO system in operation in the skies over Vancouver, BC, Canada.

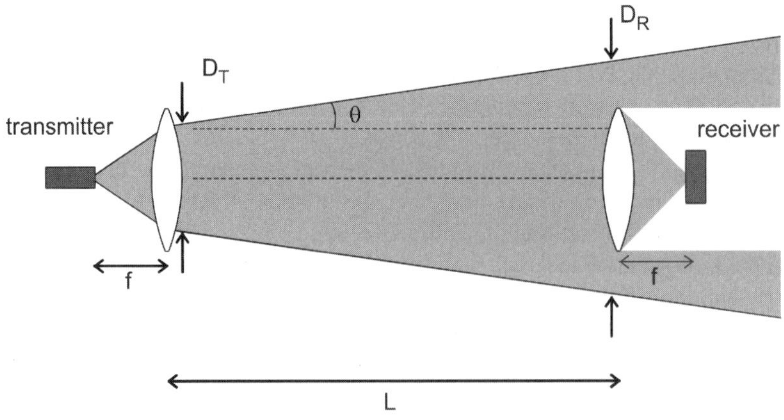

Figure 24-23 In free-space optics, light from a laser diode transmitter is collected by a lens and collimated into a beam of initial diameter D_T. The beam diverges with half-angle θ, and after propagating a distance L has diameter D_R. A portion of the broadened beam is collected by the receiver lens and focused onto a photodetector.

Figure 24-24 A rooftop, free-space optics system operates in the skies over Vancouver, BC, Canada. (Source: fSONA Systems.)

Beam attenuation due to fog is still present even with an expanded beam. However, if the system is designed with a sufficiently high margin, there will still be adequate power at the receiver for reliable detection, even with dense fog. Redundant transmission pathways in the system can also be used, so that if one pathway becomes blocked, there are other alternative pathways that can be used.

FSO is also possible in space-based applications, where it has some advantages over radio frequency (RF) communications. Because optical frequencies are much higher than radio frequencies, data can be sent much more rapidly. Also, the antenna size needed for good collimation is much smaller for optical frequencies, an important issue for spacecraft, for which reducing the size and weight of components is a high priority. The narrower and more directional nature of an optical beam does mean, however, that sophisticated pointing-control systems are needed. FSO in space has real potential, but it is not yet clear how practical it will be.

PROBLEMS

24.1 An analog signal can be more faithfully reproduced if the sampled voltage is placed into one of 4096 "bins," rather than one of 256. (a) How many bits are required for each sample in this case? (b) What would be the corresponding bit rate required to transmit a voice signal of bandwidth 4 kHz?

24.2 In a conventional television tube, the image is formed by an electron beam that makes a series of 525 horizontal sweeps across the tube face. During each sweep, the beam intensity is modulated in time to create a pattern of dots with varying intensity. The spacing between these dots corresponds to the horizontal resolution of the image. The traditional analog-broadcast TV image contains 330 resolvable dots per horizontal line, and the entire pattern of 525 lines (one frame) is repeated at a rate of 30 frames per second. (a) Estimate the bandwidth required for an analog TV signal, by taking the reciprical of the time between adjacent dots in a horizontal scan. (b) If this analog waveform is sampled at the NyQuist rate, at 8-bit (256 bin) resolution, what bit rate is required to send one video signal? (c) Using the bit rate calculated in part b, determine the number of separate video signals that can be multiplexed into one wavelength channel operating at the OC-48 rate. (d) Repeat parts a–c for an HDTV picture with 720 lines and 1280 resolvable dots per line. Note: the actual bit rate needed in practice is considerably less than you have calculated here, because there is much redundancy in adjacent frames, and this allows the data to be compressed.

24.3 The channel spacing in a DWDM system is 100 GHz. Calculate the corresponding wavelength spacing at each end of the C band (1530 and 1565 nm).

24.4 (a) If the channel spacing in a DWDM system is 50 GHz, and each channel operates at the OC-192 standard rate, determine the spectral efficiency. (b) Another communications system has a channel spacing of 10 GHz and maximum spectral efficiency of 0.4 (bits/s)/Hz. Which European/International standard rate should be used if the highest data rate is desired?

24.5 The L band extends from 1565 to 1625 nm. Assuming that the channels within this band transmit data with a spectral efficiency of 0.3 (bits/s)/Hz, determine the maximum total data rate that is possible.

24.6 (a) If the maximum tolerable bit error rate is 10^{-12}, determine the minimum average number of photons per bit for a quantum-limited receiver. (b) How much does this increase the required signal power compared with the usual bit error rate criterion of 10^{-9}? Express your answer as a ratio and also in terms of dB.

24.7 A receiver has sensitivity –60 dBm at a bit rate of 100 Mb/s. Determine the sensitivity of this same receiver at a bit rate of 400 Mb/s.

24.8 A LAN fiber optic link uses an LED source with wavelength 670 nm, modulated at 20 Mb/s. The multimode fiber has a loss of 6 dB/km at this wavelength, and an average power of 0.05 mW is coupled into the fiber. The receiver is a PIN photodiode requiring 5000 photons/bit on average for an adequate BER. The length of the fiber is 3.5 km, and the total loss from all splices and connectors is 5 dB. (a) Determine the system margin. Is this likely to be adequate? (b) Repeat this analysis if the LED is modulated at 100 Mb/s.

24.9 A long-haul fiber optic link uses a diode laser source operating at 1550 nm, which couples 0.8 mW into the core of a single-mode fiber. The receiver is an APD requiring 500 photons/bit. The fiber has an attenuation of 0.25 dB/km at this wavelength, and splice and connector losses in the system add up to 4 dB. The fiber length is 188 km, and the system operates at 2.5 Gb/s. If the system margin is to be 8 dB, determine the required gain of an optical amplifier that is inserted in the link. Express your answer in dB and as an amplification ratio.

24.10 For the fiber link of the previous problem, assume that the chromatic dispersion is 15 ps/(nm · km). Determine the spectral width of the laser source required so that the pulse spreading by dispersion is tolerable.

24.11 Eq. (24-17) gives the relation between the maximum fiber length and bit rate when limited by chromatic dispersion in single-mode fiber. The lower limit on the wavelength spread $\Delta\lambda$ is determined by the Fourier transform of the modulated waveform, which is termed the modulation bandwidth. (a) Show that at a bit rate BR, the maximum fiber length in this case is given by

$$L_{\max} \simeq \frac{c}{2D_c\lambda^2 BR^2}$$

(b) Use this to calculate the maximum fiber length for a system operating at a wavelength of 1550 nm and bit rate of 10 Gb/s, when the dispersion is 15 ps/(nm · km). (c) Repeat the calculation when the bit rate is 40 Gb/s.

24.12 Consider the three lowest levels of Er^{3+} shown in Fig. 24-14. Assume that level 3 decays primarily to level 2 by rapid nonradiative relaxation. (a) Write the rate equations for levels 2 and 3 in the case of 980 nm pumping, and show that the population of level 2 is approximately given by

$$N_2 \simeq N\frac{W_p\tau_2}{1+W_p\tau_2}$$

where N is the total number of Er ions per unit volume, and

$$W_p = \frac{P_p\sigma_p}{A_c h\nu_p}$$

(b) Using this result, show that nearly complete population inversion can be obtained on the $2 \rightarrow 1$ transition for sufficiently large pump power.

24.13 Consider the Er doped fiber amplifier that is modeled in Fig. 24-15, but now assume it is pumped at 980 nm where the absorption cross section is 3×10^{-21} cm^2. (a) Determine the pump power for signal transparency in this case. (b) Develop an expression for the gain coefficient as a function of pump power, and plot this on a graph in the manner of Fig. 24-15. (c) Explain the differences between the two graphs qualitatively.

24.14 An Er-doped fiber has length 20 cm, core radius 2.5 μm, and doping concentration 2×10^{19} cm^{-3}. The fiber is pumped at 980 nm, where the absorption cross section is 3×10^{-21} cm^2. (a) For weak pumping ($N_2 \ll N$), determine the fraction of pump light that is absorbed in the fiber. (b) Using the results of Problem 24.12, determine the incident pump power that will result in an excited state population $N_2 = 0.95\,N$ at the beginning of the fiber. (c) For the pumping rate of part b, determine the fraction of pump light absorbed in the fiber. (d) At the pumping rate of part b, what fiber length would be required to absorb most of the pump light?

24.15 An EDFA has length 1.2 m, doping level 1.5×10^{19} cm^{-3}, and core radius 2.5 μm. It is pumped at 980 nm with a power such that most of the ions in the fiber are highly excited, and also most of the pump light is absorbed in the fiber. Determine the small-signal gain in this fiber for signal light of wavelength 1550 nm, where the stimulated emission cross section is 3.2×10^{-21} cm^2. Give the gain in dB.

24.16 One potential problem with using fiber Raman amplifiers is that the strong pump light at λ_p required for amplification at one signal wavelength λ_s may interfere with a signal at another wavelength $\lambda_{s'} = \lambda_p$. (a) Assuming silica fiber with the gain spectrum shown in Fig. 24-22, is there a range of pump wavelengths that can be used to amplify the entire C and L bands without any such interference? (b) What if the entire S band is to be amplified as well?

24.17 A 2.5 km length of silica fiber carries signal light at a wavelength of 1310 nm, with mode field diameter 6 μm. The light is amplified by stimulated Raman scattering, pumping with 500 mW at the appropriate wavelength. (a) Determine the pump wavelength that will most efficiently amplify the 1310 nm light. (b) Calculate the gain of this fiber amplifier in dB.

24.18 A free-space optics system is designed for interplanetary communications. It uses a 10 W beam of 1 μm light with 50 cm initial beam radius (Gaussian beam waist). The signal is modulated at a rate 1 Mb/s, and after propagating as a Gaussian beam through space it is detected by an APD receiver that requires 500 photons per bit on average for reliable detection. The mirror that collects the light and focuses it on the detector has an area of 1 m^2. Calculate the maximum allowed distance between transmitter and receiver, and compare this to the earth–sun distance (1.5×10^{11} m).

Bibliography

Agrawal, G. P. (1995), *Nonlinear Fiber Optics,* 2nd ed., Academic Press.

Agrawal, G. P. (1997), *Fiber-Optic Communication Systems,* 2nd ed., Wiley.

Becker, P. C., N. A. Olsson, and J. R. Simpson (1999), *Erbium-Doped Fiber Amplifiers,* Academic Press.

Bhattacharya, P. (1997), *Semiconductor Optoelectronic Devices,* 2nd ed., Prentice-Hall.

Birks, T. A., J. C. Knight, and P. St. J. Russell (1997), "Endlessly single-mode photonic crystal fiber," *Opt. Lett.* **22,** 961. Reprinted by permission from the Optical Society of America.

Boyd, R. W. (1992), *Nonlinear Optics,* Academic Press.

Bube, R. H. (1960), *Photoconductivity of Solids,* Wiley.

Di Bartolo, B. (1968), *Optical Interactions in Solids,* Wiley.

Couny, F., H. Sabert, P. J. Roberts, D. P. Williams, A. Tomlinson, B. J. Mangan, L. Farr, J. C. Knight, T. A. Birks, and P. St. J. Russell (2005), "Visualizing the photonic band gap in hollow core photonic crystal fibers," *Optics Express* **13,** 558. Reprinted by permission from the Optical Society of America.

Cuisin, C., A. Chelnokov, J.-M. Lourtioz, D. Decanini, and Y. Chen (2000), Reprinted with permission from "Submicrometer resolution Yablonovite templates fabricated by x-ray lithography," *Appl. Phys. Lett.* **77,** 770. Copyright 2000, American Institute of Physics.

Hawkes, J., and I. Latimer (1995), *Lasers: Theory and Practice,* Prentice-Hall.

Hecht, E. (2002), *Optics,* 4th ed., Addison-Wesley.

Hecht, J. (2002), *Understanding Fiber Optics,* 4th ed., Prentice-Hall.

Jeunhomme, L. B. (1990), *Single-mode Fiber Optics,* 2nd ed., Marcel Dekker.

Joannopoulos, J. D., R. D. Meade, and J. N. Winn (1995), *Photonic Crystals: Molding the Flow of Light,* Princeton University Press.

Kaminow, I. P. (1974), *Introduction to Electrooptic Devices,* Academic Press.

Kashyap, R. (1999), *Fiber Bragg Gratings,* Academic Press.

Keck, D. B. (1981), in *Fundamentals of Optical Fiber Communications,* 2nd ed., C. L. Tang (ed.), p. 18, Academic Press. Reprinted with permission from Elsevier.

Lidgard, A., J. R. Simpson, and P. C. Becker (1990), Reprinted by permission from "Output saturation characteristics of erbium-doped fiber amplifiers pumped at 975 nm," *Appl. Phys. Lett.* **56,** 2607. Copyright 1992, American Institute of Physics.

Lin, S. Y., J. G. Fleming, D. L. Hetherington, B. K. Smith, R. Biswas, K. M. Ho, M. M. Sigalas, W. Zubrzycki, S. R. Kurtz, and J. Bur (1998), "A three-dimensional photonic crystal operating at infrared wavelengths," *Nature,* **394,** 251.

Marcuse, D. (1977), *Bell Syst. Tech. J.* **56,** 703–718.

Marcuse, D., D. Gloge and E. A. J. Marcatili (1979), "Guiding properties of fibers," in *Optical Fiber Telecommunications,* S. E. Miller and A. G. Chynoweth (eds.), pp. 71–72, Academic Press.

McCumber, D. E, (1964), "Einstein relations connecting broadband emission and absorption spectra," *Phys. Rev.* **136,** A954.

Meade, R. D. (1992), Reprinted with permission from "Existence of a photonic band gap in two dimensions," *Appl. Phys. Lett.* **61,** 495. Copyright 1992, American Institute of Physics.

Milonni, P. W., and J. H. Eberly (1988), *Lasers,* Wiley.

Mortensen, N. A., J. R. Folkenberg, M. D. Nielsen, and K. P. Hansen (2003), "Modal cutoff and the V parameter in photonic crystal fibers" *Opt. Lett.* **28,** 1879. Reprinted by permission from the Optical Society of America.

Nyquist, N. (1928), "Thermal agitation of electric charge in conductors," *Phys. Rev.* **32,** 110.

Othonos, A., and K. Kalli (1999), *Fiber Bragg Gratings,* Artech House.

Palais, J. C. (1998), *Fiber Optic Communications,* 4th ed., Prentice-Hall.

Palik, E. D. (1985), *Handbook of Optical Constants of Solids,* Academic Press.

Pedrotti, F. L., and L. S. Pedrotti (1993), *Introduction to Optics,* 2nd ed., Prentice-Hall.

Saito, K., A. J. Ikushima, T. Ito, and A. Itoh (1997), "A new method of developing ultralow-loss glasses," *J. Appl. Phys.* **81,** 7129.

Saleh, B. E. A., and M. C. Teich (1991), *Fundamentals of Photonics,* Wiley.

Senior, J. M. (1992), *Optical Fiber Communications: Principles and Practice,* 2nd ed., Prentice-Hall.

Siegman, A. E. (1986), *Lasers,* University Science Books.

Silfvast, W. T. (2004), *Laser Fundamentals,* 2nd ed., Cambridge University Press.

Shur, M. (1990), *Physics of Semiconductor Devices,* Prentice-Hall.

Stolen, R. H., J. P. Gordon, W. J. Tomlinson, and H. A. Haus (1989), "Raman response function of silica-core fibers," *J. Opt. Soc. Am.* B **6,** 1159. Reprinted by permission from the Optical Society of America.

Sutherland, R. L. (1996), *Handbook of Nonlinear Optics,* Marcel Dekker.

Svelto, O. (1998), *Principles of Lasers,* 4th ed., Plenum.

Sze, S. M. (1981), *Physics of Semiconductor Devices,* 2nd ed., Wiley.

Talneau, A. (2002), Reprinted with permission from "Photonic-crystal ultrashort bends with improved transmission and low reflection at 1.55 μm," *Appl. Phys. Lett.* **80,** 547. Copyright 2002, American Institute of Physics.

Tsuchiya, H., H. Nakagome, N. Shimizu, and S. Ohara (1977), "Double eccentric connectors for optical fibers," *Appl. Opt.,* **16,** 1323.

Verdeyen, J. T. (1995), *Laser Electronics,* 3rd ed., Prentice-Hall.

Weber, M. J. (1979), Reprinted from *Methods of Experimental Physics,* Vol. 15, part A, p. 175, Academic Press, with permission from Elsevier.

Wysocki, P. F., J. B. Judkins, R. P. Espindola, M. Andrejco, and A. M. Vengsarkar (1997), "Broadband erbium-doped fiber amplifier flattened beyond 40 nm using long-period grating filter," *IEEE Photon. Tech. Lett.* **9,** 1343. Reprinted with permission from IEEE.

Xia, Y., B. Gates, and Z. Y. Li (2001), "Self-assembly approaches to three-dimensional photonic crystals," *Adv. Mater.* **13,** 409. Reproduced by permission from John Wiley & Sons.

Appendix A

Solid Angle and the Brightness Theorem

SOLID ANGLE

The solid angle Ω is the 2-D analog of the conventional 1-D angle θ, as illustrated in Fig. A-1. Just as the angle θ is defined as the distance along a circle divided by the radius of that circle, so the solid angle Ω is analogously defined as the area on the surface of a sphere divided by the radius squared of that sphere. The units for θ and Ω are radians (r) and steradians (sr), respectively, although it should be noted that both of these measures of angle have no actual dimensions. Since the total surface area of a sphere is $4\pi R^2$, the total solid angle in one sphere is 4π sr.

For situations with symmetry about an axis (such as an optical fiber or the normal to a plane surface), the two types of angles can be easily related. Figure A-2 shows a differential area dA on the surface of a sphere, in the form of a thin ring centered about the symmetry axis. This ring can be thought of as the intersection of the spherical surface with two cones, one of half-angle α, and other of half-angle $\alpha + d\alpha$. The width of this ring is $R\, d\alpha$, and the radius of the ring is $R \sin \alpha$. The differential solid angle is then

$$\begin{aligned} d\Omega &= \frac{dA}{R^2} \\ &= \frac{(2\pi R \sin \alpha)(R\, d\alpha)}{R^2} \\ &= 2\pi \sin \alpha\, d\alpha \end{aligned} \tag{A-1}$$

The solid angle inside a cone of half-angle θ can be determined by integrating

$$\begin{aligned} \Omega &= \int d\Omega = \int_0^\theta 2\pi \sin \alpha\, d\alpha \\ &= -2\pi \cos \alpha \big|_0^\theta \\ &= 2\pi (1 - \cos \theta) \end{aligned} \tag{A-2}$$

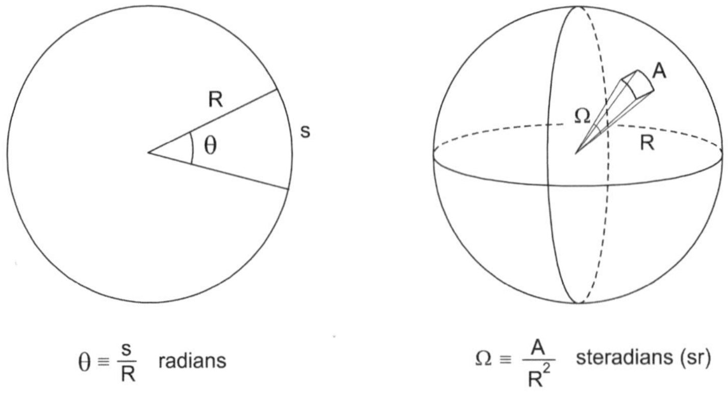

Figure A-1 The solid angle Ω is defined analogously to the conventional angle θ, using the fraction of a sphere's area rather than the fraction of a circle's circumference.

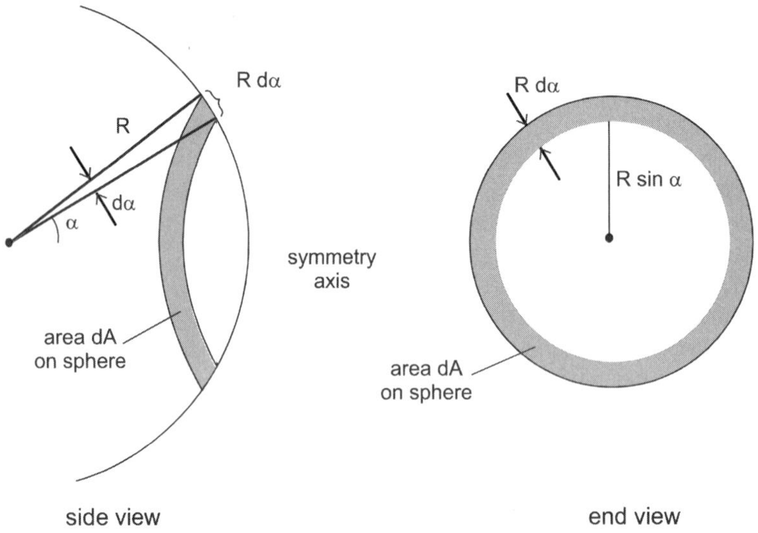

Figure A-2 Geometry for relating differential solid angle $d\Omega$ to the differential change in cone angle $d\alpha$ around an axis of symmetry.

It is often of interest to consider the small-angle approximation, where $\theta \ll 1$. In this limit, $\cos\theta \approx 1 - \theta^2/2$, so

$$\Omega \approx \pi\theta^2 \qquad \text{(cone with small half-angle } \theta\text{)} \tag{A-3}$$

BRIGHTNESS THEOREM

The solid angle is a useful concept in describing the degree of directionality for light emitted by an object. The *brightness* of a source of light quantifies this directionality, and is defined as the optical power emitted per unit solid angle, per unit area of the emitting surface. The SI units are W/(m^2 sr). Although the proper SI term for this quantity is actually *radiance* (denoted by L), the term brightness is still commonly used in photonics and laser work. Since the word "brightness" conveys a more intuitive understanding of its meaning than the technical term radiance, we will use it throughout this book.

A source has a high brightness when it emits light in a narrow range of angles (small solid angle) from a small surface area. Lasers have a much higher brightness than conventional light sources, because they are (or can be made to be) highly directional. One might think that the brightness of a conventional source could be improved by simply focusing with a lens, to create an image source with smaller surface area. However, a lens also changes the angular distribution of the light according to geometrical optics, and this tends to counteract the apparent increase in brightness.

To see how this works, consider a square light source of side h_1 being transformed by a lens into an image of side h_2, as shown in Fig. A-3. This image can be thought of as a new source of light, with a different surface area and solid angle for emission. The brightness B_1 and B_2 of the original source and its image are given by

$$B_1 = \frac{P}{A_1\Omega_1} = \frac{P}{h_1^2 \pi \theta_1^2}$$

$$B_2 = \frac{P}{A_2\Omega_2} = \frac{P}{h_2^2 \pi \theta_2^2} \tag{A-4}$$

where θ is the maximum angle from the optical axis for which light is collected by the lens, and we have made the usual paraxial approximation of small angle, $\theta \ll 1$. From Chapter 2, we have

$$\frac{h_1}{s_1} = \frac{h_2}{s_2}$$

and from the geometry of Fig. A-3 we have

$$\frac{D}{2} = \theta_1 s_1 = \theta_2 s_2$$

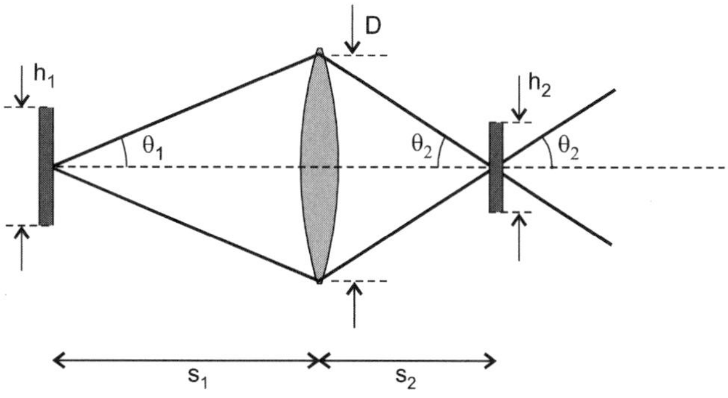

Figure A-3 A lens forms a light source image of size h_2 from a light source object of size h_1.

Combining the above two equations yields

$$h_1\theta_1 = h_2\theta_2$$

which when substituted into Eq. (A-4) gives

$$B_1 = B_2 \qquad \text{(brightness theorem)} \tag{A-5}$$

The brightness of the image is seen to be identical with that of the original object, independent of the degree of focusing by the lens. This is an example of a general principle known as the *brightness theorem,* which states that the brightness of a light source cannot be increased with passive optical components such as lenses, mirrors, or waveguides. A laser or optical amplifier is considered an "active" optical component, and the brightness theorem does not apply while light is being modified by the amplifying medium. It does apply again, however, once laser light has been generated and is freely propagating through a passive optical system.

Appendix B

Fourier Synthesis and the Uncertainty Principle

At several points in this book, we encounter the so-called *uncertainty relation,* which relates the minimum uncertainties in time and frequency, or in position and wavelength. A complete description of this involves the *Fourier transform,* the mathematical treatment of which is beyond the scope of this book. However, we can obtain an intuitive understanding of this concept by considering qualitatively what happens when we add together sinusoidal waves of different frequencies. We take this approach here, and also obtain an exact expression relating the two uncertainties for one important special case.

FOURIER SYNTHESIS

The fundamental idea of *Fourier synthesis* is that any arbitrary waveform can be constructed by adding together an infinite number of pure sinusoidal waves. This concept can be applied to either the time dependence or the position dependence of the wave, but to be concrete we will emphasize here the time dependence. When the waveform is periodic in time, it can be written as a *Fourier series,*

$$y(t) = \sum_n A_n \sin(n2\pi\nu_0 t) \quad \text{(Fourier series)} \tag{B-1}$$

where $\nu_0 = 1/T$, T is the repetition time of the waveform, and n is an integer ranging from 0 to ∞. In general, there are also $\cos(n2\pi\nu_0 t)$ terms, but for this discussion we can neglect them. The sum is over a set of discrete frequencies, including the fundamental at ν_0 and higher harmonics at $n\nu_0$. For example, a square wave can be constructed by choosing coefficients $A_n = 1/n$ for all odd values of n, and $A_n = 0$ for all even values. Figure B-1 shows the calculated $y(t)$ using the four lowest-frequency terms, along with the component waves that are added.

Even with this small number of terms, the basic square wave shape is evident in the constructed waveform. A higher number of terms would more faithfully reproduce the square wave, with the sharpness of the edges (the rise time) depending on the highest-frequency component.

When the waveform is not repetitive, a series of discrete frequencies cannot be used to construct it. This can be understood by considering a nonrepetitive waveform to be a repetitive waveform with $T \to \infty$, so that $\nu_0 \to 0$. Since ν_0 is the spacing between frequen-

Photonics and Lasers: An Introduction. By Richard S. Quimby

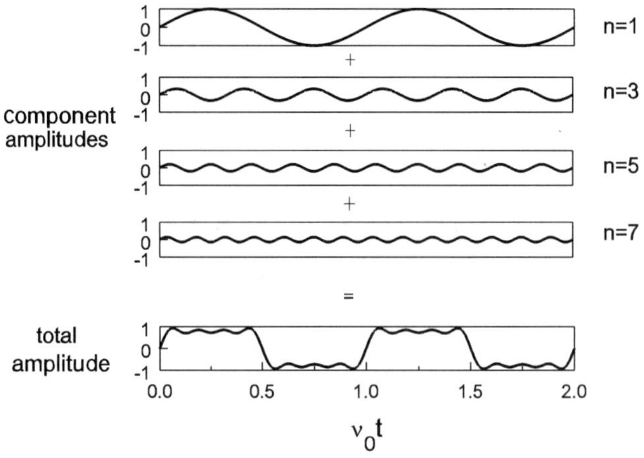

Figure B-1 Fourier synthesis of square wave from the four lowest-frequency terms of frequency $n\nu_0$. The top four waveforms are added together to give the bottom waveform.

cy components, this means that a continuous distribution of frequencies is needed. The synthesis is then described by the Fourier transform,

$$y(t) = \int A_\nu(\nu)\sin(2\pi\nu t)\,d\nu \quad \text{(Fourier transform)} \tag{B-2}$$

where $A_\nu(\nu)$is an amplitude per unit frequency (a distribution function). The product $A_\nu(\nu)$ $d\nu$ plays the same role here that the coefficients A_n do in the Fourier series.

UNCERTAINTY RELATION

To see how the uncertainty relation follows from the Fourier transform, consider the form for $A_\nu(\nu)$ shown in Fig. B-2. This "flat-top" frequency distribution has a constant value $A_\nu(\nu) = A_0$ over the range $\nu_0 - \Delta\nu/2 < \nu < \nu_0 + \Delta\nu/2$, and $A_\nu(\nu) = 0$ elsewhere. The spectral width $\Delta\nu$ might correspond to the bandwidth in a communications system, for example. The waveform $y(t)$ that results from this frequency distribution is also shown in Fig. B-2, calculated according to Eq. (B-2). The time dependence consists of a main pulse, surrounded by a series of smaller subsidiary pulses. If the pulse width Δt is defined as the full width at half maximum (FWHM) of the main peak, we find that $\Delta t \simeq 1.3/\Delta\nu$. This result contains the essential feature of the uncertainty relation: *To create a pulse of shorter duration, a greater bandwidth is required.*

The exact relationship between $\Delta\nu$ and Δt depends on the shape of the frequency distribution function $A_\nu(\nu)$, and also on the way that the pulse width is defined. If the edges of $A_\nu(\nu)$ vary more smoothly than the abrupt steps shown in Fig. B-2, the side peaks are suppressed. A Gaussian function for $A_\nu(\nu)$ has the special property that its Fourier transform is also a Gaussian function in time. The pulse width can be defined by the FWHM, the half width at half maximum (HWHM), or by the $1/e$ or $1/e^2$ points. We will not generally be concerned with these details in this book, and will write the uncertainty relation as

$$\Delta\nu\Delta t \sim 1 \quad \text{(uncertainty relation)} \tag{B-3}$$

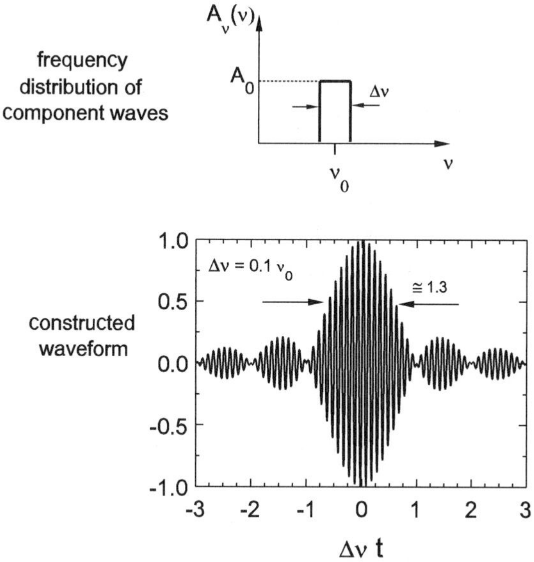

Figure B-2 Waveform synthesis using frequency components spread continuously in a range $\Delta\nu$ around ν_0. The width Δt of the main pulse depends only on $\Delta\nu$, and not on ν_0.

We have developed the uncertainty relation by simply considering the mathematics of adding waves. The results then apply to any situation that is described by a wave. This includes quantum mechanics, in which material objects with energy E have an associated wave, of frequency $\nu = E/h$. Multiplying Eq. (B-3) by Planck's constant h, we obtain the *Heisenberg uncertainty principle* for energy and time,

$$\Delta E \Delta t \sim h \quad \text{(Heisenberg uncertainty principle)} \tag{B-4}$$

This states that if measurements are made during a time interval Δt, the energy of a system will be uncertain by the amount $\Delta E \sim h/\Delta t$. For example, we can derive the minimum uncertainty in energy for photons emitted in a radiative transition. The lifetime τ of the transition gives a measure of the uncertainty in emission time, and the corresponding uncertainty in energy is $\Delta E \sim h/\tau$.

A wave can be localized in space as well as in time, and there is another form of the uncertainty relation that relates the uncertainties in wavenumber $k = 2\pi/\lambda$ and position x. This is usually written as

$$\Delta k \Delta x \sim 1 \tag{B-5}$$

and implies that for a wave packet to be highly localized in space, it must consist of a wide spread of k values. In quantum mechanics, a particle with momentum p has an associated deBroglie wavelength $\lambda = h/p = \hbar k$, where $\hbar = h/2\pi$. The corresponding uncertainty principle in quantum mechanics is then

$$\Delta p \Delta x \sim \hbar \tag{B-6}$$

EXPONENTIAL TIME RESPONSE

In many situations, the time response of a system is nearly exponential, and it is useful to know the exact constant to use in the uncertainty relation for this case. We will determine this by considering a specific example, that of the LCR electrical circuit shown in Fig. B-3. The approach will be to solve first for the transient response, and then for the steady-state response. Relating the two solutions will then yield the appropriate uncertainty relation. The resulting expression has a broader application than this particular example, because the same mathematics applies to other resonances as well, such as in mechanical and atomic systems.

(1) Transient Response

To analyze the circuit of Fig. B-3, we add up voltages around the loop, obtaining

$$iR + \frac{Q}{C} + L\frac{di}{dt} = V_s(t) \qquad \text{(B-7)}$$

where Q is the charge on the capacitor, and $i = dQ/dt$ is the current. Taking the derivative

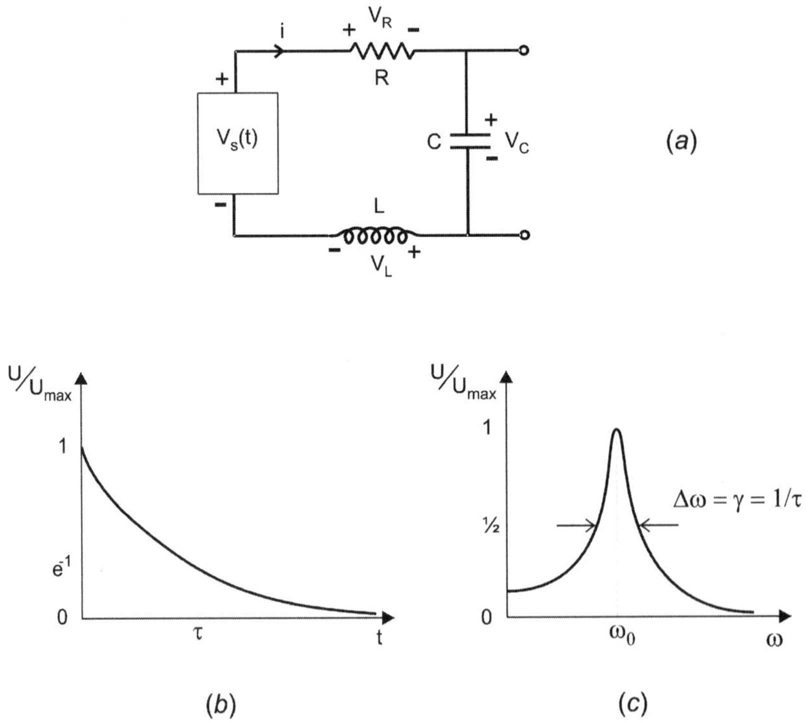

Figure B-3 (a) LCR circuit with an exponential time response. (b) Stored energy versus time when $V_s = 0$. (c) Stored energy versus driving frequency. Relating the two viewpoints gives the uncertainty relation $\Delta\omega_{1/2} = \gamma = 1/\tau$.

of this equation with respect to time, and using the dot notation for time derivatives (dot = first derivative; two dots = second derivative), it can be written

$$\ddot{Q} + \frac{R}{L}\dot{Q} + \frac{1}{LC}Q = \frac{1}{L}V_s(t) \tag{B-8}$$

To simplify notation, we define $\gamma \equiv R/L$ and $\omega_0 \equiv 1/\sqrt{LC}$, so this becomes

$$\ddot{Q} + \gamma\dot{Q} + \omega_0^2 Q = \frac{1}{L}V_s(t) \tag{B-9}$$

To obtain the transient response, we set $V_s = 0$, and look for solutions of the form

$$Q(t) = Q_0 e^{pt} \tag{B-10}$$

In general $Q(t)$ will be complex, and it is assumed that the real part is to be taken. Substituting this into Eq. (B-9) with $V_s = 0$ results in

$$Q_0[p^2 + \gamma p + \omega_0^2]e^{pt} = 0$$

$$p^2 + \gamma p + \omega_0^2 = 0 \tag{B-11}$$

Solving for p gives

$$p = -\frac{\gamma}{2} \pm \tfrac{1}{2}\sqrt{\gamma^2 - 4\omega_0^2} \tag{B-12}$$

We will assume a weak damping, so that $\gamma \ll \omega_0$. In that case,

$$p \simeq -\frac{\gamma}{2} \pm j\omega_0 \tag{B-13}$$

where we use $j = \sqrt{-1}$ rather than the usual i, to avoid confusion with the current. The solution for $Q(t)$ is then

$$Q(t) = Q_0 e^{-\gamma t/2} e^{j\omega_0 t} \tag{B-14}$$

which is the form of a weaky damped harmonic oscillator. The energy stored in the circuit is $U \propto |Q|^2$, which decays in time as

$$U(t) = U_0 e^{-\gamma t} \tag{B-15}$$

Defining the lifetime τ for energy decay by $U(t) = U_0 e^{-t/\tau}$, we identify

$$\gamma = \frac{1}{\tau} \tag{B-16}$$

In the context of the photon lifetime discussed in Sec. 16-2, we associate τ with τ_c.

(2) Steady-State Response

For the steady-state solution, we assume the circuit is driven harmonically with a voltage $V_s(t) = A \exp(j\omega t)$, where $\nu = \omega/2\pi$ is the driving frequency. Now we assume that $Q(t)$ oscillates at the driving frequency, with

$$Q(t) = \tilde{Q}e^{j\omega t} \tag{B-17}$$

where $\tilde{Q}$ is a complex amplitude that contains amplitude and phase information about the oscillation. Substituting this and the driving term $V_s(t)$ into Eq. (B-9) gives

$$\tilde{Q} = \frac{A/L}{\omega_0^2 - \omega^2 + j\omega\gamma} \tag{B-18}$$

The energy stored in the oscillator is then

$$U \propto |\tilde{Q}|^2 = \frac{A^2/L^2}{(\omega_0^2 - \omega^2)^2 + \omega^2\gamma^2} \tag{B-19}$$

According to this result, the stored energy is a maximum at resonance, $\omega \simeq \omega_0$, where the demoninator takes on its minimum value of $\omega_0^2\gamma^2$. The frequency at which the stored energy decreases by a factor of two is a measure of the resonance width, and can be determined by setting

$$(\omega_0^2 - \omega^2)^2 + \omega^2\gamma^2 = 2\ \omega_0^2\gamma^2$$

and solving for ω. Using the approximation $\omega \approx \omega_0$ near resonance, we find

$$\omega - \omega_0 = \pm\frac{\gamma}{2} \tag{B-20}$$

The stored energy is therefore at least one-half the maximum value over the frequency range $\omega_0 - \gamma/2 < \omega < \omega_0 + \gamma/2$, which corresponds to a FWHM of $\Delta\omega_{1/2} = \gamma$. Combining this with Eq. (B-16) gives an uncertainty relation of the form

$$\Delta\omega_{1/2}\tau = 1 \qquad \text{(uncertainty relation)} \tag{B-21}$$

for a system with an exponential time response. This result is used in Eq.(16-16) in connection with the photon lifetime.

List of Symbols

a	fiber core radius
a	spacing between atoms in solid
A	area of junction, detector, laser rod, or optical beam
A_c	area of fiber core
A	amplitude of electric field in light wave (Eq. 9-4)
A_{21}	Einstein *A* coefficient
b	electrooptic coefficient (Chapter 9)
B	brightness
B	bandwidth of detector (Eq. 14-28)
B_r	coefficient for electron–hole radiative recombination
B_{21}	Einstein *B* coefficient
BR	bit rate (pulses per second in digital communications)
BR_{phone}	bits per second required to send one phone conversation (Eq. 24-1)
BER	bit error rate (probability that bit will be read incorrectly)
c	speed of light (= 3×10^8 m/s)
C	capacitance
CB	conduction band
d	thickness of planar waveguide
d	grating spacing
d	width of depletion region (Eq. 10-20)
d	quantum well width
D	beam diameter
D_m	material dispersion coefficient (Eq. 6-10)
D_c	chromatic dispersion coefficient (Eq. 6-11)
D_C	transverse coherence length (Eq. 15-3)
D_w	waveguide dispersion coefficient (Eq. 6-13)
D^*	"dee star" figure of merit for detectors (Eq. 14-44)
e	magnitude of electron charge (= 1.6×10^{-19} C)
E	electric field
E_g	energy gap
f	focal length of lens
f	oscillator strength of transition (Eq. 18-44)
f_e	3 dB electrical bandwidth (Eq. 11-13)
f_a	frequency of acoustic wave

f_v	vibrational frequency (Raman scattering)
F#	F number for lens (= f/D)
$\mathcal{F}$	finesse of cavity (Eq. 16-20)
$g(\nu)$	atomic lineshape function (Eq. 18-3)
g_1, g_2	stability parameters for laser cavity (Eq. 17-16)
g_R	Raman gain function (Eq. 24-26)
G	thermal conductance (Eq. 13-1)
G	detector gain (Eqs. 13-14, 13-19)
G	amplifier gain (Eq. 19-15)
G_{th}	single-pass gain at lasing threshold (Eq. 23-10)
G_{dB}	amplifier gain expressed in dB
h	Planck's constant (= 6.63×10^{-34} Js)
h	height of object or image when imaging with lens
i	current
i_N	rms noise current
i_λ	photocurrent (Eq. 14-2)
i_{th}	threshold current for diode laser
i_0	reverse-saturation current ("dark" current in photodetector)
I	light intensity (power per unit area)
I_1, I_2	input and output intensities in optical amplifier (Fig. 19-5)
I_s	signal saturation intensity (Eq. 19-9)
I_{ps}	pump saturation intensity (Eq. 23-22)
I_π	light intensity that gives a phase shift of π radians
j	integer
J_{th}	threshold current density for diode laser
k	wave number
K	dB loss due to fiber connections and splicing (Eq. 24-7)
k_B	Boltzmann's constant (1.38×10^{-23} J/K)
ℓ	integer
L	length of fiber, Bragg grating or optical cavity
L_c	coherence length (Eq. 15-1)
m	integer (labels mode)
m	mass of sensor element in thermal detector (Eq. 13-2)
m	mass of electron (= 9.1×10^{-31} kg)
m^*	effective mass of electron or hole
M	multiplication factor (gain) in APD
M	system margin in dB (Eq. 24-7)
M^2	"m squared" figure of merit for optical beam (Eqs. 15-4, 17-22)
n	integer (labels quantum state)
n	refractive index
n_0	refractive index outside fiber
n	number of electrons per unit volume (Eq. 10-11)
n	number of photons in cavity mode (Eq. 18-12)
n_1	number of photons in a digital "one" pulse (Fig. 24-10)
n_j	number of photons in the jth cavity mode
n_{th}	theshold electron density for lasing
n_{eff}	effective refractive index for waveguide (Eq. 3-13)
n_2	nonlinear refractive index (Eq. 9-26)
N	number of refractive index undulations in Bragg grating

N	number of lasing modes in mode locking (Eq. 22-9)
N	atoms per unit volume
N_i	number of atoms per unit volume in energy state i
$N_{2,th}$	value of N_2 at lasing threshold (Eq. 20-14)
N_D	number of donor atoms per unit volume in n-type semiconductor
$\mathcal{N}$	total number of electrons in recombination region
NA	numerical aperture (Eq. 4-3)
NEP	noise equivalent power (Eq. 14-40)
p	momentum
p	number of cavity modes within laser transition bandwidth
p	holes per unit volume
P	optical power
P_p	pump power
P_{p0}	pump power incident on fiber
P_{th}	threshold pump power
P	polarization density (dipole moment per unit volume)
$P(n)$	Poisson distribution (Eq. 13-23)
$\mathcal{P}_T$	transmitter power in dBm
$\mathcal{P}_R$	receiver sensitivity (required signal power) in dBm
P_R	receiver sensitivity (signal power required for adequate BER) (Eq. 24-12)
q	principle mode number for Hermite–Gaussian beam (Eq. 17-20)
Q	total charge detected during current pulse
Q	quality factor of resonance (Eq. 16-18)
r_1, r_2	radii of curvature of laser mirrors 1 and 2 in laser cavity
R	reflection coefficient (fraction of light reflected from boundary)
R	radius of curvature of wavefront for Gaussian beam (Eq. 17-3)
R	resistance
R_L	load resistance
R_{sh}	shunt resistance (Eq. 14-10)
$\mathcal{R}$	detector responsivity (Eqs. 13-8 and 14-18)
$\mathcal{R}$	total number of atoms pumped to excited state per unit time (Eq. 19-4)
$\mathcal{R}_{th}$	threshold excitation rate (Eq. 20-18)
s	distance of object or image from lens
SNR	signal-to-noise power ratio (Eq. 14-39)
t_{tr}	transit time for electron (Eqs. 13-20, 14-33)
t_r	rise time (Eqs. 14-27 and 14-29)
T	fraction of light transmitted through boundary
T	period of wave (= $1/\nu$)
T	repetition time for pulses in digital communications
T	time between pulses in mode-locked pulse train (Eq. 22-17)
T	absolute temperature (in Kelvin)
T_c	coherence time (Fig. 15-4)
U	energy stored in resonator
v_p	phase velocity of wave
v_g	group velocity of wave
v_s	sound velocity
V	Vee parameter for fiber (Eq. 4-9)
V	volume of laser cavity
V_B	bias voltage in photodetector circuit

V_d	voltage across photodiode
V_p	normalized film thickness (for planar waveguide) (Eq. 3-8)
V_T	voltage equivalent of temperature (Eq. 14-8)
V_π	voltage across Pockels cell that gives phase shift of π radians
V_0	built-in junction potential
VB	valence band
w	beam radius for Gaussian beam (Eq. 17-2)
w_0	beam waist (minimum beam radius) for Gaussian beam
$w_{0,\text{eff}}$	beam waist (minimum beam radius) for multimode beam
W^{spont}	spontaneous transition rate (probability per unit time that a transition occurs)
W^{ind}	induced transition rate (probability per unit time for stimulated emission or absorption)
W_p	pump rate (probability per unit time that atom absorbs photon) (Eq. 23-16)
W_{pe}	pump emission rate (Eq. 23-17)
W	work function
z_0	Rayleigh range (Eq. 17-4)
α	angle that incident ray outside fiber makes with fiber axis (Fig. 4-1)
α	attenuation coefficient
α_p	absorption coefficient for pump light
α_R	attenuation coefficient due to Rayleigh scattering
β	propagation constant (z component of wave vector **k**)
β	diode ideality factor (in Eq. 10-21)
β_ν	number of cavity modes per unit frequency interval per unit volume (Eq. 16-9)
β_s	slope of output power vs. current graph for diode laser
χ	dielectric susceptibility (Eq. 9-2)
χ	electron affinity (energy to remove electron from bottom of conduction band)
χ_2	second order nonlinear susceptibility
χ_3	third order nonlinear susceptibility
δ	penetration distance of evanescent wave (Eq. 2-21)
δ	fraction of light lost in one round-trip (Eq. 20-28)
δ	index difference $n_{\text{eff}} - n_2$, in optical fiber (Eq. 6-12)
$\delta\omega$	angular frequency spacing between lasing modes
$\delta\nu$	frequency spacing between lasing modes
Δ	fractional index difference $(n_1\text{-}n_2)/n_1$
Δ	energy separation between thermally occupied levels (Fig. 23-8)
ΔN	population inversion $N_2 - N_1$
Δt	time spreading of optical pulse
Δt_p	duration of mode-locked pulse (Eq. 22-16)
ΔN_{th}	threshold population inversion (Eq. 20-6)
$\Delta\nu$	frequency width of optical source
$\Delta\nu_{1/2}$	frequency width of optical mode (FWHM) (Eq. 16-17)
$\Delta\nu_R$	Raman frequency shift
$\Delta\lambda$	spread in wavelengths for optical source
ε	effective energy in McCumber relation (Eq. 18-38)
ε	permitivity of medium
ε_0	permitivity of free space ($= 8.85 \times 10^{-12}$ F/m)

ε_r	relative dielectric permitivity (= $\varepsilon/\varepsilon_0$)
ϕ	phase of wave
ϕ	temperature increase in thermal detector
ϕ	quantum yield (or quantum efficiency) for transition (Eq. 18-46)
ϕ_r	phase shift upon reflection (Eq. 2-22)
γ	gain coefficient (Eq. 18-33)
γ_{th}	theshold gain coefficient (Eq. 20-2)
γ_0	unsaturated gain coefficient (Eq. 19-11)
η	coupling efficiency (Fig. 7-2)
η	efficiency of converting incident photons into charge carriers (Eq. 13-7)
η	efficiency of converting absorbed pump power into signal power (Eq. 19-23)
η_{abs}	fraction of incident light absorbed in photodetector material (Eq. 13-15)
η_c	efficiency for coupling light source into fiber (Eqs. 12-3, 12-11, 12-17)
η_i	radiative efficiency (Eq. 10-13)
η_s	slope efficiency of laser (Eq. 20-26)
η_{sp}	spectral efficiency (Eq. 24-6)
κ	attenuation constant for Bragg reflection in sinusoidal grating (Eq. 8-16)
λ	wavelength
λ_0	wavelength in free space
λ_c	cutoff wavelength (Eq. 4-14)
λ_B	Bragg wavelength (for strong reflection) (Eq. 8-7)
Λ	refractive index periodicity in Bragg grating or photonic crystal
μ_e	electron mobility (Eq. 13-16)
μ_h	hole mobility
ν	frequency of optical wave
ν_0	center frequency of atomic transition
ν_m	mode frequencies
θ	angle that beam makes with normal to surface
θ	divergence angle of beam (half-angle) (Eq. 17-5)
θ_B	Brewster's angle (Eq. 2-17)
θ_c	critical angle (Eq. 2-18)
ρ	energy density (energy per unit volume) (Eq. 2-8)
ρ_ν	spectral density (energy per frequency interval per unit volume) (Eq. 18-2)
σ	cross section (Eqs. 18-35, 18-36)
σ_p	absorption cross section for pump light
σ_{se}	stimulated emission cross section
τ_i	excited state lifetime of level i
τ_{21}	lifetime for spontaneous emission on $2 \rightarrow 1$ transition (Eq. 18-42)
τ_c	cavity lifetime (also called photon lifetime) (Eq. 16-13)
ω	angular frequency (= $2\pi\nu$)
Ω	solid angle (sr)

Index